Handbook of Research on Expanding Business Opportunities With Information Systems and Analytics

George Leal Jamil
Informações em Rede Consultoria e Treinamento Ltda, Brazil

A volume in the Advances in Business Information Systems and Analytics (ABISA) Book Series

IGI Global
DISSEMINATOR OF KNOWLEDGE

Published in the United States of America by
IGI Global
Business Science Reference (an imprint of IGI Global)
701 E. Chocolate Avenue
Hershey PA, USA 17033
Tel: 717-533-8845
Fax: 717-533-8661
E-mail: cust@igi-global.com
Web site: http://www.igi-global.com

Library of Congress Cataloging-in-Publication Data

Names: Jamil, George Leal, 1959- editor.
Title: Handbook of research on expanding business opportunities with information systems and analytics / George Leal Jamil, Editor.
Description: Hershey : Business Science Reference, [2018]
Identifiers: LCCN 2018004008| ISBN 9781522562252 (hardcover) | ISBN 9781522562269 (ebook)
Subjects: LCSH: Information technology--Management. | Technological innovations--Management. | Knowledge management.
Classification: LCC HD30.2 .H364255 2018 | DDC 658.4/038--dc23 LC record available at https://lccn.loc.gov/2018004008

This book is published in the IGI Global book series Advances in Business Information Systems and Analytics (ABISA) (ISSN: 2327-3275; eISSN: 2327-3283)

British Cataloguing in Publication Data
A Cataloguing in Publication record for this book is available from the British Library.

For electronic access to this publication, please contact: eresources@igi-global.com.

Advances in Business Information Systems and Analytics (ABISA) Book Series

Madjid Tavana
La Salle University, USA

ISSN:2327-3275
EISSN:2327-3283

Mission

The successful development and management of information systems and business analytics is crucial to the success of an organization. New technological developments and methods for data analysis have allowed organizations to not only improve their processes and allow for greater productivity, but have also provided businesses with a venue through which to cut costs, plan for the future, and maintain competitive advantage in the information age.

The **Advances in Business Information Systems and Analytics (ABISA) Book Series** aims to present diverse and timely research in the development, deployment, and management of business information systems and business analytics for continued organizational development and improved business value.

Coverage

- Forecasting
- Strategic Information Systems
- Algorithms
- Information Logistics
- Business Systems Engineering
- Statistics
- Geo-BIS
- Management Information Systems
- Business Decision Making
- Business Models

IGI Global is currently accepting manuscripts for publication within this series. To submit a proposal for a volume in this series, please contact our Acquisition Editors at Acquisitions@igi-global.com or visit: http://www.igi-global.com/publish/.

The Advances in Business Information Systems and Analytics (ABISA) Book Series (ISSN 2327-3275) is published by IGI Global, 701 E. Chocolate Avenue, Hershey, PA 17033-1240, USA, www.igi-global.com. This series is composed of titles available for purchase individually; each title is edited to be contextually exclusive from any other title within the series. For pricing and ordering information please visit http://www.igi-global.com/book-series/advances-business-information-systems-analytics/37155. Postmaster: Send all address changes to above address.

Titles in this Series

For a list of additional titles in this series, please visit: www.igi-global.com/book-series

Social Network Analytics for Contemporary Business Organizations
Himani Bansal (Jaypee Institute of Information Technology, India) Gulshan Shrivastava (National Institute of Technology Patna, India) Gia Nhu Nguyen (Duy Tan University, Vietnam) and Loredana-Mihaela Stanciu (University Timisoara, Romania)
Business Science Reference • copyright 2018 • 321pp • H/C (ISBN: 9781522550976) • US $215.00 (our price)

Harnessing Human Capital Analytics for Competitive Advantage
Mohit Yadav (BML Munjal University, India) Shrawan Kumar Trivedi (Indian Institute of Management Sirmaur, India) Anil Kumar (BML Munjal University, India) and Santosh Rangnekar (Indian Institute of Technology Roorkee, India)
Business Science Reference • copyright 2018 • 367pp • H/C (ISBN: 9781522540380) • US $215.00 (our price)

Corporate Social Responsibility for Valorization of Cultural Organizations
María del Pilar Muñoz Dueñas (University of Vigo, Spain) Lucia Aiello (Sapienza University of Rome, Italy) Rosario Cabrita (University Nova de Lisboa, Portugal) and Mauro Gatti (Sapienza University of Rome, Italy)
Business Science Reference • copyright 2018 • 328pp • H/C (ISBN: 9781522535515) • US $215.00 (our price)

Contemporary Identity and Access Management Architectures Emerging Research and Opportunities
Alex Chi Keung Ng (Federation University, Australia)
Business Science Reference • copyright 2018 • 241pp • H/C (ISBN: 9781522548287) • US $175.00 (our price)

Research, Practices, and Innovations in Global Risk and Contingency Management
Kenneth David Strang (State University of New York, USA & APPC Research, Australia) Maximiliano E. Korstanje (University of Palermo, Argentina) and Narasimha Vajjhala (American University of Nigeria, Nigeria)
Business Science Reference • copyright 2018 • 418pp • H/C (ISBN: 9781522547549) • US $245.00 (our price)

Modeling and Simulation Techniques for Improved Business Processes
Maryam Ebrahimi (Azad University, Iran)
Business Science Reference • copyright 2018 • 246pp • H/C (ISBN: 9781522532262) • US $185.00 (our price)

Value Sharing for Sustainable and Inclusive Development
Mario Risso (Niccolò Cusano University, Italy) and Silvia Testarmata (Niccolò Cusano University, Italy)
Business Science Reference • copyright 2018 • 398pp • H/C (ISBN: 9781522531470) • US $245.00 (our price)

IGI Global
DISSEMINATOR OF KNOWLEDGE

701 East Chocolate Avenue, Hershey, PA 17033, USA
Tel: 717-533-8845 x100 • Fax: 717-533-8661
E-Mail: cust@igi-global.com • www.igi-global.com

George Leal Jamil dedicates this book to his sister, Beatriz Leal Jamil, fellow reviewer, for all love and help to our works' success.

List of Contributors

Table of Contents

Detailed Table of Contents

in times of crisis, this is a sector with constantly expanding business whether by incorporating new technologies, new players in research and development, as well as adjustments to public health policies, regulatory issues, and global outbreaks of disease. These facts lead to constant adjustments of business in companies, universities, and government actions. This statement is aligned in the knowledge-based strategy advocated by Etzkowitz. In 2017, three lists of new strategic products for the Brazilian health system had already been changed. Using new intelligence systems, the government adopted new strategic business partnerships and were conceived in 2017 (others replaced) with budgets of more than 6 billion reais. In this scenario, the revision of the chapter will feature new author and a scope of approach will be expanded to the new policy "Technological Platform" that replaces the old policy called partnership for productive development.

A relationship between project management and knowledge management was observed with a detailed level of analysis in this chapter, as analytics tools and methods were presented to define new perspectives for these dynamics. After a theoretical review that advanced the level reached by a previous paper on the same topic a new theoretical background was completely worked, resulting in a base where a deeper way of analysis allowed, at the end, to study practical cases of rich association for PM and KM in practical, ready to apply situations. As a trend for next competitive cycles, tools, methods, and techniques that focus knowledge production for decision making are to be increasingly defined and applied, on one hand enabling organizations to propose new competitive structures and positioning, and on the other hand, presenting a more aggressive, faster, and demanding competitive environment.

The digital economy is creating disruptions in traditional industries and markets. Industrial business associations (IBAs) may face serious challenges in a near future to meet the needs and requirements of their members, particularly in supporting their growing international trade activities and internationalization processes. Digital platforms are already transforming different types of businesses across all markets. An IBA may use a digital platform, not only to keep up with the current technological trends of markets, but also to improve the internationalization support provided to their associate small and medium enterprises (SMEs). Therefore, the aim of this chapter is to present the view of these potential digital platforms' managers, by presenting the results of an exploratory field research based on 24 interviews with IBAs from Portugal, France, and the UK. Another goal is to identify current digital platforms that are being used by IBAs and to critically evaluate their potential for supporting internationalization processes of SMEs. By using these findings, a set of requirements and features for digital platforms supporting SME internationalization in the context of IBAs are derived in this chapter. These results can be used by platform designers and by IBAs for designing and developing more effective digital platforms that can meet the specific internationalization needs of their users and managers.

This chapter discusses the issues of the needs and satisfaction of the customers, in terms of information, as a basis for the practice of marketing in information management. It enhances the arguments of the relationship between marketing and information science. The chapter stresses that practicing marketing cannot be done without information about customers and to customers and their relationship with information management, in information science. It stresses the importance of the studies and research on marketing of information as a philosophical approach for the information management process. The structure of the chapter synthesizes the existing academic work, seeking to generate new knowledge. It presents the promotion and communication of information in organizations from the evolution of the concept of marketing, in an integrated manner, in order complete the implications for future research.

Information, a tool to reduce uncertainty and develop knowledge in organizations, is an important aid in the decision-making process. There are several characteristics that describe the quality of information that will allow the analysis of its value. Quality information warrants us best results when competing with other organizations. Its value is related to the results that it will allow us to obtain, and it depends on the context. Patent information must be of high quality to permit the search and retrieval of the documents needed to solve a problem or stimulate new ideas and solutions. Old inventions can generate new ideas; technologies for one application can be introduced for a new domain and can be applied in ecologically sustainable solutions. The current high number of patent applications reduces the quality of patent information due to the time needed to filter and search for all the prior art available. Some standards, together with machine translation, have been set up to avoid this situation and improve the quality of the patent information retrieved by the interested public. A comprehensive survey of the relevant literature available made us aware that commercial databases supply some value-added information to help the researchers and improve the efficiency of the search queries. Some of these features could be applied by the national and international intellectual property offices.

This chapter approaches the context of healthcare, observing how information system conceptual background can help to comprehend the evolution and adoption of emerging technologies on the applicable solutions. As a critical sector, both from economic and life quality points of view, healthcare is an excellent environment to observe how these tools and its associated infrastructure make it possible for new business organizations to be proposed. The chapter is aimed to develop a theoretical and practical comprehension around concepts through isolated and integrated analysis.

In order to subsidize the public policy planners, identifying innovative sectorial nuclei that support the design of public policies to support innovation in companies, the chapter proposes and constructs an indicator of innovation potential with characteristics that fill an important gap, where it is possible to reach the groups of innovative companies in each sector (and their territories in which they are inserted), with a process of data collection focused on secondary sources and yet of high effectiveness and assertiveness. In this way, it is possible to make a detailed analysis of what is most directly relevant to decision makers, with the need for a low investment and the possibility of constant updating of the information. Therefore, this chapter will discuss the construction and demonstration of practical examples of the innovation potential indicator and their respective subindexes.

Interaction between devices, systems, and services is provided in internet through a new concept: internet of everything (IoE). This new concept should present numerous opportunities for the creation of new devices and applications in the years that follow and can dramatically affect the day-to-day life of all of us. New possibilities of information access, transmission, analysis, and interaction are practically infinity. With so many benefits, it is expected to come with several challenges. System structure and architecture and information confidentiality, integrity, and availability are aspects that someone should keep in mind to design an IoE application. This chapter offers to the reader an overview of the evolution of devices connectivity until it reaches the IoE. Some IoE concepts and applications, together with some challenges, are presented.

Many researchers have explored the knowledge management theory. However, to the author's knowledge few were interested in the tacit knowledge construct, whether it is gained inside or outside the organization. This chapter has a challenge to analyze in-depth the embedded knowledge gained from the customer, especially as it sheds light on the role of customer experiential knowledge by defining the customer experiential knowledge. It follows an emphasis on the customer experience and its close relationships with innovation management. Hence, a thorough theoretical background is presented progressively in order to define a new organizational competence labeled CEKMC. The first part presents an overview of the knowledge status, fundamental knowledge views, the evolutionary theory to the tacit knowledge construct. The second part stresses the definition of customer tacit knowledge related to customer experience. Finally, the conclusion defines a new organizational competence relative to this knowledge while discussing its contribution, especially to the experiential innovation type.

In this chapter, the implementation of an information strategy project in a Portuguese real estate company is analyzed. This involves its framing—historical and socio-economic—and a brief description of the activity sector in which the company operates. Several well managed projects have been developed to improve the competitive position of the company, but without focus all the activities lose strength because they might not reach their proposed targets. Some tools to identify the information needs of business activity developed are described, as well as the role of information as a promoter of competitive advantages. Social media tools were utilized and proved to be a great strategic decision. To conclude, a few reminders of the factors to consider in developing the information strategy to implement and that information management without a strategy could result in several diversified decisions without any positive consequence for the organization.

In the present context, the term intelligence has marked the economic and social fields. The increase in computational power of the technologies, associated with the activities of the people and organizations, economic know-how development needs of the market, the imperative of participation in economic dynamics established, among other things, is critical to the survival and for the sustainability of organizations. This is the dimension of competitive intelligence. But the term intelligence is not limited only to the external side of organizations, where this concept has gained broad meaning. This should also encompass the internal domain of organizations, organizational intelligence, and in this sense, refers to the centrality of functioning, flexibility of the articulation of resources, the indispensability of the systems architecture, and to the management insight. The ability to see important issues in the chain of value is crucial to identify the most important factors for success, to anticipate the competition, and to have pro-active management instead of reacting to market pressures.

The more than 100 million patents registered in the European Patent Office provide an unprecedented source of scientific and technological information in the history of mankind. The technological management of this information is exploited to develop technological advances in scientific, technological, and educational organizations and companies. New standards of product and process safety and effectiveness have been introduced across the world, and public and private business strategies are under constant review to comply with the prevailing paradigm. The health sector releases more than 1 million papers a year on scientific progress, while technological (patents) advance 10% per year. Therefore, updating the contribution of the information science through scientific and technological knowledge in intellectual property, a case study, will provide a contribution to reflection for the business in research, development, and innovation in health. These facts lead to constant adjustments of business in companies, universities, and government actions. In 2017, three lists of strategic products for the Brazilian Health System were changed. Using new intelligence systems, the government has adopted new strategic partnerships with the private sector, and were conceived in 2017 (others replaced) with budgets of more than US$ 2 billion. This chapter explores the contribution of information science through scientific and technological knowledge in intellectual property.

The challenges of the information society require frequent strategic reorientations that, in the context of the relational economy, emphasize the importance of IS/ICT in adapting to change, competitiveness, and economic sustainability. The easy adoption of the new technologies by the people, the paradigm of the availability of the new distribution channels, the reduction of the time-to-market, among other examples, have demanded to the responsible ones of the organizations and the SI dynamic of management, and adequate framework of the stated challenges. The preparation and management of the associated changes presuppose a coordinated and coherent combination of four vital axes for organizational success: business performance, information management, systems architecture and information technologies, and relational positioning. In this area, the development of an urban view of the value chain and its participants and the adoption of a logistics perspective on the momentum appear to be the primary adjustment in the support of organizational models to business.

A collaborative space for stimulating creativity is a place of learning through the exchange and sharing of knowledge and experience among its members. It allows the leveraging of innovation through the use of technological resources available in the space, stimulating the creativity of its participants, enabling the development of products and solutions based on personal projects—do it yourself (DIY)—from ideation, or the construction supported on knowledge developed by other elements together, collaboratively, enhancing the final result—do it with others (DIWO). A research project is being held to create a new lab, or transform and adapt one of the existing lab's, in a Fab Lab or a Maker Space to let students, teachers, and staff give wings to their imagination and develop innovative solutions to solve real problems while they interact and exchange tacit knowledge, making it explicit after concluding their projects when they share their research reports.

Revisiting a concept is a mandatory approach when new technologies emerge. This chapter presents a reviewed study, based on a previous publication, about information quality. As it happened in the first attempt to study and develop this concept, it remains challenging, as this context contemplates both tangible and intangible observations for two unlimited parts—information and quality—researched isolated, with its own alternatives, methods and objectives for several years and from different points of view. This chapter's approach is to bring additional work published since the original one and develop conclusions reached at that time, providing a new level of comprehension, which allows the continuity of this debate of information quality, considering the emerging strong trend of analytics processing and informational service offering.

With the evolving understanding of the role of the supply chain management and its potential to add value, integrated business planning is a key concept in any modern organization. That brings complexity to supply chain management, requiring companies aiming to operate a world-class process to have a strong coordination between internal functions, which is only possible with a highly efficient information management framework. This chapter discusses how companies can extract competitive advantage from the use of available information on the supply side. For that, as based on the supply-side methodology, a new analytical layer is built and applied based on a multivariate statistical approach that makes possible the creation of groups according to characteristics of the capacity to supply goods and services. As analytical locus, the authors selected the Minas Gerais territory, Brazil.

This chapter begins with an introduction to the telecom sector, communication using mobile phones, and evolution of wireless communication. It presents the current scenario in this sector, the challenges faced by the telecom industry, especially with ever increasing data network traffic. Finally, the possibilities of harnessing the power of big data analytics, new techniques, and technologies that drive innovation in telecom are presented to help service providers make better decisions and react quickly to threats on the competitive horizon.

Following the reflection around the emergence of "research university" in the context of the slow but progressive increase in value of science and technology and research and development in Portugal, a study applied to the knowledge transfer and the process of innovation in the university, in the context of a master dissertation in information science (IS), study area of information management, is presented. The university is one of the most important institutions in the context of the national innovation system (SNI), being part of its mission the creation and transfer of knowledge. At the University of Porto (U.Porto), projects, such as the University of Porto Innovation unit (U.Porto Inovação) and the Science and Technology Park of the University of Porto (UPTEC) seek to support the university's innovation value chain, promoting the reinforcement and solidification of knowledge transfer and of the relations between the university and companies, as well as the promotion and support to the creation of companies with a technological, scientific, and creative base, and the attraction of numerous innovation centers of national and international companies. This chapter points out an informational perspective on I&D+i (research and development and innovation) and entrepreneurship, based on the systemic theory and the quadripolar method, as theoretical and methodological guidance tools, and an information management/knowledge management approach of innovation models for the knowledge economy, the national and international referents, and corresponding set of indicators. An exploratory study, which allowed the identification of internal and external agents, the resources, the relations between actors and institutions, the processes and flows, and the main inputs and outputs, is presented. The most relevant result is embodied in a model of innovation indicators in an academic context and applied to the University of Porto.

Preface

It was another glad event, in my relationship with IGI Global, when the invitation came to edit this work, *Handbook of Research on Expanding Business Opportunities With Information Systems and Analytics*. The proposed challenge took us to the approach of several already publishes papers and chapters, welcoming some new works, promoting a better understanding of new technologies fronts, presented by the "analytics" strong trend on automated data collection to produce information and knowledge and their integration with the classical conceptual context of information systems.

Generally, it was understood that this new age of data collection will affect all aspects of businesses, improving and bringing complexity to management. Also, the demand for integration of all resources, managerial principles, processes and people development fit in the theoretical base of information system, a classical, traditional background that was invoked frequently by different scientific fields, as Economy, Management, Production Engineering, Computing Science, Information Science and Technology, among many others, to improve the contextual and complete, integrated studies. As those research fronts were confronted and, eventually, put together to become an analytical prism, this book could receive contributions of researchers from different fields, producing a multidisciplinary content, highly desirable to interpret new waves of technology and management, as those we now face.

This result in a high expectation for contributions of the chapters here presented, when some of older studies were improved, updated – not excluding the old ones, that was an editorial directive – and new ones were added, resulting in a literature that can advance overall comprehension on information and knowledge management. From traditional sectors to new startups arrangement, from classical levels of technological resources, to modern internet-of-the-things components and integration, passing through big data trends, analytics and information systems are two strong concepts which must be addressed carefully, to result in potential competitive advantage offer and managerial knowledge generation. This, among other trends, can affect and develop new ways for decision-making and implementation soon.

In the following, the nineteen chapters of this book are presented.

In Chapter 1, George Jamil, Leandro Rocha and Cecília Carvalho approached market intelligence, developing former studies they already published and practiced for competitive, market sectors. Market intelligence presents itself as a powerful cyclic process for commercial industries, also becoming a process to integrate data and information, spread in market arrangements and sectors, to produce consistent knowledge for decision-making in strategic marketing planning effort. As a robust process for knowledge generation, market intelligence is related to information system concept and its related developments and observed together with analytics, composing an interesting conceptual background which advanced from the previous studies, an evolution for this important and applicable technology-drive process.

Jorge Magalhães, Marlene Menezes, Zulmira Hart and Adelaide Antunes discussed, in Chapter 2, the relationship of information system conceptualization for healthcare sector. This is strong trend for several years, as this is an undisputable critical sector which presents a constant evolution, based on technology, process management, specialization, continuous development and facing strong challenges, as commercial and regulatory issues. In this study, authors evaluated the modernization of managerial techniques, introducing a new policy as a methodological aspect, improving a previous study, also amplifying the outcomes concerning sectorial capabilities on producing knowledge for strategic decision-making in the new era, facing those growing challenges.

An important area of study which has becoming progressive more impacted by the adoption of new technologies is Project Management. A project, with classical concepts and managerial aspects, is continuously being classified, promoted, integrated, executed and kept with growing usage and adoption of big data, analytics, IoT, among other emerging technologies. George Jamil and Luiz Fernando Magalhães, in Chapter 3, promoted an opportune update in a previous study around these issues, integrating it from knowledge management point of view, adopting information system approach. The association of these conceptual arenas – project management, technology, information system, data science and analysis – results in a powerful and detailed method to understand precisely how projects can be better understood and produced soon, being this chapter a contribution to develop this comprehension.

Eric Costa, Antonio Soares and Jorge Pinho de Sousa, researchers associated with InescTec of Portugal, produced a remarkable view for digital economy in Chapter 4, analyzing the introduction of international business association (IBA) theoretical conceptualization, aiming to understand disruptive events and their associated implications. As a result of a field research held in Portuguese, French and British organizations, this study presents how digital platforms, a context where information is produced and managed dynamically, can promote better conditions for internationalization of small and medium enterprises (SMEs). Another production of this study is a method to evaluate how these platforms can be designed in the future, allowing application of its results by analysts and developers.

In Chapter 5, José Rascão approached one of the richest contexts for applied information management research: marketing planning. Observing the customer satisfaction factor, the author proposed information marketing context for an observation arena, aiming to understand the contours of information management application for marketing decisions, implicating also in assessing of communication, commercialization and overall value positioning and offer through information systems at times of emerging technologies and methods. It is important to develop, as another important result of his reflections, the comprehension on information as a valuable product, both for organizations and for final customers.

As for decision-making processes, Sérgio Lopes, Sérgio Oliveira and Paulo Melo studied, in Chapter 6, how information management can promote a better way of scenarios choosing, approaching a competitive regulatory aspect: patent and licensing systems. As analytics and other emerging technologic resources can be applied, a better condition to promote a precise, efficient and qualitative mechanism can be reached, allowing to analyze the immense content of patents to identify potential industrial standards. This chapter develops a contextual analysis of actual literature, with implications for improved methodologies in adoption by international intellectual property offices and managers.

As discussed in other chapters of this book, healthcare sector became a dramatic scenario for emerging technologies adoption. As this organizational arrangement aims to improve our quality of life and, essentially, our living, technologies and the associated managerial implications must be applied to answer regulation, demand for new medical and related interventions, knowledge development for strategic management, among several other perspectives. Liliane Jamil, Augusto Vieira and Antonio

Xavier offered a study where reflections produced in a previous analysis are updated, taking analytics and the information system context in addition, to evaluate how information management can benefit healthcare business and research in the future. In Chapter 7, authors aimed to develop these thoughts, allowing to compose a base where this discussion can be further researched and practiced in various fields of healthcare services.

Chapter 8, produced by Brenner Lopes and Luander Falcão, assessed the need of a methodology to support public policy planners, to promote strategic investments to foster innovative entrepreneurial arrangements. This study proposes a set of indicators, associated with a collection and analysis process, oriented to produce results that will help evaluate competitive contexts, basic infrastructural issues, capabilities, competences and demands, from primary and secondary sources of information. This way, it gathers innovation, innovation management, information and knowledge composition and its consolidation for practical, public strategic decision-making criterion. At the end, some practical cases are worked, illustrating how this method can be efficiently applied.

Internet of things can become an important data collection and production supporting technology, innovating to bring automation and integrative approach, dynamically configuring networks and objects autonomous connections. To discuss better the introduction of this already successful technology, Cláudio Pessoa, Cássio Bastista and Marco Elísio Marques wrote Chapter 9. Producing a basic conceptual review, a discussion about market perspectives and their overall integration to other systems, from isolated devices to be applied to our routine living actions, ranging to sophisticated cloud supporting connections, this chapter contributes on leveling the knowledge about IoT fundamentals and promotion of a platform for new studies.

Dhouha Jaziri presents a potential association of knowledge management theories with the emerging scenario of data science, emerging technologies, analytics and organizational arrangements, proposed by the book project definition. The author, in Chapter 10, approaches marketing theories, from the customer point of view, who is taken as an active, systemic agent in interactions with value offer positioning by strategic and competitive market players. In her new study, professor Dhouha emphasizes the relationship of customers experiences regarding value offers when producing a set of organizational competencies, relating this continuum to knowledge generation, potentially emerging from new technological scenario.

Researchers Sérgio Lopes, Sérgio Oliveira and Paulo Melo wrote Chapter 11 to contribute about information strategy, as the overall and essential planning abilities to understand and project a set of future actions regarding information management in modern organizations. Authors approached the relevant context of social media usage for business in the digital era with the innovative promising support of emerging technologies. After a contributive association on theoretical fundamentals, which produces a fair base to promote new studies, they also presented a good approach regarding a practical case of a Portuguese company.

In Chapter 12, authors Pedro Anunciação and Antonio Peñalver analyzed the context of organizational intelligence. This remarkable association with a wide and prospective contextual background, announced by intelligence concept, helps define precisely this concept, around some of its practical observations, like competitive intelligence. This presentation of OI relates also to several other conceptual contexts, such as information systems and information architecture, producing a consistent base for conceptual definition, done in the perspectives of technological waves as those studied for the book proposition.

Information science is considered as an active background to observe and produce comprehension around information management phenomena. This lemma is the base of Chapter 13, where authors Jorge Magalhães, Flávia Mendes, Adelaide Antunes and Zulmira Hartz observed the potential of information science methodological analysis on an interesting and peculiar organizational case: the proposition of new criterion to select and classify patents from huge public knowledge bases, allowing their application for decision-making in the organizational context where huge industrial complexes compete for healthcare markets and knowledge-propelled profits.

An innovative proposition to comprehend the immense context of information systems, when applied to practical decision situations, is addressed by Pedro Anunciação, Francisco Esteves and Fernando Gonçalves in Chapter 14. In an interesting study, these authors relate the business observation of information systems to organizational urbanism, proposing it as the way of analyzing the integration of information management to all areas of one organization. This conceptual background is developed taking four main axial aspects of study – business performance, information management, systems architecture and information technology positioning – building an interesting and powerful alternative to improve understandings of information science and its methods when comprehending how information can be generated, collected, produced, stored and delivered to main users in a value chain.

In Chapter 15, Sérgio Lopes and Joberto Martins evaluated the rich contemporary arena of cooperative spaces of ideation, regarding innovation production around new projects. Associating the aspects of DIY (do it yourself) to DIWO (do it with others), authors proposed an integrated overview of one evolutionary path towards collaborative project management based on new technologies, information and knowledge management. It is another interesting approach, presented in this book, that observes how innovation concept can be detailed, improved and developed when related to emerging technologies introduction in organizational competitive arrangements.

Chapter 16, authored by George Jamil, reviews the concept of information quality, both as a static pre-definition, to be attributed to informational application of a content, and a dynamic process which will define perspectives for optimal information usage in any organizational process. This chapter also observes how information technology and other emerging trends can potentialize new systems and value offers based on information services and production, which, eventually, can result in quality loss, posing difficulties and disappointments for the applied information management principles. As this last process implies in better technological specifications and adoption, quality becomes a critical issue, deserving the observation and study presented by the author.

Information applied to decision-making is one of the main scenarios intended for the book project development. Investment plans, because of information availability, became a complex decision context, illustrated by Brenner Lopes and Luander Falcão in Chapter 17. Authors analyze a potential method to promote understanding, through information system and management in the state of Minas Gerais, in Brazil. Their approach encompasses one sectorial approach regarding optimal supply chain management in this competitive area of Brazilian territory, serviced for many different players, connecting industrial and service facilities and attractive markets. Information management perspectives are directly related to perspectives of strategic decision-making in a demanding context.

Chapter 18, written by Vishal Mahajan, Renuka Mahajan and Richa Misra, studied how analytics tools and methods are applied in the Indian telecommunications market to prevent customer churn. Telecom industry became a tough competitive market, facing all threats of competition, approached by strategy and marketing authors for several years – as "war on prices", customer volatility, increasing of customer bargain power – among several others. The application of new, emerging technologies, as analyzed and presented by these authors, relate the new resources, all of them applied to emerging informational trends, to produce advantage in a traditional, competitive sector.

Finally, Chapter 19, proposed and authored by Fábio Gonçalves, Maria Manuela Pinto and Alexandra Xavier, observed how information management can benefit knowledge transferring aspects, from universities to markets, specifically studying the Portuguese business environment. With experience in sponsoring programs and researches regarding innovative industries, startups arrangements and promotion and new technologies application for business, authors analyzed the perspective of comprehension models – as the Quadripolar method – to understand how information can help optimally accomplish this relationship on knowledge transfer, critical for economic systems.

This book enabled us to witness a special moment when emerging technologies start to be intensively applied in business contexts, for decisions that encompass simple, operational, daily tasks, to critical, complex and rare strategic alternatives. During this project, the continuous interaction with authors, themes, topics and several questions that emerge from this fascinating and turbulent context of new technologies and its associated methods, processes and tools, it was possible to observe and understand the complexity around the attractive and risky implementation of new ideas.

Will this new wave of resources result in competitive advantage for the pioneers? Will it be worthy, when observing the huge investments not only for technology adoption, but also with its related supporting structure – personnel, organizational structures, market definitions, and many others? Will information management remain ethical, effectively contributing for humanitarian quality of life promotion, or will it produce a new wave of untreated unemployment, with many undesirable, complex and socially risky factors? Is there any identifiable kind of methodological base to be applied in the adoption of these new technologies? Will the useful, actual and well-constructed set of principles of information system management for organizations be comprehended to allow controlled, planned and qualified plans for emerging alternatives adoptions?

These questions and many others follow the apparent excitation related to analytics, big data, internet of things and other information and knowledge management tools these days. This book intends to promote the questions, scientifically raise precise doubts and initially propose a path for some answers.

Good reading!

George Leal Jamil
Informações em Rede Consultoria e Treinamento Ltda., Brazil

Chapter 1
Market Intelligence as an Information System Element:
Delivering Knowledge for Decisions in a Continuous Process

George Leal Jamil
Informações em Rede Consultoria e Treinamento Ltda., Brazil

Leandro Henrique Rocha dos Santos
In3 – Informação, Inteligência e Inovação Ltda., Brazil

Cecília Carvalho Jamil
Institute Français de la Mode, Brazil

ABSTRACT

The market intelligence process has been studied from several points of view and different approaches. In this chapter, the authors discuss the perspective of market intelligence as a component of information system projects and specifications. As the results show, both from theoretical perspectives and studies, and from successful practical applications, this better comprehension results in a possibility of higher levels of knowledge production, allowing the integration of marketing strategic, tactical, and operational perspectives with also a collateral benefit to integrate other organizational processes, such as competitive intelligence. The chapter text approaches conceptual development and relationship and practical cases observation, which will compose the final scenario where market intelligence, treated as an information system active component, is a vital and strategic tool to implement competitive advantage alternatives.

INTRODUCTION

As market intelligence (MI) is an evolutionary concept, in the sense it is regarded to strategic marketing planning, a process which is being re-evaluated in modern, technologically-driven new contexts, understanding its perspectives and applications become an opportune development. In this chapter, as

DOI: 10.4018/978-1-5225-6225-2.ch001

an update of Jamil (2015), new trends of analytics, big data and the wide coverage promoted by the information system concept will be taken to reevaluate marketing intelligence conceptual background and its applicable unfolding.

Market intelligence performs what can be called a multidisciplinary context. From various contributions, as Soares and Carvalho (2013) and Jamil (2015), among many others, this process has been discussed for many years, as it integrates aspects and facts that emerge for many scientific and practical areas, such as Marketing, Strategy, Information Technology, Information Science, Market research and others. Interestingly, its results also promote a wide repercussion over market chains, producing several strategic opportunities that can be evaluated for traditional and new companies, organizational propositions.

It is opportune to affirm that, in the view of this chapter, market intelligence is a process for one stance of knowledge management in organizations, aiming to productively deliver knowledge for decisions. Proposing an initial view, MI can be regarded as a cyclic process to provide knowledge for strategic marketing planning decisions, for groups of homogeneous organizations or "sectors". Organizations must not only react to external factors or phenomena, but also act effectively, trying to lead their sector proposing and executing plans and differential strategic positioning, approaching Marketing tactics and operational formulations as one of the most relevant alternatives (Boblitz, 2006; Porter, 2008; Mintzberg, Ahlstrand & Lampel, 2009; Kim & Mauborgne, 2015). Market intelligence is affirmed as a process resulting in an organizational continuum that aims to answer typical decision problems faced by firms when competing in actual business environments (Van Kesteren, 2012; Jamil, Santos, Alves, & e Furbino, 2012).

As a provocative practical research opportunity, market intelligence, this way, can be comprehended as a component of an information system. As conceptualized by Stair and Reynolds (2008) and O´Brien and Marakas (2008), information systems are a set of components, adjusted and implemented towards processing information for one organization. There are several points of mutual interest when relating these concepts – IS and MI – as this chapter will search for this potential correspondence to understand better this perspective. According to the study of Mac Innis (2011), it is possible to understand how conceptual studies are still needed to understand and promote the evolution of Marketing principles, applied to organizational actions. At a theoretical level, information systems compose a correlated set of hardware, software, skilled and prepared people, communication resources, process definitions and, overall, definition of data availability in all organizational flows. In a practical view, information systems produce arrangements of information, potentially providing knowledge for several corporative, organizational implementations, such as decision-making (Turban, Mc Lean & Wetherbee, 2002; Turban, Rainer Jr., Potter, 2007; Stair & Reynolds, 2008; O´Brien & Marakas, 2008).

Market intelligence, in its way, is a continuous process that aims to produce knowledge for strategic marketing decisions, such as planning, which will coordinate both tactical and operational marketing efforts and implementations, according to the formal orientation regarded to strategic plans (Albescu & Pugna, 2014; Jamil, 2012; Jamil, 2015).

Potential contacts amongst market intelligence, competitive intelligence, data analytics, big data analysis and information systems concepts will be studied in this chapter, to reflect around the possibilities of better results of its application towards an "intelligent organization" design and performance. As those concepts and its relationship can enhance substantially an organizational structure, design and development, they can also provoke additional level of complexity, resulting in more confusion for entrepreneurs and planners. This chapter intends to contribute, with an additional study regarding a practical view of this conceptual background and its relationships, bringing light, in an orderly way, turning possible better implementation of applicable management instruments, decisions and choices around information and knowledge management.

Here we aim to detail MI concept, working from theoretical point of view through literature and adds a practical approach on studying real cases of market intelligence potential applications. As MI is a multidisciplinary context, its conceptualization also provokes perspectives for research from other scientific fields, as Information Science, Systems and Management, Computing Science, Human resources management, Strategy, Marketing, among many others.

BACKGROUND

According to the goals of this chapter, a theoretical review must be conducted, enabling to understand and promote the study of a conceptual background relationship. This relationship presents an additional goal for this chapter, as it, *per se*, constitutes in an expressive goal, as it is useful for several researches which will approach MI concept dynamically, in potential relations with other organizational processes, integrating the boundaries of what can be called and correctly identified as an information system.

Analytics

A powerful process can be observed when we understand data production as one of the essential facts towards knowledge generation for any organization. As knowledge is critical to innovate, decide, communicate and implement decisions, data generation, as one of its components, must be perceived correctly to understand the limits to be reached with the knowledge produced. As we face new communication and technological trends, such as Internet of things and big data, we also are facing situations where data will be generated in a more massive way, bringing another level of knowledge generation perspective, but also presenting challenges to knowledge production (Jamil & Magalhães, 2015).

Analytics trends emerge as this massive data arises from one objective source: internet interactions. As users interact through internet with the multifaceted characteristics of applications, such as mobile systems and "apps", software, pieces and components of software and hardware, domestic utilities such as TV sets and other apparatus which, indeed, are computers that produce data around users´ commands and actions, data about these activities is potentially produced all the time (Chau & Xu, 2012). Observing, for example, the case of Google, as a noticeable and reputable company, that offers solutions from e-mail services, open communication platforms, social media, mapping services (with support for traffic management, at a corporative and personal level), internet browser and operational system for desktops, mobile computers and telephones / smartphones. Google, this way, can absorb data from several different sources, recording it to user behavior studies, such as market profiling, interests, reactions and others. Google can promote a deep level of understanding toward users´ comprehension and behavior prediction, which allow this company and its associates to better decide around a future offer or interaction with some specific group of users (Choo, 2003).

As it can be understood from SAS (2014) and SAS (2017), analytics is a general expression that identifies software utilities, analytical methods, graphical interfaces and various other supporting tools that will provide contents of information and knowledge about users' interactions with specific sites, such as news, social media and electronic commerce. As it can be perceived from Google (2017), typical information produced, in a very first level of analysis, reports times, actions and preferences adopted by users when they are doing regular actions in internet, such as browsing electronic catalogs of products, posting comments over a previous initiative of other user, producing a visual interaction in a social media platform, consulting – with variable level of interest – a content in internet, like a summary report of a banking account, as many others.

With these reports, one analyst can, for example, refine his studies about segmentation and user profiling, user reaction predictions and forecasts, user demands for information and responsive actions in one commercial transaction, among many other alternatives (El-Gayar & Timsina, 2014). Some major contributions from analytics applications are:

- Fast, dynamic, responsive reactions can be made, as instantaneous analysis is easier to implement, when compared to more rigid, strict systems. This way, a commercial organization can analyze the acceptance of one of their sales offer through internet, enabling managerial actions in few minutes, as to implement already-coded changes in one website, through a nicer interface, and changing its business model basics, as one user can receive a more provocative offer, modifying ways of payment and additional value through additional products.
- Organizational memory of external events, as suppliers and customers, can be tracked, by the analysis provided by analytics tools, enhancing the perceptions of the external competitive environment, an essential approach to strategic planning methodologies. This can result in better, more precise goal definitions and actions plans.
- Completely absorb users' behavior around news, offers, announcements and promotions, producing a better way to specify communication campaigns and more precise advertisements.
- When thinking of one company different from the commercial arena, as in Healthcare, Governmental public services (as safety or transportation), analytics can also provide subsidies to understand citizens needs, such as services information, reaction to campaigns and other cultural, social and market manifests. This will enable the design of better management for public services.

One opportune thing to observe, around analytics, is that there are many companies offering contents, services, researches and tools to promote their applications, toward corporative decisions, plans and market approaches, as it is possible to verify from big players, such as Google, Facebook, SAS, Oracle and many others, who owns some huge databases and well-designed tools to process analytics, and other small, agile, flexible and focused companies, or startups, that will provide some complementary tools and processes for collection, analysis and knowledge production (Jamil & Magalhães, 2015; Yonce, Taylor, Kelly, & Gnau, 2017).

Analytics continuum is, undoubtfully, a strong trend in the forthcoming years, as it is becoming a real level of software to be applied for knowledge generation for business strategic decisions, or, as it will be discussed in the next section, an important part of an organizational information system.

Information Systems

The concept of information system shows how one practical evidence can eventually impose some restrictions over the scientific definition of a research object. Sometimes confused with information technology, software or even to academic contexts, such as software engineering, information systems can be defined as a set of coordinated components put together to produce knowledge from data and information. Although the main target of an information system can be regarded as being information, it is opportune to associate this concept to knowledge production (Stair & Reynolds, 2008; Jamil, 2014). It is obvious that a relationship between information systems and market intelligence (and other concepts referring to intelligence) emerges from this first observation.

It is interesting to define, initially, that information system is not associated to contexts, sectors that will use information, organization types and, mainly, time lapses or opportunities. It is a broad concept that will encompass all these definitions, concepts and conceptual relationships, bringing an opportune background, allowing a potential discussion around capabilities and strategic resources that can emerge from this relationship (Jamil, 2013).

According to Jamil (2012), an information system project will have to address these components:

- **Hardware:** Not restricted to computers, but also covering all hardware pieces, such as sensors, transmitters, network connection components, etc. that will perform the data transmission path from the evidence, collection point, to other components of the IS.
- **Software:** Conceptualized as the active parts that will process, automatize operations, analyze and inform productions of the information system. As software invades our daily lives, embedded even in small domestic appliances, it becomes more valid as an instrument of human interaction. Some usual software pieces are not related nowadays to computers, as they are used isolated, but interfaces to usual, trivial instruments such as TV sets, bathroom utilities, kitchen appliances, smartphones, etc.
- **People:** As data and information providers, information and knowledge users, processors, active agents and evaluators, people must be identified, classified and understood, to completely domain our interactions that will demand all the processing designed for one information system. Also, it is important to state that ethical aspects of knowledge production and application, emerge when "people" element is precisely addressed.
- **Communication Resources:** As results from one information processing must be communicated to their users to be applied for some work, activity or decision. In these communication resources, social media interfaces, displays connected to internet of things servers (such as glasses, panels, etc.), among many others, can be considered.
- **Organizational Structures:** To understand precisely where data and information is produced, stored and handled in one organizational array, noticing that, sometimes, this arrangement extends from the boundaries of a strict organizational design.
- **Process Definitions:** As one information system will interact and produce contents from and to organizational processes, such as supplying knowledge to the operational level, allowing to adjust machinery (a robotic array, for example) performance and sequence or, in another perspective, for strategic level, contribute to a more precise and objective strategic decision. This way, the interaction and comprehension about all these processes are extremally useful to compose the IS project design.
- **Data "Understanding":** Denoting where is produced, how it is produced, the perspective on composing data for information, and so on. Data and information must not be only "mapped", but understood by the information system engineer or analyst, consolidating their roles as sources to be processed to produce the desired knowledge.

With this wide and complete definition, we intend to produce a context for the big data, analytic techniques and market intelligence services inter-relate, composing one information system definition and one information system practical specification. Next, it is interesting to understand, from the conceptual points of view, how an information system will interact to these other items to produce the results demanded by organizations.

Organizational Strategy and Knowledge Demands

It is possible to understand the requirements of knowledge for strategic planning phases (Kaplan & Norton, 2007; Johnson, 2012; Dimitrios, Sakas & Vlachos, 2013). From these and other authors, it is evident that, for strategy formulation, design, planning, execution and monitoring, how knowledge about various aspects, factors, historical events and processes is essential and critical for fair levels of strategic planning (Schiffman & Kanuk, 2010; Ferrel & Hartline, 2010). A detailed observation should be paid to strategic marketing decisions, as market intelligence has the fundamental contribution for these planning processes.

Strategic marketing decisions are the main market intelligence process objective, as the knowledge produced can be used for studies and decisions with clearer risk delimitations and definitions for implementations, with better specifications for customer aggregated value factors, answering to the basic organizational marketing demands (Kotler & Keller, 2005; De Man, 2012).

As typical decisions of strategic marketing processes that can be targeted by market intelligence process results are: product line configuration, chain adjustment and distribution, pricing (and financial management), advertising and general communication responses, differentiation as a value-based strategy and marketing channel comparative analysis. As examples of knowledge needed in usual marketing decisions, it is possible to cite: consumer behavior details, demographic perspectives and constraints, customer reaction to distribution forms and communication, distribution channels performance and financial performance for all productive chain components. Knowledge provided for marketing strategic planning will allow any organization to develop its practical, tactical-to-operations plans, which will, at the end, specify the real work to be executed, aligned to those strategic views and propositions (Kotler & Keller, 2005; Schifman & Kanuk, 2010).

One important perspective is to understand strategic marketing planning as a real connection of Marketing to Strategy (Soares & Carvalho, 2013; Jamil, 2013). Strategic goals are defined because of the strategic planning process, corresponding to the composition of a final scenario to be reached in a period and scope of resources applied, resulting in the overall planned situation (Kaplan & Norton, 2007; Porter, 2008). Through these details, tactical and operational activities are expected to be developed with connection, alignment and adjustments, promoting the strategic governance of one company or organization. This potentially implicates in a coordination that guarantee better usage of resources, monitoring of future positioning, continuous chances for planning, strategic learning and, finally, a predicted, planned implementation of actions, with more understanding, even at the operational level, to the basic question of "what is being done", concerning the future of the company.

Data, Information, and Knowledge

Knowledge can be understood as composed from essential elements, data and information. This proposed structure is approached by several authors, and allow the comprehension and validity of market intelligence process. Approaching from a managerial viewpoint, we can consider the works of Davenport and

Prusak (2000), Tuomi (2000) and Lucas Jr. (2005), who defined not only the concepts of data, information and knowledge but also their relationship and integrated application views.

Data can be considered as an absolute value that can be obtained directly from a measurement activity or collected from an automated source (Tuomi, 2000; Jamil, 2005). This definition presents how data is flexible to communicate, as it can be easily generated, collected and processed by automated instruments and machinery, an important perspective when dealing with the new context of technological implementations. However, data lacks more meaning, as it is almost impossible to deduce more from the context, development or process from it was created, being just an instantaneous observation of a reality.

Information is conceptualized, from these and other authors, such as Inmon, Strauss, and Neushloss, (2008), as a collection of homogeneous data collected in a context, providing a better condition for deciding, but offering more complexity to be treated or processed – stored, shared and standardized. Information increases decision capabilities, but demands additional, sophisticated and expensive work to be applied, as additional work must be conducted to assure its homogeneity and cohesion (as one example: measurements are collected, corresponding to data from one specific event, in homogeneous, identical conditions – composing information about this phenomenon).

In development, knowledge can be understood as composed through the collection of information, including details and delimitation of the processes which produced that information. Knowledge allows maximum decision capabilities, for example, enabling even prediction capabilities which are essential for planning (Davenport, 2000; Jamil, 2005; Nonaka, 2008, Badia, 2014). But on the other hand, knowledge is difficult – complex, expensive – to manage, communicate, being questionable to be stored and critical to be shared, resulting in a need of a specific process to treat it, which is defined as knowledge management (Jamil, 2001; El-Bashir, Collier & Sutton, 2011).

It is opportune to define, for example, the new trend of "big data" studies, as one area that will receive potential contributions of knowledge management research, as this fundamental aspect of conceptual relationship is being significantly treated, increasingly perceived by entrepreneurs and socio-political actors. Additional attention must be paid to the process of "knowledge generated from data" which proves how important is to observe data, information and knowledge in an organizational context for decisions such as strategic marketing planning (Ohata & Kumar, 2012; Park, Huh, Oh & Han, 2012).

Market Intelligence

Intelligence is a concept that provoked many scientists and practitioners of various sectors, nowadays calling additional attention due to the emerging trend of several new technologies and innovations, some of them addressed in this chapter and in this book. Some of most remarkable contributions from other scientific fields are discussed here, to improve market intelligence comprehension. This approach was also applied by several authors, as Makadok and Barney (2001), Markovitch, Steckel and Yeung (2005) and Albescu and Pugna (2014), as the contextual dynamics and common-sense comprehension, sometimes produce confuse, imprecise views about intelligence.

As approached before, in a first glance synthesis, market intelligence composes internal, sectorial (external, but correlated by internal organizational configurations), and external sources to result in knowledge to be applied mainly for strategic marketing contexts and decision-making oriented to implement aggregated value positioning for products and services in competitive contexts (Jamil, Santos, Alves, & e Furbino, 2012; De Man, 2012, Jamil, 2014).

Some concepts and theoretical fundamentals are now re-emerging, such as artificial intelligence. Disseminated as a powerful service, applied in mobile applications, electronic commerce services, among many others, artificial intelligence is a remarkable announcement as a concept, to be observed to build the desired comprehension about MI. "Artificial intelligence" (AI) was proposed in 1950 decade, as an academic subject, sometimes related to Philosophy or, alternatively, understood as a computational engineering support for automated or robotic system, as it was presented by John McCarty in 1956, in a conference at Dartmouth College. Along with futuristic views, of robots replacing human operators and decision-makers, it turned out to be a provocative area that motivated computing, mathematic, physics and specialists of other areas to develop theoretical and practical studies to comprehend its applications and limits.

It is possible to define artificial intelligence, improving perceptions and comprehensions about this vision of intelligence fundamentals, as defined by Russel and Norvig (2009): a continuum that promotes progressive and adaptive learning absorbed from the reality, through modelling and explicit relations, transforming it in rules as it is produced from accumulative knowledge created.

As one integrative view, artificial intelligence can be related to knowledge management exactly when it is perceived how this process of accumulation of knowledge results in composition of a set of rules that could serve for further implementations, decisions and automation of tasks. In this path, it is possible to understand how phases of the "knowledge spiral" cycle, proposed by Nonaka (2008), occur, as an approach to the same phenomenon. As a fundamental process of knowledge management, the spiral is so correspondent to one cycle of knowledge accumulation. In advance, comprehending the process of this progressive learning, it is possible to adhere a conceptualization for artificial intelligence, bringing light to the conceptual and theoretical perspective of association of these two relevant processes. As it was suggested for robotic and automated or controlled environments, artificial intelligence can be considered as a form to codify predictable knowledge for repetitive application, with accumulation of rules, perceptions, exact domains and contexts, that can occur from continuous cycles of process (Hubber, 1990; Weiss & Verma, 2002; McAfee & Brynjolfsson, 2012).

Another association, which usually occurs at the practitioners' side, is to confuse MI with competitive intelligence (CI). CI is a process that results in complementarity to market intelligence, although problems occur due to the lack of delimitation in some definitions of these two concepts. For competitive intelligence, it is adopted the statements of a reputable source, SCIP (2012), who defines it as: *A process of monitoring the competitive environment and analyzing the findings in the context of internal issues for decision support.*

Based on previous studies from Kahaner (1998) and Miller (2002), competitive intelligence is defined as a process related to strategic problems solution, configuring CI as an organizational process that has the goal to produce better decision support, specifically for higher executive levels, when addressing strategic problems. It is important to remark that, as a process primarily designed to produce knowledge from external environment interpretations, CI offers a potential objective response to a precise problem, usually proposed in a formulation, revision or sudden decision of a strategic planning task. Competitive intelligence can be defined, this way, as a cycle which is started by a strategic problem perception, translated into an objective question that must be answered to allow the development of strategic planning process, resulting in some reviews of practical actions to be implemented. It is so, a process designed from sources of past occurrences, which demands interventions from specialists, allowing the projection of actions and positioning in one strategic advance. It can, also, constitute a learning process, an implementation alternative for strategic planning advance (Makadok & Barney, 2001; Jamil, Santos, Alves, & e Furbino, 2012).

From information technology (IT) and computing science fields, another conceptual base is "business intelligence" (BI). This approach emerges as a relevant and broadly discussed concept, for several applications, also facing indeterminations from the application scenario. It defines many interpretations for set of tools, techniques, methods and integration of resources proposed to produce objective knowledge from disperse sources of information and data, as fundamentally analyzed by Inmon, Strauss and Nishloss (2008) and Kimball and Ross (2010). The Computing Science literature mainly focus on technical issues and details, such as implementation models, physical storage, database structuring and integration of BI to other IT platforms, such as mobile applications (Kimball & Ross, 2010). Thinking this way, intelligence is produced from technological infrastructural components together with its correspondent management, which offers not only the mechanisms for its development, integration between logical models to physical implementations, but also for knowledge codification, process modeling and implementation and knowledge communication and distribution, aspects understood as basics of a knowledge management process (McAfee & Brynjolfsson, 2012; Courtney, 2013; Jamil, 2014).

Another special aspect is integration, as the business intelligence usually is based on an infrastructure that is implemented through an associated arrangement of organizational resources, not regarding aspects, such as cyclic updates or evolution, but thinking it as a solution-driven knowledge provider (Cao, Zhang & Liu, 2006; Kimball & Ross, 2010; El-Bashir, Collier & Sutton, 2011; Hoffman, 2013).

The joined view of these concepts can be observed at a strategic level, where MI concept can be studied and reaffirmed. Both processes – CI for strategic decision, such as those in planning and executing strategies and MI for marketing strategic decisions – have the related strategic contribution, with the paradox of an eventual overlap, due to its incomplete, one-side view that not consider limits and real implementation details. It also allows to understand how they can be complementary and interrelated, reiterating that market intelligence has a more practical, applicative "to the market" objective, as it is designed to support a connection between organizational strategy formulation and answers provided by strategic marketing decisions (Johnson, 2012; Soares & Carvalho, 2013).

It is important to define, at the end of this conceptual background work, that viewpoints from market intelligence and Marketing intelligence are different. Although it seems obvious, this comparison must be addressed to carefully allow the comprehension discussed in this chapter. Market is a competitive context, where several pressures, forces, regulations and strategies are put in action, aiming to dynamic commercial and social interactions and exchanges among economic and social actors (Porter, 2008). Marketing, as it was conceptualized before, is a set of disciplines and contexts applied to study on how to aggregate value for products and services (Kotler & Keller, 2005). So, market intelligence, in our viewpoint for this work, is a continuous process that allows the production of knowledge from data and information to be applied for strategic market decisions, or, better defining, strategic marketing planning (Chen, Chiang & Storey, 2012; Vavra, 2012; Dimitrios, Sakas & Vlachos, 2013). This relationship structures a precise way of thinking, as market intelligence process helps one organization to practice a better Marketing planning, at the strategic level, becoming competitive in a market.

Setting the final adjustment for market intelligence process, it is important to define the main components of the MI process itself. This way, it must be considered that data and information are spread over a value-aggregated chain, as MI produces knowledge to be applied for sectorial strategic marketing decision on plans. Based on these definitions, market intelligence process is defined as the following continuous process, in a cycle definition (Jamil, 2014; Jamil, 2015):

Phase I: Preparation:

- **Aggregated Chain Value Analysis:** A comprehension step, which enables to understand precisely how the organizational arrangement where the MI cycle will occur and have its results applied.
- **Data and Information Diagnosis:** Where business sector contents of data and information can be detected and found, to constitute a mapping component that represents what is available for initial process design.
- **Knowledge Needs Diagnosis:** To conceive, initially, what are the main contents of knowledge demanded by typical market competitors, for decision-making in their strategic marketing planning work.
- **Observation:** These two last steps can be exchanged and even be practiced in a cycle, which interacts until the final users – for example, in focus group series of sessions (Bazeley, 2015) – converge about data and information locus and knowledge needs, as their final objectives, to proceed to the cyclic, continuum phase, detailed below.

Phase II: MI Continuum - Process (modeling and implementation):

- **Planning:** As a start / stop point of every cycle, it is a step of planning next cycle execution – resource allocation, pre and post requirements, deliveries, risks, etc. – based on the last cycle execution, gathering good practices experiences and planning schedules, budgets, human resources delegations, etc.
- **Collection:** It is the automated or manual gathering of data and information from organizations of the sector analyzed, from sources indicated by the mapping effort in the first phase. External sources, as government and agencies, research institutes and other partners, if diagnosed as essential or opportune, are also collected.
- **Validation:** Normalization, review and correction of data and information collected, as those sources can produce different (identification, formats, updates, etc.) or mistreated contents that must be leveled to become understandable as similar for cyclic analysis.
- **Processing:** Usage of business rules, logic modeling, statistical and mathematical analysis applied to data and information to produce specified knowledge contexts, usually with intervention of sectorial specialists and analysts, to assure the best level of results were produced to be sent to final users.
- **Sharing:** Results sharing of the processing that are delivered, presented, transmitted to interested market players for further application in their decision scenarios, through training sessions, social and technological implementations (as corporative social media, etc.), public information, etc.

This definition, encompassing the first conceptual affirmation and the structure depicted above, constitutes the theoretical background of market intelligence, being defined by the Figure 1.

In the following, an analysis of practical cases is held, validating this framework.

As a practical and final remark, it is usually recommended for market intelligence users, as we recall about those results are produced for a sector, a set of companies and market actors in a competitive context, that each company, economic actor would, thereafter, produce their own tactical and operational

Figure 1. Market intelligence cycle
Source: Author.

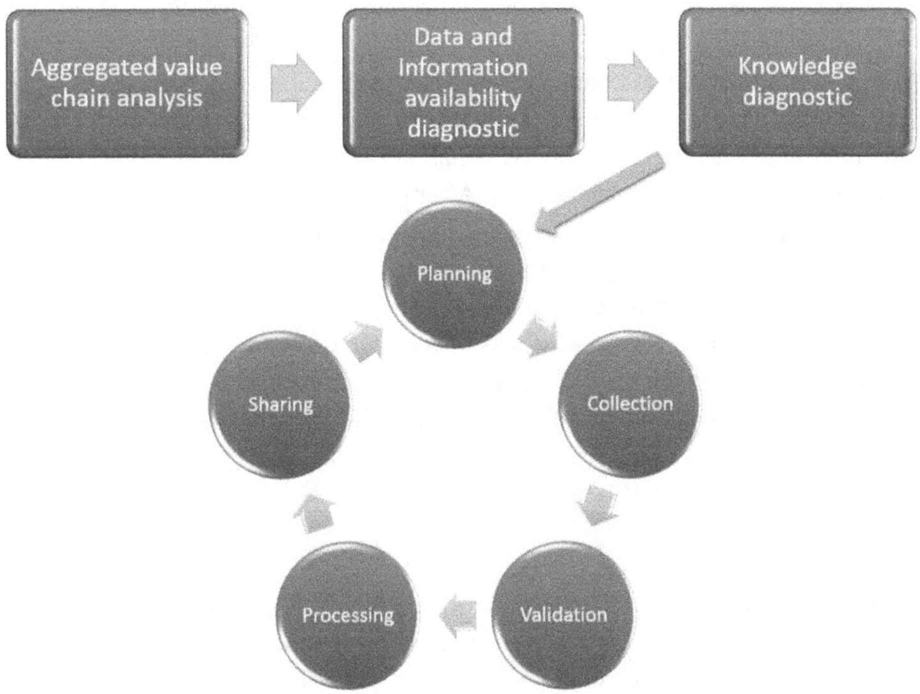

plans. So, the overall strategic coordination, designed by market intelligence production, is to be used as the guiding strategic base for further development of subsequent plans, as each agent receives these results, from MI and can, in advance, produce their own plans to implement better designed decisions for product and services to be strategically positioned in the competitive context.

MARKET INTELLIGENCE AS AN INFORMATION SYSTEM COMPONENT

In this section, some practical issues and cases will be approached, to promote the complete understanding of marketing intelligence process, as a component of an organizational information system. It is our main goal to promote an integral view, understanding how this completion can occur in competitive scenarios.

As defined in the original study about market intelligence, one basic, practice-oriented concept that must be observed is the informational association, and its related services, as it was presented by Marchand, Kettinger and Rollins (2001). As announced by these authors, this competitive grouping must be thought as a group of organizations that presents similar informational behaviors in terms of their problems, needs, infrastructures, research methods and processes and, overall, same knowledge requirements for planning decisions, especially around Marketing, which was the main focus of MI process.

The first case to be analyzed is a market intelligence real consulting case, held by a third-party consultancy for tourism organizations. In this case, the informational association, which will represent the economic agents who will be benefitted from the MI knowledge provision, is composed by companies, such as hotels (medium and small, considering the number of potential guests serviced per month and year), transport companies (for this research, just terrestrial transportation as mini-buses and private trains were considered) and tourism agencies, ranging from small to medium companies, considering their financial incomes and number of customers serviced per month and year.

Initially, a motivational work was conducted, as some representatives – managers and top-level employees – participated in a training session, held mainly through Internet, as an online course, to prepare for market intelligence perspectives, restrictions (what the cycle will not do for business), schedule, efforts, aspects to data and information privacy, etc. After this preparation, it was expected that all agents were prepared about the schedule, resources, delegations and objectives for the MI process.

Then, Phase I procedures – Value-aggregation chain modelling; Data and information sources identification and Knowledge needs according to simulations of strategic marketing planning decisions – were detailed, scheduled and planned. The collection of these sources was started, and overall conditions to check for each step validation were reinforced, as, at the final of Phase I, market intelligence analysts were satisfied over the results obtained, as the mapping for MI cycle was sufficiently designed.

From the information systems theoretical background, it is possible to identify how several informational components can be set for the MI process:

- Data and information contexts and contents.
- Processes designed, according to the production of data and information and knowledge needs. Also, the main phases of the market intelligence cycle are reported and scheduled, as the process is planned at the end of Phase I and the beginning of Phase II.
- People, with skills defined in profiles, trained and prepared to provide needed contents and receive the benefits of the MI cycle.
- Information technology associated plans – for hardware, software at all levels needed to implement the MI cycle.
- Organizational structure, as the design of inter-relations for organizations and agents is set, to allow all MI process to be structured and work.

For the Phase II, the plan step of the cycle was detailed, allowing the complete cycle to be precisely planned. In this step, forms and associated formatting specifications were designed and implemented, associated with simple and accessible tools, such as Google Documents spreadsheets and documents facilities. Guidance about the filling of these documents was provided through e-mails and associated small videos, uploaded to specific Youtube channel.

With the plan set, the collection step was started, as all interested users and agents provided their mapped data and information through those documents, filling the needs for further analysis. Typical data collected was several salaries per hour, for routinely and managerial tasks, costs to receive each type of guest (classification done by the agents´ representatives in the planning step of the cycle), maintenance costs, infrastructure (furniture, food, catering services, etc.) costs, technology needed, among several others. These data and information were collected from the sources detected and identified in the first Phase.

Validation, as the next step of Phase II, analyzed the data to review and correct discrepancies, strange values for data (for example, a cost associated to a typical worker payroll, which seems to be very high or very low), check data dependencies (as if the data referring to one property was corrected registered, as the dependent data was present), and so forth. This important step assures that all data and information collected will be supplied in the same format, standards, processing terms, names and identification, allowing it to be processed.

Advancing, the data and information were processed, to obtain, as typical analysis, maximum, minimum, average values for some critical variables. Also, trends were calculated and projected, for some variables. Interestingly, a "council" of specialists – formed by some top-level managers, experienced employees and even frequent tourists – analyzed and discussed the knowledge production, enabling to obtain several additional conclusions, to help inform all involved agents about the best way to implement some market decisions.

As typical tactical and operational marketing decisions, it was possible to identify:

- Association with a Cruise touristic company, offering discounts to some passenger profiles who will have the potential to contract additional services from hotels – such as tours, massages / spas, healthcare orientation, etc.
- Offer free or cheap transportation services for downtown, touristic areas, airports, bus stations and other attractions or facilities for tourists.
- Form a "club for purchase" of some supplies, such as food, beverage, cleaning materials, infrastructure, maintenance equipment, etc.
- Form also a "cooperative" of employees, allowing companies to share services, such as electricians, carpenters, waiters, concierge, etc.

These decisions are to be detailed, planned and implemented in specific plans, which will derive from MI knowledge provided, as this cycle will have the goal to produce strategic marketing planning orientation to be applied as a base for these subsequent phases and plans.

Concluding the MI cycle for this case, there was the knowledge sharing phase, as the results produced – strategic marketing planning knowledge – were sent to all involved agents through reports, which were convened by all participants in a standard accorded, previously, with a small, objective course on how to interpret some typical results. Consulting services for subsequent planning, at tactical and operational levels, was recommended by the consulting company, as, for ethical reasons, the company that acted as a MI consultancy can not work as specific consultant for one involved company because it accessed data, information and knowledge of all involved actors, being able to force decisions and interfere with the complete sector.

Finally, the cycle, in each execution or turn, prescribes a review, sometimes in an auditory style, to check if it performed as planned, resources were applied and adopted, answers were produced, users were satisfied, and if any risk emerged or correction action must be implemented in the next round. This way, the process, as designed for robustness, had the specification for its own perfection in each turn.

Again, it was completely possible to understand this cycle as an information system-style process, as, according to those definitions previously worked, market intelligence process produced knowledge, from data and information, based on an optimal structure formed by an adjusted set of organizational components. This example shows the complete alignment of the principles of information systems, and the powerful and wide process known as market intelligence. Other examples are discussed below, in a more superficial approach, as to illustrate this context application for other realities, with practical, commercial, marketing decisions being potentially implemented.

In the healthcare area, in another example, one association of research institutes, composed mainly by pharmaceutical and related companies, which aim to exchange knowledge, regarding laboratory drugs for human tests, produced an interesting proposition for market intelligence process. As this organization has an associative set of heterogeneous corporations – such as transnational research and commercial companies, small focused research firms and other agents, as distributors, regulators and implementation-oriented companies – knowledge sharing was a challenge. Justifying this affirmation, this association used to hold six different seminaries and congresses in a typical yearly schedule which gathered few interested people, not producing the desired level of discussion and organizational impact.

Knowledge collection, production and sharing, in this case, emerged as an excellent suggestion, as an alternative for this association to produce a more efficient way to reach their affiliates, delivering the desired answers and experiences. After a period of negotiations, a contract was registered, and the market intelligence process was defined. Starting a comprehension on how an information system concept was taking place, the composition of firms and associated objective for knowledge sharing, set by goals explicitly defined in the contract, along with data and information definitions to allow its collection, were finally addressed.

MI pre-cycle phase I was negotiated and defined with some internal sub-cycles to refine the data and information availability definitions and knowledge needs. For this last effort, it was opportune the adoption of a simulation, storytelling approach for modelling needs for a typical affiliate in a two-day training journey. After this step, the Phase I was completed, with another essential element of an information system project correctly addressed: people. After the interaction, people (some experts in medicine manufacturing, healthcare issues and regulations) not only could be considered trained, but also involved, motivated and gathered around the information system supported by market intelligence cycle.

Phase II steps, the MI cycle performed with some restrictions and difficulties in the start, until it reached a perfection level, correcting some assumptions proposed in the negotiation phase that not corresponded effectively to real situations on the field. After two years, it was possible to understand this cycle as a correctly proposed market intelligence implementation, which allowed also the association to perform its duties and responsibilities. Nowadays, in an example of productive change due to the success of this MI project, this association held, yearly, a bigger event with seven preparatory meetings, re-structuring its previous confuse, expensive and ineffective way of knowledge sharing journeys, along with other methods – as a blog and a social-media-like forum – to distribute the results of an information system adopted to reach its desired productions.

Finally, as other perspective of market intelligence application for an information system that encompass several users, it was possible to check the following experiences:

- Integration with competitive intelligence cycles, in information technology area. As a group of companies contracted a marketing intelligence consulting service and, after its successful knowledge production, a strategic opportunity arouse. This was identified as the need for a certification on software quality production, as major software development companies would demand this certificate as a capability demonstration, regarding the producer competence and ability on adhere to required quality standards. Software development companies who adhered to MI cycle received all knowledge that could promote their evolution towards quality higher levels – costs, schedules, technological demands, etc. – and could plan for this certification. As a strategic demand, two major players among those market intelligence users group started the evaluation on if and how to obtain the quality level certification, as a typical competitive intelligence approach. As those two cycles were concluded, a precise plan was detailed, and these two competitors got the certification. Although in different levels, after two years of efforts, they were not only better prepared when compared to their competitors, but also more qualified regarding contractor´s formal criteria, getting contracted for more sophisticated and profitable works.

- Another case happened in the civil engineering sector, as one company, from the group of a market intelligence users, sponsored a private research around a new material for thermal coverage, which was expensive but could result in better performance and recognized level of further satisfaction by their final customers. As this new material had all aspects of its physical purchase, storage, implementation and maintenance studied, it was more precisely applied in practical solutions, resulting in a proved better performance for the habitational units built by this company, offering better value perception for customers. Here, a marketing plan, that was proposed after strategic design, was conducted to enable the application of this new material with objective expectations on its financial returns when being installed in the final production units.

It is interesting to perceive, in these and other several examples, on how market intelligence process offered a good perspective of integration with other topics, themes and managerial demands, such as information technology implementation, tactical and operational marketing approaches, commercial researches and overall strategic alternatives. This was done, as a valid assumption, because MI integrated a genuine information system which operated organically, integrating several companies and economic agents. This way, it is correct to affirm that MI cycles and process, as an information system component, produced as much as the desired integration, as qualified results for business. Knowledge was produced from data and information at the expected level.

CONCLUSION

At the end of this chapter, an inevitable conclusion regards the integration of market intelligence to other concepts, and, indeed, practiced services and applications. It approaches and cooperate with all marketing planning processes, such as those selected to be applied at the managerial, top-level strategy studies, to those at the operational level, where value is firmly negotiated with final users, effectively producing the intended results as previewed. So, studies of these combined concepts, observing market intelligence with other topics, such as information systems, all marketing planning background, strategy, information technology management and several others can result in good perspectives for practical application of scientific knowledge, regarding data and information being generated to produce market knowledge, implicating in a better decision ability.

In these times where technology is emerging, new business models are proposed and immediately implemented, process data collected and produced every time, not only by human agents, but also from machines and regular devices and apparatus, are becoming strategic. New scenario propositions, fast decision-making, high level risk management, among other needs and trends must compose manager´s routine, demanding to be supported by effective and efficient processes, such as market intelligence, which must be understood relating to other planning and implementation-oriented processes.

REFERENCES

Akbar, H. (2003). Knowledge Levels and their Transformation: Towards the Integration of Knowledge Creation and Individual Learning. *Journal of Management Studies*, *40*(8), 1997–2021. doi:10.1046/j.1467-6486.2003.00409.x

Albescu, F., & Pugna, I. B. (2014, July). *Marketing intelligence: The last frontier of business information technologies. Romanian Journal of Marketing*, 55–68.

Badia, A. (2014). Data, information and knowledge: An information science analysis. *Journal of the Association for Information Science and Technology*, *65*(6), 1279–1287. doi:10.1002/asi.23043

Bazeley, P. (2015). Mixed Methods in Management Research: Implications for the Field. *Electronic Journal of Business Research Methods*, *13*(1), 27–35. Available online at www.ejbrm.com

Boblitz, M. C. (2006). Looking out the window: Market intelligence for a view of the real world. Healthcare financial management, 46-53.

Chau, M. & Xu, J. (2012) Business intelligence in blogs: understanding consumer interactions and communities. *MIS Quarterly, 36*(4), 1189-1216.

Chen, H., Chiang, R. H. L., & Storey, V. C. (2012). Business intelligence and analytics: from big data to big impact. MIS Quarterly, 36(4), 1165-1188.

Choo, C. W. (1996). The knowing organization: How organizations use information to construct meaning, create knowledge and make decisions. *International Journal of Information Management*, *16*(5), 329–340. doi:10.1016/0268-4012(96)00020-5

Choo, C. W. (2003). *The knowing organization: how organizations use information to construct meaning, create knowledge and make decisions.* Oxford, UK: Ed. Oxford University Press.

Clark, T. D. Jr, Jones, M. C., & Armstrong, C. P. (2007). The dynamic structure of management support systems: Theory development, research, focus and direction. *Management Information Systems Quarterly*, *31*(3), 579–615. doi:10.2307/25148808

Courtney, M. (2013). Puzzling out Big Data. Engineering & Technology, 7(12), 56-60.

Davenport, T. H., & Prusak, L. (2000). *Working knowledge: how organizations manage what they know* (2nd ed.). Harvard Business Press.

De Man, D. (2012). A platform for market intelligence. *International Trade Forum,* (4), 21-21.

Dimitrios, N. K., Sakas, D. P., & Vlachos, D. S. (2013). The role of information systems in creating strategic leadership model. *Procedia: Social and Behavioral Sciences*, *73*, 285–293. doi:10.1016/j.sbspro.2013.02.054

Dotzel, T., Shankar, V., & Berry, L. (2013, April). Service Innovativeness and Firm Value. *JMR, Journal of Marketing Research*, *50*(2), 259–276. doi:10.1509/jmr.10.0426

El-Bashir, M. Z., Collier, P., & Sutton, S. G. (2011). The Role of Organizational Absorptive Capacity in Strategic Use of Business Intelligence to Support Integrated Management Control Systems. *The Accounting Review*, *86*(1), 155–184. doi:10.2308/accr.00000010

El-Gayar, O., & Timsina, P. (2014). Opportunities for Business Intelligence and Big Data Analytics In Evidence Based Medicine. *Annals of 47th Hawaii International Conference on System Science*. 10.1109/HICSS.2014.100

Ferrel, O. C., & Hartline, M. (2010). *Marketing Strategy*. South Western College Publications.

Google. (2017). *Google Analytics*. Available at https://www.google.com/analytics/

Hoffman, L. (2013). Looking back at the big data. *Communications of the ACM*, *56*(4), 21–23. doi:10.1145/2436256.2436263

Huber, G. P. (1990). A theory of the effects of advanced information technologies on organizational design, intelligence and decision making. *Academy of Management Review*, *15*(1), 47–71.

Inmon, B., Strauss, D., & Neushloss, G. (2008). *DW 2.0: The Architecture for the Next Generation of Data Warehousing*. Morgan Kaufmann.

Jamil, G. L. (2001). *Repensando a TI na empresa moderna*. Rio de Janeiro: Axcel Books do Brasil.

Jamil, G. L. (2005). *Gestão da Informação e do conhecimento em empresas brasileiras: estudo de múltiplos casos*. Belo Horizonte: Ed. Con / Art.

Jamil, G. L. (2013). Sistemas de informação e a tomada de decisões: estudos de caso sobre a inteligência de mercado e a gestão de marketing. In P. Campos & P. Q. Brito (Eds.), *Novas tendências em Marketing Intelligence*. Ed. Actual.

Jamil, G. L. (2014). Inteligência de Mercado como um processo de gestão da informação e do conhecimento: proposta de oficinas de capacitação setoriais. Post-doctoral Project at Faculdade de Letras from Universidade do Porto, Porto University, Portugal.

Jamil, G. L. (2015). Market intelligence. In *Encyclopedia of information Science and Technology*. Hershey, PA: IGI Global. doi:10.4018/978-1-4666-5888-2.ch508

Jamil, G. L., & Magalhães, L. F. C. (2015). Perspectives for big data analysis for knowledge generation in project management contexts. In G. L. Jamil (Ed.), *Handbook of research on effective project management research through the integration of knowledge and innovation*. Hershey, PA: IGI Global. doi:10.4018/978-1-4666-7536-0.ch001

Jamil, G. L., Santos, L. H. R., Alves, M. L., & Furbino, L. (2012). *A design framework for a market intelligence system for healthcare sector: a support decision tool in an emergent economy. In Handbook on research of ICTs for Social Services and Healthcare: developments and applications*. Hershey, PA: IGI Publishing.

Jeston, J., & Nellis, J. (2007). *Business Process Management: practical guide to successful implementations* (2nd ed.). Butterworth-Heinemann.

Johnson, J. E. (2012, July). Big data + Big Analytics + Big opportunity. *Financial & Executive*, 51-53.

Kahaner, L. (1998). *Competitive Intelligence: How to gather, analyse, and use Information to move your business to the top*. New York: Touchstone Books.

Kaplan, R., & Norton, D. (2007). *The execution premium: linking strategies to operation for competitive advantage*. Harvard Business School Press.

Kearns, G. S., & Lederer, A. L. (2003). A resource based view of IT alignment: How knowledge sharing creates a competitive advantage. *Decision Sciences, 34*(1), 1–29. doi:10.1111/1540-5915.02289

Kim, W. C., & Mauborgne, R. (2015). *Blue Ocean Strategy: Expanded Edition*. Harvard Business Review Press.

Kimball, R., & Ross, M. (2010). *Relentlessly Practical Tools for Data Warehousing and Business Intelligence*. John Wiley and sons.

Kotler, P., & Keller, K. (2005). *Marketing Management* (12th ed.). Prentice Hall.

Kremer, M., Mantin, B., & Ovchinnikov, A. (2013). *Strategic consumers, Myopic retailers*. Darden School of Business at University of Virginia. Available at http://www.darden.virginia.edu/web/uploadedFiles/Darden/Faculty_Research/Research_Publications/Ovchinnikov_StrategicConsumer_MyopicRetailers.pdf

Laudon, K., & Laudon, L. (2009). *Management Information Systems* (11th ed.). Prentice Hall.

Lucas, H. C. Jr. (2005). *Information technology: strategic decision making for managers*. Hoboken, NJ: John Wiley and Sons.

MacInnis, D. (2011). A framework for conceptual contributions in Marketing. *Journal of Marketing, 75*(4).

Mahrt, M., & Scharkow, M. (2013). The value of big data in Digital Media research. *Journal of Broadcasting & Electronic Media, 57*(1), 20–33. doi:10.1080/08838151.2012.761700

Makadok, R., & Barney, J. B. (2001, December). Strategic Factor Marketing Intelligence: An application of information economics to strategy formulation and competitor intelligence. *Management Science, 47*(12), 1621–1638. doi:10.1287/mnsc.47.12.1621.10245

Marchand, D., Kettinger, W., & Rollins, J. (2001). *Making the invisible visible: how companies win the right information, people and IT*. Wiley.

Markovitch, D. G., Steckel, J. H., & Yeung, B. (2005). Using Capital Markets as Market Intelligence: Evidence from the Pharmaceutical Industry. *Management Science, 51*(10), 1467–1480. doi:10.1287/mnsc.1050.0401

McAfee, A., & Brynjolfsson, E. (2012). Big data: the management revolution. Harvard Business Review, 90(10), 60-68.

Miller, S. (2002). Competitive Intelligence - an overview *Competitive Intelligence Magazine, 14*(3), 43-55. Retrieved from http://www.sci.org/library/overview.pdf

Mintzberg, H., Alhstrand, B., & Lampel, J. (2009). *Strategy Safari: the complete guide through the wilds of strategic management* (2nd ed.). Pearson Education Limited.

Nonaka, I. (2008). *The knowledge creating company*. Harvard Business Review Classics.

O'Brien, J., & Marakas, G. (2008). *Management Information Systems*. Irwin: Mc Graw Hill.

Ohata, M. & Kumar, A. (2012, September). Big Data: A Boom for Business Intelligence. *Financial Executive*.

Park, S. H., Huh, S. Y., Oh, W., & Han, S. P. (2012). A social network-based inference model for validating customer profile data. MIS Quarterly, 36(4), 1217-1237.

Porter, M. (2008). *On Competition*. Harvard Business School Press.

Russel, S., & Norvig, P. (2009). *Artificial intelligence: A modern approach* (3rd ed.). Prentice Hall.

SAS. (2017) *What is Analytics*? Available at https://www.sas.com/en_us/insights/analytics/what-is-analytics.html

SAS. (n.d.). *SAS Enterprise Miner – SEMMA*. Available at http://www.sas.com/technologies/analytics/datamining/miner/semma.html

Schiffman, L., & Kanuk, L. (2010). *Consumer behavior*. Prentice Hall.

SCIP. (2012). Strategic and Competitive Intelligence Professionals. *What is competitive intelligence*. Retrieved from http://www.scip.org/content.cfm

Soares, C., & Carvalho, A. C. P. L. F. (2013). *Novas tendências em Marketing Intelligence* (P. Campos & P. Q. Brito, Eds.). Ed. Actual.

Stair, R., & Reynolds, G. (2009). *Principles of information systems*. Course Technology.

Tuomi, I. (2000). Data is more than knowledge: Implications of the reversed knowledge hierarchy for knowledge management and organizational memory. *Journal of Management Systems*, *16*(3), 103-117.

Turban, E., Mc Lean, E., & Wetherbee, J. (2002). *Information technology for management: transforming business in the digital economy* (3rd ed.). Hoboken, NJ: John Wiley and Sons.

Turban, E., Rainer, R. K. Jr, & Potter, R. E. (2007). *Introduction to information systems*. Hoboken, NJ: John Wiley and Sons.

Van Kesteren, J. (2012). How a market intelligence system can add value to the decision making process of NDP and sales: a design science approach. Master Theses of School of Industrial Engineering, Eidhoven University, Netherlands.

Vavra, T. G. (2013). The advantage of aftermarketing. *American Salesman, 58*(5), 13-18.

Weiss, S. M., & Verma, N. K. (2002). A System for Real-time Competitive Market Intelligence. *Proceedings of of the eighth ACM SIGKDD international conference on Knowledge discovery and data mining*, 360-365. 10.1145/775047.775100

Yonce, C., Taylor, J., Kelly, N., & Gnau, S. (2017). BI Experts´ perspective: Are you ready for what´s coming in Analytics. *Business Intelligence Journal, 22*(3), 36-42.

Chapter 2
Information System for Knowledge Management of the Technological Platforms in Brazil Healthcare

Jorge Lima Magalhaes
Oswaldo Cruz Foundation, Brazil & NOVA University of Lisbon, Portugal

Marlene Menezes
Oswaldo Cruz Foundation, Brazil

Zulmira Hartz
NOVA University of Lisboa, Portugal

Adelaide Antunes
National Institute of Industrial Property, Brazil & Federal University of Rio de Janeiro, Brazil

ABSTRACT

In the information era, great challenges in healthcare are now intensifying. Regarding challenges for public health, they are intensified not only in developing and less developed countries on account of parasitic diseases, but also for developed countries because of emerging and reemerging diseases. When these diseases are associated with chronic or acute diseases, they intensify causing a serious public health problem as drugs resistance, neglected disease, and those that perpetuate poverty. Thus, it is important to find out new tools for management information. In health area, the financial turnover in the world was US$ 1 trillion dollars only in medicines sector. In Brazil, about US$ 40 billion. Even in times of crisis, this is a sector with constantly expanding business whether by incorporating new technologies, new players in research and development, as well as adjustments to public health policies, regulatory issues, and global outbreaks of disease. These facts lead to constant adjustments of business in companies, universities, and government actions. This statement is aligned in the knowledge-based strategy advocated by Etzkowitz. In 2017, three lists of new strategic products for the Brazilian health system had already been changed. Using new intelligence systems, the government adopted new strategic business partnerships and were conceived in 2017 (others replaced) with budgets of more than 6 billion

DOI: 10.4018/978-1-5225-6225-2.ch002

reais. In this scenario, the revision of the chapter will feature new author and a scope of approach will be expanded to the new policy "Technological Platform" that replaces the old policy called partnership for productive development.

INTRODUCTION

The information can be seen as a condition for survival, given that it extends the communication context and rescuing and preserving social memory. Its value is intangible and resists all the mechanisms of destruction and oblivion since the collection of information. This is due, on the grounds that allows reconstructing the cognitive and knowledge assessment of a given situation in question (Lawrence & Giles, 2000).

However, in the 21 century, the capacity to generate speed information is modern and unprecedented in the world because of the Internet. Data created are practically instantaneous. Technological per-capita capacity of the world to store information nearly doubled every 40 months since the 1980s. From 2012, every day 2.5 quintillion (2.5×10^{18}) bytes of data are created (Lynch, 2008). This phenomenon growing is called Big Data (Lawrence & Giles, 2000; McKinsey Global Institute, 2011).

Quantity of daily data posted on the Web has led to a constant training of new professionals in all science areas. New activities have appeared, such as the Data Scientist - one who constantly seeks the best way to deal with the phenomenon of Big Data. The Big Data term has been consolidated within the scientific community due to the set of technological solutions capable of dealing with the ongoing accumulation of data that are unstructured and are captured from several sources. They are presented in order of petabytes, i.e.; quadrillion bytes of stored data. It's challenging the way to deal with these issues both for conducting scientific projects and any type of business organizations (Lynch, 2008).

At the institutional level decisions made without the right information lead to inaccurate decisions and sometimes disastrous. Decisions based on facts and reliable information are more likely to generate good results, thereby enabling decision maker's subsidies to meet the everyday challenges. Proper and timely information can develop effective strategies and acts proactively. This action can be called a competitive strategy when it involves business approach, which maximizes the value of the capacity of the organization to distinguish the company from its competitors (Porter, 2008).

Any scientific research area or an organization to lead to the advancement in technology, i.e., to improve to the management of large volumes of data, must apply the concept to extract data in order to have a more consistent view for decision making of managers.

It is globally recognized that Science is an intensive data information but the scale with which it presents itself in recent times is exponential in any science area. Even considering their correlations in a globalized context. Therefore, it requires new tools for extracting, analyzing and an informational treatment. One approach that has enabled this activity is that the information science which have focused on information enables multidisciplinary interface with several areas (Magalhães, Quoniam, & Boechat, 2013; Quoniam, 2011; Trigo, Gouveia, Quoniam, & Riccio, 2007).

In this sense, this work considered a case study of management information in the "Public Health" area. The choice of this area occurred due whereas the term "health" is present in about 50% of Big data. This data total, 47% are related to the "Public Health". Therefore, it is urgent to seek studies aimed at helping better management of this science area (Magalhães & Quoniam, 2013).

INFORMATION AND KNOWLEDGE MANAGEMENT

Information

Information is a data or set of data articulated to construct a meaningful message. The way to organize the data is an intention. It is therefore partially subjective. The information implies a transmitter and a receiver but also a media whose nature is far from neutral (Kira Tarapanoff (org.), 2015).

These poles suppose the existence of an aptitude, in the form of selective understanding, to extract the meaning of the information from the surrounding noise. Information is, therefore, a set of data placed in a context, mainly organizational for what interests us here, and bearing a particular meaning. Again, the concept has given rise to considerable developments in the information system and the development of protocols to make information systems as efficient as possible. Three notions are generally associated: that of information system (with the question of its economic performance), that of quantity of information (with the tension that appears between the amount of information made available and the super information) and ambiguity (which opens the question of relevance, a question linked to the imprecise notion of "quality" of information) (Pesqueux, 2005).

Information is a condition for survival, given that it expands the context of communication, rescuing and preserving social memory. Its value is intangible and resists all mechanisms of forgetfulness and destruction, since the collection of information reconstruction allows the cognitive evaluation and the knowledge of a certain reality in question (Gavirneni, Kapuscinski, & Tayur, 1999).

In the 21st century, thanks to the evolution of the Web, the speed of information generation is unprecedented and unprecedented also in the world. The world's per capita technological capacity to store information has almost doubled every 40 months since the 1980s. The Internet evolved from a purely static platform and assumed to be dynamic and an interactive role (Minelli, Chambers, & Dhiraj, 2013), allowing users to exchange large amounts of information instantly.

Once the proliferation of data and information through the internet and the flow of information is established and there are no restrictions on distance and availability, the big question that arises is the ability to screen, interpret and convert this volume of information into knowledge. The identification and analysis of the amount of information of a given area, whether scientific, technological or business; identify their state of the art and their respective correlations have become hard work (Aleixo & Duarte, 2015; Magalhães & Quoniam, 2013).

Therefore, it is necessary a management analysis to integrate actions and to provide the effectiveness actions to generate technological innovation and, consequently, wealth of the country. Thus, a better management for information architecture, also known as business architecture, encourage a new understanding (Ross, Weill, & Robertson, 2006).

Knowledge Management

Knowledge Management requires the involvement and support of all the company's stakeholders to preserve, transmit and develop knowledge. Indeed, it is the individuals who are at the center of the creation of value and who hold the keys to the success of such a project. The management of the knowledge and know-how of the company is therefore not universal, it depends strongly on the culture of the country in which it is practiced (Balmisse, 2006).

In order to motivate individuals to share their knowledge, it is important to consider cultural and human factors, and not just tools and procedures. Pesqueux et al (2011) define "culture" as the set of values, norms, habits and customs specific to a society that will influence the personality of the individual (Pesqueux & Ferrary, 2011).

Therefore, the Knowledge to Action Process conceptualizes the relationship between knowledge creation and action, with each concept comprised of ideal phases or categories. As shown in Figure 1, a knowledge creation "funnel" conveys the idea that knowledge needs to be increasingly distilled before it is ready for application. The action part of the process can be thought as a cycle leading to implementation or application of knowledge. In contrast to the knowledge funnel, the action cycle represents the activities that may be needed for knowledge application (Canadian Institutes of Health Research, 2016).

According Morin (1982), it is necessary to elevate the concept of "system" of this theoretical level to the paradigmatic level. A paradigm, for the author, is the set of fundamental relationships of association and/or opposition between a small number of "master concepts" that command/control all existing knowledge in all the speeches and theories. Thus, to sustain the action cycle (application) of the knowledge

Figure 1. Knowledge to action progress
Source: (Government of Canada, 2010).

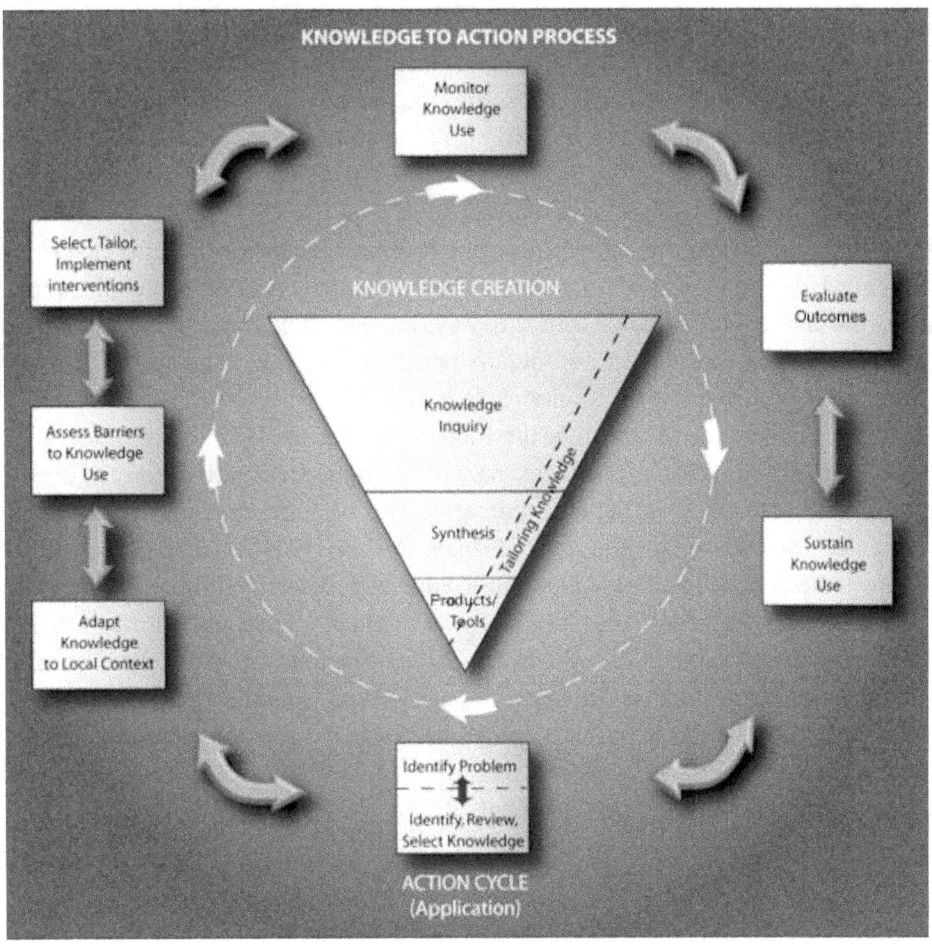

to the action progress, it is necessary to try to identify/define what are these notions, featuring the new paradigm and to resume as the (re)definition of the system where the logic of complexity in modeling the actual performance interact with a posture of the researcher and the actors involved in evaluative process (Hartz, 1997; Morin, 1982; Piaget & Garcia, 1983).

In this sense, to identify, classify, measure, disseminate and training competencies are essentials to aid strengthen to the R,D&I in Management of Science, Technology and Innovation in health. According to Hartz (1997), evaluation and health are characterized by phases. The first, is the "systemize" and refers to a "structuralism and cybernetic" practice supported on mathematical theories which stands the analysis of hard-systems (Le Moigne, 1994). This approach is still *analytic in nature and positivist in attitude* (...). Thus, knowledge needs to be thought as a process that takes place within a complex system of interactions between researchers and knowledge users which may vary in intensity, complexity and level of engagement, depending on the nature of the research and the findings, as well as the needs of the particular knowledge user – knowledge translation (Brouwers, Kho, Browman, Burgers, Cluzeau, & Feder, 2010).

Knowledge translation increasingly assumes critical importance for health research, since it is recognized that the creation of new knowledge, often alone, does not lead to its application or to effective impacts on the health of populations (Canadian Institutes of Health Research, 2016; WHO I Special Programme on Research & Training in Tropical Diseases, 2006).

There are several critical aspects in translating research results into best practice. These include, among others, the discrepancy between the knowledge needs identified by the communities and the work, done by the researchers during the "research" process and the gap between the knowledge produced and its incorporation into good health practices and policies (Pearson, Jordan, & Munn, 2012). Thus, there is a growing interest in the research on the mechanisms underlying the translation of knowledge, seeking to maximize and maximize the results of this process.

The active involvement of the various sectors of society has been instrumental in ensuring the production of useful knowledge that meets real needs as well as their application in problem solving, especially in the context of vulnerable populations. In this way, the adoption of participatory and intersectoral approaches has the potential to promote community participation, individual and community empowerment, and to reduce health disparities (Koelen, 2001).

In the context of collaborative research, the translation of knowledge is an innovative and multifaceted practice that allows multidirectional exchanges of knowledge, skills and resources, and the co-production of knowledge among academics, representatives of communities, professionals and policy makers (Brouwers, Kho, Browman, Burgers, Cluzeau, & Feder, 2010). In addition, the translation of knowledge allows the development of shared knowledge and understanding among universes initially foreign to one another, but which, in turn, gradually reconfigures themselves through their interactions (Carvalho, Bodstein, Hartz, & Matida, 2004; Hartz, 2012). The innovation character of this approach is in the establishment of connections or alliances that are formed from the interests and capacities of the individual actors. During the formation of these connections, the relations between actors are defined and negotiated and, therefore, the network and its connections are dynamic, fluid and unstable (Boland, Phillips, Ryan, & McPhee-Knowles, 2012; Hippel, 2005).

OVERVIEW ON PUBLIC POLICIES IN BRAZIL'S HEALTHCARE

Development in science, technology and innovation are key components for socio-economic growth of the Brazil. In this context, the health sector plays a key role as a space for technological innovation, new jobs and income and the promotion and protection of health for the people´s welfare. It is the duty of the State to ensure the health to the entire population, through the formulation and implementation of economic and social policies aimed at reducing the risks of diseases and other health problems, and to established conditions that ensure universal and equal access for all. Similarly, to promote actions and services for protection and recovery of the society (de Carvalho, Souza, Shimizu, Senra, & Oliveira, 2012; De Negri & Kubota, 2008; Gonçalves, 2006).

The importance of health and development should be emphasized as an economic good, capable to generate jobs and income. Innovation processes are not socially natural, recognizing that the definition of technological trajectories is conditioned by the global productive and competitive structure, whose dynamics are not necessarily directed to social demands. In this scenario, it is essential that national State acts as an inductor and articulator of the various interests that affect the generation of new technologies, giving a direction to the objectives to a universal health system and welfare´s State.

Brazil, as well as the global trend, has undergone processes of demographic, epidemiological and nutrition transition with significant reduction of fertility and birth rates besides a steady increase in life expectancy. This process results in changes in the patterns of occurrence of pathologies and increased significantly the prevalence of chronic diseases. This changing in epidemiological profile of the country causes an increased demand for new medical technologies to ensure continuous improvement in the health population (Gadelha, Carvalho, & Pereira, 2012c).

In this context, health products are expanding and occupying an important space in the Brazilian economy. It consists of an overload of items with different levels of complexity, ranging from a simple infrared lamp until the MRI equipment to a culture medium to a reagent kit for detection of HIV and medicines. Therefore, these products are used in conducting medical, dental and physiotherapy procedures, as well as the diagnosis, treatment, rehabilitation or patient monitoring.

It is one of the most vital and dynamic sectors of the economy. According to the Ministry of Health (MoH), in Brazil, there was a 72% growth in the number of devices in health services in the National Health System (SUS - Brazilian term) over the past five years. Sector revenues reached more than 100%, from 2005 to 2013, reaching the order of US$ 5.5bn.

The expansion of access to health care has led to a significant increase in Industrial Economic Health Complex (CEIS) trade deficit in the last decade, from a level of US $ 3 billion in 2003, to US $ 11.5 billion in 2014 (Gadelha, 2015).

The Ministry of Health announced, in 2016, that it will guarantee investments of almost US$ 2 billion to stimulate national production of medicines, inputs and technologies in health. It is worth remembering that national spending on the acquisition and distribution of medicines was almost US$ 5 billion in the last year.

Basic Health Care is a priority for the Ministry of Health, whose budget forecast, until the end of 2017 is U $ 5.8 billion, an increase of 10.4% compared to last year. The Ministry of Health has said to increase the pass-through to purchase drugs which part of the Basic Component of Pharmaceutical Assistance (CBAF). The annual value will increase from R $ 5.10 to R $ 5.58 per inhabitant, which should increase the amount of products available to the population. Only in the state of Rio de Janeiro, the investment in basic care will be US$ 37 million.

Public and private investments in health research in the country are expected to reach US $ 4 billion over the next four years - equivalent to almost 0.30% of Brazil's GDP.

According on the Monthly Survey of Physical Production of IBGE, the sector growth of 8.5% in production of medical and dental instrumentation equipment, from January to September 2013, is compared to the same period in 2012. Meanwhile, sales in retail trade of pharmaceutical, medical and orthopedic goods grew at around 9.5% in the same period (Gadelha; Costa & Maldonado, 2012b; Gadelha; Quental & Fialho, 2003; IMS Institute for Healthcare Informatics, 2012; Ministério da Saúde, 2014).

The growing of the market in Brazil is maintained for years to come. However, according to industry experts it is necessary to maintain this growth for Brazil´s progress. In other words, prioritizing policies and actions that can increase access by incorporating new technologies and eliminate any regulatory barriers, aiming to meet and discuss the scenario of health care products in Brazil.

Another factor which contributes to market expansion and access to medication is the advancement of research in the development of innovative products. In this sense, it must be stimulated the development and domestic production of medicines and drugs. So, Ministry of Health (MoH) in Brazil increased purchases of medicines and supplies within the SUS. MoH was responsible for create the Program for the Development of the Health Industrial Complex - PROCIS and the Partnership for Productive Development (PDP). These actions aim to promote joint technology development and exchange of knowledge for innovation within public and domestic private producers, making them competitive and skilled (Cartaxo, 2011; Centro de Gestão e Estudos Estratégicos & Ministério da Ciência, Tecnologia e Inovação, 2013).

The PDP is considered one of the most important actions of the government in health care that is part of plan produced by the Production Development Policy. This is part of the government's industrial policy entitled "Greater Brazil" ("Brasil Maior" Plan – Brazilian term) - http://www.brasilmaior.mdic.gov.br/). The focus is on innovation and growth of Brazilian industry, using the transfer of technology for the production of medicines, vaccines and equipment from private companies to public laboratories. Offset by the guarantee of the purchases made by the government around of US $ 12 billion/year. In addition, to generate significant savings for the Ministry of Health (MoH) and decrease the trade deficit of health, which reached US $ 11 billion in 2012. This initiative also benefits the population because it ensures the supply of essential drugs to the Public Health System (SUS - Brazilian term)[1].

The strengthening of the Industrial Complex has become a strategic axis of the Federal Government with the expectation of creating technological solutions in health, generate national technological protagonism and generate economic growth. In 2016, a new policy called "Policy of Innovation for the Economic Complex of the Health" was launched, changing the focus of the production of a specific product by a platform that manages this product.

Aligned to the purpose of the macro plan "Greater Brazil" the National Health Plan (NSP) 2012-2015 and the National Policy on Science, Technology and Innovation Management (PNCTIS) is an integration of policies aimed at local production of medicines considered high value-added and essential in the successful treatment of chronic and complex diseases, i.e., strategic medicines for the country. Thereby, it is possible to see, in Figure 2, the National Strategy for Science, Technology and Innovation.

The government has been directing the development of PDP with strategic products as well as the high cost medicines or biologic drugs produced and distributed from living cells. MoH published a list of strategic products, including 52 medicines. Biological products production in Brazil, included on the list, besides expanding the assistance, would impact directly on the budget of the Ministry of Health. This is because, despite representing only 4% of the quantity distributed by SUS, these products consume 51% of the available resource for the purchase of medicines. Eculizumab, for example, is among the ten

Figure 2. National strategy for science, technology, and innovation 2010-2015
Source: http://www.mct.gov.br/ (Access: September 26, 2014).

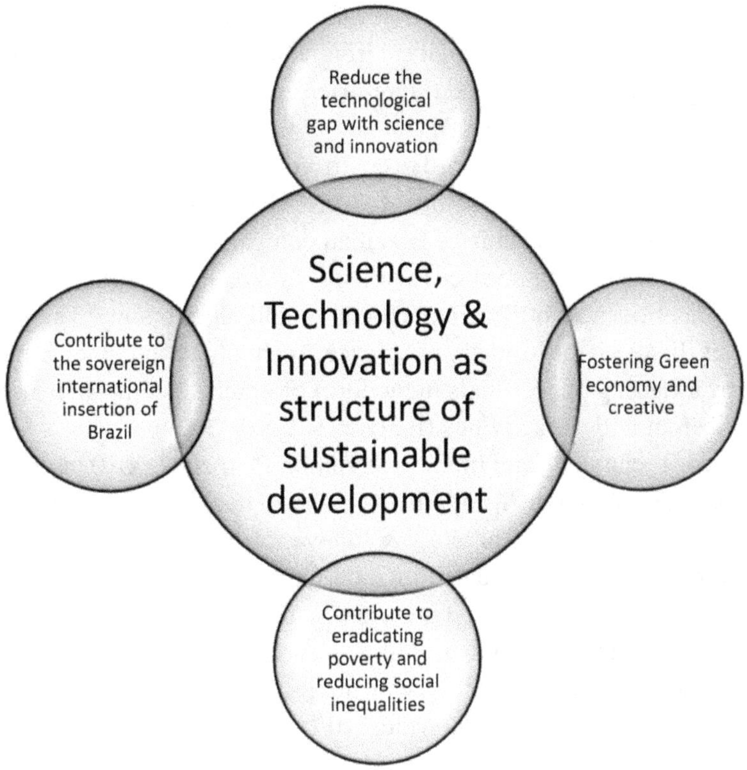

most judicialized medicines in the country, and demanded, in 2016, a cost of $ 391.8 million to serve 336 patients. Also included in the list of priorities of the Ministry are two other drugs that are among the ten most judicialized in the country: Galsulfase and Alfagalsidase.

Since 90s, the federal government has given expression to its interest in developing the pharmaceutical sector by introducing the Forum for Competitiveness in the Pharmaceutical Production Chain, mainly the creation of the Law for Generic Medicines (Law No. 9,787, February 10, 1999). It was set up as a space for discussing government policies for the pharmaceutical production chain and was jointly coordinated by the Ministry of Development, Industry and Foreign Trade and the MoH.

More expressively in the 2000s, the recognition of this relationship of development and health has supported the formulation of new public policies, with the marked need to develop the productive base of health and its capacity to generate innovations.

Initially, in 2003, the Industrial and Technological and Foreign Trade Policy (PITCE, Brazilian term) was launched, which represented the resumption of two central issues in the current macro policy agenda: the defense of the need for an industrial policy, and the importance of selecting strategic sectors to foster national development, including health.

The PITCE had targeted the pharmaceutical industry as one of its strategic objectives, recognizing the importance of this sector globally, its dynamic, knowledge-and innovation-intensive nature, and the high demand for global investments in Research, Development and Innovation (R,D&I). The PITCE was conceived with the aim of strengthening and expanding Brazilian industry by improving businesses' innovative capacity. With its long-term strategic vision, the central pillar of PITCE was to innovate and add value to the processes, products and services of national industry. This policy operated on three aspects: horizontal action lines (technological innovation and development, exports / entry to foreign markets, industrial modernization, institutional environment), strategic sectors (software, semiconductors, capital goods, drugs and medicines) and future-oriented activities (biotechnology, nanotechnology and renewable energy) (Gadelha & Costa, 2012a).

BNDES's Profarma program was developed to contribute to the implementation of the PITCE, and is now used to finance PDPs. It is divided into three sub-programs which provide funding for investments in production, RD&I, and its strengthening of Brazilian companies.

The government also created the Innovation Economy Funding Program, aimed at boosting innovation and improving the competitiveness of Brazilian company, and the economy. The program is funded by FINEP funding agency, linked to the Ministry of Science, Technology and Innovation (MCTI – Brazilian term) and operates by awarding public funds in the form of grants to companies with which they share the costs and risks inherent to these activities (Furtado & Urias, 2009).

In 2007, the Science, Technology and Innovation Action Plan 2007-2010 of the Ministry of Science and Technology, under the heading of "health inputs", placed health as one of the strategic areas, guiding financing and National Science, Technology & Innovation System.

Brazil's industrial policy was reformulated in 2008 into the Production Development Policy. It involves different ministries and has integrated goals and actions, clear financing and concrete mid-term responsibilities. The Production Development Policy's actions are divided into three levels: systems, structuring programs and strategic highlights. There are programs for a variety of industries, including the health industry (Costa, Gadelha, & Maldonado, 2012; Gadelha & Costa, 2012a).

In 2008, Brazil created the Health Industry Executive Group (GECIS) under the MoH. Its purpose is to take concrete actions and measures to further the creation and implementation of regulations for the government's development strategy for health sector, based on the national guidelines for strengthening the health industry, as well as other related measures.

From 2011, PDPs initiative has gained scale, perhaps representing the most ambitious policy for the implementation of the constitutional precepts that the market is part of national patrimony, that scientific and technological development should be stimulated by the state, and that the health is a right, for which national technological training is an essential condition. This fact required an extension of its regulatory framework in order to give it a compatible institutional base.

PDPs had their first criteria defined in 2012, with the objective of rationalizing the state's purchasing power, encouraging technological development in the country, and focusing on local production of strategic products for the SUS. From that point on, standards and criteria were established for these agreements between public institutions and private entities for the transfer of technologies.

Lastly, the priority of the drug segment and the health-care production complex was reaffirmed by the National Science, Technology and Innovation Strategy launched in 2012, which emphasized the need to promote mechanisms to stimulate innovation in health and intensify technology transfer for the national public laboratories, while pointing out a series of gaps in the Brazilian production base that needs to be overcome. Another development concerns support for the production of public institutions, aimed at the supply of medicines, vaccines, diagnostic reagents, biopharmaceuticals and, more recently, potential manufacturers of medical equipment and devices, in the context of the Industrial Complex Investment Program of Health (PROCIS).

About PDPs, even before the full transfer of technology, partnerships for productive development generate an economy for the Federal Government of 30% in value of imports. That is because, the patent holder starts supplying the product to Brazil and to the partner public laboratory with a discount on the price practiced in the international market. And, over time, it will transfer the technology until the product is produced entirely in Brazil, also implying a reduction of 5% per year.

In 2016, was launched Policy of Innovation for the Economic Complex of the Health. Under this theme, there were promises of investment in manufacturing infrastructure of synthetic and biological drugs. It is a reorganization to strengthen the national strategy to foster Brazilian industrial production, creating technological solutions in health. The new policy highlights eight platforms, including therapeutic targets (rare diseases and neglected diseases) and technological and productive routes (biotechnology, chemical synthesis, phytotherapy, blood derivatives, radiopharmaceuticals/nuclear medicine production and health products). The aim is to accelerate the production of new products, other therapeutic targets, such as oncological, immunosuppressive and antiretroviral drugs, reducing costs and increasing access.

The Innovation Law (#1097 from December 2, 2004) provides incentives for science and technology research and innovation in industry. This law aims to capacity building, technological autonomy and industrial development. The law recognizes the different social actors involved and encourages the participation of science and technology institutions into innovation process, regulating partnerships and alliances, and the sharing of infrastructure and intellectual property.

It is worth noting that patenting is a strong stimulant for RD&I in the health industry[2]. Brazil's researchers are done by public research institutions. So the way is interacting with the business world and the evaluation criteria of the nation's academic system must be taken into account. Within the scope of the Industry, Technology and Foreign Trade Policy (PITCE), the Innovation Law aims, among other things, to make it easier for public institutions and universities to grant licenses for their patents to private companies, while encouraging greater efficiency in this system by means which include making it an evaluation criterion (Antunes & Mercado, 2000; Magalhaes, JL, Antunes, AMS, & Boechat, 2012c; "Parceria para o Desenvolvimento Produtivo - PDP", 2014c).

In addition, it should be noted that one of the aspects of Brazilian policy was the search for the internalization of the production of health products in the country (medicines and drugs, equipment and materials, vaccines and diagnostic products), through the establishment of partnerships for the productive development (PDPs) between public producers and private companies with national and foreign capital.

Nonetheless, the sector for medical and in-vitro diagnostic materials and equipment (also known as health products) is one of the most dynamic. Brazil is the second largest market of the emerging countries, with sales of US$ 12 billion of these products, or 5% of world demand. Global sales are estimated at around US$ 250 billion, and are growing at around 6% a year. The Brazilian market exceeds this growth rate, rising at 10% a year. However, the full potential of this market in Brazil is not being harnessed because, depending on the product, purchases for the SUS account for 35-65% of the existing market (BRASIL/MS/FIOCRUZ/IPEA, 2012; BRASIL/TCU, 2014d).

One of the Brazilian government's key incentive policies is the Greater Brazil Plan for 2011-2014 (replacing the PITCE), which organizes inter and intra-sectorial initiatives. The former is designed to improve the efficiency of domestic production. Meanwhile, the intra-sectorial measures are devised according to the key features, challenges and opportunities of the industries in question, and are organized into five blocks for the formulation and implementation of programs and projects. These cover areas, such as strengthening production chains, building new technological and business competencies, energy supply chains, export diversification and internationalization, and competencies in the natural knowledge economy (BRASIL/MS/FIOCRUZ/IPEA, 2012).

In 2016, Ministry of Development, Industry and Foreign Trade, in partnership with the National Confederation of Industry (CNI), launched the Brazil More Productive (B + P) Program. The goal is to substantially increase the productivity of participating business units from rapid corrections in the production process. The methodology was already tested in 2015 by Senai in a pilot project involving 18 medium-sized companies, and the results were significant: a 42% increase in productivity, and a 21% reduction in production costs. Highlights of this initiative are its excellent cost-benefit ratio and the concrete possibility of measuring results.

Brazil More Productive Program aims to help the country's industrial companies to improve their production processes, with implementation of fast, efficient and low-cost actions. The project consists of 120-hour on-site technology consulting on the spot, for the diagnosis of processes, proposals for improvements to obtain productivity gains, reduction of production costs, and monitoring of implementation and results.

The policies introduced in recent decades have sparked the need to integrate the knowledge produced in alignment with the previously defined strategies. To assure success and the effective absorption of knowledge in Brazil for the production of strategic medicines from the SUS list, the Brazilian government through the MoH, has prepared partnership agreements under the Production Development Policy. The partnerships must include, at least, one foreign pharmaceutical laboratory, one official pharmaceutical laboratory, and one Brazilian pharmaceutical laboratory or pharmaceutical company.

It is worth mentioning the creation of the Patent Act. In 1994, it was signed by members of the World Trade Organization (WTO) Agreement on Aspects of Intellectual Property Rights (TRIPS) which is known as an international agreement administered. This agreement establishes that the entire production of the countries member must granted patents for chemical and pharmaceuticals. The patents in the field have become more intense and controlled in member countries of the WTO. This agreement establishes that the production and sale of drugs only can be made by the holder of the patent, for a period, at least, 20 years. In this way, it prevents the production and marketing of pharmaceutical products for other pharmaceutical companies, encouraging investment in innovation. In Brazil, it was regulated by Law No. 9.279 of May 14, 1996.

Therefore, pharmaceutical patents have become so important, once that contain novelty and reveal the details of the invention, if process or product. By the other hand, the researchers and / or managers consider patents as an important indicator of innovation. Studies related that patent of the company, countries and owners of the invention can reveal the technological dynamism of a particular industry, and provide information about the direction of technological change. As regards the field of health, the pharmaceutical industry is essential for the R,D&I. This industry is growing every year, and it had generated about U$ 33 billion, between April 2015 and March 2016, a growth of 10% compared to the same period a year earlier, according to IMS Health.

The Production Development Policy became the government's new industrial policy, under the legal provisions of the Innovation Law 10.973/2004. Laws 11.196/2005 providing tax benefits for technology (RD&I); Law 11.105/2005 (Biosecurity) and 6.041/2007 (Biotechnology Development Policy) have main aims to expand access to strategic technologies and to reduce the vulnerability of the public health system (SUS) by strengthening the health industry. The first PDPs were created in 2009 as part of a broad policy linking health with development for the welfare of the population, and to aid consolidate the national production of strategic technologies for the SUS (Cartaxo, 2011).

In Figure 3, synthetically it can be seen the evolution of health policies in Brazil.

Figure 3. Evolution of some policies in Brazil for public health
Source: Magalhães, Hartz, and Antunes, 2016, Adapted by authors.

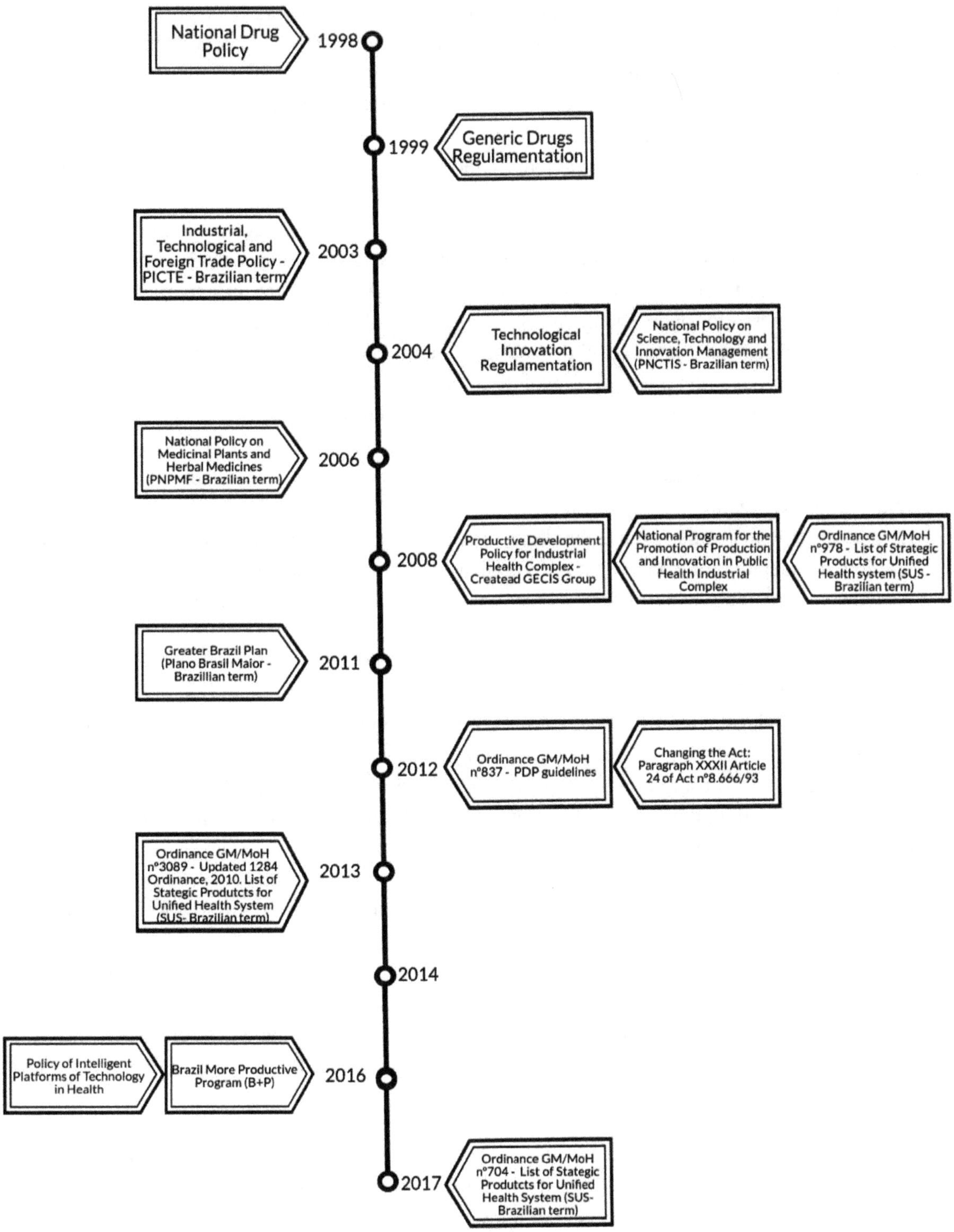

A BLUEPRINT TO MEASURE THE KNOWLEDGE TRANSLATION

Public Policy is in the context of the architecture of knowledge. Therefore, is pressing management analysis tools to integrate actions and to provide the effectiveness actions to generate technological innovation and, consequently, wealth of the country.

Certo and Peter (2005) regard an analysis of the environment, which indicates that this context (business architecture requires information architecture) is been monitored by entire organizational environment to identify opportunities, challenges and threats regarding current assess and future risks (Certo; Peter, 2010).

Under this perspective, it can be concluded that the information architecture, linked to business for innovation in given segment, is also linked to the organizational logic for business processes, technology transfer and infrastructure of the Information Technology (IT). This reflects the integration and standardization required for operation of an organization model. So, this "architecture" requires a long-term view of the processes, systems and technologies, in order to be built in the interim organizational capabilities that permeate three stages of evolution, namely (Centro de Gestão e Estudos Estratégicos & Ministério da Ciência, Tecnologia e Inovação, 2013; Certo & Peter, 2010):

1. Architecture with business silos, where the functional needs or the needs of individual business units are maximized;
2. Architecture with standardized technology, which aims to provide IT efficiency through standardization and technology via increasing centralization of management technology;
3. Architecture with modular business, where preserving global standards while enable local differences with customization.

Considering the global technological advancement in the knowledge era and even the public health, arises the need to find better treatments and quality of life to people. Then, to cogitate technology transfer in healthcare by PDP can be understood as a customization for the local reality of a nation second model, practiced by the Special Programme for Research and Training in Tropical Diseases (TDR) of WHO. This model promotes the knowledge translation which redefines the scope of the search, as shown in Figure 4 (Carvalho, Bodstein, Hartz, & Matida, 2004; WHO | TDR, 2013a; Zicker, 2007):

1. Endemic countries involved in setting the research agenda, and also on how to respond to their priorities;
2. Strengthen the research capacity of individuals, institutions and networks in the countries, whether endemic or low-income;
3. Promote the application of innovations and strategies in political and health practices; and,
4. Serve as a platform to establish consensus on priorities, funding and management processes in research.

There are several Public Policies in healthcare. These policies provide an NHS that meets their goals as well as to produce drugs and medicines in the country. Nowadays, the incorporation of new technology with efficient tools to assess the implementation from Policy to impact is the "Knowledge Translation" (KT) (Carvalho, Bodstein, Hartz, & Matida, 2004; De Negri & Kubota, 2008).

Figure 4. Redefinição do Escopo de Pesquisa – modelo TDR – OMS
Source: Special Programme for Research and Training in Tropical Diseases (TDR) WHO – Geneva.

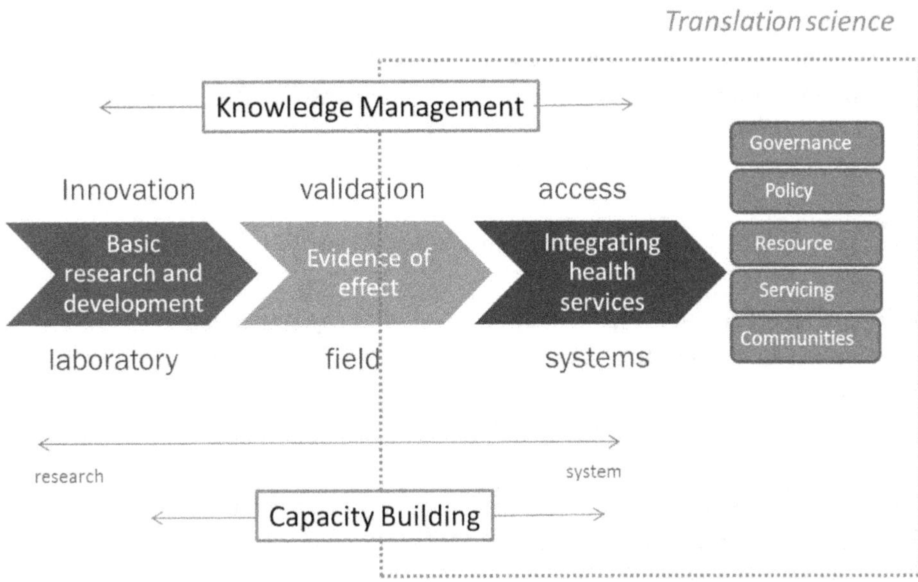

This Policy in Brazil, included not only new generic products, but also the internalization of new technologies. Actually, Brazil is totally dependent on imports, making it vulnerable for the National Defense Drugs and Medicines. In this context, the PDP can contribute to minimize this fact. Note, in the 90's, Brazil lost nearly 2000 factories pharm chemical after opening markets. In this sense, became dependent on trivial products, such as acetylsalicylic acid (Aspirin) and Amoxicillin (antibiotic), with imports from China and India (Antunes & Mercado, 2000; Magalhaes, Antunes, AMS, & Boechat, 2012b).

These analyzes are not used properly by scarcity of information, either by research about methodologies of KT, or by prospecting and monitoring in health care whose value is to assess the effectiveness of the policy, since its implementation to impact on society (appropriation of value to society and innovation) (Hartz, 2012; Zicker, 2007).

Evaluation studies of government policies and programs allow researchers and implementers of this program which are able to objectively make decisions with better quality and optimizing public spending on various activities. Thus, it's possible to identify and overcome bottlenecks or points that lead to success. Consequently, there are new perspectives to implement rational policies with greater capacity, such as better results for the political in Science, Technology and Innovation in Health (De Negri & Kubota, 2008; Ross, Weill, & Robertson, 2006; Santos, Toledo, & Lotufo, 2009).

Fostering for research is understood as programmed actions to improve and innovate policies to promote and protect health. The evaluation should provide better understanding and organizational learning on the scope of these results for decision making and greater transparency in the management of public resources (De Negri & Kubota, 2008).

In this overview, initiatives to identify, classify, measure, disseminate and train human resources methodologies are essentials to help strengthen the R,D &I in Management of Science, Technology and Innovation in health. According Hartz (1997), "Evaluation and Health" is characterized by phases. The first is the "systemize" and refers to a "structuralism and cybernetic" practice supported on mathematical theories which stands the analysis of hard-systems (Le Moigne, J.L., 1994). This approach is still *analytic in nature and positivist in attitude (...). Things were reduced the whole (holism) for parties (atomism), revealing himself actively reductionist.*

Melese (1990) also demonstrates the need to redefine "the system" because there are those wanting to escape the crippling reductionism, who use the term in this restricted form, and run the risk of forgetting that it is possible to reduce both the "whole" for the part. This approach is one that considers "the system" a concept of three faces (Melese, 1990; Morin, 1982a):

1. This "whole" combined with the macro unit where the parties have their own identity and a common identity;
2. Interactions or set of relationships that are created in the system;
3. The organization and the partners, expressing a constitutive nature of the interactions and with idea of giving your core system.

A DOUBLE-QUICK EVALUATION OF THE PARTNERSHIP FOR PRODUCTIVE DEVELOPMENT IN BRAZIL

The quality of life of the Brazilian people is on the government's agenda. Its public health programs are designed to foster the healthy ageing of the population, with economic development being linked to improved health conditions, fostering a sustainable lifestyle. Brazil is seeking to reduce its health sector deficit, and to boost its technology and RD&I, including a variety of initiatives and investments targeting academia, the private sector and government. These include a government effort to build technological competency in the production of drugs and medicines back to the relative levels seen in the mid-1980s (Magalhaes, Antunes§, & Boechat, 2012a).

One of the key policies in this area is the Production Development Policy launched by the government in mid May 2008, which replaces the Industry, Technology and Foreign Trade Policy (PITCE, acronym in Portuguese), introduced in March 2004. The 2008 policy had a broader scope, greater depth, and focuses on increasing coordination, controls and targets. Like its predecessor, is structured in a way to align the public and private sectors, with the former being responsible for facilitating business ventures through tax/fiscal incentives, credit lines, reduced bureaucracy and regulatory adjustments.

Production Development Policy is a more complex set of measures designed to expand access to strategic technologies, and reduce the vulnerability of the country's public health system by strengthening the health industry. The objectives, which are designed to foster could, if in practice, raise the production capacity and competitiveness of Brazilian industry, producing a ripple effect throughout society (Ministério da Saúde, 2014a).

However, it should be recalled that, although the PITCE did not fully attain all the desired outcomes, it left a legislative legacy that favors certain sectors of the economy, providing them with certain specific lines of credit from the Brazilian Development Bank (Banco Nacional de Desenvolvimento Econômico e Social, BNDES). Meanwhile, PDP aim to foster technology transfer between public and private institutions and technologies capable of meeting the public health system's (SUS) need for strategic products as set forth in MoH directive 3039/2013, which was derived from directive 1284/2010.

The directive in question lists several priority products for use in health programs, such as vaccines and medicines for cancer, women's health, mental health, and chronic diseases. For instance, Imatinib

Mesylate was the first Brazilian cancer drug, produced by a PDP for chronic myeloid leukemia which, alone, produced a real saving of US$ 15.5 million in 2013 vis-à-vis 2012. In the middle of 2014, there were 102 PDPs for over 75 products. There are over 17 public pharmaceutical laboratories, and 53 private ones involved in these PDP (Gadelha & Costa, 2012a; Macedo, 2014a; Ministério da Saúde, 2014a, 2014b).

The Secretariat of Science, Technology and Strategic Supplies, under the MoH, is responsible for managing the PDP in the pursuit of making Brazil a leading player in innovation and strategic development. The list in directive 3039 does not only contain products the MoH has determined as strategic, but also indicates to the country's leaders what products should be prioritized in individual projects designed to boost production, innovation and technology transfer in the country, as well as regulatory mechanisms. By this means it provides strategic support for the health industry (Gadelha & Costa, 2012a; Gadelha, 2012).

Some PDP objetives ("Parceria para o Desenvolvimento Produtivo - PDP", 2014c):

- To streamline the State's purchasing power by selectively centralizing health expenditures, to reduce the acquisition costs of the SUS and enable the production in the country of innovative products of extreme importance to health, focusing on improving the population's access to strategic supplies;
- To foster joint technological development and knowledge exchange for innovation;
- To prioritize the local manufacture of products of a high cost or of public health and social impact; and
- To significantly and progressively reduce prices as technology is transferred and developed in Brazil.

The plan for the execution of the Production Development Policy contains targets known as "systemic actions", which, once implemented, will have an impact on costs in Brazil thanks to their broad scope. The aim is to:

1. Increase annual fixed investments/GDP;
2. Increase private RD&I expenditure/GDP;
3. Increase the market share of Brazilian exports;
4. Support small and medium-sized businesses by fostering capacity building for competition in foreign markets, thereby increasing the number of such businesses which export goods.

The absorption and integration of knowledge fostering more aligned actions and strategies is designed to replicate in the field of health the success achieved in other sectors of the economy. Some examples ("Parceria para o Desenvolvimento Produtivo - PDP", 2014c):

1. **World Leadership:** Goal achieved in mining and steel, aeronautics and ethanol production;
2. **Penetration of Markets:** Durable consumer goods and standardized capital goods. However, for health the aim is also to position or maintain the production system amongst the leading global exporters;
3. **Specialization:** To build and consolidate competitiveness in technology-intensive areas. The strategy covers, not just the health industry, but also information technology and custom-made capital goods;

4. **Brand Differentiation or Enhancement:** To position Brazilian brands and businesses amongst the top five in the world;
5. **Greater Access by the Population to Goods and Services:** Not only health services, but also broadband internet, durable and non-durable consumer goods and civil construction.

This superstructure of broad goals and specific targets is set by the Production Development Policy, where one sector which has been benefitted can interlink and interact with others. Thus, support health industry and other industries. Action lines of the Production Development Policy are designed to enhance competitiveness and encompass a variety of technologies, including nanotechnology and biotechnology.

Another consequence of the Production Development Policy is to attract multinationals to make investments in health in Brazil. One example is an agreement between multinational Merck Serono and Brazilian firm Bionovis, mediated by the MoH. The aim is to further the PDPs underway for the Brazilian production of biological drugs. Merck Serono, as a partner, provides to build a new factory to produce six biological medicines. The current partnership allows investments worth US$ 250 million for construction, technological product development and technology transfer. Meanwhile, Bionovis will invest US$ 0.5 billion in the production of biological medicines in Brazil in the coming five years (BRASIL/MS/FIOCRUZ/IPEA, 2012; BRASIL/TCU, 2014d; "Parceria para o Desenvolvimento Produtivo - PDP", 2014c).

This example has a high-cost, but in counterpart cutting-edge products will be developed and produced by a private Brazilian company, Bionovis, and public institutions (Fundação Oswaldo Cruz and Instituto Vital Brasil) for the treatment of cancer, rheumatoid arthritis and other conditions, namely: etanercept, rituximab, bevacizumab, cetuximab, infliximab and trastuzumab. Like the other PDPs, the aim is to make Brazil autonomous in the production of these medicines, while helping to reduce the country's trade deficit ("Parceria para o Desenvolvimento Produtivo - PDP", 2014c).

Some other products have been registered in Brazil by PDPs: a two-in-one antiretroviral, tenofovir disoproxil fumarate 300 mg + lamivudine 300mg; and cabergoline 0.5 mg (treatment of the excess production of the female hormone prolactin or hyperprolactinemia). The antiretroviral, alone, should bring about a saving of around US$ 110 million for the MoH, in five years. When all the PDPs now signed are implemented, Brazilian government expenditure will be cut by around US$ 2 billion a year, while the impact on imports, by the end of the fifth year, will reach around US$ 1.8 billion.

Use of the Information System for Knowledge Management in PDP Through the Patents Platform

The process of updating the strategic medicine list should be annual, in accordance with the recommendations issued by the Executive Group of the Industrial Health Complex (GECIS), and coordinated by Science, Technology and Strategic Inputs Office - SCTIE with the other offices of the Ministry of Health, other offices that integrate GECIS, and Entities of the Competitiveness Council of the sector (Brasil, 2014a).

The current list was published in March 2017, where it contemplates 56 medicines (Brasil, 2017). Of the 56 drugs provided, a small contribution of antibiotic class. According to Figueiredo (2016), in 2013 list, some diseases and therapeutic classes come to appear in large numbers, as is the case of antibiotics (12), monoclonal antibodies (11) and cancer drugs. Although the high number of antibiotics as strategic drugs for domestic production, none of them was selected for PDP. The following year, only 3 drugs in the antibiotic class were present, while the current list (2017) contemplates only 1 drug in the class, doxycycline.

Antibiotics appears in a situation with important gaps, referring not only to diagnostic and therapeutic tools, but also to intervention in the health services and prescription of medicines, as well as in the national production of them, making them strategic, taking in view of increasing bacterial resistance. It is worth remembering that some of these diseases and conditions, besides having a high burden in Brazil, are present in WHO's report on priority medicines as conditions where therapeutic gaps have been identified (WHO, 2013).

This scenario has apparently not changed since there are currently no active PDPs for this class. Most active PDPs are intended for the production of high-cost drugs.

Doxycycline drug, present in the current list, is a tetracycline class drug, incorporated into the SUS in 2014 through Ordinance n°16, dated May 15, 2014, through the evaluation of the CONITEC organ, for treatment of Brazilian spot fever and other rickettsial diseases in the Unified Health System - SUS (Brazil, 2014b). Doxycycline is the drug of choice for the treatment of all tick-borne diseases regardless of severity in children and adults. Doxycycline is the tetracycline of choice because of its pharmacokinetic advantages, lower toxicity and cost-effectiveness (Chapman, 2006).

There are still few producers of injectable form of doxycycline. A brief survey on the Canadian drugbank database (https://www.drugbank.ca/) reveals a worldwide production centered on fewer than 10 companies worldwide.

Evaluating the compound in its technological context, using the online tool Aulive - Patent Inspiration, there are 665 patents related to doxycycline, 518 of them are patents related to the A61K code (codification internationally assigned in order to classify different groups; A61K patents are related to preparations for medical, dental or toilet purposes). However, only 164 patents are related to the injectable pharmaceutical form.

Although the number of patents is not expressively high, there is an increase in the number of deposits from the date of their publication, as noted in Figure 5.

It is worth to note that only two countries account for almost 50% of total patents worldwide. United States holds about 26% of total patents, and 23% of China (Figure 6). Other countries, such as Germany and Switzerland, contribute only 5%.

CONCLUSION

Over the last decade, a set of public policies have been formulated in Brazil, aiming to progress in overcoming the vulnerability of the health productive base, and mitigate its consequences on the sustainability of the health system. Although the identification of the potential of the health productive base to leverage the economic and technological development of the country and its insertion in the political agenda goes back to the 1970s, it is from the 2000s onwards that a new generation of public policies to support strengthening of the Industrial Economic Health Complex (CEIS).

The Productive Development Partnerships (PDPs) were inaugurated in 2008, and are a set of partnerships between public and private institutions, aimed at reducing the vulnerability of SUS and the prices of strategic health products, by internalizing and developing high value strategic technologies aggregate. They are an instrument of public policy based on a new perception of health policy, which takes into account both the structural characteristics of the international markets for health inputs and the capacity

Figure 5. Activity by publication date
Source: Aulive – PatentInspiration.

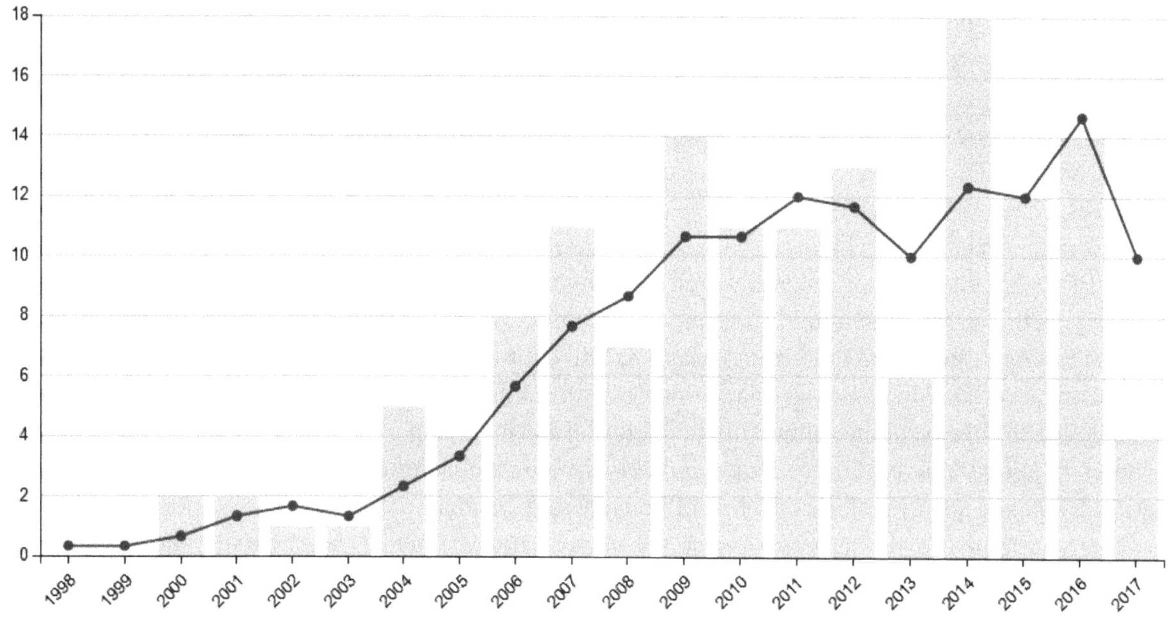

Figure 6. Global distribution of patents about injectable pharmaceutical medicines of Doxycycline
Source: Aulive – PatentInspiration.

of the national productive base to insert themselves in them. Thus, the central objective of the PDPs is to technologically increase the health production base in the country, giving more autonomy to the SUS and encouraging economic growth, by strengthening the productive capacity of the national public and private laboratories associated with the transfer processes of involved in these partnerships.

Strategic alliances resulting from PDP can be used as a way to stimulate the promotion of R,D&I in another areas. So, the time is ripe for new partnerships and investments in healthcare.

Considering the political, industrial and academic context surrounding the field of health, there is clearly potential for continuous success for PDP and technological platform. The SUS list of strategic medicines could be transformed in opportunities for national investments and the introduction of measures for the short, medium and long term for knowledge translation, and, consequently, ownership of innovation by the Brazilian population. The management of the patent information system as a subsidy to the decision makers is efficient. This process detects where the technology is present and their respective trends. Concerning doxycycline, it is highlighted as a priority for the Brazilian Government. In this sense, of the number of patents available during this study, 32% are related to pharmaceutical formulations. In addition to identifying the US and China as the main players holding the technology.

REFERENCES

Aleixo, J. A., & Duarte, P. (2015). Big data opportunities in healthcare. How can medical affairs contribute? *Revista Portuguesa de Farmacoterapia*, *7*, 230–236.

Antunes, A., & Mercado, A. (2000). *A aprendizagem tecnológica no Brasil: a experiência da indústria química e petroquímica*. Editora E-papers.

Balmisse, G. (2006). *Guide des Outils du Knowledge Management: panorama, choix et mise en oeuvre*. Recuperado de http://www.leslivresblancs.fr/informatique/applications-pro/knowledge-management/livre-blanc/outils-du-km-panorama-choix-et-mise-en-oeuvre-62.html

Boland, W. P., Phillips, P. W. B., Ryan, C. D., & McPhee-Knowles, S. (2012). Collaboration and the Generation of New Knowledge in Networked Innovation Systems: A Bibliometric Analysis. *Procedia: Social and Behavioral Sciences*, *52*, 15–24. doi:10.1016/j.sbspro.2012.09.437

BRASIL. MS., D. (2014). *Parceria para o Desenvolvimento Produtivo - PDP*. Accessed may 14, 2014, de http://portalsaude.saude.gov.br/index.php/o-ministerio/principal/leia-mais-o-ministerio/581-sctie-raiz/deciis/l2-deciis/12090-parceria-para-o-desenvolvimento-produtivo-pdp

BRASIL. MS. (2014a). Portaria nº 2.531, de 12 de novembro de 2014. Redefine as diretrizes e os critérios para a definição da lista de produtos estratégicos para o Sistema Único de Saúde (SUS) e o estabelecimento das Parcerias para o Desenvolvimento Produtivo (PDP) e disciplina os respectivos processos de submissão, instrução, decisão, transferência e absorção de tecnologia, aquisição de produtos estratégicos para o SUS no âmbito das PDP e o respectivo monitoramento e avaliação.

BRASIL. MS. (2014b). Portaria nº16, de 15 de maio de 2014. Torna pública a decisão de incorporar a doxicilina injetável e o cloranfenicol suspensão para terapêutica da febre maculosa brasileira e outras riquetsioses no Sistema Único de Saúde - SUS.

BRASIL. Ministério da Saúde. Portaria nº 704, de 08 de março de 2017. Define a lista de produtos estratégicos para o Sistema Único de Saúde (SUS), nos termos dos anexos a esta Portaria.

BRASIL/MS/FIOCRUZ/IPEA. (2012). *A Saúde no Brasil em 2030: diretrizes para prospecção estratégica para o Sistema de Saúde Brasileiro*. Brasil. Recuperado de http://www.isags-unasursalud.org/biblioteca_interna.asp?idArea=33&lang=1&idPai=3024

BRASIL/TCU. (2014). *FISCSAÚDE (No. Relatório Sistêmico de Fiscalização da Saúde - FISCSAÚDE)*. Brasil: TCU - Tribunal de Contas da União. Accessed by http://portal2.tcu.gov.br/portal/page/portal/TCU/imprensa/noticias/noticias_arquivos/032.624-2013-1%20Fisc%20Saude.pdf

Brouwers, M. C., Kho, M. E., Browman, G. P., Burgers, J. S., Cluzeau, F., Feder, G., ... Zitzelsberger, L. (2010). AGREE II: Advancing guideline development, reporting and evaluation in health care. *Canadian Medical Association Journal*, *182*(18), E839–E842. doi:10.1503/cmaj.090449 PMID:20603348

Canadian Institutes of Health Research. (2016). *Knowledge Translation and Commercialiation*. Recuperado 11 de abril de 2017, de http://www.cihr-irsc.gc.ca/e/29418.html#1

Cartaxo, R. J. (2011). *Metodologia de Priorização para Produção Nacional dos Medicamentos Pertencentes à Lista do Sistema Único de Saúde* (Dissertação de Mestrado). National Institute of Industrial Property (INPI - Instituto Nacional de Propriedade Industrial). Accessed by http://www.inpi.gov.br/portal/

Carvalho, A. I., Bodstein, R. C., Hartz, Z., & Matida, Á. H. (2004). Concepts and approaches in the evaluation of health promotion. *Ciência & Saúde Coletiva*, *9*(3), 521–529. doi:10.1590/S1413-81232004000300002

Centro de Gestão e Estudos Estratégicos, & Ministério da Ciência, Tecnologia e Inovação. (2013). *Parcerias Estratégicas nº 36* (36º ed, Vol. 18). Brasília: CGEE. Accessed by http://www.cgee.org.br/parcerias/parcerias.php

Certo, S. C., & Peter, J. P. (2010). Administração Estratégica: planejamento e implantação de estratégias (3o ed). Prentice-Hall.

Chapman, A. S.Tickborne Rickettsial Diseases Working Group. (2006). Diagnosis and Management of Tickborne Rickettsial Diseases: Rocky Mountain Spotted Fever, Ehrlichioses, and Anaplasmosis - United States. A Practical Guide for Physicians and Other Health-Care and Public Health Professionals. Morbidity and Mortality Weekly Report (MMWR). *Recommendations and Reports*, *55*(RR04), 1–27. PMID:16572105

Costa, L. S., Gadelha, C. A. G., & Maldonado, J. (2012). A perspectiva territorial da inovação em saúde: a necessidade de um novo enfoque. *Revista de Saúde Pública*. doi:10.1590/S0034-89102012005000066

de Carvalho, A. L. B., Souza, M. de F., Shimizu, H. E., Senra, I. M. V. B., & de Oliveira, K. C. (2012). SUS management and monitoring and evaluation practices: Possibilities and challenges for building a strategic agenda. *Ciência & Saúde Coletiva*, *17*(4), 901–911. doi:10.1590/S1413-81232012000400012

De Negri, J. A., & Kubota, L. C. (2008). *Políticas de Incentivo à Inovação Tecnológica*. Brasília: Ipea - Instituto de Pesquisa Econômica e Aplicada. Accessed by http://www.ipea.gov.br/portal/index.php?option=com_content&view=article&id=5569

Figueiredo, T. A., & Pepe, V. L. E. (2016). *A seleção de medicamentos estratégicos para a produção nacional e a saúde coletiva. Saúde Amanhã. Textos para Discussão; n. 21*. Fundação Oswaldo Cruz.

Gadelha, C. A. G. (2015). Reflexões sobre a saúde na agenda contemporânea do desenvolvimento: Saúde, desenvolvimento e inovação. Rio de Janeiro. *Cepesc*, *1*, 43–69.

Gadelha, C. A. G., & Costa, L. S. (2012a). Saúde e desenvolvimento no Brasil: avanços e desafios. *Revista de Saúde Pública*. doi:10.1590/S0034-89102012005000062

Gadelha, C. A. G., Costa, L. S., & Maldonado, J. (2012b). O Complexo Econômico-Industrial da Saúde e a dimensão social e econômica do desenvolvimento. *Revista de Saúde Pública*. doi:10.1590/S0034-89102012005000065

Gadelha, C. A. G., Quental, C., & Fialho, B. de C. (2003). Saúde e inovação: Uma abordagem sistêmica das indústrias da saúde. *Cadernos de Saude Publica*, *19*(1), 47–59. doi:10.1590/S0102-311X2003000100006 PMID:12700783

Gadelha, P., Carvalho, J. N., & Pereira, T. R. (2012c). *A Saúde no Brasil em 2030: diretrizes para a prospecção estratégica do sistema de saúde brasileiro*. Editora FIOCRUZ. Accessed by https://portal.fiocruz.br/sites/portal.fiocruz.br/files/documentos/capa_saude_brasil_2030.pdf

Gavirneni, S., Kapuscinski, R., & Tayur, S. (1999). Value of Information in Capacitated Supply Chains. *Management Science*, *45*(1), 16–24. doi:10.1287/mnsc.45.1.16

Gonçalves, A. (2006). Problema de Saúde Pública: Caracterizando e avaliando aplicações. *Revista Brasileira de Epidemiologia*, *9*(2), 253–255. doi:10.1590/S1415-790X2006000200013

Hartz, Z. M. de A. (1997). *Avaliação em saúde: dos modelos conceituais à prática na análise da implantação de programas*. Publisher FIOCRUZ. Accessed by de http://books.scielo.org/id/3zcft

Hartz, Z. M. de A. (2012). *Avaliação em saúde: dos modelos conceituais à prática na análise da implantação de programas*. Editora FIOCRUZ. Recuperado de http://books.scielo.org/id/3zcft

IMS Institute for Healthcare Informatics. (2012). *The Global Use of Medicines: outlook through 2016*. Recuperado de www.imshealth.com

Kira, T. A. R. A. P. A. N. O. F. F. (org.). (2015). Analise da Informacao para Tomada de Decisao Desafios e Solucoes (Vol. 1). Brasil: Editora Intersaberes. Recuperado de http://www.estantevirtual.com.br/b/kira-tarapanoff/analise-da-informacao-para-tomada-de-decisao-desafios-e-solucoes/158550962

Koelen, M. A. (2001). Health promotion research: Dilemmas and challenges. *Journal of Epidemiology and Community Health*, *55*(4), 257–262. doi:10.1136/jech.55.4.257 PMID:11238581

Lawrence, S., & Giles, C. L. (2000). Accessibility of information on the Web. *Intelligence*, *11*(1), 32–39. doi:10.1145/333175.333181

Le Moigne, J. L. (1994). *La théorie du Système Géneral: théorie de la modélisation*. France.

Lynch, C. (2008). Big data: How do your data grow? *Nature*, *455*(7209), 28–29. doi:10.1038/455028a PMID:18769419

Macedo, B. B. (2014a). *Parcerias para o Desenvolvimento Produtivo (PDPs) como Estratégia de Acesso aos Medicamentos*. FIOCRUZ/Farmanguinhos, Especialização de Gestão da Inovação em Medicamentos Fitoterápicos.

Macedo, B. B. (2014b, outubro). *Parcerias para o Desenvolvimento Produtivo (PDPs) como Estratégia de Acesso aos Medicamentos*. FIOCRUZ/Farmanguinhos, Rio de Janeiro. Accessed by www.far.fiocruz.br

Magalhaes, J. L., Antunes, A. M. S., & Boechat, N. (2012a). *Technological Trends in the Pharmaceutical Industry: the matter of neglected tropical diseases – An overview of the Research, Development & Innovation in Brazil*. Synergia Editora. Accessed by http://www.livrariasynergia.com.br/livros/M39700/9788561325732/tendencias-tecnologicas-no-setor-farmaceutico-a-questao-das-doencas-tropicais-negligenciadas-edicao-bilingue.html

Magalhaes, J. L., Antunes, A. M. S., & Boechat, N. (2012b). *Tendências Tecnológicas no Setor Farmacêutico: a questão das doenças tropicais negligenciadas - uma perspectiva da P,D&I no Brasil* (Vol. 1). Synergia Editora. Accessed by http://www.livrariasynergia.com.br/livros/M39700/9788561325732/tendencias-tecnologicas-no-setor-farmaceutico-a-questao-das-doencas-tropicais-negligenciadas.html

Magalhães, J. L., Hartz, Z., & Antunes, A. M. (2016). An Evaluation of Partnership for Productive Development in Brazil's Healthcare: Measuring the Knowledge Translation from Implementation to the Impact. In Handbook of Research on Information Architecture and Management in Modern Organizations (pp. 525-544). IGI Global.

Magalhães, J. L., & Quoniam, L. (2013). Perception of the Information Value for Public Health: A Case Study for Neglected Diseases. In *Rethinking the Conceptual Base for New Practical Applications in Information Value and Quality* (p. 345). IGI Global. Recuperado de http://www.igi-global.com/chapter/perception-information-value-public-health/84218

Magalhães, J. L., Quoniam, L., & Boechat, N. (2013). Web 2.0 Tools for Network Management and Patent Analysis for Health Public. *Revista de Gestão Em Sistemas de Saúde*, *2*(1), 26–41. doi:10.5585/rgss.v2i1.48

McKinsey Global Institute. (2011). *Big Data: The Management Revolution - Harvard Business Review*. Recuperado de http://hbr.org/2012/10/big-data-the-management-revolution/ar/1

Melese, J. (1990). *Approches Systémiques des Organisations: Vers l'Entreprise à Complexité Humaine*. Paris, France: Les Éditions d'Organisation.

Minelli, M., Chambers, M., & Dhiraj, A. (2013). *Big Data, Big Analytics*. John Wiley & Sons, Inc. Recuperado de https://books.google.com.br/books/about/Big_Data_Big_Analytics.html?hl=pt-BR&id=Mg3WvT8uHV4C

Ministério da Saúde. M. (2014a). *Parceria para o Desenvolvimento Produtivo - PDP*. Accessed in November 2, 2014. http://portalsaude.saude.gov.br/index.php/o-ministerio/principal/leia-mais-o-ministerio/581-sctie-raiz/deciis/l2-deciis/12090-parceria-para-o-desenvolvimento-produtivo-pdp

Ministério da Saúde. M. (2014b). *SCTIE*. Accessed in November 2, 2014 by http://portalsaude.saude.gov.br/index.php/o-ministerio/principal/secretarias/sctie

Morin, E. (1982a). Le Système: Paradigme ou/et Théorie. In *Sicence avec Conscience*. Paris, France: Libraire Arthème Fayard.

Morin, E. (1982b). Le système: Paradigme ou/et théorie. Science avec Conscience.

Pearson, A., Jordan, Z., & Munn, Z. (2012). Translational Science and Evidence-Based Healthcare: A Clarification and Reconceptualization of How Knowledge Is Generated and Used in Healthcare. *Nursing Research and Practice*, *2012*, 1–6. doi:10.1155/2012/792519 PMID:22474583

Pesqueux, Y. (2005). *Management de la connaissance: un modèle organisationnel?* Management De La Connaissance: Un Modele Organisationnel? Recuperado de https://halshs.archives-ouvertes.fr/halshs-00004005/document

Pesqueux, Y., & Ferrary, M. (2011). Management de la connaissance (2o ed). Economica. Recuperado de /Entreprise/Livre/management-de-la-connaissance.

Piaget, J., & Garcia, R. (1983).. . *Psychogenèse et Histoire des Sciences*, *19*(1), 304.

Porter, M. E. (2008). *Competitive Strategy: Techniques for Analyzing Industries and Competitors*. Simon and Schuster.

Quoniam, L. (2011). *Competitive Intelligence 2.0*. ISTE Ltd and John Wiley & Sons Inc. Accessed by http://www.iste.co.uk/index.php?f=a&ACTION=View&id=409

Ross, J. W., Weill, P., & Robertson, D. (2006). *Enterprise Architecture as Strategy: Creating a Foundation for Business Execution*. Boston: Harvard Business Review Press.

Santos, M. E. R., Toledo, P. T. M., & Lotufo, R. A. (2009). *Transferência de Tecnologia: estratégias para a estruturação e getão de Núcleos de Inovação Tecnológica*. Komedi. Accessed by http://www.pucrs.br/ett/Livro%20Transferencia%20de%20Tecnologia.pdf

Trigo, M. R., Gouveia, L. B., Quoniam, L., & Riccio, E. L. (2007). *Using competitive intelligence as a strategic tool in a higher education context* (D. Remenyi, Org.). Nr Reading: Academic Conferences Ltd.

von Hippel, E. (2005). Democratizing innovation: The evolving phenomenon of user innovation. *Journal Für Betriebswirtschaft*, *55*(1), 63–78. doi:10.100711301-004-0002-8

WHO. (2006). *Bridging the "Know-Do" Gap: meeting on knowledge translation in Global Health* (No. WHO/EIP/KMS/2006.2). WHO. Recuperado de https://www.measureevaluation.org/resources/training/materials/high-impact-research-training-curricula/bridging-the-know-do-gap.pdf

WHO. (2013a). *TDR | TDR Performance assessment framework*. Special Programme for Research and Training in Tropical Diseases (TDR). Recuperado de http://www.who.int/tdr/publications/about-tdr/strategy/framework/en/

WHO. (2013b). *Priority Medicines for Europe and the World Update Report*. WHO.

Zicker, F. (2007). *TDR'S Research and Capacity Development Strategy*. WHO | TDR. Accessed by http://www.edctp.org/forum2007/presentations/Workshop_on_Enhancing_capacity_-_Fabio_Zicker.pdf

ENDNOTES

1 The SUS was constituted in the 1988 Federal Constitution to serve the whole population, from the poorest to the wealthiest, in equal measure. Socially speaking, it was a great step forward for a developing country like Brazil. However, it still falls short in its goal of providing comprehensive healthcare, with less months-long waiting lists for doctor's appointments and shortages or lacks essential medicines for free distribution.

2 According to the Industrial Property Law #9279 of 1996, a patent is "a privilege conferred by the State that assures its holder exclusivity in exploiting the patented technology, safeguarding the holder's right to prevent third parties from exploiting (producing, selling, buying, stocking) the protected object." Filers of patent applications do so with the expectation of being awarded rights over the object of the application, but these rights can only be fully exercised after they are effectively awarded the patent. (BRASIL, PRESIDÊNCIA DA REPÚBLICA, 1996).

Chapter 3
Improving Project Management Decisions With Big Data Analytics

George Leal Jamil
Informações em Rede C e T Ltda., Brazil

Luiz Fernando Magalhães Carvalho
Banco Itaú, Brazil

ABSTRACT

A relationship between project management and knowledge management was observed with a detailed level of analysis in this chapter, as analytics tools and methods were presented to define new perspectives for these dynamics. After a theoretical review that advanced the level reached by a previous paper on the same topic a new theoretical background was completely worked, resulting in a base where a deeper way of analysis allowed, at the end, to study practical cases of rich association for PM and KM in practical, ready to apply situations. As a trend for next competitive cycles, tools, methods, and techniques that focus knowledge production for decision making are to be increasingly defined and applied, on one hand enabling organizations to propose new competitive structures and positioning, and on the other hand, presenting a more aggressive, faster, and demanding competitive environment.

BACKGROUND

This chapter was produced adopting Jamil and Carvalho (2015) as a base, producing a review of its findings, adding studies from big data and analytics perspectives.

Project management (PM) is a remarkably dynamic area, where data, information and knowledge are systematically and empirically produced and, as well, demanded, in fast-change scenarios. As projects are to be detailed, proposed, negotiated, executed and maintained, several contents of these fundamental assets are to be faced by the project manager and his associated team, users and other project stakeholders, as decisions need to be produced with precision every time. Poor decisions, taken from indefinite or problematic data and information, will lead to more risky and precarious implementations.

DOI: 10.4018/978-1-5225-6225-2.ch003

This chapter reviews the previous findings of Jamil and Carvalho (2015), detailing how big data and analytics, as a potential trend to produce more qualified and applicable knowledge, can result in better decisions, producing a challenging, but worthy picture for project management.

PROJECTS, STRATEGIC DECISIONS, AND DEMANDS FOR KNOWLEDGE

The basic definition of a project already motivates to think about knowledge and its strategic applications. It is possible to check, from PMBoK (2013), that projects are ways to achieve organizational strategic goals. By analyzing this initial concept, it is possible to understand that a view of the future for this organization was projected, configuring goals to be achieved. As this new situation is faced from organizations, knowledge can be one of the most critical assets to be applied to produce the design of this reachable description. But it is also relevant to understand the role of projects in this process, how its execution and implementation, as final results, were based on data, information and knowledge, as these contents were also generated during the project management phases. It is an important relationship, which defines the background of this research.

It is important to remind that strategy is a discipline oriented to coherently propose future positions for one organization in its social, business and competitive environment. One strategy is stated through a set of consistent decisions, expressed in a base that constitutes the organizational strategic plan, as a coherent path of subsequent decisions (Porter, 2008). In this context, strategic planning is a process that continuously study organizational resources, capabilities, perspectives and competencies, with addition of determination of internal and external factors, aiming to monitor and design implementable actions to reach established goals (Kotler & Keller, 2005; Porter, 2008; Mintzberg, Ahlstrand & Lampel, 2009).

Jamil *et al.* (2012) conceptualized strategic planning, as the main organizational process, as it is a "knowledge demanding" cohesive set of tasks, which needs, for its success, some factors, components and contents, such as:

- Studies of external business environment, considering data and information from competitors, suppliers, customers, legal aspects, business rules, market reactions to events, etc. (Porter, 2008).
- Internal business environment analysis, observing corporate resources, motivations and restrictions and its correlated control and management perspectives, competitive advantage positioning, etc. (Mintzberg, Ahlstrand & Lampel, 2009).
- Data and information about a market history, as a "knowledge base of best practices", which can be revoked for simulations, business models and dynamic studies for competitive design (Choo, 2005; Nonaka, 2008).
- Details of connected and interrelated projects, assuming they are data, information and knowledge producers, as the practices cited above and other productions, that will enable favorable conditions to plan and execute a new project.
- Definition of measurements and methods to measure, or "indicators", as financial quantitative demonstrations, reputation, performance and many other factors that can be set by strategy staff and followed through a plan execution for strategic monitoring that will allow its following and management in the workplace (Kaplan & Norton, 2007).

It is opportune, from the definitions pointed above, to notice the importance of data, information and knowledge sources. This fact is fundamental to comprehend how Big Data contents, as announced in the following text, are relevant for internal, external and historical studies to serve as bases for strategic planning, as it results in goals prediction for new reachable and challenging points.

Among typical strategic planning tasks conducted during the phases of project, planning and execution, it is possible do describe:

- Definition of strategic goals and correspondent ways of monitoring, defining a reachable scenario as a success for the specific planning process;
- Intermediate and final results evaluation or plan monitoring;
- Project resources management methods and goal reaching perspectives.

In these cases, strategic planning and its associated execution demand data, information and knowledge to plan, review, modify and design business dimensioning and positioning. The "meaning" or "sense" about competitive scenarios, which will result in more precise strategic propositions are affirmed by various authors as intermingled with structured knowledge availability (Choo, 2003; Kearns & Lederer, 2003; Prado, 2006).

KNOWLEDGE MANAGEMENT FOR PROJECT MANAGEMENT-AN IMPORTANT RELATIONSHIP

The relationship of PM and knowledge management (KM) must be observed initially for the proposed objective of this study. As defined by Project Management Institute (PMI, 2014), projects are a *temporary group of activities designed to produce a unique product, service or result.* As an additional conceptualization from the same source, project management is:

... the application of knowledge, skills and techniques to execute projects, effectively and efficiently. It's a strategic competency for organizations, enabling them to tie project results to business goals — and thus, better compete in their markets.

Project, project management and its associated perspectives produce an interesting research object to understand its correspondence with organizational knowledge.

Reich, Gemino and Sauer (2008; 2012) tell that knowledge management (KM) practices have a contributive interaction with project management planning capabilities, which result in production of knowledge for strategic advantage, as projects can be related to strategic planning, as its goals executive plans and procedures. The analysis held by the authors, which evolved from one approach to the older, through information and knowledge contents, shows a detail level further explored in this section when considering the cooperation of KM practices to PM planning. From these works, also, it is possible to understand the complexity and richness of project management contexts when observing these tasks as critical consumers and producers of organizational knowledge, interacting with all organizational levels. It is important to address this relevant aspect, as analytics tools, as dynamic processors of data and information in all organizational contacts, will help decisively to produce knowledge that will be managed in advance, according to the perspectives of a continuous process, the knowledge management activities. It will, for example, refer to the major definition at PMBoK (2013) for "knowledge areas", such as financial, risk, executive and operational communication, human resources, acquisition, cost, time and quality project management.

It is possible to explore this critical relationship of KM and PM, as it was observed in several theoretical and practical analysis. For example, in Landaeta (2008) we find an evaluation on how the transfer of knowledge among projects can benefit its design and management. In this study, there is an observation on how a set of KM practices can improve a project portfolio management, serving as an important strategic tool for companies that compete in several markets or with diversified product lines that always demand new projects for its marketing and commercial positioning.

Recurring to Reich, Gemino and Sauer (2012), it is possible to understand how risks management practice, another critical PM task, is improved by the relation with applied KM actions, findings that are consolidated on the final model presented by these authors. Gasik (2011) researched a model that also sought to associate KM and PM principles, resulting in an analysis of the importance of knowledge assets in the project management organizational roles, from the planning according to strategy to implementation. For Müller *et al.* (2013), it is possible to understand how the modern approaches of project management correspond to some observed through the KM conceptual and theoretical base. For example, project management offices (PMOs) propositions and actions, along with management tasks distributed over companies' networks are approached by these authors as evidences of these interactions.

Regarding to the works of Barney (1991) and Porter (2008), it is possible to affirm that PM and KM practices alignment appear as a critical strategic relation, with clear potential to generate competitive advantage, as a valuable, rare and irreplaceable resource (Barney, 1991). Projects can be defined as methodological ways to produce results according to strategic planned goals and actions, and it is opportune to understand how knowledge is applied on decisions taken during its planning and execution (Barney, 1991; Porter, 2008; Nonaka, 2008). In addition, the association of a corporative and inter-company projects, known as project management *portfolio* (PMBoK, 2013), brings another level of demand for knowledge, with additional degree of complexity to produce it, as those contexts where these projects are held also reveal, produce and apply data, information and knowledge.

As these and other works reviewed, it is possible to affirm the perception of KM and PM integration, and the role of analytics in its development in modern contexts as the rationale for this chapter. This proposition constitutes a valid and opportune research objective. The introduction of new fronts of study, such as risk management, financial project structure, project portfolio management, among many others, represent new trends for applied studies. It is possible also to think Big Data and its associated phenomenon, analytics, as new areas to be studies, producing several essential, opportune and trending perspectives for this potential integration. For project management base, as it was also approached by PMI (2013), the potentials for data analysis, information and knowledge generation are immensely opportune at this moment, facts that will be explored in the forthcoming analysis.

DATA, INFORMATION, AND KNOWLEDGE

The relationship and conceptualization for these basic concepts is interesting, as they are useful components for business activities and plans. Nowadays, the comprehension of the dynamics of these contents constitutes in opportunities to offer services and products, as they present opportunities by themselves and in business plans that relate to each other, as information systems, information technology applications and implementations, among several others. With the insertion of new technology-based services and new methods of industrial production, dynamical collaborations among companies and events, such as the Analytics and Big Data phenomena, this integrative and complementary view demands more study (Ohata & Kumar, 2012; McAfee & Brynjolfsson, 2012; SAS, 2017).

It is possible to understand data, information and knowledge as isolated concepts, but also in terms of their relationships, which already advance on the perspective of information and knowledge management (Jamil, 2005). Davenport and Prusak (2000) affirmed a set of definitions and contours for these concepts that will contribute for the theoretical base adopted in this chapter. For them, these basic concepts keep its definition, but also reveal relationships with each other, a dynamic characteristic. Tuomi (2000) studied the producing of data from knowledge conception. It is an apparent opposite way as usually treated by literature, but it indeed reinforces the concepts of each item and its correspondent relationship. Also from Lucas Jr. (2005), Turban, Mc Lean, and Wetherbee, 2002, Turban, Rainer and Potter, 2007 and Stair and Reynolds (2009), it is possible to understand how Information Technology (IT) and communication resources applied such concepts in themes as technological infrastructure, information systems design and web-based applications development, nowadays impacting the design of applications, or, as it is commercially described, the "apps".

Jamil (2005) consolidated these definitions, producing a conceptual base that states data as an absolute value that can be obtained directly from a measurement, observation or collected from an automated source and is expressed in usual measuring standards (as currency, metering system units, etc.). Although showing flexibility and easiness to generate, storage and communication, data lacks context, or immediate support for more complex decisions, being an understandable building block for more consistent concepts.

Information was conceptualized as a collection of homogeneous, correlated data, added with notions from the context where it was produced. With this wider conception, it is possible to expect results of better contribution for predictions, as decision-making, planning activities, prospection and several others with more complex managerial results. Information increases organizational decision prospects, but it demands more resources to be precisely applied, as the support from information systems and its linked infrastructure, demanding investments and adequate planning abilities (Turban, Rainer & Potter, 2007; Stair & Reynolds, 2008).

Based on these concepts, knowledge can be produced from a collection of homogeneous information, including definitions of the processes that generated that set of information. Knowledge allows the maximum level of decision capabilities, description and analysis. For example, it enables even competitive predictions such as market dynamics, product negotiation evolutions, demands, etc. (Davenport, 2000; Akbar, 2003; Jamil, 2005; Nonaka, 2008). But on the other hand, knowledge shows many challenges for its management, as to store, convert, communicate and share, are not simple, straightforward activities, resulting in the need of a specific process to treat it, as it is defined by authors as knowledge management (Choo, 2003; Kearns & Lederer, 2003; Jamil, 2005; El-Bashir, Collier & Sutton, 2011).

To conclude this initial conceptualization about these fundamental concepts, it is relevant to point out, from the authors cited above and many others (for example, for those that study marketing relationships such as customer-behavior knowledge and strategic planning implementations), that it is possible to obtain knowledge from data and from information, as it is also possible to develop information from data. This relationship prescribes a managerial cycle, where the contents are applied to produce each other, eventually resulting in a process description that can be found and evidenced in productive actions, chains and organizational trivial tasks. This result is adequate to define how knowledge management practice is more frequent and simpler than it appears when approaching the theme just for strictly theoretical origin.

BIG DATA AND ANALYTICS

Some years ago, it was possible for any Internet web-based service manager to understand how interactive his service was: a lot of commercial, interactions, customer data flown through servers, web transactions, orders and other simple dynamics. Lately, these findings called attention when data mining services were offered in simple, cheap way, aiming to decipher behaviors, reactions and preferences, enabling web-site and electronic commerce managers to understand and even to predict some trends – as positive reactions by customers, how he will potentially react to an offer, what are the reasons that will take him to analyze a competitor´s offer and so forth. Then, we reached the "big data" wave, as unstructured communication tools invaded our lives forever (Dodge, 2010).

Although not new, "Big data" phenomenon has been perceived by entrepreneurs and socio-political actors, as a process of "knowledge generation from data". This new wave of scientific and practical evidences proves how important is to observe data, information and knowledge in organizational and competitive context for decisions, such as strategic planning (Ohata & Kumar, 2012; Park, Huh, Oh & Han, 2012; SAS 2017). It is a new area of scientific and applied research, which must be adopted for future studies about information and knowledge management, along with several other contributive fields. It calls special attention due to the huge amounts of data generated randomly and spontaneously throughout the world, in numberless ways. It is also supported by the fast introduction of technology and the continuous reduction of IT infrastructural costs, such as storage, transmission, and many other components. Along with these remarkable features, it is possible to observe how big companies, such as Google, Microsoft and Facebook, or small, fast-action, innovative ones, reacted to this data production, offering easy-to-access tools for analysis, scenario detailing and prediction, known as "analytics" (Rayome, 2016).

As a referential source, McKinsey (2011) described big data as *datasets whose size is beyond the usual ability of typical database software tools to capture, store, manage, and analyze*. Obviously, as it seems, to use these data sets, huge computing resources, optimized algorithms and other technologic resources and structures must be applied in a planned way to extract knowledge from big data. IT market and scientific communities have been studying tools and associated techniques to capture, collect, store, classify, analyze and produce deeper results or applicable outcomes from datasets (McKinsey, 2011; Chen, Chiang & Storey, 2012; Ohata & Kumar, 2012; Courtney, 2013; SAS, 2017).

Additionally, analytics emerged as both a demand from market agents, regarding that massive data generation and technology availability, as new resources are presented everyday by IT players (De Roche, 2016; Jain, 2016; SAS, 2017). As typical analytics services, it is possible to identify big players offers, such as Google and Facebook, when they put available for this service customers, solutions as:

- Number of users on-line and their geographical distribution.
- Information technology infrastructure used for access.
- Frequent sites accessed, based on user groups.

- Some typical customer reactions, as, for example, to marketing advertisements, announcements and offers.
- Conversion Rates: How many users advanced negotiation with refer to offer announcements.
- Selling performances, for each offer through internet.
- Segmentation information, as customer identifications and definitions can be used to analyze their adjustments to each offer (Mehmod, 2017).

As a typical analytics application, one business analyst can evaluate how a product offer prescribed a trend, aiming to understand its perspective adjustment to statistical models and projections, which may allow adjust better this offer. As a comprehensible detailing from this commercial study, several small and medium projects can be started, as commercial points of sales promotions and discounts, commercial campaigns, logistics operations, human resources training and preparation, among several others. It is a simple, clear way to relate the conceptual base, the practical context that motivate this chapter writing: projects and project management that emerge from analysis produced from big data generation and analytics projections.

Informational products can now be regarded as definitions presented by Manyika *et al.* (2011) and IBM (2014), as they are created from big data and analytics, when authors cited contents to produce strategic and tactical plans for various industries around the world, and how they connect to strategic project management, enabling projects that will not only follow, but execute strategy actions on the field. This factor allows pondering the effort trying to relate such production of knowledge to strategic marketing planning, and perceive it as critical tasks conducted by competitive firms specially when applied for long time plans. In one of these cases, it is opportune to refer about market intelligence as one process improved by such a correlation of data treatment and knowledge generation for critical tactical and strategic decisions, more specifically, for strategic marketing planning (Jamil, 2013).

As one of the interesting organizational processes to understand, taken as an example of a knowledge producer of sectorial data, market intelligence (MI) is exposed by Jamil *et al.* (2012) as a cyclic set of activities coordinated to provide strategic marketing plan knowledge for a group of companies who compete in the same sector. MI has a proposed structure of phases such as value chain modelling, knowledge demands diagnosis, data and information availability and a cycle of collection, validation, classification, processing, communication and end-cycle review. MI is a process oriented to produce knowledge for market sectors, as any of the companies who receive this produced knowledge take advantage of it in its specific strategic and tactical plans – such as marketing, logistics, project management, etc. MI deals directly with data, encompassing those huge sets identified as Big Data, as it is treated properly through any analytical process (Jamil, Santos, Alves, & e Furbino, 2012; Johnson, 2012; De Man, 2012; Dimitrios, Sakas & Vlachos, 2013).

Big data and analytics studies and practical applications brought a new way to observe how knowledge can be produced and applied in new entrepreneurial contexts, such as those put available by systems interconnectivity, new software design and social media applications, as they can be used to collect and treat some customer, citizens, agent reactions to any event. Classical and new methods, such as statistical and heuristics, has been suggested for studies that will allow, for example, the precise application for information technology. This way, it is possible to define that we have a specific condition to depict precise targets and reachable goals for future projects. As this section of literature review showed, these concepts can define already perspectives of the basic relationship sought in this chapter, where project management is benefited by knowledge management which, in its way, is produced and processed through new processes and services, as big data and analytics.

KNOWLEDGE MANAGEMENT, BIG DATA, AND ANALYTICS

Knowledge management is conceptualized as a process, a continuum, composed by a set of predicted and manageable tasks that deal with organizational knowledge (Jamil, 2005). Clearly, it is a simple definition, also supported by an objective concept of knowledge, as exposed earlier, presenting the perspective referring to practical aspects, allowing further discussions about the KM goal and practices.

Nonaka (2008) defined knowledge management as a more complex process, with wider results, designed to answer several organizational needs for planning and execution. As this process is approached as a more complex theoretical construction, it was consolidated in reviews produced by Christea and Capatina (2009) and Haslinda and Sarinah (2009), where, typically, several KM research models were described and analyzed. Jamil (2005) studied basic knowledge oriented activities, taking business organizations as examples and its correlation, resulting in a composed set of tasks, such as searching, generating, registering, sharing, valuating and strategic application monitoring, which was proposed also as an integrative KM process model, designed to receive additional contributions in future studies.

Clark Jr., Jones and Armstrong (2007) and El-Bashir, Collier and Sutton (2011) also studied knowledge management systems in an integrative perspective. These and other studies that adopt integrative focus propose KM process integration to other organizational factors and processes, becoming not only a function, but a set of principles which can be applied to improve knowledge application in all organizational activities. From these sources is also observed the importance of information and technological systems as components of such processes, to allow understanding and corroboration of the intended conceptual relationship between knowledge and project management practices. This implementation-oriented contribution allows us to understand how IT can be a supporting technology for information systems, which, by this way, can also encompass social media, big data and analytics processing for project management perspectives.

As we aim to discuss how knowledge management can occur in this new scenario of massive data generation, it is also important to relate classical concepts, as the organizational business intelligence (BI) (Cao, Zhang & Liu, 2006). It was defined as an important set of tools, its related methods and analytical procedures exposed mainly by information technology (IT) market. Defining this way, BI is now a receptive conceptual and practical field for the application of big data and analytics resources (Inmon, Strauss & Nishloss, 2008; El-Gayar & Timsina, 2014). These peculiarities encompass technological issues and information generation concept, when the first group of authors dedicate a special effort to identify BI as a set of components, such as data warehouses (as storing resources), "mining" tools (specifically for pattern recognition processes, relationship and frequency analysis) and result production, as customized report tools facilities.

As Kimball and Ross (2010) detail, BI depends not only on the definition of this set of components, but also in its integration, planned application and overall conditions to be used in organizational situations to produce its complex and usually synthetic results. Taking their definition of pre-requisites for a BI effective application, it is useful to recall, from studies of information systems as from Stair and Reynolds (2009) that, among these basic elements to be used, an organization needs skilled, prepared people with fair conditions to interpret results and implement the BI outcomes for decisions, as those discussed earlier in this text. For project management, for instance, as it will be better debated in the following text, a BI-oriented set of IT tools can produce several results as simulations for risk management, financial values, human resources allocation, etc., allowing a deeper study for PM planning and execution.

BI tools and associated techniques assume a relevant situation in this study, as they represent a set of organized components to process data to produce a synthetic, specific knowledge for decision-making, exactly aligned to what is observed for a conceptual relationship between project and knowledge management practices, which will be expanded in this study, regarding the introduction of big data, analytics. Interestingly, it will perform an integrated informational environment, as presented by information system researchers.

Finally, it is important to discuss information systems. Information systems (IS) are arranged sets of several components oriented to treat information to produce desired results for pre-defined organizational goals (Stair & Reynolds, 2009; Jamil, Santos, Alves, & e Furbino, 2012). This concept represents an important functional and strategic arrangement of IT hardware, software, interconnection infrastructure and skilled people, which will implement processes, tasks and methods that solve information-driven organizational problems. Several authors from project management area exemplify ISs that impact expressively works as financial, risk, people, logistics and other administrative PM tasks.

Information systems, as a theoretical objective, is an opportune way to understand in a broader perspective, how information will be dealt in one organizational arrangement, how it can be treated to be applied in its value-aggregation chain and, finally, how information is processed and flows through organizational networks. IS will also be a connection point for the present study, relating some of our theoretical contexts in one complete, dynamic and interrelated research scenario.

THEORETICAL BACKGROUND

As an integrated approach is adopted for this chapter, an opportunity to study several concepts arise and will be, at least partially, addressed in this section. Among these concepts, as a focus, we can explore big data, project management, knowledge management and information systems theoretical references. This study intends to set a first level of relationship, allowing to understand this immense potential and enabling further discussions about KM, Big Data and PM practices.

APPROACHING KNOWLEDGE DISCOVERY AS A PERSPECTIVE OF KNOWLEDGE PRODUCTION FROM PROJECTS

Projects can be regarded, firstly, as a source of knowledge. As PMI (2013) conceptualized project management as a continuous process, there is a potential arena where data and information are continuously processed and will allow, by consequence, the possibility to produce knowledge. In these terms, the theoretical delimitation of knowledge discovery in databases (KDD) can be an essential base to propose the relationship among big data, analytics and project management (Fayad, Patesk-Shapiro & Smyth, 1996).

As stated by (Fayad, Patesk-Shapiro & Smyth, 1996; Yonce, Taylor, Kelly, & Gnau, 2017), the generation of potential value for users and customers of information and knowledge systems depend or are related to analysts' abilities on extracting useful answers from datasets, as, for example, business reports, trends, sales reports, customer-oriented driver's information and many others. This view relates the main value, not exactly to the amount of data, but in users´ ability and conditions to analyze it, according to final business rules and applications. These abilities can be significantly improved, if analytics and big data tools and resources can be applied. In the following, we aim to discuss some resources, as examples.

From Chapman et al. (2000), SEMMA, and from Hampton (2011) and SAS(2014), CRISP-DM are proposed solutions of methodologies for knowledge discovery processes, which became almost industrial standards, as they are widely applied for knowledge generation from datasets, such as those that compose the functional array of big data assets and resources. It is also notable the popularization of tools which enable knowledge discovery in massive data, such as the MapReduce models and techniques, cited by Jeffrey and Ghemawat (2008) and the Hadoop framework, presented by White (2009). According to Azevedo and Lourenço (2008), each step of CRISP-DM and SEMMA can be mapped to KDD steps and, by this way, regarded as actions of a typical KM process. Dodge (2010) and Chen, Roger and Storey (2012) studied also the application of mining techniques over big data contents, discussing how the adaptation of the classical mining methods for knowledge generation in this new environment can produce applicable knowledge for business decisions.

As Berry and Linoff (1997) defined that data mining as the analytical step of the KDD process in which prepared data is used to generate applicable knowledge, it is possible to understand how this opportune process is reaffirmed in SAS (2014) and SAS (2017), this last reference clearly connecting to project management actions and plans. This production of knowledge can be performed using various approaches and algorithms for purposes, such as marketing support (Rygielskk, Wang, & Yen, 2002; Hu & Liu, 2004; Yonce, Taylor, Kelly, & Gnau, 2017), stock marketing perspectives (Enke & Thawornwong, 2005; Bollen, Mao & Zeng, 2011) and fraud detection (Chan, Fan, Prodromidis, & Stolfo, 1999; Phua, Lee, Smith, & Gayler, 2010). Kremer, Mantin and Ovchinnikov (2013) showed how a process that could coordinate and prepare information and knowledge about the external competitive environment favors strategic marketing planning, connecting it to the organizational strategic process.

As cited before, a critical aspect for information management is customer behavior. Various authors described how an alignment of internal resources could occur to fulfill customers' expectations, attributing the term "myopia" to identify the correspondent difficulty for a company to understand correctly its market. A strategic alignment, for example, using an information system, will define a relationship of internal forces and resources to this external context, to implement effective strategic marketing. This perspective integrates big data datasets, processed through KDD resources, through an information system, to allow project management tasks implementations.

Approaching, generally, KDD alternatives, we find from the sources cited: Classification, Regression, (cluster) analysis, rules generation and outlier detection. Summarizing these possibilities, we can describe *Classification task* as labeling a set of data using a decision model created from a set of known data that can allow prediction of new data generated through the same system. *The regression task* is also based on known records, but besides "labeling" new records, as it was proposed by Classification, approximated numerical values could be calculated, or inferred, another trend or prospective prediction possibility.

Big contents of datasets can be split using *Cluster analysis* or *clustering*. Dividing the data into groups, can minimize instances (or occurrences) differences, maximizing the analytical views of several groups, understanding how data can reveal evidences from a big sample, which may vary from one situation to another. This way, it is possible to generate *association rules* to verify the co-occurrence between observations from data generated from past activities. Finally, *Outlier detection* is related to identifying patterns in data that do not conform to expected behavior (Varun, Banerjee & Kumar, 2009).

RELATING KNOWLEDGE MANAGEMENT, ANALYTICS, BIG DATA AND PROJECT MANAGEMENT

It is possible to understand several proposals that examine project management as a dynamic context, where knowledge can potentially flow among its activities, being expressively understood as a knowledge management evolution. Considering the SECI model, discussed by Nonaka (2008), allow to stress the relation and definitions for KM practices for project management tasks. From this study, it is opportune to define that there is a big comprehension of knowledge management contribution for PM, also that this contribution was not sufficiently addressed at the time of its writing. In this analysis, producing explicit knowledge from tacit level, and then combining it to other explicit knowledge, resulting in a permanent cycle condition where new tacit can be absorbed, reminds completely situations of projects being proposed, planned and executed, starting from improvised, practical compositions to more structured, explicit, process-oriented basics, illustrating how knowledge management processes can be investigated in this example.

For Reich and Wee (2006), several explicit knowledge-based components for these basal directives for project management, such as the project charter, project scope statement and the project plan can be depicted as KM. In this work, the authors also call attention to the knowledge production and demands for the 44 processes defined by PMI for PM, calling a special attention for tacit knowledge management needs.

It is possible to extend Landaetta (2008) findings to project portfolios, where a set of projects can be managed together, in one level, and then be observed in further detail. His conclusions showed the relevant impacts on the fostering of this knowledge flow among projects, implicating in better project execution performance. He also obtained some results, which may lead to better levels of capabilities formation, starting a positive cycle of knowledge management. Alternatively, Paramkushan and Gordon (2013) analyzed the inhibiting factors for this knowledge flow. It was studied a group of six information technology projects, where authors also reported their findings to project performance, but this time observing how the lack of the flow will impact negatively on the project final and significant factors. As those studies show, knowledge production, coming from expressive sets of data, impacts positively, when efficiently managed, in various project aspects, as those critical final-user points, such as performance, quality and communication.

As it was defined for analytics, it is opportune to understand that, nowadays, its propositions arise from commercial, simple offers, as those produced by analysis of popular, public platforms, as Google apps – Mehmod (2017) – to more sophisticated and complex solutions, that allow planners and strategists to develop additional comprehension regarding business decisions, trends and contributions (Deloitte, 2017; Linke, 2017). This way, analytical studies, arising from big data contents, populated and shared through project management activities, could result in potential knowledge to improve further decisions for PM planning actions. This opportune cyclic process will be approached in the last section, where some study cases will be analyzed. It can, initially be perceived from the findings of Müller *et al.* (2013), where authors concluded that earlier collaborators play a decisive role in knowledge management for projects, when considering organizational clusters.

With all these examples of association of KM and PM, it is interesting to appreciate how big data sources are generated, stored, managed and applied in the context of projects. Approaching those references cited before, this association is explored in the following. From these studies, it was noticed that:

- Sometimes, data is not predicted and organized in the dynamic environment of projects. It is always available and intensively produced, resulting in a perception of an almost unlimited source of signals linked to all project management events. Informal datasets are also produced, as it can be identified as tacit knowledge sources.
- Knowledge is intuitively produced also, sometimes, through informal processes related to cultural aspects of each organization, being applied without any structured perception, as tacit knowledge contents. This situation produces a continent that is difficult to apply in new situations, as it is related to environmental aspects – such people´s conditions, organizational tacit communication, competitive pressions, etc.
- Knowledge demands also grow in the field, on the application situations, as projects usually are proposed in such an evolving, more complex ways, to achieve new and higher value goals.
- On the strategic level, it is also interesting to point out that the generated knowledge can be further applied for KM itself improving, as it also receives more structured knowledge production to improve PM practices as well.
- Knowledge is managed in contexts of multiple projects and various organizations, the *portfolios,* bringing the complexity described as those scenarios related by the literature with big data.

This last observation identifies the situation where it is possible to understand big data as referring to contexts of huge amounts of available data in dynamic conditions. Additionally, the need of knowledge production from this data and the correlated need for knowledge management process for the production itself potentially implies in better project management practices and results.

Some aspects important to assess, from this broad conceptual relationship perceptions are:

- It is possible to describe a cycle, where, progressively, organizations that improves KM practices can improve their PM practices also allowing, in result, better KM practices. It recalls a "spiral" model, as that announced by Nonaka (2008), generally identified as SECI model.
- Big data contents can be analyzed and treated by some tools, methods and techniques, as those exemplified above and many others. It can be recognized as those sets produced by interactions in multiple, concurrent projects in any organizational contents.
- KM practices remain very challenging for organizations, as efficient and coherent application of IT tools, infrastructure, information systems projects and implementations, planned managerial tasks and skills.
- There are several models and recommendations proposed by the literature at this time, with no predominant trend at all, as this context is still new to the market, various practices and empirical research are in exercise now, that must be investigated further for a potential description for this relationship.
- Big data contexts, as projects for strategic purposes and knowledge management practices are continuously being proposed, studied and applied. In their way to be precisely, conceptually affirmed, they enable to better implement tools - as information systems, for example.

Rajamaran and Ullman (2011) present an interesting provocation, announcing that instead of "supercomputers", these new approaches demand also a technological infrastructure specially customized to produce knowledge in an effective and objective way, where, for example, resides project management usual demands. This factor is sometimes forgotten in theoretical approaches and practical applications, as immense amount of data – characterized as Big Data – must be continuously processed, sometimes, under the application of complex algorithms, providing knowledge results to be further used in plans and associated actions. Along with these processing capabilities, storage is also an infrastructural issue that must be considered in information systems projects, and its related implications. Distributed file systems tools are techniques to be applied for efficient performances in these contexts of knowledge production.

Some authors, as Boyd and Crawford (2011), argue about the application of "networked datasets", allowing the adaptation of technological infrastructure methods for knowledge production. For them, these new technologies can improve knowledge management, as *Ford changed the way we made cars*. It must be recalled, eventually, that Toyota process-driven and oriented systems and production design also modified Ford initial ideas on manufacturing.

The following section produces some study cases, to exemplify how this conceptual network can really be applied.

CASE STUDIES

The dynamic conceptual relationship among data generation, big data-based knowledge production, decision making, overall performing knowledge management in project management contexts is to be approached in this section, oriented to produce a practical view that elaborates the final composition for the intended goal of this chapter. As it happens in multidisciplinary observations, this consolidation could be produced in various ways, adopting several techniques and different methodological propositions. Here, it was adopted a genuine multidisciplinary observation, supported by the previous theoretical background developed in the previous sections of this text, regarding practical cases of infrastructure, industrial and services projects.

For the first case, a usual infrastructure project is to be analyzed. It is a basic study, adapted from a real case, which is an intention of a local government to improve a traffic interstate road in a region of an emerging country. It is a complex decision, as it contemplates two different views for the results of this project, after the cited road and its associated supporting structures – bridges, tunnels, services areas, etc. – are delivered to the population:

- For some citizens, it is a real opportunity of economic growth, presenting the improvement of passengers and load transportation, along with jobs created along the road, both to keep it working and others derived from its main objectives. These additional jobs could correspond to the service people, hired for food catering, hotels, transportation itself, etc. As the local government impose some transportation taxes, the public budget can also be benefited from business improvement.
- On the other hand, a significant part of the population will prefer not to have this road system increment, as more traffic will go through their cities and region. Pollution, heavy traffic, potential occurrences of accidents and other facts would be possible, bringing new security issues. Along with these unwished events, this part of population argues about heavy public investments on an old-fashioned system, which should be effectively replaced by modern systems, such as electrical trains (clean energy), for example.

As it happened to some private companies, the govern, with the help of their Public Relation and Communications secretariat, starts a campaign in social media, radio and television to inform citizens about their intention on improving, augmenting this traffic system, implementing some benefits and increments to the old one. This campaign is not merely informative, as several monitoring services, based on an information system, created specifically to follow citizens reaction (which be a base to be applied in the future for other processes, as elections, other public projects, overall public relations with citizens, international government actions, etc.) through some channels – social media, mainly, but also from manifests, newspapers publications, political reactions, among many others – intending to develop a "trend topic" model, that shows, in practical terms, how local citizens react, influence each other, etc.

Data is collected from structured sources – as, for example, money spent for some citizens and political parties' movements in advertisements pro or against the project, number of reactions – by type, projections of investment and return or payback values – and from unstructured, mainly characterized by social media reactions – Facebook posts, tweets, blogs interactions, public services answers to citizens – to compose a big data content, which will be analyzed. This analysis, which typically produces knowledge from heterogeneous data (as big data promises), will return the first level of analysis to local government managers.

As we are thinking about an information system, analytics are applied, in a continuous way, to understand how people will react to some simulations produced by local authorities, testing practical hypothesis. In some tests, local government managers inform that only a part of the road will be expanded – it is not a technical recommendation, indeed, but a test to produce another level of communications exchange, interaction. These new, unstable announcements generate some additional events which will be analyzed, producing new waves of knowledge from data and information, designing specific decision scenarios, where it is possible to identify main groups of citizens that want to react, how they react, what are their preferences and how they will potentially think about the final governmental decision, if the negotiations for this project are to be started.

In this case, big data and analytics are to be applied to understand citizens´ reactions to one initiative from the local government, as is to be inserted in a real process of one information system. This IS, which involves information technology, people, data collected and compared, organizational structure, communication resources and process designed, will result not only for this specific decision, which is already important, but in an information and knowledge process that can be repeated or adapted for other public and private projects, reinforcing our conceptions over the richness for this interaction in practical cases. It can generate, for instance, some public governmental actions and plans, based on predictions made over citizens previous behavior. More knowledge, methodologically generated, applied to better decision about projects.

In another practical case, an industrial food processing corporation wants to adopt an "industry 4.0" based system in one of its plants and distribution line. The alternative of industry 4.0 is a modern, competitive investment, which proposes to automatize production lines, with the implementation of robots, adjusted production schemes based on knowledge generated from the commercial fronts and, overall, optimizing industrial production. As it usually happens in projects of this size and impacts, it also presents an irreversible situation where human jobs will be lost forever, as operators are to be changed by these programmable machines and optimized level of process controlling.

This project contemplates a decision scenario, as it also will take a long time – typical estimated for some years – to payback the complete investment made in this change. As this company has to replace human operators and managers for automatized infrastructure, optimization must be achieved faster, at the beginning of this new working business model, assuring this modernization to return the investment as fast as it is possible.

Thinking this way, the company starts a research about its production lines, using social media contents and customer face-to-face promotions in supermarkets, to understand several issues regarding customers preferences and habits. Some tests and researches to be conducted will try to measure satisfaction levels, brand retention, product lines diversification potential strategies, customer expectations, etc. These tests results are to be analyzed consolidating two types of data – structured, formal, defined and unstructured, informal, spontaneous – as it will be collected from commercial documents, spreadsheets, invoices, budgets and schedules (structured) and blog posts, interviews, reactions perceptions (unstructured). For this, first, a big data processing system is implemented, receive these contents and depicting some details about customers´ reactions. This system is so added by analytics processing, where a comprehension about the evolution of these reactions can occur if some changes are to be implemented in the productive chain.

It is important to say that we have several different signals coming from customers reactions. At a first glance, it can show, in a similar way as we understood the first case, of the traffic road project, the continuity of the customer satisfaction, even if the processed food package he or she is acquiring now was produced by a robot, not by a human operator. But in another level, it can show, objectively, how products are really acquired by customers, how we develop our preferences towards a specific purchase – ways, information needed, amount paid, etc. – allowing this company to precisely decided what production line will be automatized and when. This is a typical strategic decision for project management, which will impact directly on the investment plan for its food processing plant automatization, industry 4.0, project.

So, this way, big data and analytics will produce knowledge allowing the first level of adjustment for this project, or an association of smaller projects, a *portfolio*, which can be coordinated. It is not a one-way knowledge generation for project management indeed, but a continuous cycle, a knowledge management market intelligence scenario, where commercial data is produced to adjust projects implementation in a period.

In another typical case, that are impacting a complete sector nowadays, it is analyzed how a communication agency can implement all this informational infrastructure to bring more automation to its daily routines of process and works. Analyzing in a simple way, just observing its more typical service for a regular customer, one agency has the goal to provide an advertising campaign for a customer – typically a commercial company – as to inform the final customer about an added value in a product or service. This is a frequent relevant part of the communication plan, which is an indispensable service for a marketing strategic and tactical plan.

Thinking "digitally", this agency, in the future, could receive a draft document from its customer, as the business owner will fill a form, through internet, with its potential demands. After automatically identified – by a big data tool – basic key-words detailed in that document are to be analyzed by analytics software and systems in Internet, identifying usage, frequency, trends, resulting in some pictures that will inform the publisher about scenarios for efficient advertising.

For example, we can consider the agency customer if it is a clothe shop, which mainly sells sports materials and related items, from worldwide known brands, as a multi-brand strategy. From this analysis, the agency can react to an intention from the shop owner, as he informed, in his draft document, that he wants to open a new space in his shops, focusing new sports modality (for example, as it happened few years ago in Brazil, where NFL was not well known – now, it is a successful case in this country).

The agency, producing a big data analysis, can research internet contents, data collected from press agencies channels and other structured and unstructured ways, producing a picture about the actual market situation, future trends, expected transaction volumes, customer segments that will be candidates for promotions, etc. These knowledge components will generate a series of marketing – related project plans, to be detailed, developed and implemented in the future.

Interestingly, the first level of advertisement can also be digitally produced and digitally disseminated, using social media channels, communicating with potential customers, associated to press and institutional channels of that specific sportive modality (local confederation, clubs, fans clubs) about their offers of sports materials.

As additional marketing strategies, customers can have "loyalty" and other retention marketing strategies applied, when – maybe automatically – studying their consumption levels, media reaction, interactions, integration to the advertisements, to evolve and improve their participation and levels of adhesion to marketing campaigns. Maybe this more sophisticated agency performance can reach other level of service, transforming it into a marketing agency, who will produce, with the help of this information system (big data, analytics, social media and related infrastructure) in an effective advisor on customer capture, development and retention strategies, in typical service-sector projects.

CONCLUSION

Project management emerges as a turbulent, dynamic, controversial context where data, information and knowledge are generated massively and demanded for better decision processes and structures. In this chapter, aiming to expand previous developments, regarding a publication where it was studied the relationship of project management and knowledge management, new components for these potential contributions were addressed: analytics and information systems architecture. As these two concepts prescribe modern possibilities to produce knowledge for application in organizational decisions, the association of PM and KM is improved, together with its perspectives of new competitive forces.

A theoretical review was conducted, to define how analytics will compose with big data, as the former work assessed this important resource as a trend of knowledge production. After this, knowledge management and project management were proposedly studied together, to observe their correlation. Interestingly, this relation appears both as a dynamic process, oriented to solve business issues and complex situations, but also as a potential generator for new strategic positioning, to be structured by strategic planning process in organizations. Information systems, a concept that composes with every informational source, regarded to align several active elements – such as people, organizational structure, process definitions and so for – promote the final key part to be added to the theoretical background.

As we discussed final examples for this dynamic challenging solution, attention has been called for its complete integration when simple, frequent actions of actual market competitive methods are observed with more detail. Finally, it is possible to affirm that this chapter observed and promoted a study where knowledge management and project management expressive relationship can be improved in modern value-aggregated chains, enhancing their abilities, perspectives of modern trends adoption (such as automation, for example) and strategic solutions, with the quick responses demanded by nowadays market competition.

REFERENCES

Azevedo, A., & Lourenço, A. (2008). *KDD, SEMMA and CRISP-DM: a parallel overview*. Available at http://hdl.handle.net/10400.22/136

Barney, J. (1991). Firms Resources and Sustained Competitive Advantage. *Journal of Management, 17*(1), 99–120. doi:10.1177/014920639101700108

Berry, M. J., & Linoff, G. (1997). *Data mining techniques: for marketing, sales and customer support*. New York: John Wiley and Sons, Inc.

Bollen, J., Mao, H., & Zeng, X. (2011). Twitter mood predicts the stock market. *Journal of Computational Science, 2*(1), 1–8. doi:10.1016/j.jocs.2010.12.007

Boyd, D., & Crawford, K. (2011). Six provocations for Big Data. *Annals of "A decade in Internet time: Symposium on the dynamics of the internet and society"*.

Cao, L., Zhang, C., & Liu, J. (2006). Ontology-based integration of business intelligence. *Web Intelligence and Agent Systems. International Journal (Toronto, Ont.), 4*, 313–325.

Chan, P. K., Fan, W., Prodromidis, A. L., & Stolfo, S. F. (1999). Distributed data mining in credit card fraud detection. *IEEE Intelligent Systems and their Applications,* 14(6), 67-74.

Chapman, P., Clinton, J., Kerber, R., Khabaza, T., Reinartz, T., Shearer, C., & Wirth, R. (2000). *Step-by-step mining guide*. SPSS.

Chau, M., & Xu, J. (2012). Business intelligence in blogs: understanding consumer interactions and communities. MIS Quarterly, 36(4), 1189-1216.

Chen, H., Chiang, R. H. L., & Storey, V. C. (2012). Business intelligence and analytics: from big data to big impact. MIS Quarterly, 36(4), 1165-1188.

Choo, C. W. (2005). *The knowing organization: how organizations use information to construct meaning, create knowledge and make decisions*. Oxford, UK: Ed. Oxford University Press. doi:10.1093/acprof:oso/9780195176780.001.0001

Chou, J. S., & Yang, J. G. (2012). Project management knowledge and effects on construction projects outcomes: An empirical study. *Project Management Journal, 43*(5), 47–67. doi:10.1002/pmj.21293

Clark, T. D. Jr, Jones, M. C., & Armstrong, C. P. (2007). The dynamic structure of management support systems: Theory development, research, focus and direction. *Management Information Systems Quarterly, 31*(3), 579–615. doi:10.2307/25148808

Courtney, M. (2013). Puzzling out Big Data. Engineering & Technology, 7(12), 56-60.

Cristea, D. S., & Capatina. (2009). *A. Perspectives on knowledge management models*. Analls do "Duranea de Jos". *Universidade de Galati, 15*(2), 355–366.

Davenport, T. H., & Prusak, L. (2000). *Working knowledge: how organizations manage what they know* (2nd ed.). Harvard Business Press.

David, E., & Thawornwong, S. (2005). The use of data mining and neural networks for forecasting stock market returns. *Expert Systems with Applications*, *29*(4), 927–940. doi:10.1016/j.eswa.2005.06.024

De Man, D. (2012). A platform for market intelligence. *International Trade Forum,* (4), 21-21.

De Roche. (2016). Five critical aspects to applying data analytics. *Teammate solutions.* Available at http://www.teammatesolutions.com/five-critical-steps-to-applying-data-analytics.aspx

Dean, J., & Ghemawat, S. (2008). MapReduce: Simplified data processing on large clusters. *Communications of the ACM*, *51*(1), 107–113. doi:10.1145/1327452.1327492

Deloitte. (2017). *Five questions about applying analytics to risk management*. Available at https://www2.deloitte.com/content/dam/Deloitte/au/Documents/risk/deloitte-au-risk-risk-angles-applying-analytics-risk-management-250215.pdf

Dimitrios, N. K., Sakas, D. P., & Vlachos, D. S. (2013). The role of information systems in creating strategic leadership model. *Procedia: Social and Behavioral Sciences*, *73*, 285–293. doi:10.1016/j.sbspro.2013.02.054

Dodge, J. (2010). Mining for Gold. *Information Management*, *10*(5), 38-40.

El-Bashir, M. Z., Collier, P., & Sutton, S. G. (2011). The Role of Organizational Absorptive Capacity in Strategic Use of Business Intelligence to Support Integrated Management Control Systems. *The Accounting Review*, *86*(1), 155–184. doi:10.2308/accr.00000010

El-Gayar, O., & Timsina, P. (2014). Opportunities for Business Intelligence and Big Data Analytics In Evidence Based Medicine. In *Annals of 47th Hawaii International Conference on System Science*. IEEE. 10.1109/HICSS.2014.100

Fayyad, U., Piatetsky-Shapiro, G., & Smyth, P. (1996). The KDD process for extracting useful knowledge from volumes of data. *Communications of the ACM*, *39*(11), 27–34. doi:10.1145/240455.240464

Gasik, S. (2011). A model of project knowledge management. *Project Management Journal*, *42*(April), 23–44. doi:10.1002/pmj.20239

Hampton, J. (2011). *SEMMA and CRISP-DM: Data mining methodologies*. Available at http://jesshampton.com/2011/02/16/semma-and-crisp-dm-data-mining-methodologies/

Haslinda, A., & Sarinah. (2009). A review of knowledge management models. *The Journal of International Social Research, 2*(9), 187-198.

Hu, M., & Liu, B. (2004). Mining and summarizing customer reviews. In *Proceedings of the tenth ACM SIGKDD international conference on Knowledge discovery and data mining (KDD '04)*. ACM. Retrieved from http://doi.acm.org/10.1145/1014052.1014073

Inmon, B., Strauss, D., & Neushloss, G. (2008). *DW 2.0: The Architecture for the Next Generation of Data Warehousing*. Morgan Kaufmann.

Jain, C. L. (2016). How to Use Big Data and Predictive Analytics to Improve the Success of New Products. *Review of Business*, *37*(1), 48-55.

Jamil, G. L. (2005). *Gestão da Informação e do conhecimento em empresas brasileiras: estudo de múltiplos casos*. Belo Horizonte: Ed. Con / Art.

Jamil, G. L. (2013). Approaching Market Intelligence concept through a case analysis: Continuous knowledge for marketing strategic management and its complementarity to competitive intelligence. *Annals of Centeris 2013 – Conference on Technology and Information Systems*.

Jamil, G. L., Santos, L. H. R., Alves, M. L., & Furbino, L. (2012). A design framework for a market intelligence system for healthcare sector: a support decision tool in an emergent economy. In *Handbook on research of ICTs for Social Services and Healthcare: developments and applications*. Hershey, PA: IGI Publishing.

Johnson, J. E. (2012, July). Big data + Big Analytics + Big opportunity. *Financial & Executive*, 51-53.

Kaplan, R., & Norton, D. (2007). *The execution premium: linking strategies to operation for competitive advantage*. Harvard Business School Press.

Kearns, G. S., & Lederer, A. L. (2003). A resource based view of IT alignment: How knowledge sharing creates a competitive advantage. *Decision Sciences*, *34*(1), 1–29. doi:10.1111/1540-5915.02289

Kimball, R., & Ross, M. (2010). *Relentlessly Practical Tools for Data Warehousing and Business Intelligence*. John Wiley and sons.

Kotler, P., & Keller, K. (2005). *Marketing Management* (12th ed.). Prentice Hall.

Kremer, M., Mantin, B., & Ovchinnikov, A. (2013). *Strategic consumers, Myopic retailers*. Darden School of Business at University of Virginia. Available at http://www.darden.virginia.edu/web/uploadedFiles/Darden/Faculty_Research/Research_Publications/Ovchinnikov_StrategicConsumer_MyopicRetailers.pdf

Landaetta, R. R. (2008). Evaluating benefits and challenges of knowledge transfer across projects. *Engineering Management Journal*, *20*(1), 29–38. doi:10.1080/10429247.2008.11431753

Linke, R. (2017). Applying analytics – Ideas and examples from leading companies and researches. *MIT Management Newsroom*. Available at http://mitsloan.mit.edu/newsroom/articles/reading-list-applying-analytics/

McAfee, A., & Brynjolfsson, E. (2012). Big data: The management revolution. *Harvard Business Review*, *90*(10), 60–68. PMID:23074865

Mehmod, H. (2017). *A big list of what Google Analytics can and can not do*. Available at http://marketlytics.com/blog/list-of-things-google-analytics-can-and-cannot-do

Mintzberg, H., Alhstrand, B., & Lampel, J. (2009). *Strategy Safari: the complete guide through the wilds of strategic management* (2nd ed.). Pearson Education Limited.

Müller, R., Glücker, J., Aubry, M., & Shaun, J. (2013). Project management knowledge flows in network of project managers and project management offices: A case study in the pharmaceutical Industry. *Project Management Journal*, *44*(2), 4–19. doi:10.1002/pmj.21326

Nonaka, I. (2008). *The knowledge creating company*. Harvard Business Review Classics.

Ohata, M. & Kumar, A. (2012, September). Big Data: A Boom for Business Intelligence. *Financial Executive*.

Paramkusham, R. B., & Gordon, J. (2013, Fall). Inhibiting factors for knowledge transfer in information technology projects. *Journal of Global Business and Technology*, *9*, 2.

Park, S. H., Huh, S. Y., Oh, W., & Han, S. P. (2012). A social network-based inference model for validating customer profile data. MIS Quarterly, 36(4), 1217-1237.

Phua, C., Lee, V., Smith, K., & Gayler, R. (2010). *A comprehensive survey of data mining-based fraud detection research.* Available at http://arxiv.org/ftp/arxiv/papers/1009/1009.6119.pdf

PMBoK. (2013). *Project Management Body of Knowledge* (5th ed.). Project Management Institute.

Porter, M. (2008). *On Competition*. Harvard Business School Press.

Prado, D. (2006). *Gerenciamento de programas, portfólios e projetos nas organizações*. Belo Horizonte: INDG.

Rajamaran, A. & Ullman, J. D. (2011). *Mining of massive datasets.* Cambridge Editors.

Rayome, A. D. (2016). *Seven ways to build trust and analytics at your company*. Tech Republic Blog. Available at https://www.techrepublic.com/article/infographic-7-ways-to-build-trust-in-data-and-analytics-at-your-company/

Reich, B., Gemino, A., & Sauer, C. (2008). A temporary model of information technology project performance. *Journal of Management Information Systems*, *24*(3), 9–44.

Reich, B., Gemino, A., & Sauer, C. (2012). Knowledge management and Project-based knowledge in its projects: A model and preliminary results. *International Journal of Project Management*, *30*(6), 663–674. doi:10.1016/j.ijproman.2011.12.003

Reich, B., & Wee, S. W. (2006). Searching for knowledge in the PMBoK Guide. *Project Management Journal*, *37*(2), 11–27.

Rygielski, C.; Wang, J. C. & Yen, D. C. (2002). Data mining techniques for customer relationship management. *Technology in Society, 24*(4), 483-502. 10.1016/S0160-791X(02)00038-6

SAS. (2017). *What is Analytics?* Available at https://www.sas.com/en_us/insights/analytics/what-is-analytics.html

SAS. (n.d.). *SAS Enterprise Miner – SEMMA*. Available at http://www.sas.com/technologies/analytics/datamining/miner/semma.html

Stair, R., & Reynolds, G. (2009). *Principles of information systems*. Course Technology.

Tuomi, I. (2000). Data is more than knowledge: Implications of the reversed knowledge hierarchy for knowledge management and organizational memory. *Journal of Management Systems*, *16*(3), 103-117.

Turban, E., Mc Lean, E., & Wetherbee, J. (2002). *Information technology for management: transforming business in the digital economy* (3rd ed.). Hoboken, NJ: John Wiley and Sons.

Turban, E., Rainer, R. K. Jr, & Potter, R. E. (2007). *Introduction to information systems*. Hoboken, NJ: John Wiley and Sons.

Varun, C., Banerjee, A., & Kumar, V. (2009). Anomaly detection: A survey. *ACM Computing Surveys, 41*(3), 2–15.

Yonce, C., Taylor, J., Kelly, N., & Gnau, S. (2017). BI Experts´ perspective: Are you ready for what´s coming in Analytics. *Business Intelligence Journal, 22*(3), 36-42.

Chapter 4
On the Use of Digital Platforms to Support SME Internationalization in the Context of Industrial Business Associations

Eric Costa
University of Porto, Portugal & INESC TEC, Portugal

António Lucas Soares
University of Porto, Portugal & INESC TEC, Portugal

Jorge Pinho de Sousa
University of Porto, Portugal & INESC TEC, Portugal

ABSTRACT

The digital economy is creating disruptions in traditional industries and markets. Industrial business associations (IBAs) may face serious challenges in a near future to meet the needs and requirements of their members, particularly in supporting their growing international trade activities and internationalization processes. Digital platforms are already transforming different types of businesses across all markets. An IBA may use a digital platform, not only to keep up with the current technological trends of markets, but also to improve the internationalization support provided to their associate small and medium enterprises (SMEs). Therefore, the aim of this chapter is to present the view of these potential digital platforms' managers, by presenting the results of an exploratory field research based on 24 interviews with IBAs from Portugal, France, and the UK. Another goal is to identify current digital platforms that are being used by IBAs and to critically evaluate their potential for supporting internationalization processes of SMEs. By using these findings, a set of requirements and features for digital platforms supporting SME internationalization in the context of IBAs are derived in this chapter. These results can be used by platform designers and by IBAs for designing and developing more effective digital platforms that can meet the specific internationalization needs of their users and managers.

DOI: 10.4018/978-1-5225-6225-2.ch004

INTRODUCTION

Today, we live in a world with constant developments in technology and with a society increasingly connected. In a rising generation of micro-entrepreneurs, freelancers working anyplace at any time, and new business models, we are witnessing to at least three important changes in the business world: (1) a more shared economy, where collaboration is the emerging paradigm; (2) the use of digital information and communication, social media and new technology platforms to improve businesses and management processes; and (3) an increasing competition environment for all types of companies and organizations. This can also be a challenge for industrial business associations (IBAs).

Globalization and the development of the European Union (EU) have put pressure on traditional business associations to re-invent themselves for remaining relevant institutional actors in representing the interests of companies and organizations (Streeck, Grote, Schneider, & Visser, 2006). IBAs are characterized by their collective and member-based nature, where membership is most of the times voluntary, and members have a common interest (Bennett, 1998; Traxler & Huemer, 2007). These institutional entities will need to adapt to the new requirements of their members, in order to remain an active support for their activities. This can specifically occur in the case of satisfying the needs of small and medium enterprises (SMEs).

SMEs have been gaining more power and representativeness in international markets by assuming different manifestations as international new ventures and born-global firms, meaning that many of them now start their international operations right from inception or soon thereafter (Knight & Liesch, 2016). According to Giudici and Blackburn (Giudici & Blackburn, 2013):

It is often asserted that the advent of the digital economy has substantially reshaped how SMEs conduct their sensing and seizing activities across international markets. Therefore, new requirements are needed by IBAs to keep up with the evolution and changes that might occur with their associate SMEs.

In fact, IBAs have been already facing increasing difficulties, mainly related with problems of low membership and high opting out, instability over time, and inequalities due to the existence of associations of different types or sizes of businesses (Bennett, 1998; Traxler & Huemer, 2007). In addition to that, many companies have multiple association memberships, both at the individual/staff and business levels (Boléat, 2003), which naturally creates some sort of competition environment among the different types of associations. Other potential challenges faced by IBAs are related with limited revenue channels, limited audience, declining trade show attendance, not demonstrating value and benefits of membership, not aligning industry issues with member needs (Schutzius, 2016).

Notwithstanding the challenges and the wide variety and number of organizations that make up the business associative system, the common denominators for these institutional entities is the fact that most of them are surviving through membership fees (increasingly smaller due to the high number of competing associations), are dependents on communitarian funds and provide services and training (Bennett, 1998; Boléat, 2003; Fries, 2008; Mikamo, 2013).

Internationalization processes require accessing, processing and organizing large amounts of information (Xie & Amine, 2009). However, the lack of information and the lack of suitable tools for managing information and knowledge are identified as the main obstacles for SME internationalization (Costa, Soares, & Sousa, 2016a; Hsu, Chen, & Cheng, 2013; Rodriguez, Barcos, & Álvarez, 2010; Sommer & Haug, 2011). Given their limited resources and difficulties in reaching markets, SMEs tend to rely more extensively on external networks to operate in international markets than large enterprises (Hessels & Terjesen, 2010). SMEs and large enterprises also differ in their need for services from the IBAs. In the case of SMEs, due to their resource constraints, they tend to be more dependent than large firms on the services, information and contacts generated through business associations (Gashi, Hashi, & Pugh, 2013). From the other side, large enterprises can easily make or buy these services independently of associations (Traxler & Huemer, 2007).

Previous studies on international business show that, although existing different opinions about the utility of their services, IBAs are identified as one of the export information sources (Costa, Soares, & Sousa, 2017a; Leonidou & Theodosiou, 2004) and SMEs can use their institutional network resources for overcoming barriers to operate in international markets (Costa, Soares, & Sousa, 2017b; Gashi, Hashi, & Pugh, 2013). Such institutional network support by IBAs is provided through (Costa, Soares, & Sousa, 2017b): promotional activities; counselling, training, and technical and legal support; information sharing; and cooperation with other institutional entities.

Nevertheless, previous research also show that IBAs can have a more active role in supporting SME internationalization by fostering more collaborations between members (Costa, Soares, & Sousa, 2015). This support can also be enhanced by improving their information management role to better meet companies' informational needs (Costa, Soares, & Sousa, 2016b; Costa, Soares, Sousa, & Jamil, 2017). Therefore, to improve their role in supporting SME internationalization, one important development is for IBAs to evolve towards collaborative networks by sharing more organized and valued information with their associates and also by fostering information flows and more collaborations between them (Costa, Soares, & Sousa, 2017b).

Many information communication technologies (ICT) have been designed and developed to improve business and management processes and activities of companies and organizations, as well as to support collaboration in business networks (Bellini, Ascenzo, Ulskaia, & Savastano, 2016; Carneiro, Soares, Patrício, Azevedo, & Pinho de Sousa, 2013; Durugbo, 2015). Regarding the international business, some studies have been discussing the importance of digital technologies to support the international activities of SMEs (Brouthers, Geisser, & Rothlauf, 2016; Giudici & Blackburn, 2013). This makes even more sense when dealing with information, as internationalization processes require to manage large amounts of information about foreign market conditions, available funding and opportunities and partners for collaborations (Costa, Soares, & Sousa, 2016a; Xie & Amine, 2009). However, previous research points to the need for extending our understanding on how and whether technology is suitable and viable for supporting SME internationalization and collaboration (Dutot, Bergeron, & Raymond, 2014). In addition, the effect of the use of digital platforms on management processes is relatively unexplored (Spagnoletti, Resca, & Lee, 2015).

Therefore, we hypothesize in this paper that collaborative digital platforms may substantially help the management processes of IBAs for improving their support to SME internationalization. In particular, it can contribute to the effectiveness of information and knowledge management processes and to the creation of new collaborative networks among their members (Costa, Soares, & Sousa, 2016a). We believe that such transition to technology is crucial for the survival of IBAs, as it can avoid the loss of more associates, or even prevent the worst scenario of the IBAs' services to no longer make sense for companies.

Accordingly, this paper aims to understand the perspective of IBAs on the use of collaborative digital platforms for supporting the internationalization of their associate SMEs by recognizing their problems, needs and opinions about this subject. Another goal is to identify current digital platforms that are being used and to critically evaluate their potential for internationalization support. The research questions that guide this study are:

RQ1: What are the current digital platforms used by IBAs to support SME internationalization?
RQ2: How do IBAs perceive the use of digital platforms to improve the support to the internationalization processes of their associate SMEs?

To answer these research questions, and based on empirical evidence, this paper presents a qualitative exploratory field research, providing insights from interviews performed with 20 IBAs in Portugal, 4 IBAs in the UK, and 1 IBA in France. By using these findings, the objective is to obtain a preliminary set of requirements and features for digital platforms supporting SME internationalization, managed by IBAs.

The structure of this book chapter is as follows. After this introduction, literature on IBAs supporting SME internationalization is presented, as well as some background on collaborative networks and digital platforms. After that, the research methodology is described and the results are presented and discussed. Finally, the main conclusions and directions for further research are presented in the last section of the chapter.

RESEARCH BACKGROUND

Industrial Business Associations Supporting SME Internationalization

There are thousands of business associations, trade associations, industrial associations, federations and chambers of commerce in the European Union (EU). While different in nature, each type of business association represents the interests of companies and organizations (Streeck, Grote, Schneider, & Visser, 2006). An IBA is a business and trade association specifically focused on industry, representing firms rather than individuals, assuming two different types (Costa, Soares, & Sousa, 2017b; Costa, Soares, Sousa, 2017): (1) sectoral IBA, by representing firms of a specific industrial sector; or (2) multi-sectoral IBA, by acting as intermediary within a network of companies from different industrial sectors. On top of that, they act as intermediaries between business sectors and governments (Traxler & Huemer, 2007), and they provide services for their members (e.g. newsletters, technical support, information services, conferences and exhibitions and training) (Boléat, 2003).

Research on the role of IBAs in internationalization processes of SMEs is still in its infancy. Gashi et al. (2013) and Patel-Campillo & DeLessio-Parson (Patel-Campillo & DeLessio-Parson, 2016) suggest that the propensity of SMEs to export is positively influenced by their membership in business and trade associations. The study of Costa et al. (Costa, Soares, & Sousa, 2017b) goes deeper on this topic, by developing a conceptual model that explores the activities and the use of institutional network resources by IBAs to support and facilitate internationalization processes of SMEs. Findings show that SMEs can benefit from the institutional network resources and support of IBAs to start or continue their internationalization strategy. This support is mainly provided through promotional activities, such as trade fairs, trade missions, promotional actions and organization of events. Services such as counselling, training and information sharing also help associate companies to successfully approach foreign markets. In addition, SMEs can benefit from the network of contacts that IBAs have with other institutional entities such as governments, other associations and IBAs, embassies, foreign companies, consulting firms, funding bodies and regulatory bodies. In this same study, the future vision of IBAs in terms of topics of improvement is also presented. Results show that collaboration and the use and development of new ICT are the main components of their future strategy for improving the services and activities for helping companies in their international endeavors.

Finally, based on the perspective from some companies that are members of IBAs, Costa et al. (2017) show that, although recognized as a useful source of information, the IBAs' information management role can be improved to better meet their internationalization needs:

- Improved information organization e.g., by filtering and organizing information according to countries in which each associate is present or have sales;
- More effective information dissemination e.g., by more selectively disseminating information about more specific subjects such as legal aspects (e.g. legislation and customs rules, patents);
- Promoting information sharing among the members e.g., experiences and lessons learned;
- Improved information quality, by providing information specifically tailored to the needs of SMEs.

Another conclusion from this study is that IBAs can also improve their role in the promotion of collaboration in internationalization. Accordingly, arguments from previous studies (Bell, 2006; Wang & Gooderham, 2014) state that a successful business association must be able to strengthen and foster collaborations and collective actions among members. However, the engagement of these institutional entities as collective action facilitators is still limited (Pieth, 2012).

Collaborative Network Perspective of IBAs

Costa et al. (Costa, Soares, & Sousa, 2017b) suggest that, in a network perspective, IBAs may improve their support to internationalization by acting as a broker among associate SMEs to create collaborative networks (CNs) with the aim of facing internationalization challenges. A CN is formed by independent organizations that interact and cooperate most of the times through collaborative ICT to share risks and benefits (Camarinha-Matos & Afsarmanesh, 2005; Carneiro, Soares, Patrício, Azevedo, & Pinho de Sousa, 2013; Parung & Bititci, 2008; Romero, Galeano, & Molina, 2009). One of the main interlocking themes related to an effective network implementation is the network governance (Popp, MacKean, Casebeer, Milward, & Lindstrom, 2014). Provan and Kenis (Provan & Kenis, 2008) define network governance as using institutions and structures of authority and collaboration, with the aim of allocating resources and managing joint actions across the whole network. Three different types of governance structures within organizational networks are proposed in the literature (Popp, MacKean, Casebeer, Milward, & Lindstrom, 2014; Provan & Kenis, 2008):

- **Shared Governance:** A participant-governed network without a formal administrative entity, where all members contribute to the network management and leadership, requiring a high level of trust;
- **Lead Organization:** A network where a hub organization or key member act as the manager and administrative entity by keeping the network partners aligned;
- **Network Administrative Organization:** A network where an independent, neutral and separate administrative entity is established to manage the network by acting as a network broker and by managing conflicts.

Therefore, IBAs can be classified in this study according to their network governance as a Network Administrative Organization (NAO). Most of the times, this means that the flows of control and information are essentially dyadic, centered on the governing role of the IBA (Figure 1).

Carneiro et al. (2013) point two important assumptions from previous studies on CNs. The first one is that collaboration is a fundamental process for successful organizational networks. The second one is that technologies that have the specific purpose of supporting collaborative processes are a key fostering factor for CNs. In a similar way, Fedorowicz and Sawyer (Fedorowicz & Sawyer, 2012) present some recommendations for implementing CNs that encompass both the technology and the organizational setting. One of those recommendations is also to leverage technology for advancing a collaborative network, including the design of management processes and technologies of collaboration and information sharing to support routine use and to increase system usage. Therefore, the ideal situation for organizations is to have an integration of technological tools that support three primary collaboration dimensions (Carneiro, Soares, Patrício, Azevedo, & Pinho de Sousa, 2013; Rabelo & Gusmeroli, 2008): communication; information sharing; coordination. Accordingly, in most cases, a well-designed and well-developed digital platform may allow to achieve this integration.

Figure 1. IBAs as network administrative organization
Adapted from (Provan & Kenis, 2008).

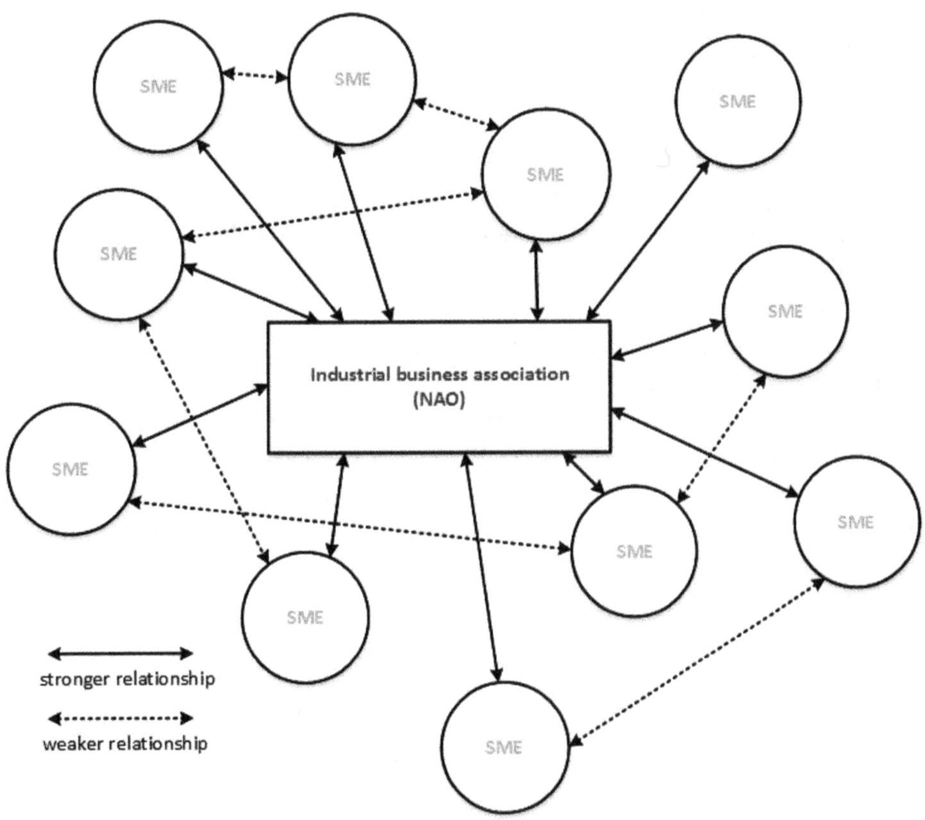

Digital Platforms

There are various conceptualizations for the term "digital platforms". Sun et al. (Sun, Gregor, & Keating, 2015) and de Reuver et al. (de Reuver, Sørensen, & Basole, 2017) provide extensive lists of digital platform definitions, which can be derived either from management researchers or from information systems (IS) researchers. The presented digital platform definitions also look at this term from different perspectives, for example by considering digital platforms from a technical point of view as an extensible codebase to which complementary third-party modules can be added, or from a more sociotechnical perspective, as technical elements (of software and hardware) and associated organizational processes and standards (de Reuver, Sørensen, & Basole, 2017). Both studies end up by recommending that researchers and scholars need to provide clear definitions of the term "digital platform" to avoid misinterpretations and to clarify and make explicit whether the technical or sociotechnical view is being considered. Nevertheless, some of the most complete or most used definitions are:

- *Digital platforms are software-based external platforms consisting of the extensible codebase of a software-based system that provides core functionality shared by the modules that interoperate with it and the interfaces through which they interoperate* (Ghazawneh & Henfridsson, 2015; Tiwana, Konsynski, & Bush, 2010);
- *A platform is a building block that provides an essential function to a technological system and serves as a foundation upon which complementary products, technologies, or services can be developed* (Gawer, 2009; Spagnoletti, Resca, & Lee, 2015);
- *An IT-platform is defined as comprised of a technological base on which complementary add-ons can interoperate, following standards and allowing for transactions amongst stakeholders, within the platform-centric ecosystem* (Sun, Gregor, & Keating, 2015).

Many digital platforms have been designed and developed for many purposes in a variety of industries. The disruptive power of platforms has been demonstrated by the rapid dominance of platform businesses over traditional industries (Parker, Alstyne, & Choudary, 2016). Digital platforms have become important for the activities of organizations, facilitating the communication inside work groups (Mansour, 2009), supporting online communities (Spagnoletti et al., 2015), creating knowledge (Tseng & Johnsen, 2011), managing knowledge in clusters of firms (Cremona, Ravarini, & Sutanto, 2014), and establishing collaborations (Carneiro, Soares, Patrício, Azevedo, & Pinho de Sousa, 2013). Parker et al. (2016) provide a detailed and comprehensive analysis on the rapid rise of the platform model, explaining that this has already transformed many major industries and will transform many others in the next few years. In this way, digital industry platforms are a technological trend that is having a profound impact on companies. According to Mergel (Mergel, 2017), digital transformation requires a holistic effort of rethinking and changing the main processes of organizations, beyond the traditional digitization efforts performed in the past, requiring cultural, managerial, process and developmental changes by the organization as a whole.

Due to the variety of the growing user base, and with the constant development and addition of new IT capabilities and complements, the design context of digital platforms is subject to wide range of change (Spagnoletti, Resca, & Lee, 2015). Therefore, efficient platform designs must meet the needs and requirements of their community of users. Nonetheless, according to Spagnoletti et al. (2015), there is still a lack of understanding on how organizations can effectively design digital platforms supporting online communities.

Some digital platforms have been developed to support the internationalization of SMEs in the EU, mainly to facilitate cooperation between European clusters (European Commission, 2014). For instance, some more active and developed IBAs also gained interest on this kind of technology and are currently using digital platforms to support members with information about markets and by allowing large companies to subcontract smaller ones for specific international operations (Costa, Soares, & Sousa, 2017b). Nevertheless, it seems that most of the existing digital platforms to support SME internationalization are only focused on information dissemination, failing to integrate other key collaboration components, such as communication and coordination. In addition, many IBAs are still using basic forms of communication and information dissemination with their members, such as phone calls and emails, and are not active enough to support their international activities (Costa, Soares, & Sousa, 2017b; Costa, Soares, Sousa, & Jamil, 2017). As stated by Sedoglavich (Sedoglavich, 2012), the success of international activities is mainly based on firms' technological capabilities and technological knowledge. Therefore, to face the competition from other institutional and private entities, this shift to technology can be crucial for IBAs to remain as the first choice for SMEs in what regards the internationalization support.

RESEARCH METHODOLOGY

This study comes in the sequence of previous exploratory studies performed by the authors. In a first exploratory field research, we had the perspective from 5 Portuguese companies associated with IBAs, regarding the use of digital platforms managed by IBAs to support their information management and collaboration activities in internationalization processes. The results of this exploratory study can be found in (Costa, Soares, Sousa, & Jamil, 2017). After that, also based on empirical evidence, we extended the previous study by adding new interviews performed with 20 IBAs in Portugal. The objective of this new study was to explore the activities and the use of institutional network resources by the selected IBAs to support and facilitate internationalization processes of SMEs. The findings of this second exploratory study can be found in (Costa, Soares, & Sousa, 2017b).

Therefore, with the current chapter, the aim is to bring the view of prospective digital platforms' managers (i.e. the IBAs), by explaining the factors that influence the adoption of digital platforms to support SME internationalization. Another goal is to identify current digital platforms that are being used and to perform an evaluation of their potential for supporting internationalization processes. This is important to understand what is the current offer in the market.

Therefore, this exploratory study started in Portugal and the sample was large enough to reach theoretical saturation (Corbin & Strauss, 2015; Eisenhardt, 1989). Nevertheless, to increase the validity of our findings, we decided to make a few more interviews in other countries, specifically for this new study. So, 1 IBA from France was interviewed (because one of the authors had the opportunity to interview it in a specific international event), as well as 3 IBAs in the UK (as the same author also had a research visiting period in this country). Therefore, the sample for this study is composed by a total of 24 IBAs. Figure 2 shows the research design considered for this exploratory filed research.

Accordingly, a qualitative design is used to help specifying a preliminary set of requirements and features derived from the interviews, for digital platforms managed by IBAs to support SME internationalization. The qualitative approach is adopted to better fit with this exploratory-oriented study. Qualitative research has been continuously growing and spreading across disciplines (Golden-Biddle & Locke, 2007), where the aim is not to generalize the results, but its main value is the specific description

Figure 2. Research design of the exploratory filed research

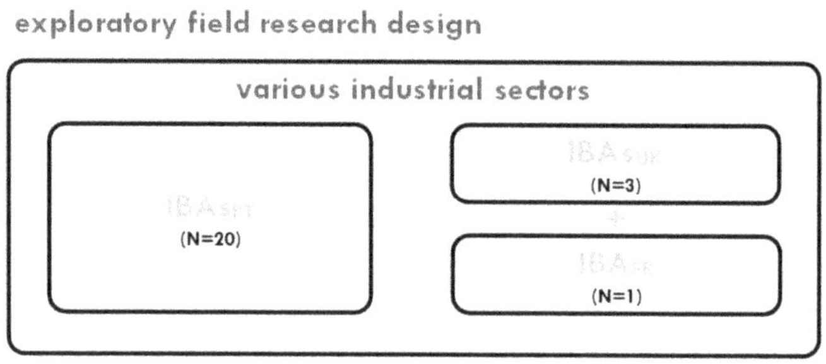

related to the particularity of a specific context (Merriam, 1998; Yin, 2009). Table 1 shows the main information about the selected IBAs.

Data was collected through interviews with IBAs' managers and people involved in international activities. Some of the procedures for the interviews were: (1) using an informed consent form; (2) using a semi-structured interview guide; (3) recording the conversations; and (4) request documentation to be used as secondary sources for data triangulation (Yin, 2009). The main questions of the interview guide considered for this study are:

Q1: Do you think that the management of information, knowledge and collaboration related to internationalization processes could be a success factor if mediated by a digital platform?
Q2: Do you think that the internationalization support of the IBA can be improved with the use of technology such as digital platforms?
Q3: What are the most important requirements and features to be considered in such digital platforms?

Afterwards, the interviews were transcribed as soon as possible to ensure a constant comparison (Eisenhardt & Graebner, 2007), and were coded using the qualitative data analysis software MAXQDA. Finally, data was analyzed by carrying out within-case and cross-case analysis, until reaching a point of data saturation (Eisenhardt, 1989). In addition, the identified digital platforms used by IBAs were also analyzed to understand the main purpose and features to support SME internationalization.

FINDINGS AND DISCUSSION

Digital Platforms Used by IBAs to Support SME Internationalization

Results from the interviews with IBAs show that most of them use basic forms of digital communication and information dissemination with their members, such as emails and websites. Some of them also use social networks like Facebook, Twitter or LinkedIn to disseminate more specific information content. Nevertheless, some of the IBAs show a motivation to change that and are now developing and adopting new ways and new ICT for improving the support to their members. In this chapter, we have special interest in knowing more about digital platforms that are specifically being used for enhancing the internationalization activities of SMEs. Therefore, the list presented in Table 2 refers to digital

Table 1. Main information about the selected IBAs

ID	Country	Industry Sector	Members	No. of Interviewees	Position of the Interviewee(s)	Duration of the Interview (Min.)
IBA_{PT1}	Portugal	Automotive	70	1	Technical Secretary	72
IBA_{PT2}	Portugal	Construction	7000	3	Secretary-general; Economist; Board Advisor	52
IBA_{PT3}	Portugal	Metals	900	1	Internationalization Project Manager	46
IBA_{PT4}	Portugal	Wood and furniture	700	1	Director-general	40
IBA_{PT5}	Portugal	Food	400	1	General Coordinator	61
IBA_{PT6}	Portugal	Metallurgical and electromechanical	500	1	Director-general	54
IBA_{PT7}	Portugal	Electrical and electronic	90	1	Economics and Association Service	43
IBA_{PT8}	Portugal	Cork	270	1	Director-general	63
IBA_{PT9}	Portugal	Chemicals	40	1	Director-general	18
IBA_{PT10}	Portugal	Footwear	400	2	CEO; Project Manager	71
IBA_{PT11}	Portugal	Furniture	200	1	Marketer-internationalization	49
IBA_{PT12}	Portugal	Mining	200	2	Director of Communication and Advice for International Relations; CFO and International Relations	45
IBA_{PT13}	Portugal	Multisector	800	1	Director of International Relations Department	52
IBA_{PT14}	Portugal	Multisector	100	1	Secretary-general and Director of International Relations	49
IBA_{PT15}	Portugal	Defense and weaponry	70	1	Director-general	44
IBA_{PT16}	Portugal	Aerospace	40	1	General Executive Manager	66
IBA_{PT17}	Portugal	Multisector	280	3	President of the Administrative Council; Board Administrator; Technical Administrator	39
IBA_{PT18}	Portugal	Agro food	130	1	Executive Coordinator of Markets Division	54
IBA_{PT19}	Portugal	Multisector	370	1	Project Manager	49
IBA_{PT20}	Portugal	Multisector	2000	1	Information Manager	36
IBA_{FR1}	France	Textile	50	1	Operational & International Director	47
IBA_{UK1}	UK	Multisector	1800	1	Director Commercial	24
IBA_{UK2}	UK	Multisector	1050	2	International Trade Manager; Head of Corporate Partnerships and Membership	61
IBA_{UK3}	UK	Multisector	4800	1	International Trade Coordinator	24

platforms developed by IBAs alone, or by other entities with the partnership of the IBAs. To maintain the confidentiality of the IBAs, the name and web links of these digital platforms are kept anonymous.

P1 is an open digital platform that was developed by a governmental entity, in collaboration with other institutional entities, including two of the interviewed IBAs. The main features of this digital platform are related with information dissemination about internationalization topics to facilitate the international expansion of companies of the sector.

Table 2. Digital platforms developed by IBAs to support SME internationalization

ID	Purpose	Main Features
P1	Promote capacity building for internationalization in the sector	• Products and markets: documents and information retrieved from other websites to present statistical data about imports and exports, by product sector and by country. • Export facilitation: information produced to serve as guidelines and roadmap to exportation. • Investment and trade: information and links to instruments and support programs (national programs, EU Horizon 2020, etc.). • Strategies and policies: information about public policies in support of internationalization. • Other information: information about news and events of the sector.
P2	Promote the SME transfer market and its importance and provide companies and investors a single point of contact for national and international business opportunities	• Opportunities: add new opportunities; edit opportunities; map opportunities; eliminate opportunities; approval of opportunities; • Manage documentation; • Search by ID, keyword, title or location; • Contact new partners; • Search for investors.
P3	Support companies in getting relevant information for an adequate international expansion	• Databases of companies of the sector; • Industrial parks of the sector; • General information about these markets; • Internationalization opportunities in these countries; • News about the sector; • Support guides for internationalization to these markets; • Partners in each of these countries.
P4	Offer a range of business information to help companies in defining their internationalization strategies and in increasing their visibility in foreign markets	• Spotlight: area where news about events, opportunities and other issues related to internationalization are shared. • Company Profile: area where the participating companies are presented through a brief description. • Business Matching: placement area for offers and opportunities for partnerships between Portuguese companies and foreign companies. • Tenders: a section intended for dissemination of project opportunities, funding opportunities, and consultations in foreign markets. • Marketplace: a space reserved for the exclusive promotion of products and services of companies in a showroom environment. • Benchmarking: a section designed for the publication of case studies promoting the sharing of successful practices.
P5	Systematize information about markets, and identify an international business opportunity to group companies in collaborative networks	• Business: information on public national and European tenders for each area of activity of the sector; opportunities resulting from public procurement; receive inquiries from companies and individuals; request budgets; receive business alerts daily summaries of information by email. • Accreditation: have the accreditation seal of the IBA visible in the directory. • Business Directory: detailed access to information about all companies available in the directory.
P6	Create collaborations between associates	• Search for associates of the IBA using an intelligent search to consult information classified for example by type of organization, industry, product or service, technologies, scientific domains, and customers; • Discussion forums; • Challenges to create synergies between the associates; • Ideas sharing; • Working groups.

P2 is a Portuguese business transfer digital platform for SMEs and was launched by one of the IBAs in partnership with a City Council. It allows an access to a *personalized, flexible and confidential support service to find a new business opportunity.*

P3 is a digital platform developed by one of the IBAs, in cooperation with international associations and other institutional entities in Portuguese-speaking countries. The information disseminated in this platform is only available for the members through registration. The main features are to present a set of information about the sector in Portugal and in Portuguese-speaking countries. The information in this digital platform is provided by IBAs of these countries and by other public entities linked to the industry.

P4 is an open digital platform, also available as mobile platform, which aims to support internationalization of companies by disseminating good practices and the promotion of international results, offering a set of tools for the internationalization activity. This digital platform was developed by one of the IBAs in cooperation with Enterprise Europe Network. This IBA has access to a set of databases with information about foreign companies, and disseminates market studies about foreign countries and about concrete companies that are abroad.

P5 is a digital platform developed by one of the IBAs, where companies and individuals of the sector need to register to have access to a range of information and features. In the future, this IBA pretends to establish links between companies through the digital platform by launching specific opportunities and create the contact between companies for a specific internationalization project, aiming to be the main business center in the sector in Portugal.

Finally, P6 is a digital platform developed by another IBA with reserved access to its members. The main objective of this digital platform is to create collaborations between the members. Although not specifically focused in internationalization processes, the creation of collaborations can trigger the development of partnerships for specific internationalization opportunities.

During the interviews, the participants also pointed out to the use of additional digital platforms that are focused on international trade but that are managed by other institutional entities such as governments, business clusters, or export agencies. Table 3 presents those additional digital platforms. In this case, the names of the platforms are revealed as it does not affect the confidentiality agreements established with the interviewees, and those institutional entities and digital platforms are well-known by most of the business people that deal with internationalization issues in Portugal and in the UK.

The External Trade and Investment Agency of Portugal (AICEP) provides a digital support for international trade activities of companies. Its main digital platform - Portugal Global (P7) - is mainly concerned with having information for support the international expansion of Portuguese companies. Nevertheless, it presents some modularity by being connected and by providing additional services using other external applications and add-ons, such as the Portuguese suppliers' digital directory and the digital platform 560pt. The latter intends to integrate information about Portuguese products and services, aiming at stimulating the internationalization of the respective companies. However, it has a restricted access by only making the information available for registered users.

Another example of a digital platform to support SME internationalization is the one developed by the Department for International Trade (DIT) of the UK Government - Great (P8). This is a very recent tool that was designed to provide services for increasing the exports of British companies. In this digital platform, the type of support available is divided according to the international trade experience of a company: companies with no export experience; companies that occasionally sell abroad; and experienced exporters. There are also features that allow to find buyers in foreign markets and other ones that encourage foreign investment in the country by providing information on a wide range of investment opportunities in the UK.

The last digital platform considered in this study, Export Britain (P9), presents similarities with the previous one. The aim is also to support the international operations of companies in the UK, but here the platform managers and the support is provided by the British Chambers of Commerce (BCC). In this digital platform, it is possible to find a variety of services and information that try to cover each stage of the internationalization process, i.e. the pre-firm internationalization phase, the initial internationalization

Table 3. Digital platforms developed by other institutional entities to support SME internationalization

ID	Name, Purpose, and Main Features
P7	Portugal Global (AICEP - External Trade and Investment Agency of Portugal) Encourage foreign companies to invest in Portugal, and contribute to the success of Portuguese companies abroad in their internationalization processes or export activities. Features: • Exporter's Guide: guide for SMEs that want to know how to export successfully. Explanation on the various stages of the export process and ways to overcome risks. • Internationalization Guide: guide to facilitate the path of business expansion to external markets, especially SMEs; case studies and success stories. • Events and Internationalization Actions: diverse internationalization actions to empower and promote companies by bringing their offer of products and services to the demand in external markets. • Buy From Portugal: Portuguese suppliers' digital catalogue/directory. • 560 pt: digital platform that integrates information to promote the internationalization of companies and the exports of national products and services, and create business opportunities for Portuguese entrepreneurs. • Support: financial support, export shop, account manager, financial multilateral, external network.
P8	Great (DIT - Department for International Trade, Government UK) Help British companies to export to new markets. Features: • Help to Sell Abroad: the type of support available is divided and applies to any type of company, i.e. new to exporting, selling abroad occasionally, experienced exporters. • Export Opportunities: find and apply for overseas opportunities from businesses looking for products or services; get alerts; get support along each stage of exporting. • Find a Buyer: allow companies to create a profile; let international buyers get in touch with companies' sales team; showcase companies' projects and experience; give companies international credibility by displaying independent data. • Selling Online Overseas: reach consumers online around the world; get exclusive deals negotiated on behalf of UK businesses. • Get Finance: discover how to make the export business of companies more competitive while managing the payment risks; access additional working capital, credit insurance or bank guarantees; understand the competition. • Events: list of different types of events that allow companies to export to new markets.

continued on following page

Table 3. Continued

ID	Name, Purpose, and Main Features
P9	Export Britain (BCC - British Chambers of Commerce) Help businesses to make new connections overseas and provide relevant information for internationalization. Features: • Export Services: export readiness; export finance; getting goods to market; international market access; sales & marketing; market promotion; market entry; market development; regional expansion. • Export documentation: make arrangements for legal documents to be apostilled or signed by a notary public; provide an online document processing service; export documentation certificates. • Export training: export training courses within the British Chambers of Commerce National Training Program for exporters. • Export events & missions: international trade events and trade missions to the world's biggest and fastest-growing markets. • Export publications: publish reports and surveys into aspect of international trade.

phase and the continued internationalization phase (Melén, 2009). Training and events on international trade are also promoted through this tool.

The next section of this paper presents an evaluation and a discussion on the identified digital platforms.

Evaluation of Current Digital Platforms

Results from the interviews show that only a few IBAs are actually using digital platforms to provide an internationalization support for their members. From the 24 interviews, we have only identified 6 digital platforms developed by IBAs. This situation shows that these institutional entities are still not following the current digital transformation and disruption of markets.

In a first glance, results of the evaluation show that most of the identified digital platforms are mainly focused on information dissemination, where the purpose is to organize international trade information into different topics to be accessed and used by its users. However, the process of information sharing (i.e. information shared by members with the IBAs or information shared between members) is not so evident. In many cases, there are no room for promoting the sharing of information between users, only allowing the platforms' managers to upload and distribute the information content. Regarding the communication, the level is basic in most of the cases. Some communications are also developed through the users' registrations in the digital platforms and the respective communication with the IBA. Nevertheless, P2, P5, P6 and P8 allow users to communicate with each other and share some information. None of the digital platforms allow a full digital coordination of international trade activities of companies.

All these digital platforms have resulted from recent projects of the IBAs that were financed by national or European funds. So, it is still soon to measure the real impact that is currently causing in the SMEs' international operations. Nevertheless, one thing is to comply with well-developed design rules and features to obtain a well-designed digital platform, but another thing is on how to create value and on how to attract users for its proper development. According to de Reuver et al. (2017), there is still a need for a comprehension on why some digital platforms succeed, while others fail. With this current study, it is difficult to infer about that aspect. But what we can conclude from the interviews is that, from one side, if the digital platform is new, it is still early to understand its real impact and success because it has not reached a proper level of maturity. From other side, we found two cases that have been unsuccessful, i.e. the case of P6 and the 560pt platform, integrated in P7. Those two digital platforms were developed more or less 5 years ago, but there has been a low traffic and a reduced interest from companies to access them. Many of the Portuguese IBAs recognize the existence of the 560pt platform that was developed by the Portuguese government, together with AICEP and other institutional entities. The problem is that they also consider that this project has been a flop, and that this specific digital platform does not have the capacity for providing a proper support to internationalization. The features presented are rudimentary, the information uploaded is not updated, and is still very generic information about some products and services from Portuguese companies. The IBAs are also aware that the majority of their members do not use this tool. Although still accessible, this digital platform has not suffered major modifications from the moment of its creation.

In the case of P6, it is the own IBA's manager that considers the failure of their collaborative digital platform. In this case, the digital platform seems well-designed. However, many investments were made to create P6, but the problem here was the lack of traffic from its members, mainly explained by, among other reasons, the fear of companies for sharing their information with others, or by the lack of time, or even lack of willingness for using such tools. These problems, together with other challenges on the use of digital platforms, will be presented and discussed in more detail in the next sub-section of this chapter.

Another topic of this discussion is related with the competitive environment and the overlaps of support between IBAs and governmental entities. This situation was clearer in the UK, where P8 is managed by the Department for International Trade, and P9 by the British Chambers of Commerce. It seems that everyone wants to be the main hub for international trade businesses in the country, justified by the clear similarities regarding the services provided by those two specific digital platforms, as well as their main objectives: P8 has the purpose of *Help British companies to export to new markets*, while the main aim of P9 is to *Help businesses (in the UK) to make new connections overseas and provide relevant information for internationalization.* This situation shows that there is a lack of cooperation between institutional entities that provide support to SME internationalization, so more collaborations are needed to facilitate and to improve the international expansion of businesses.

We work very closely with government agents like DIT but we work separately from them at the same time. So, we support each other but we don't always work alongside them. And the reason for that is that I personally believe that there is a very big difference between business-to-business relationships and business-to-government department relationships... That's not decrying the government agencies, the DITs, etc. ... we have a very healthy respect for them and they for us but we do just tend to keep... where we can work together we absolutely do, but most of the time we are working independently. (IBA_{UK2})

Some questions that may arise for future research are: Do we need different digital platforms for every institutional entity or do we need to aggregate efforts of gathering information and contacts in a unique platform with the collaboration of everyone?

Obviously, the answer to this question is challenging because, by having only one digital platform to support internationalization in each country, there is the problem of defining its governance: who will be the main responsible? who will manage its content? who will provide funding? what type of information content each player needs to share? etc. Therefore, in our opinion, digital platforms from governments and other institutional entities should exist for more general information about international trade activities, across all types of sectors. In the case of the IBAs, as they have more specific knowledge about their respective business sectors and as they already have a kind of network established with their members, they should develop their own digital platforms to provide a more detailed and focused information on the respective sector, as well as to facilitate the establishment of real collaborations among their members. This view is also reflected in the opinion of most of the interviewees, as there is a general agreement that the digital platforms from governmental entities and from agencies are useful to have access to statistics, studies and more general information about internationalization. But, for more specific information about each sector, it is difficult to find valued information that can be used by companies in their decision-making processes.

Nonetheless, insightful discussions on the use of digital platforms were held with the IBAs involved in this study. In these discussions, the interviewees emphasized many important aspects to have in consideration when developing digital platforms. The opinion from IBAs that are not using such platforms and digital tools was also crucial to understand the reasons for not using it. Therefore, the set of interviews that were performed in this study allowed to obtain a rich content about the factors that influence the adoption of digital platforms by IBAs to support SME internationalization. These results were fundamental to help specifying some requirements and features for digital platforms that can improve information and collaboration management processes of IBAs in internationalization contexts. The next section presents the detail of these requirements and features.

REQUIREMENTS AND FEATURES FOR DIGITAL PLATFORMS DERIVED FROM THE INTERVIEWS

Requirements and Features From Companies (First Exploratory Study)

The interviews that were performed in this study allowed to obtain different needs and important considerations regarding the use of digital platforms for supporting the SMEs' internationalization. Before presenting the results of the IBAs' side, we retrieve some of the main points that were mentioned in our first exploratory study with 5 companies associated with IBAs (Costa, Soares, Sousa, & Jamil, 2017), which are important to have in consideration in this present chapter. In this previous study, the focus was on understanding the perspective from those companies, regarding the use of collaborative digital platforms managed by their respective IBAs for supporting their internationalization processes. Table 4 presents the main considerations from this first exploratory study, divided by features, requirements and possible difficulties. We differentiate requirements and features, considering the first one as something the digital platform must do or a quality it must have, and the second one as a set of related requirements that allows the user to satisfy a business objective or need.

The considerations described in Table 4 give us an overview of what companies expect and desire from digital platforms for create the interest in using such tools for their international operations. The

Table 4. Main considerations from companies associated with IBAs from our previous study

Requirements	
International Trade Information	The information disseminated by the IBAs through digital platforms could be about market conditions, countries' attractiveness, internationalization strategies and funding projects and programs.
Participation of Academic/Research Entities	The possibility of using a collaborative digital platform only makes sense if entities, such as Universities or research institutes have some participation in its development, in order to connect the business world with the academic world.
Added Value	All parts and all users must share something meaningful in this digital platform, not only a few players. Companies must feel that there is a real need for them.
Flexibility in Information Sharing	A digital platform may work if it has not generalized information, but features that can allow to create and direct information within groups and market segments.
Collaboration	Create collaborations among members can be relevant for making joint strategies and for sharing resources and risks.
Features	
Information Management (In General)	Digital platforms could improve information sharing and facilitate the search for partners and opportunities. IBAs could disseminate legal information (e.g. about patents, legislation and customs rules) and filter information according to each member (e.g. information about the countries in which each associate is present or has sales).
Opportunities Sharing	The IBA could disseminate funding and research opportunities, as well as some more institutional information.
Matchmaking (For Approaching Markets)	The IBA could align members with same strategies, resources and philosophies by creating synergies to approach specific markets.
Matchmaking (For Projects Opportunities)	A possible feature of a digital platform is to have a matchmaking process between European projects and the research interests of companies.
Information from External Sources	A digital platform could retrieve information from different websites and external digital sources of interest and re-organize it for the interests of the members.
Collaboration Features	The IBA could foster collaborations using digital platforms by disseminating the skills and competencies of each associate company, increasing their visibility within foreign markets and industrial sector.
Experiences Sharing	A digital platform could allow companies to share previous experiences in foreign markets with other members, in order to understand the rules and specificities of each market, as well as to reduce the possibility of making same mistakes.
Difficulties	
Information Management	Transposing all the information management into a digital platform can be very difficult because most of the companies have no desire in sharing their information for confidential reasons and also due to the lack of time for structuring it in a simple and pleasant way to read. Another reason is that people may prefer to exchange experiences in more informal ways, through personal contact.
Platform's Governance	The governance of digital platforms to support internationalization can be difficult due to the lack of resources (human, financial), and the lack of time of both the IBAs and their members.
Content Management	To avoid failures, a digital platform needs great human power mostly dedicated to content management.

Costa, Soares, Sousa, & Jamil, 2017.

processes of matchmaking, and of having access to more organized and valued information, are the main considerations from the interviewed companies. In Table 4, it is also presented the main problems that also may arise. Here, the lack of resources and the possibility of companies for not having time or motivation to share or to access information, represent the main pointed difficulties by the companies.

Requirements and Features From the IBAs: Information

After analyzing the current offer in terms of existing digital platforms, and with the perspective from potential users (companies associated with IBAs), it is now possible to present the perception from the possible platforms' managers, which, in this case, are the IBAs. Accordingly, the 24 interviews considered for this present study were transcribed and coded in different categories to help the presentation of the main requirements and features for digital platforms.

The first category is "INFORMATION", which is distributed by different subcategories of what should be the information accessible in a digital platform supporting SME internationalization:

- Synthetize international trade information;
- International trade documents;
- Financing and credits;
- Detailed information about trade fairs and trade missions;
- Calls for projects / tenders;
- Detailed information about members / database of members;
- Databases of distributors and agents;
- International contacts / databases of companies in foreign markets;
- Legal information and market conditions;
- Policy making information;
- Publications and market studies.

Therefore, findings from the interviews with the IBAs show that a good digital platform should allow to have access to a set of important information that can help companies in their international activities, such as international trade documents, information about trade fairs and legal information. Different databases should also be included in the design of digital platforms supporting internationalization with detailed information about the IBAs' members, accompanied by specific information on distributors, agents, foreign companies and other international contacts. According to the opinion of the interviewees, although recognized as a difficult process, a digital platform could be useful if it allows to gather information from different sources, organize it and synthetize it for a better treatment and management, in order to be shared with the IBAs' members.

Requirements and Features From the IBAs: Collaboration

The second category defined is "COLLABORATION", which shows some requirements and specific features that could allow to use information for creating collaborative processes of internationalization:

- International business opportunities disclosure;
- Digital platform dynamic and interactive;
- Integrate external entities of interest to contribute with information;
- Foster business collaboration;
- Logistics sharing;
- Find partners for projects' consortiums.

Regarding the creation of digital collaboration, the IBAs think that a digital platform could have an area of business opportunities where all the offers, regarding the opportunities for internationalization, could be inserted and presented for the members. The digital platform should also be dynamic and interactive, where companies could insert information and share their stories and experiences with other members. The integration of information from external entities, such as funding entities, clusters, governments, embassies, international trade agencies and foreign IBAs, could also be a success factor to have access to different types of information to be used by the IBAs and their respective members. Other specific collaboration features of a possible digital platform of support to internationalization could be to allow a member to contact and subcontract another member of the IBA, to insert information for sharing commercial agents, or even to share specific logistics activities such as sharing destinations for products, sharing transportation modes, or sharing warehouses.

Requirements and Features From the IBAs: Communication

Another category resulting from the transcriptions of the interviews is "COMMUNICATION". The subcategories defined for this category are:

- Facilitate general communication;
- Contact memos;
- Social networks;
- Media and social communication;
- Promote the image of the sector / country;
- Newsletters.

A digital platform should facilitate the general communication of the IBA with its members or among members. A feature could be to choose specific people or specific department of a member to communicate a certain internationalization subject, i.e. the information should be more targeted and more classified. Another feature described by one of the interviewees is the possibility to create contact memos, to understand, for example, that in a certain day, at a specific hour, person X from company Y called to ask for clarifications about a particular subject. Many IBAs also consider that a digital platform should be integrated with the current IBAs' social networks (Facebook, LinkedIn, and Tweeter) for a proper communication about international trade subjects, as well as by including the clipping of news form the internet, television, journals, etc. All those ways of communication, together with the multimedia resources of the IBAs (e.g. videos), can help the IBA to promote the image of the sector or of the country for the outside world.

Requirements and Features From the IBAs: Other

The last category defined includes all other requirements and features that the IBAs consider as relevant for increasing their interest on using digital platforms for supporting SME internationalization, "OTHER". The list of requirements for this category is:

- Added value / incentive for using a digital platform;
- User-friendly;

- Updates of platform's versions;
- Quality management;
- Flexibility in sharing the information;
- Summarized information;
- Language / idiom;
- Different types of support for different types of companies;
- Companies need technology / technological capital;
- Follow the needs of companies;
- Iterative process of design.

Accordingly, most of the IBAs consider that a digital platform only makes sense if it has an added value for the members. It must allow to reach a specific objective or to meet a specific need. It must be something objective, to not only represent an information portal. Therefore, platform designers should consider how to create an incentive for companies to use such tools, and how to create demand and traffic on the platform. This is an important requirement because if companies see the utility of using digital platforms, they will share information and collaborate with other members. Another requirement is for the digital platform to be user-friendly. Some IBAs consider that adopting an interface leader, such as the ones from Facebook or LinkedIn, could also potentiate more usage and more acceptance from the users. If developed, a digital platform should also allow easy updates in its different versions, as well as allow the possibility to create add-ons and to adapt and upgrade new software and new servers, without incurring in huge investments for companies for maintaining it updated and functional. Other requirements by the interviewees is to have flexibility in disseminating the information, allowing to select the type of information and the targets and members to send the information within the platform. This situation should avoid overloading companies with information that they have no interest on it, as there are different kinds of information that are needed by each company. The information should also be summarized because companies have no time and availability to access and read large content of text. The digital platform should also be multi-language to reach a higher number of people of interest in foreign markets. According to the interviewees, the information and support provided could be divided according to the type of company, in what regards its level of international experience. This is a relevant aspect because, naturally, a company that just started to export have different needs from a company that is used to all the procedures and to all the existing international trade information. Finally, it was possible to retrieve some other important requirements from the interviews, which are related with the need for technological capital of companies and regarding the process of platform design. Some IBAs defend that companies should modernize themselves by adopting technology in their processes of internationalization, to getting to know people, have online face-to-face meetings, and to even create business. Another issue is that technology facilitates information management processes, as well as can allow IBAs to think about and to create new instruments and tools to support their members. Some IBAs defend that the most successful companies are the ones where there was an intervention of technological capitals that entered and gave great potential to companies in international markets. For the platform design, based on the experience of some IBAs, the whole design process should always follow the needs of companies, using an iterative process of design. Companies should be involved in the design process, to understand their needs, expectations and objectives, allowing to raise the interest on the utilization of the digital platform. Therefore, one recommendation from the interviews is to *create a solution according to what companies want and not according to what we think that might interest them* because, at the end, companies will do what they want, and not what the IBA wants. So, before starting to mount the first piece of the platform software, the opinion is that platform designers should listen companies and ask for feedback to meet their goals. The interface of the digital platform should also be simple and have only the essential features to work well for companies.

Besides all these new requirements that were derived from the performed interviews for the category "OTHER", some additional features were also mentioned:

- Restricted access and levels of access;
- Creation of forms;
- Information on the impact of the initiatives developed;
- Market selection tool;
- Add-ons;
- Translation services;
- Integration with existing tools;
- Forums.

Most of the IBAs think that some of the digital platform features should have a restricted access to their members, in order to allow them to collaborate and share experiences with each other without constraints. Another suggestion is to have different levels of access to the information and to the main features of the platform, i.e. distinguish between what is management of content for members and for non-members. The interviewees also mention the creation of various digital forms could also be of interest. Examples of such forms are purchase requisitions, billing orders, price orders, budgeting requests, procurement requisitions, satisfaction survey, general surveys about companies' performance indicators (e.g. export volume of the company), evaluation forms for trade fairs, etc. So, there are many documents and forms that the IBAs consider that could be on digital platforms. Another feature of interest for the IBAs is the possibility of having a market selection tool that can look at the profile of companies and the criteria they are looking for in the market, and help look at criteria like people in market, imports, exports and other indicators, and then select the market and show the highest potentials and risks. The provision of online translation services to companies is also referred as being relevant for translating international trade documents. Finally, the creation of online forums for potentiate the communication and collaboration among members, and the integration of the digital platform with existing tools in the market (from governments, trade agencies, banks, etc.) are also some other features of interest for the IBAs.

Difficulties Considered by the IBAs

The interviews with the 24 IBAs allow us to understand some of the potential problems and difficulties on the use of digital platforms for internationalization purposes:

- Too many digital platforms;
- Lack of human resources;
- Lack of financial resources;
- Companies not willing to share information;
- Companies not willing to use such tools;
- Lack of time of companies;

- Lack of involvement / engagement of companies;
- Personal contact will be always needed;
- Platform governance;
- Platform maintenance;
- Companies wanting information that the IBA cannot provide;
- Different IBAs for the same sector;
- Too many IBAs and institutions of support (multiple membership of companies);
- Market studies are expensive;
- Companies not aware of the services of the IBAs;
- Lack of support from governments;
- Problems of information standardization;
- Companies face difficulties for screening project opportunities;
- Cyber-crime and data protection;
- SMEs and the fear of collaboration.

The first problem is referred by some IBAs about the existence of too many attempts to develop digital platforms for internationalization purposes (and many have failed) by different kinds of entities (trade agencies, banks, governments, etc.), as it seems that everyone wants to have their own digital platform to support companies and all have the same expectation of wanting information forwarded to their own digital platforms. So, the first difficulty that an IBA can face is on how to guarantee that its platform will be the one chosen by the companies, and on how to stand out from the competition. Another major problem is on how to motivate companies for sharing information and for making use of the features of a specific digital platform for helping them in their international processes. Some of the problems related with that is the lack of human and financial resources. Some IBAs, who have a small team working inside the association, recognize that a well-developed digital platform, with different kinds of features and information, could require many resources (both human and financial) to preserve it, resources that some could not possess at the moment. Others consider that it is nearly impossible to have one person from the IBA specifically dedicated to feeding and updating a digital platform. The initial costs for developing such tools, as well as the costs of maintenance are also some impediments for its adoption and acceptance. Although considered as potential useful, the cost of diffusing a new digital platform is also an important factor to have in consideration and that may cause additional difficulties. According to some IBAs, the problem of information is to collect it and systematize it.

Another potential problem considered by most of the interviewees is the lack of time and lack of willingness for companies to use such tools, as they are always inserted in pressuring environments and many are not open for adopting new tools or changing habits. Many IBAs complain about the lack of involvement of companies and that, most of the time, it is hard to get information from them. Therefore, asking them to make the share of information and experiences with other members (including competitors) a part of their day-to-day activities, represents an issue that will be very difficult to solve. Many interviewees point the culture and the fear of sharing their information as major problems faced by most of the companies. Therefore, the IBAs recommend to always maintain the informal and personal contact, as it is easier and more acceptable for companies to share their experiences and thoughts, for example, in trade fairs, events, or in a lunch/dinner. Registering this virtually and asking them for writing and uploading information on a digital platform is something that the IBAs think that will be, at least, challenging. Another IBA considers that from an international perspective, online instruction and assistance is often more confusing and they find that companies like to always speak to somebody.

Other difficulties on the use of digital platforms are concerned with the platform governance and platform maintenance. Who will upload and update the information? How to motivate and attract users to keep the platform "alive"? How will the platform be updated? Will other institutional entities contribute and upload their information? Who will fund and develop the digital platform? Those are examples of questions asked by the IBAs in the interviews. In the opinion of a few IBAs, a product can be very good and well developed. But, if the IBA cannot do a proper maintenance, with updates in information and updates in features, there is no point for using such tools. Therefore, before designing and developing any digital platform, platform designers must carefully think about its governance model for meeting the reality of each IBA. Another issue defended by the IBAs is that the digital platform should facilitate a determined process or activity of a company by not creating more complications for them, and by not requiring more work and effort.

Continuing the enumeration of potential difficulties, in some situations the members of IBAs are looking for information that the IBA is not able to provide or, in some other cases, companies are not aware of the full range of services provided by the IBAs. Market studies are also expensive to obtain or to produce. It would also be difficult to standardize the information in a digital platform, due to the various sub-sectors that many times an IBA represents. Each sub-sector has its main needs and nuances that can create additional difficulties in the process of standardizing the information to be of interest for everyone. Another problem is for small companies to share their ideas with others, with the fear that a bigger company, with more human power and financial resources, will just swollen them up and take the idea all the way, and then lose their ideas. Cyber-crime is another issue that may arise from the use of technology like digital platforms.

The last topic of this section of difficulties is concerned with the existence of too many IBAs and trade associations, which creates multiple memberships of companies, as sometimes it is possible to find two different IBAs for the same sector. The pointed problem is that there are too many offers for the promotion of internationalization activities, which overloads companies with large amounts of information and creates a confusion in what is the real capacity of each business association to provide a good and quality service.

CONCLUSION AND FUTURE RESEARCH

This chapter provides the results from 24 interviews performed with IBAs from Portugal, France, and the UK, on the use of digital platforms to support SME internationalization in the context of collaborative networks. The first conclusion is that most of the IBAs are not using technology, such as digital platforms, to support the activities of their members. A communication using the email and the dissemination of information through their websites are still the basic technological channels typically used. Nevertheless, a few IBAs are trying to follow the technological trends that are happening in the current digital economy, demonstrated by the use of more advanced technological means and tools. From these interviews, it was possible to identify 9 digital platforms used by IBAs to support the international trade activities of SMEs, by describing their main objectives and features. Six of them were developed by some of the IBAs, while the other three are from government entities or from trade agencies.

An evaluation of these 9 digital platforms was also performed. The main findings of this evaluation show that the existing offer of digital platforms of support to internationalization lacks to meet some important requirements for obtaining well-designed digital platforms that can allow to create collaborative networks. Nonetheless, the majority of these digital platforms are still new and it seems too early to apprehend the real impact that are causing in the international activities of SMEs. Additionally, with this evaluation, it was also possible to identify a few cases of failure, mainly evidenced by the lack of involvement of companies. There are also overlaps in terms of support activities to SMEs between sectoral IBAs, multi-sectoral IBAs, trade agencies, and governments. It looks like the different institutional entities that provide support to the internationalization of SMEs are trying to be the main hub or representative, by developing and using similar digital tools and services, without collaborating with each other to provide a proper collaborative international trade support. We also found differences in terms of the levels of international service development among the interviewed IBAs, differences that are also reflected in terms of the acceptance and adoption of technology to their support services. Therefore, we can conclude that before using a digital platform or other more developed IT tools of support to internationalization, some IBAs need to focus on improving some other basic aspects of their digital evolution, such as getting information from their members to develop a suitable database with detailed information for each member, or designing and developing new websites that reflect the current technological trend in the markets.

Besides the presentation and discussion about the current digital platforms used by IBAs, we were able to obtain a rich set of requirements and features derived from the interviews. First, a brief summary of our previous exploratory study with companies associated with IBAs was presented, giving an overview of what companies expect and desire from digital platforms to raise the interest on the use of such tools. Features such as matchmaking processes or having access to organized information distributed by different internationalization topics, were the main considerations for the companies. However, the main concerns were on the lack of resources and the lack of time to use digital platforms in their day-to-day international trade activities.

After that, the requirements and features considered by the 24 IBAs were also presented and divided by different categories: INFORMATION (e.g. upload or have access to international trade documents; calls for projects; databases of members, distributors, and agents; legal information and market conditions); COLLABORATION (e.g. international business opportunities disclosure; foster business collaboration; logistics sharing); COMMUNICATION (e.g. contact memos; social networks; media and social communication); OTHER requirements (e.g. added value; user-friendly; updates of platform's versions; flexibility in sharing the information; follow the needs of companies); and OTHER features (restricted access and levels of access; creation of forms; market selection tool; market selection tool). Therefore, the perspective from the potential platforms' managers allowed us to obtain a detailed set of requirements and features on the use of digital platforms for supporting SME internationalization. Nevertheless, this study also allowed us to understand the main problems and difficulties that can emerge when using and developing digital platforms. Different types of difficulties were mentioned by the IBAs. The main ones are related with the lack of human and financial resources by both the IBAs and the SMEs to manage and maintain such digital tool. As shown by some companies in our previous exploratory study, some IBAs also think that companies might not have time or people dedicated to access and use the information and features provided by a digital platform. Another recurrent problem is the fact that many companies are still not ready or are reluctant to use such technologies in their activities, mainly due to the fear of sharing their information and experiences, but also because it seems that the personal contact is still the preferred method to do that kind of tasks. Finally, some other difficulties were mentioned such as the platform governance, platform maintenance, the existence of too many digital platforms or too many IBAs and institutional entities providing the same kind of support.

All the information and findings of this chapter can be used by digital platform designers and by IBAs in general, to design and obtain more effective digital platforms in the future that can meet the specific needs of users and managers, always having in consideration the main problems and difficulties of their implementation. In terms of contributions to theory, this chapter brings new knowledge on how organizations can effectively design digital platforms supporting collaborative networks, having the specific context of internationalization. The results of this chapter will be used in future developments of this study, by proposing and validating a design theory, in the form of design propositions, for designing effective collaborative digital platforms managed by IBAs for supporting internationalization processes of SMEs. This will be achieved by performing case studies and focus groups for designing and validating/evaluating the design propositions.

ACKNOWLEDGMENT

This research was funded by the Portuguese funding agency, Fundação para a Ciência e a Tecnologia (FCT), through the Ph.D. Studentship SFRH/BD/110131/2015. It was also supported by the Project "TEC4Growth - Pervasive Intelligence, Enhancers and Proofs of Concept with Industrial Impact/ NORTE-01-0145-FEDER-000020" financed by the North Portugal Regional Operational Programme (NORTE 2020), under the PORTUGAL 2020 Partnership Agreement, and through the European Regional Development Fund (ERDF).

REFERENCES

Bell, S. (2006). A victim of its own success: Internationalization, neoliberalism, and organizational involution at the business council of Australia. *Politics & Society*, *34*(4), 543–570. doi:10.1177/0032329206293643

Bellini, F., Ascenzo, F., Ulskaia, I., & Savastano, M. (2016). Digital service platform for networked enterprises collaboration: A case study of the NEMESYS project. In IESS 2016, LNBIP 247 (pp. 313–326). Academic Press. doi:10.1007/978-3-319-32689-4_24

Bennett, R. (1998). Business associations and their potential contribution to the competitiveness of SMEs. *Entrepreneurship and Regional Development*, *10*(3), 243–260. doi:10.1080/08985629800000014

Boléat, M. (2003). *Managing trade associations*. London: Trade Association Forum.

Brouthers, K., Geisser, K., & Rothlauf, F. (2016). Explaining the internationalization of ibusiness firms. *Journal of International Business Studies*, *47*(5), 513–534. doi:10.1057/jibs.2015.20

Camarinha-Matos, L., & Afsarmanesh, H. (2005). Collaborative networks: A new scientific discipline. *Journal of Intelligent Manufacturing*, *16*(1), 439–452. doi:10.100710845-005-1656-3

Carneiro, L., Soares, A., Patrício, R., Azevedo, A., & Pinho de Sousa, J. (2013). Case studies on collaboration, technology and performance factors in business networks. *International Journal of Computer Integrated Manufacturing*, *26*(1–2), 101–116. doi:10.1080/0951192X.2012.681914

Corbin, J., & Strauss, A. (2015). *Basics of qualitative research: Techniques and procedures for developing grounded theory*. Thousand Oaks, CA: Sage Publications.

Costa, E., Soares, A., & Sousa, J. (2015). A new insight in the SMEs internationalization process. In M. L. Camarinha-Matos, F. Bénaben, & W. Picard (Eds.), *Risks and Resilience of Collaborative Networks: 16th IFIP WG 5.5 Working Conference on Virtual Enterprises, PRO-VE 2015, Albi, France, October 5-7, 2015* (pp. 398–410). Cham: Springer International Publishing. 10.1007/978-3-319-24141-8_36

Costa, E., Soares, A., & Sousa, J. (2016a). Information, knowledge and collaboration management in the internationalisation of SMEs: A systematic literature review. *International Journal of Information Management*, *36*(4), 557–569. doi:10.1016/j.ijinfomgt.2016.03.007

Costa, E., Soares, A., & Sousa, J. (2016b). Situating case studies within the design science research paradigm: An instantiation for collaborative networks. In H. Afsarmanesh, M. L. Camarinha-Matos, & A. Lucas Soares (Eds.), *Collaboration in a Hyperconnected World: 17th IFIP WG 5.5 Working Conference on Virtual Enterprises, PRO-VE 2016, Porto, Portugal, October 3-5, 2016* (pp. 531–544). Cham: Springer International Publishing. 10.1007/978-3-319-45390-3_45

Costa, E., Soares, A., & Sousa, J. (2017a). From data sources to information sharing in SME collaborative networks supporting internationalization: A socio-semantic approach. In L. M. Camarinha-Matos, H. Afsarmanesh, & R. Fornasiero (Eds.), *Collaboration in a Data-Rich World: 18th IFIP WG 5.5 Working Conference on Virtual Enterprises, PRO-VE 2017, Vicenza, Italy, September 18-20, 2017* (Vol. 506, pp. 478–490). Cham: Springer International Publishing. 10.1007/978-3-319-65151-4_43

Costa, E., Soares, A., & Sousa, J. (2017b). Institutional networks for supporting the internationalisation of SMEs: The case of industrial business associations. *Journal of Business and Industrial Marketing*, *32*(8), 1182–1202. doi:10.1108/JBIM-03-2017-0067

Costa, E., Soares, A., Sousa, J., & Jamil, G. (2017). Information management for network transformation in industrial enterprises associations: The case of the internationalization process. In G. L. Jamil, A. L. Soares, & C. R. M. Pessoa (Eds.), *Handbook of Research on Information Management for Effective Logistics and Supply Chains* (pp. 415–436). Hershey, PA: IGI Global; doi:10.4018/978-1-5225-0973-8.ch022

Cremona, L., Ravarini, A., & Sutanto, J. (2014). KMS in a cluster of firms: The role of a digital platform. In *Information Systems, Management, Organization and Control. Lecture Notes in Information Systems and Organisation* (Vol. 6, pp. 253–262). Cham: Springer; doi:10.1007/978-3-319-07905-9_18

de Reuver, M., Sørensen, C., & Basole, R. (2017). The digital platform: A research agenda. *Journal of Information Technology*, 1–12. doi:10.105741265-016-0033-3

Durugbo, C. (2015). Modelling information for collaborative networks. *Production Planning and Control*, *26*(1), 34–52. doi:10.1080/09537287.2013.847217

Dutot, V., Bergeron, F., & Raymond, L. (2014). Information management for the internationalization of SMEs: An exploratory study based on a strategic alignment perspective. *International Journal of Information Management*, *34*(5), 672–681. doi:10.1016/j.ijinfomgt.2014.06.006

Eisenhardt, K. (1989). Building theories from case study research. *Academy of Management Review*, *14*(4), 532–550. doi:10.5465/AMR.1989.4308385

Eisenhardt, K., & Graebner, M. (2007). Theory building from cases: Opportunities and challenges. *Academy of Management Journal*, *50*(1), 25–32. doi:10.5465/AMJ.2007.24160888

European Commission. (2014). *Supporting the Internationalisation of SMEs*. Retrieved from http://ec.europa.eu/enterprise/policies/sme/files/support_measures/internationalisation/report_internat_en.pdf

Fedorowicz, J., & Sawyer, S. (2012). *Designing collaborative networks: Lessons learned from public safety. IBM Center for The Business of Government: Collaborating Across Boundaries Series*. Washington, DC: IBM Center for The Business of Government.

Fries, L. (2008). Boundarians that shape market actors: A review of the literature on business associations. In *24th IMP-conference* (pp. 1–17). Academic Press.

Gashi, P., Hashi, I., & Pugh, G. (2013). Export behaviour of SMEs in transition countries. *Small Business Economics*, *42*(1), 407–435. doi:10.100711187-013-9487-7

Gawer, A. (2009). *Platforms, markets and innovation*. Edward Elgar Publishing. doi:10.4337/9781849803311

Ghazawneh, A., & Henfridsson, O. (2015). A paradigmatic analysis of digital application marketplaces. *Journal of Information Technology*, *30*(3), 198–208. doi:10.1057/jit.2015.16

Giudici, A., & Blackburn, R. (2013). *Digitally-enabled network intermediation in the internationalization of SMEs: A dynamic capabilities perspective*. Nemode.Ac.Uk.

Golden-Biddle, K., & Locke, K. (2007). *Composing qualitative research* (Vol. 2). Thousand Oaks, CA: Sage Publications. doi:10.4135/9781412983709

Hessels, J., & Terjesen, S. (2010). Resource dependency and institutional theory perspectives on direct and indirect export choices. *Small Business Economics*, *34*(2), 203–220. doi:10.100711187-008-9156-4

Hsu, W., Chen, H., & Cheng, C. (2013). Internationalization and firm performance of SMEs: The moderating effects of CEO attributes. *Journal of World Business*, *48*(1), 1–12. doi:10.1016/j.jwb.2012.06.001

Knight, G., & Liesch, P. (2016). Internationalization: From incremental to born global. *Journal of World Business*, *51*(1), 93–102. doi:10.1016/j.jwb.2015.08.011

Leonidou, L., & Theodosiou, M. (2004). The export marketing information system: An integration of the extant knowledge. *Journal of World Business*, *39*(1), 12–36. doi:10.1016/j.jwb.2003.08.005

Mansour, O. (2009). Group intelligence: A distributed cognition perspective. In *INCOS'09 - International Conference on Intelligent Networking and Collaborative Systems* (pp. 247–250). IEEE. 10.1109/INCOS.2009.59

Melén, S. (2009). *New insights on the internationalisation process of SMEs: A study of foreign market knowledge development. Economic Research Institute*. Stockholm School of Economics.

Mergel, I. (2017). *Digital service teams: Challenges and recommendations for government. IBM Center for The Business of Government: Using Techbology Series*. Washington, DC: BM Center for The Business of Government.

Merriam, S. (1998). *Qualitative research and case study applications in education*. San Francisco: Jossey-Bass Publishers.

Mikamo, S. (2013). Business associations and politics in the post-EDSA Philippines: Neither oligarchy nor civil society. *Philippine Political Science Journal, 34*(1), 6–26. doi:10.1080/01154451.2013.789165

Parker, G., Alstyne, M., & Choudary, S. (2016). *Platform Revolution*. New York: W. W. Norton & Company, Inc.

Parung, J., & Bititci, U. (2008). A metric for collaborative networks. *Business Process Management Journal, 14*(5), 654–674. doi:10.1108/14637150810903048

Patel-Campillo, A., & DeLessio-Parson, A. (2016). Why types of operations, trade associations, and production trends matter in the geographic branding of an emerging industry. *Journal of Wine Research, 27*(3), 242–256. doi:10.1080/09571264.2016.1202218

Pieth, M. (2012). Collective action and corruption. In *Collective Action – innovative strategies to prevent corruption*. DIKE AG; doi:10.1007/978-3-319-04480-4_6

Popp, J., MacKean, G., Casebeer, A., Milward, H., & Lindstrom, R. (2014). Inter-organizational networks: A critical review of the literature to inform practice. *IBM Center for The Business of Government*, 123.

Provan, K., & Kenis, P. (2008). Modes of network governance: Structure, management, and effectiveness. *Journal of Public Administration: Research and Theory, 18*(2), 229–252. doi:10.1093/jopart/mum015

Rabelo, R., & Gusmeroli, S. (2008). The ecolead collaborative business infrastructure for networked organizations. In Pervasive Collaborative Networks (pp. 451–462). Academic Press. doi:10.1007/978-0-387-84837-2_47

Rodriguez, V., Barcos, L., & Álvarez, J. (2010). Managing risk and knowledge in the internationalisation process. *Intangible Capital, 6*(2), 202–235. doi:10.3926/ic.2010.v6n2.p202-235

Romero, D., Galeano, N., & Molina, A. (2009). Mechanisms for assessing and enhancing organisations' readiness for collaboration in collaborative networks. *International Journal of Production Research, 47*(17), 4691–4710. doi:10.1080/00207540902847280

Schutzius, M. (2016). *Best practices for industry associations*. Retrieved from http://go.euromonitor.com/white-paper-associations-2016-Associations-Research-Strategy-Development-Member-Engagement.html

Sedoglavich, V. (2012). Technological imperatives in the internationalization process: Results from a qualitative investigation of high-tech SMEs. *Management Research Review, 35*(5), 441–459. doi:10.1108/01409171211222386

Sommer, L., & Haug, M. (2011). Intention as a cognitive antecedent to international entrepreneurship-understanding the moderating roles of knowledge and experience. *The International Entrepreneurship and Management Journal*, *7*(1), 111–142. doi:10.100711365-010-0162-z

Spagnoletti, P., Resca, A., & Lee, G. (2015). A design theory for digital platforms supporting online communities: A multiple case study. *Journal of Information Technology*, *30*(4), 364–380. doi:10.1057/jit.2014.37

Streeck, W., Grote, J., Schneider, V., & Visser, J. (2006). *Governing interests: Business associations facing internationalism*. London: Routledge.

Sun, R., Gregor, S., & Keating, B. (2015). Information technology platforms: Conceptualisation and a review of rmerging research in IS research. In *Australasian Conference on Information Systems* (pp. 1–17). Academic Press.

Tiwana, A., Konsynski, B., & Bush, A. (2010). Platform evolution: Coevolution of platform architecture, governance, and environmental dynamics. *Information Systems Research*, *21*(4), 675–687. doi:10.1287/isre.1100.0323

Traxler, F., & Huemer, G. (2007). *Handbook of business interest associations, firm size and governance: A comparative analytical approach*. London: Routledge.

Tseng, K., & Johnsen, R. (2011). Internationalisation and the internet in UK manufacturing SMEs. *Journal of Small Business and Enterprise Development*, *18*(3), 571–593. doi:10.1108/14626001111155718

Wang, J., & Gooderham, P. (2014). Institutional change and regional development in China: The case of commodity trading markets. *Environment and Planning. C, Government & Policy*, *32*(3), 471–490. doi:10.1068/c11254b

Xie, Y., & Amine, L. (2009). Social networks and the internationalization of Chinese entrepreneurs. *Global Business and Organizational Excellence*, *29*(1), 61–78. doi:10.1002/joe.20299

Yin, R. (2009). Case study research: Design and methods (4th ed.). Thousand Oaks, CA: Sage Publications.

Chapter 5
Information Marketing

José Poças Rascão
Polytechnic Institute of Setúbal, Portugal

ABSTRACT

This chapter discusses the issues of the needs and satisfaction of the customers, in terms of information, as a basis for the practice of marketing in information management. It enhances the arguments of the relationship between marketing and information science. The chapter stresses that practicing marketing cannot be done without information about customers and to customers and their relationship with information management, in information science. It stresses the importance of the studies and research on marketing of information as a philosophical approach for the information management process. The structure of the chapter synthesizes the existing academic work, seeking to generate new knowledge. It presents the promotion and communication of information in organizations from the evolution of the concept of marketing, in an integrated manner, in order complete the implications for future research.

INTRODUCTION

This article aims to make an innovative marketing approach of information from the market and business, understanding of information in organizations with or without profit. In this sense, it is necessary to clarify certain concepts in the field of information science. One of these concepts is the unit of information. Although very widespread in the literature, is not yet widely defined. Guinchat and Menou (1994) use this term to refer to the companies that have the mission: *to identify that can, as accurately as possible, information that can be useful to managers in support of decision making*. (337)

METHODOLOGY OF APPROACH

This is a descriptive and analytic approach seeking to know and analyze the existing scientific or cultural contributions on this subject, from the review of the existing literature. The survey was structured based on the systemic approach, seeking to the understanding of the problems of post modernity in the work of Erikson (1998), looking for practical, operational or troubleshooting application of "real life" organizations.

DOI: 10.4018/978-1-5225-6225-2.ch005

FUNDAMENTAL CONCEPTS

Data, Information, and Knowledge

In a global economy, information and knowledge are the greatest competitive advantages that organizations can have (Davenport & Prusak, 1998). In recent years, we have witnessed a remarkable transformation in society have quickly transitioned into a society based on industry and transport into a knowledge-based one.

One major challenge for management is to understand what information means: how to manage and interpret it and what decisions it allows to make in an era of world-wide communications, since information is the link that unites us. By being able to transmit large amounts of information quickly across continents, we transform the world into a global metropolis.

Another challenge facing managers is the wealth of information in present day society, whose most obvious distress signal in this kind of society is the combination of the production of large amounts of information, the intensive use of information technologies and communication, and the continuous learning process. The articulation of these three aspects suggests that the information society quickly transitioned into the society of knowledge. The symbolic culture of this society requires new types of learning, organization and management and, therefore, also one for information management.

In the information and knowledge society there are several hierarchical levels or progressive stages directly related to the learning process of this knowledge. Thus, we can distinguish three stages: data, information, and knowledge. The messenger on foot or on horseback gave way to the highways of information. So, what is information all about, then? Whatever resource may become valuable to be compiled, saved, duplicated, sold, stolen and even sometimes, a motive for murder.

Many people in organizations spend their workday gathering, studying and processing information. Some industries have been developed based on the information resource to produce technology (process technology - the computer, product technology - software and communication technology - communications equipment + software) in order to store, process, transmit and easily access information.

Managers cannot open a newspaper without being confronted to the term "information." Countless books contain the word "information". Lots of people in organizations perform activities related to the word "information". It seems easy to describe what it consists of. However, when we start thinking about the term "information", we experience some difficulty in finding the appropriate definition. Part of the difficulty for managers is in understanding information lies in the fact that they are so accustomed to dealing with it on a day-to-day basis, that they do not realize the complexities involved. Managers only realize difficulties when they are confronted with a new language. The potential for misinterpretation is always present.

Given the vital role communication plays within organizations, those involved in decision-making must find ways to minimize the likelihood of error. To do this end, it is necessary to understand how communication unfolds - how information is transmitted from person to person, from computer to computer, and between people and computers. The need to understand information - what it is and how it flows - is not just limited to large organizations. Whenever a person communicates with another one, we have a flow of information since communication is a means to provide information to others. Although used with much interchangeable frequency, the terms "information" and "knowledge" do not mean the same. "Information" is not the same thing as "data", although the two words are frequently confused. Hence, it is essential to make a distinction of the subtle differences between these two concepts.

Data are generally provided by newspapers, reports and computer software. For example, a list of the Stock Exchanges on a financial news page constitutes data. When data is acquired and brought into a global information structure previously attained, these data are converted into information. So, when someone reads the newspaper Exchange price list, one gets information about various companies. What allows managers to obtain information from a journal data is their previous knowledge of what these values mean and how the stock market works.

Data do not carry meaning or significance of facts, images or sounds, as they lack relational elements essential for the establishment of a complete meaning, where an internal relational structure for a cognitive purpose is absent; that structure is one of the attributes of the information. Data are transformed into information when their creator assigns meaning to them (Davenport & Prusak, 1998.)

Data are partial representations of facts, images or sounds and have no meaning by themselves, as they do not lead to the understanding of facts or situations and are converted into information when introduced into an information structure (global) already acquired (Alter, 1992; Davenport & Prusak, 1998; Devlin, 1999.) The concept of information as data is commonly seen in the field of scientific research. In this view of information as data with assigned meaning (Checkland & Howell, 1998) or simply:

Information = data + meaning (Devlin, 1999)

Information is the result of adding the data of a specific pattern of relationships establishing its format. Acting upon information, is not only to act on the data which make it up, but also act on the relationships that are established, i.e., on the collective or individual formatting standards and through them upon the way reality is perceived and the follow-up action.

Information is a quantity that measures and/or reduces the uncertainty: that is, everything that reduces the uncertainty experienced by an observer in relation to the occurrence of an event. Information is the product of the course of data transformation and relationship, or of other information operations, by means of which the significant conceptual relations between elements are summed up, considering a particular purpose of communication. Information refers to the body of facts and / or events in a convenient decision-making format within a context that defines the relationships between the data (Zikmund, 2000).

The concept of information can be understood from many different perspectives. Information is an object created by men to represent an identifiable event for them in the real world, integrating and linking a set of records or data, (Le Moigne, 1979; Alter, 1992). It is the difference that makes the difference (Bateson, 1972).

In terms of equation:

Information = Data + meaning

When information is internalized to the extent that can be used, it is called knowledge (Zikmund, 2000). It is a fluid mix of framed experiences, values, contextual information and expert insight, and provides a framework for evaluating and incorporating new experiences and information. In organizations, it is not only found in documents and reports, but also in organizational routines, processes, practices and norms. Knowledge originates and is applied in the minds of experts (Davenport & Prusak, 1998; Zikmund, 2000).

Knowledge is the information taken as valid and accepted by integrating data, actions, information and sometimes assumptions. The need to know requires that someone get, combine and interpret information.

Information can be regarded as a "substance" that can be acquired, stored and owned by a person or a group and transmitted from person to person or from group to group. Information has certain stability and it is, perhaps, the best thought of as existing at the level of society (Davenport & Prusak, 1998). In consequence, information may be feed knowledge.

Knowledge = Internalized information + ability to use it in new situations.

Thus, knowledge is a mix of information, experience and knowledge that provides a structure that can be applied to evaluating new information or new situations (Zikmund, 2000). Knowledge lies essentially and intrinsically within people who, at individual level, are far more complex and unpredictable than society as a whole. Consequently, it is not surprising that knowledge results much more difficult to attain than information. Knowledge exists primarily within people; it is an integral part of human complexity and unpredictability (Davenport & Prusak, 1998). Knowledge has a fundamental duality: it is something storable (at least sometimes we intend to do so) and something that flows (something that is communicated from person to person). Quite possibly, it is this duality of knowledge (something that flows and the storage process) what makes its handling and management difficult.

Competitive Intelligence

According to Ferreira (2004), there are two definitions of the term "intelligence": one related to the Latin word *Intelligentia*, which corresponds to College to learn, understand, and the other related to the English term which means *Intelligence* information service. These definitions are in accordance with the definition of Fuld (1995), in which intelligence is the result of an information service, through a process of understanding the context, providing information that serves to support decision.

Among the various existing definitions in the literature, the SCIP settings – *Strategic and* Competitive Intelligence Professionals and ABRAIC-Brazilian Association of competitive intelligence analysts in that competitive intelligence it is a process of management of competitive environment and analysis of results, under the prism of domestic issues, to support decision-making.

The competitive intelligence helps the managers of any organizations take the decision on information from various sources. It is a continuous process that involves legal and ethical information collection, analysis and dissemination of intelligence-controlled to the decision makers.

According to ABRAIC (2011), intelligence is a permanently specialized activity with the goal of making available information (intelligence) of interest to a particular organization, and the safeguarding of this information against adverse action of any kind. It is an activity (of intelligence) focused on the world of business, i.e. to the competitive environment, and seeks the maintenance or development of competitive advantage over competitors.

Being a service of information, competitive intelligence covers various technical skills of information professionals: understanding the information needs of the decision makers, technical collection and processing of information from several types of sources (human and documentary), analysis of this information with a focus on solving the problem of decision making, production and dissemination of information of high added value, among others.

Information Science

Borko (1968) defined:

information science as a discipline that investigates the properties and behavior of the information, the forces that govern your flow and processing to optimize its accessibility and use. It relates to the body of knowledge pertaining to production, collection, compilation, storage, retrieval, interpretation, transmission, processing and use of the information.

According to Borko (1968) the information science investigates the properties and behavior of the information, the forces governing the informational flow and means of the processing and information processing for its optimization of access and use. It is an interdisciplinary science and related to various fields derived as mathematics, logistics, Linguistics, psychology, computer technology, search operations, graphic arts, communications, library science, management and other similar fields.

It has a component of pure science, which investigates the subject without taking into consideration its application, as a component of applied science, which develops services and products. Librarianship and documentation are applied aspects of information science (Borko, 1968).

For Goffman (1970), the objective of information science is to establish a unified scientific approach to studying the various phenomena that involve the notion of information, if such features are found in biological processes in human existence or on machines created by humans. Consequently, the subject must be related to the establishment of a set of fundamental principles that govern the behavior of the whole process of communication and its associated information systems.

Saracevic (1991), studied the evolution of the information science and defined it as:

a field dedicated to scientific and professional practice issues facing the problems of effective knowledge communication and record-keeping of knowledge among human beings, the social context, institutional or individual use and information needs. The treatment of these issues is considered of particular interest the advantages of modern information and communication technologies (ICT's).

The information science was born after World War II to solve a big problem, which was also the major concern of both the Documentation and the Information Retrieval which is to gather, organize and make accessible the cultural, scientific and technological knowledge produced in the world. The information science is an exact science and was born to achieve an exact knowledge from the inspiration of mathematical and quantitative models (Bronowski, 1977).

It is based on objectivity sought to formulate universal laws of the "behavior" of information. It is strongly influenced by the empirical sciences intended to establish universal laws that represent the informational phenomenon and hence the need for resorting to mathematical models (information theory), physical (Entropy) or biological (epidemiological theory).

Saracevici (1991) studied the evolution of the problem that guides the information science and redefined it as:

a field dedicated to scientific and professional practice issues facing the problems of effective knowledge communication and record-keeping of knowledge among human beings, the social context, institutional or individual use and information needs. In the treatment of these issues, are considered of particular interest the advantages of international modern informational technologies.

MARKETING

During the first thirty years of the last century, the success was achieved by the companies that offer the lowest price. The products were generally not distinguished and the ability to produce the lowest price was the secret of success. However, in the 1930's, the demand for consumer goods was walking to the saturation and, as such, the customers/consumers have begun to be more demanding than just the basic performance.

According to Igor Ansoff (1978) in early years of the decade 1930's, General Motors has triggered a shift of production to the mentality of the market. This change symbolized the displacement of standardized products for differentiated ones. In contrast to the guideline for the production the new secret of success began to move to the "marketing orientation". Promotion, advertising, sale and other ways of influencing the consumer became primary concerns of managers.

The shift toward marketing orientation meant an offset of an introverted perspective and, turned into it to an extroverted and open perspective. Also, it meant a change in the mentality of the production managers for those with marketing mentality, which led to a struggle for power and the need to acquire new skills to find new solutions of problems, modifications of structures and systems, and the acceptance of new levels of uncertainty about the future.

It was only after the Second World War that many industrial enterprises were driven by new technologies, primarily toward a belated marketing orientation, and then towards the more turbulence post-industrial era. At this point, the total marketing concept established a balance between conflicting demands of marketing and production.

The literature review takes us from the School of Scientific Management of Taylor, in the first decades of the last century, until the evolution of the concept of marketing today, whose focus is centered on customer marketing and positioning of the company. The Delta model Hax (2001) characterizes the evolution of organizations with more product-centric focus, through a more recent step of customer orientation and market, to arrive to new forms of relationship with customers, competitors and to ensure a sustainable, competitive positioning in the context of globalization.

According to Wang Mattar (1986), is found that organizations are at different stages of development, which are in direct correspondence with the degree of application of Marketing techniques:

- **Organizations at the First Stage:** Are just producing products and / or services facilitators, and are geared little value to them;
- **In the Second Stage:** Organizations incorporate the boosters to produce goods and services and complement basic offer facilitators and are geared to the needs related to the health and well-being of clients, but have a strong dependence on brokering.
- **Organizations at the Third Stage:** Are facing a combined offer of products and or services. Are organizations that add greater value to the products and / or services.

- **Organizations in the Fourth Stage:** Are configured in both vertical and horizontal network, including suppliers, distributors, public and private organizations that seek to meet the needs of customers and retain them.

Are these globalization and being part of this phenomenon, the processes of internationalization of large organizations control almost the entire market, leading to a strong concentration and that, in practice, this network covers the major developed countries.

Intelligence Units

The term unit of information, although very widespread in the literature, it is not yet widely defined. Guinchat and Menou (1994) use the term to refer to companies that have as their mission: *to identify as accurately as possible, the information that may be useful to managers to support in decision-making.*

SA (2009, p. 17) refers to the units of information, *as a generic term to denote different types of libraries or information of business organizations.* In the author's conception, the use is justified, as the libraries and / or business organizations increase their field of action with regard to the management of information stored not only in media (books / paper), but also in electronic, digital and virtual props.

Sueli Amaral (1998) defines intelligence units like:

(...) any type of organization active in the area of information and / or documentation to work with the records of knowledge in any kind of support, regardless of their designation. They are considered units of information, all types of libraries, business units / services documentation center, according to its area of operation and extension. (15)

Rosenfeld and Morville (2002) proposed a model that represents the information architecture as an intersection of context, content and decision makers / users which is seen in Figure 1.

For authors, each element of the model has the following meaning:

- **Context:** All organizations have mission, objective, strategies and organizational resources (human, financial, material, technological and informational). It is necessary to understand the goals

Figure 1.
Adapted from Rosenfeld and Morville (2002), "Information Architecture for the World Wide Web, O'Reilly, USA.

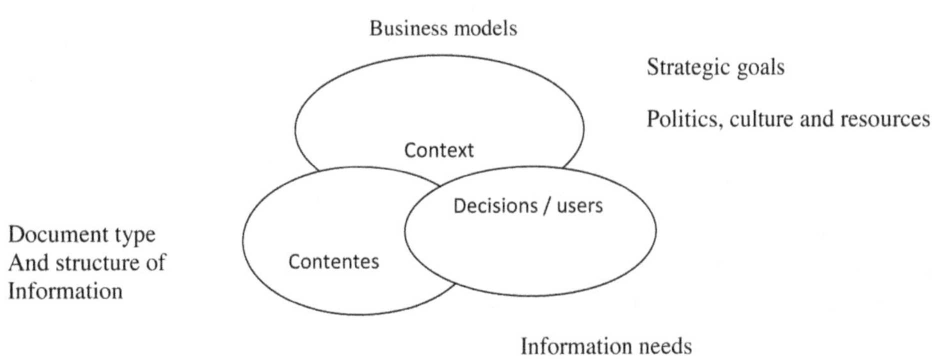

of the organization behind the informational space and resources available for the design and implementation of information architecture, This must be shaped considering the peculiarities of each context.

- **Contents:** Over the structures of representation and organization of informational content and their potential contemplates the nature, volume and available sources of content and its potential growth;
- **Managers/Decision Makers/Users:** Is needed to understand their roles and responsibilities, as well as their information needs that are extremely variable and influence behaviors of search for new information;
    - Operational managers are large consumers of information, daily of organizations;
    - Interim managers are consumers, but also already interpret and make decisions on the resources made available to their area of responsibility or business unit;
    - Top managers, are primarily interpreters of trends on the global and immediate surroundings, as well as skills and internal capabilities-which information (strong signals and or weak) in which they will support in the definition of the strategy and global objectives, monitor the implementation of the strategy and the achievement of pre-established objectives.

Gilchrist and Mahon (2004) present a more pragmatic definition focused on the organizational context, i.e. a plan whose goals are to provide relevant information to the right people at the right time. Also, include the need for integration of the production, organization and use of information from a common ontology.

According to the model proposed by Beijerse (2000), the professional information must understand that the information cannot be considered only as a commodity, since it is not limited to a single product or raw material for domestic use. The information should be studied as an essential element that allows the transformation of society.

According to Angelo, e. s., Ziviani, f., (2011), units of information are:

social enterprise, including economic, political, cultural factors and educational, involved with the management of information, in order to facilitate intellectual control and the logic of the informational assets through the collection, selection, treatment storage and distribution of information to increase the understanding of the quitrent on the statement. Also reduce their uncertainty the difference between the amount required and available for decision-making, about a particular subject.

According to the authors, the aim of the unit is to disseminate information needs and provide the customer the correct care for the effective use of their records between humans. Offer services of tangible form (printed) and / or intangible (customized, personal, virtual) to the organization that maintains and do not exist in isolation, which may or may not-for-profit (Tarapanoff, 2006).

Regardless of your level of expertise or extension, can be: any kind of library and archive, since traditional virtual and digital information analysis Center or documentation, Engineering draws, video libraries, provided that it has as its mission the provision of informational inputs to the process of evolution (Amaral, 2008; Ramos, 1996).

Information is the raw material of fundamental importance in the generation of knowledge to boost the development of society. How to develop any activity or project, whatever the stage of the work, without information?

MARKETING AND INFORMATION SCIENCE

The Marketing is present in all professional areas and in information science, although some rejection. However, it is stated that the marketing of the products and / or services is useful to solve the problem of the lack of visibility (information). Information workers seek the teachings of the Marketing to apply them in information management and to ensure better performance and the success of organizations Many reasons can explain this behavior, among them is the limitation of Marketing promotion, which is the visible part of the Marketing (Amaral, 2001).

Make the disclosure to promote the products and / or services, is not enough, because the Marketing is more than that. The Marketing is a function of management, that is, is a social function and a set of processes that involve the creation, communication and delivering value to customers, as well as the management of relationship with them, to benefit the organization and its interested public (American Marketing Association, 2004).

Although he appeared in for-profit organizations, since 1969 when Philip Kotler and Sidney l. Levy published their article in the Journal of Marketing, it is admitted the possibility of non-profit organizations adopt marketing techniques. The adoption of these techniques depends on the research and study on the adoption of Marketing in the specific context of organizations.

Borko (1968) admits that information science is an interdisciplinary science related, among other areas of knowledge and, therefore, is the business sciences. As Marketing is a process management approach and therefore is related to information science about information management. This association allows you to consider the term information Marketing.

According to Amaral (2004), the study of the information Marketing refers to the organization and to the systems involved with the processes of management of information relating to the various stages of the cycle of information, in particular, the research, collection, processing, distribution, analysis/ interpretation to support in decision-making on the part of managers.

The author believes that the organization of information and systems involved are considered units of information in organizations, with or without profit. Kotler and Keller (2006) include the information in their studies, when you consider that marketers are involved in marketing products and services, events, personal experiences, places, property, organizations, information and ideas.

If information workers do not stay awaken to the consolidation theory information science, they are losing the opportunity to highlight to society the importance of its role in information management for the featured deserved. They are not recognized by our society and the social role of the intelligence units and their managers as agents of society transformers.

Among scholars and researchers of information science, resistance to adoption of marketing techniques is a reality. But despite this strong rejection, paradoxically, it is said, too often, for example, that the marketing of library or the problem of the lack of visibility of the information resulting from the lack of marketing.

THE MARKETING AND INFORMATION SOURCES

According to Cunha (2001), information sources may include manuscripts and printed publications, as well as objects that transmit information. They can be divided into three categories: primary, secondary and tertiary documents.

Choo (1994; 2006) classifies the information sources in four categories: external and personal sources, external and internal, personal and impersonal, and internal and impersonal.

According to Pacheco and Valentines (2010), the categorization of sources of information allows you to understand the scale of each source, i.e. the primary sources express the direct interference of the author; the secondary sources make use of knowledge of primary sources, since there is a different treatment of them according to their function; tertiary information sources enables primary and secondary sources to be found.

Raj (2009) groups the sources of information on:

- **Personal External Sources:** Peers from other organizations, experts, customers, competitors, consultants, partners, scientific and commercial events (face-to-face or virtual interaction);
- **Internal Personal Contacts:** Employees/employees, coworkers, superiors, partners (face-to-face or virtual interaction);
- **Personal Electronic Sources:** Electronic mail (personal or company), forums, discussion groups on the Web, Messenger, Skype and similar;
- **Impersonal External Sources:** Documents produced out of organizations (magazines, newspapers, books, reports, periodicals, technical regulations, governmental publications, radio broadcasts and TV);
- **Impersonal Sources:** Documents produced within the internal organizations (reports, studies, reports, memos, files, etc.);
- **Impersonal and Electronic Sources:** Electronic documents in General, intranet, electronic database, site of the Organization, several Internet sites, news portals, etc.).

Basu and Barbosa (2009) share the Internet information sources in various sectors: mailing lists, emails (newsletters and Marketing), newsletters, chats, instant messages, search sites or search tools, intranets, extranets and Web sites.

Tomaél et al (2004), define ten criteria to evaluate the quality of information sources on the Internet: identification of sources of consistency in training, reliability of information, adequacy of supply, internal and external links, and ease-of-use of the information, layout, source restrictions perceived and user support.

Eppler (2006), states the quality of information sources on the Internet presents a few problems, such as information overload, misunderstanding of information, bad judgment and the misuse of the information.

According to Petro (2008), sources of information vary according to the areas and professional groups, types and degrees of need, in accordance with the purpose and informational situation. With the identification, classification, selection and organization of information sources, can develop the process of using these sources, in the different activities of the organizations.

It can be defined as information source all that generates or conveys information, that is, any object or to respond to a need for information on the part of whoever needs, including products and services of information, people or network of people, computer programs, digital media, sites and portals.

We present other conceptual contributions about sources of information in Table 1.

The lack of quality sources of information is associated with trust, security, loyalty and truthfulness of the sources and the information is useful, contextualized, integrated, and easily accessible and in vary-

Table 1.

References	Main Issues
Davenport, 2000	The sources should be as varied and complex as possible
Sugahara and Jannuzzi, 2005	The information sources can be internal and external. Indoor fountains are from internal organs. The external sources are divided in: sources close to market activities, sources of professional character and expert and institutional sources.
Pereira and Barbosa, 2007	The information sources are classified according to the source, relationship, proximity and a half.
Alvarenga Neto, 2008	The mapping of the internal and external information sources face the complexity and multiplicity of the same.
Wensing, 2010	The sources of information have evolved over time: stones, Papyrus, paper, photographs, electronic and electromagnetic means, so the sources of information have become synonymous with informational resources available in digital format.
Barreto, 2010	The information itself is something immaterial. The information while merchandise exists only when is quantified, or it was the result of a technical condition of production, such as a book, an article, a song, a printed image, a work of art, etc.

ing formats. Another important aspect is the source of the information to be objective, concise, correct, direct, present in several literatures, magazine and recognized.

The use of information involves the selection (filters), technical treatment of the sources of information in response to a question, a problem solving, decision making, negotiation or understanding of a situation.

INFORMATIONAL MARKETING

According to Kotler and Keller (2006), for most organizations is not the problem, but rather, what to say (information), and how and how often to say it. This occurrence is Informational Marketing, which work all the variety of information involving the creation, dissemination and delivery of value to the target audience, as well as the management of the relationship that will benefit the Organization and the customer.

According to Angelo, e. s., Ziviani, f., (2011), the Informational Marketing prepares a perspective for sale, avoiding scares or excessive messages by offering guided by competence (Amaral, 2004):

- Know to interpret informational anxiety of audience;
- Be receptive to criticisms, suggestions and complaints, valuing the evaluations;
- Provide timely returns and courteously;
- Be effective, giving emphasis to the results and developing the capacity to set priorities;
- The entrepreneur must reverse the roles with the client and answer, among others, issues such as:
    - *What is the useful information to the customer making the decision to buy?*
    - *Why choose our products and or services to the detriment of competitors?*

Then put in place communication actions to publicize what it does, what it offers and why should you use your products and / or services, where you can buy and service information, technical support and the organization itself. The disclosure points that the benefits that the products and / or services can offer, rather than worry about only in describing their characteristics (Amaral, 2001).

The brand supported informational campaigns integrates the reference of the target audience, since it provides the accurate information for decision making. Even if there is no immediate need, the organization appears as one that should be sought when the necessity of the product and / or service arise.

According to Kotler and Armstrong (2007), customer/consumer can feel stimulated and got the offer promptly, once the customers/consumers go through five stages of buying decision: recognition of need; search for information, assessment of alternatives, buying decision and post-purchase behavior.

RELATIONSHIP MARKETING

According to Amaral (1998), information is essential to the practice of marketing because it is an indispensable factor to boost the development of society and an input of fundamental importance of generating knowledge which, in turn, makes it possible to, efficiently, meeting the diverse needs of the population.

The generation of knowledge-dependent marketing allows the satisfaction of needs, expectations and concerns of the population, questioning the marketing dependence with regard to information seems to be out of sense.

Until 1925, what prevailed was a guide to the production, whose concentration was in manufacturing efficiency. Between 1925 and 1950, the orientation for the production gave way to guidance for sales where the concentration was about selling products. Between 1950 and 1990, the prevailing orientation was the marketing orientation, whose concentration was about the needs and desires of customers.

These three guidelines were, essentially, the conquest of new customers for organizations. There research showed that keep customers was cheaper than gain new ones. These searches seem to have been a kind of "splitter of the waters" in marketing, since they generated a paradigm shift in posture adopted by then.

Winning new customers, who scored so heavily marketing culture until 1990, gave way to customer retention, in which knowledge-based segmentation of the customer happened to be of paramount importance. Organizations have come to realize that obtaining quality information about the client could generate competitive advantage.

This new point of view gave rise to the so-called non-profit relationship. The relationship marketing is, in fact, a new philosophy of organizational management based on acceptance of customer orientation and for profit on the part of organizations and the recognition that should be searched new forms of communication, establishing a deep and lasting relationship with customers, *prospects*, suppliers and all intermediaries, in order to obtain sustainable competitive advantage (Bretzke, 2000).

The relationship marketing is based on the learning relationship between the organization and their customers, *prospects* and suppliers to obtain competitive advantages. According to Geus (2000; Kotler, 2003), *the ability to learn faster than competitors may be the only sustainable competitive weapon*. In this relationship, the more you interact with these stakeholders, more you find out what they really want, offering customized products and services.

Morin (1991; p. 19), states that (...) to know a piece we can't isolate a Word, a piece of information, it's necessary to turn it to a context and mobilize our wonder, our culture, to reach an appropriate and timely knowledge of the same. The problem of knowledge is a challenge because we can only know as Pascal, if we know the parties; we know the whole in which they are located.

Barreto (2002), advocates the role of information as a tool to modify individual and group consciousness of the individual, since, thanks to the information; the man is placed in line with the memory of its past and its future prospects. From this perspective, the author claims that he has established a relationship between information and knowledge, since the information is so perceived and accepted, *placing the individual in a better stage of development, conscious of itself and into the world where is your individual Odyssey* (Barreto, 2002, p. 49).

Barreto (2002) emphasizes that the information is assimilated in an imaginary point of the present, but with strong reference to the past of the individual informational, combined with a considerable weight of its prospects and possibilities in the future. Barreto (2002, p. 50), argues that:

production of knowledge is a reconstruction of the mental structures of the individual through his cognitive competence, that is, a change in his stock of accumulated knowledge.

The point of view of Barreto (1994) is that the information to be set to: *symbolically significant structures with the ability to generate knowledge in the individual, and in your group or society.* (3)

Can deduce that meet the customer, more than simply information about it, requires the employees and managers of organizations generating concrete actions that modify their individual and in your group consciousness, in the sense of the customer. So, you can create them a predisposition to satisfy the expectations and desires of the customers and, thus, practice marketing more suited to their theory.

The information has assumed importance as a factor of competitive differentiation, becoming increasingly studied carefully in the marketing. It is known that the fundamental condition in the practice of marketing is knowing what customers of an organization want and need, and to offer them, as far as possible, what meets these expectations and desires. To satisfy this requisite, it is necessary to "listen" with certain frequency this audience and frequency this audience, so that the principle of marketing can be practiced in its fullness.

INFORMATION MARKETING IN ORGANIZATIONS

There are several factors that can influence or delay the adoption of market techniques in information management, based on the concept of information or business units, related to the practice of marketing: unethical by some researchers labeled, the lack of focus in the units of information and or business units on priority of managers, the lack of knowledge of marketing techniques on the part of managers, the paucity of literature on the application of marketing techniques, as well as the inadequacy of the economic theory of information.

According to Amaral (1998), there is a huge shortage of documents (articles, theses, dissertations) on this topic. The focus of the information marketing is not always the main highlight of the published documents. The theoretical concepts of Marketing are little discussed in the literature.

Due to technological advances, the information can be electronically disseminated and reach a growing audience in less time, but still faced with problems related to communication primitives. The optical market leads to a better understanding of what is the "business" fundamental question on the marketing orientation. When are known and use marketing techniques in the analysis and segmentation of the market, the design of business units / all information is easier in decision-making, because the activities are developed based on market studies, (interests, needs, expectations and desires of consumers).

If in practice, the marketing concept of information units and/or business units to consolidate the marketing theory of information, we need to encourage research in this area and increase the scientific production of researchers interested in this subject. The units of information need to be seen as business units, since the information is the raw material of knowledge, and both are intangible assets of strategic relevance in the information and knowledge society in which we live. Despite the information be considered a feature of organizations, represents a particular class of other resources.

According to McGee, Prusak (1984), the differences arising from the potential of the information and the challenge to manage it, since the information is infinitely reusable and does not deteriorate or depreciates and its value, is determined solely by the consumer and, as such, it is necessary to adapt the offer of information products and services to your interests and needs.

This reinforces the concept of information marketing since it emphasizes the importance of consumer's information and consumer client need for adequacy of provision of information products and services to their interests and needs. Customers/consumers are increasingly demanding and consider the informational products and services offered in the light of the added value that those products and services you can offer.

The new information and communication technologies allow the easy storage of data on goods, services and people, as well as the easy access to them. This development combined with marketing techniques makes it possible to group the customers/consumers by similar data, enable, on the one hand, to plan the different operations and marketing activities and identify business opportunities and direct and instant contact.

The goal of the marketing managers is to determine the needs and desires of potential customers/consumers (target market), to satisfy them with appropriate *design,* communication and distribution, in addition to a competitive price for the products and services. The marketing assumes an understanding of needs, perceptions, preferences, interests and favorites of customers/consumers, to satisfy them and by the behavior of the target audience, appropriateness of messages from the media, and costs of facilities, having the purpose to maximize their activities in the area in which it is applied.

Implementing the marketing concept for the information/business units is not always comprehensively focused by the design, as the assimilation of marketing concepts is often limited to promotional aspects, more difficult to conquer and maintain markets. Despite the limitation of marketing approach, in the dimension of the competitiveness of markets, promotion and communication of information assume role in the survival of organizations.

According to the model proposed by Sueli Amaral (2008), *each organization can assign a different importance and conduct marketing activities differently, depending on its philosophy of action.* The three dimensions can be summarized as follows:

- **Philosophical Dimension:** To ensure customer satisfaction;
- **Functional Dimension:** To contribute to the achievement of the exchanges between the Organization and the market;
- **Administrative Dimension:** Study, analysis, planning and implementation of marketing activities.

The philosophy of practice in units of marketing-oriented information is service-oriented and the relationship with customers, making emphasis in the Exchange, through administrative actions, aimed at the goals of organizations. Functional function relates to the ease of trade between the organization and the market, while the administrative dimension deals with the analysis, planning, implementation and control of marketing activities

The adoption of marketing techniques by the units of information shall correspond to the vision of the information market in the context of business information, which can only be obtained when they are conceptual and philosophical principles recognized in marketing.

According to Sueli Amaral (2008), there are various social actors in the information market and schematically understand the business of organizations and intelligence units or business which can be observed in the following figure:

INFORMATION MARKET AND BUSINESS INFORMATION

Kotler (1978, p. 37-38) defines as, *a distinct group of people and/or organizations that have resources that they want or can change for different benefits*. From the point of view of organizations, market "is a potential arena for the exchange of resources". Later, Kotler (2000, p. 140) presents the market definition as, *the set of current and potential buyers of a supply to the market*. He also states that the size of the market depends on the number of buyers that may exist for a given market.

To Kotler (2000, p. 31) there are the following types of market:

- **Producers-Buy Resources (Raw Materials, Labor, Financial Resources):** That transforms us into products and services. Sell the finished products to markets intermediaries who, in turn, sell them to the final consumer markets. These sell their skills and abilities with which receive money to buy goods and services;
- **The Government (Government Market):** Taxes to acquire assets of resource markets, producers and intermediaries to offer public services based on these resources acquired;

Figure 2. Business information

Global and Iimmediate Environment

Environment Global and immediately

Market Information / Information Units

Suppliers	Interested	Media	Customers
Offer of: . Products and -Services	. Organizations . Government . Unions . Families -Workers	. Press . Radio . TV . Internet . Etc.	. Looking for: -Products and -Services . Attendance: -behavior -satisfaction . Relationship: -Receptivity -Complaints -Suggestions

Request for information

Cost of Information

Information Business / Business Units

The information market is evolving and all organizations of the industry information need to identify new market segments, develop new products and services and new relationships with their target audience, using the marketing techniques and tools to succeed in a paradigm shift of society (Rowley, 2006). The paradigm shift reflects the changing nature of information services and the theoretical and practical perspectives of the marketing due to the emergence of the Internet.

According to Reboredo (2003, p. 3), the market information may be perceived in a comprehensive manner to the extent that the information can be:

- Recorded/encoded in various ways;
- Duplicated and reproduced endlessly;
- Transmitted through various media;
- Preserved and stored in various media;
- Measured and quantified;
- Added other information;
- Organized, processed/treated according to various criteria;
- Retrieved whenever necessary according to certain rules.

According to Wurman (1995, p. 47-48), the information marketing can be understood as follows:

- **Cultural Information Market:** Covers the areas of Arts, humanities and social;
- **Information of Market Average:** Covering the areas of communication and information science;
- **Information Market in Science:** Covers the area of science in all its fullness;
- **Information Market in Technology:** Covers the area of technology in all its breadth;
- **Information Markets of Reference:** Covers all areas of human knowledge, presented as a reference or support or vital statistics.

According to Valentim (2000), information markets are mutually supportive in the production of information in the different areas of knowledge to which they are linked, establishing a value chain that will grow whenever the information is produced.

Barreto (1996; 1999) does not accept the transposition of marketing concepts because, according to him, the information market has characteristics which are peculiar, as, in the studies already carried out it is the provision that determines the search of information.

However, based on market definitions presented by Kotler (1998; 2000), analysis of information laws proposed by Moody and Walsh (1999), and the vision of Rowley (2006), about information marketing, you can understand that the information market is that market where there is actual or potential interest in its own information. This market consists of all the people and organizations that have an interest in information, also have certain resources and needed to be exchanged for it. The information market is present in those who offer information (offer) as those that need and seek it (seek).

We need to reflect that the raw material of everyday work, information and knowledge are complex, changeable and diverse. The complexity may explain the lack of consensus on the outcome of the epistemological studies about the concept and definition of information and knowledge.

PROMOTION AND COMMUNICATION OF INFORMATION

One of the roles of marketing in organizations is the promotion; to make known the same and their products and/or services by potential customers/consumers making them attractive, as well as showing how they may be used where to buy them and highlight the main benefits in its use, as well as keep customers/consumers constantly informed.

According to Le Coadic (1996), the communication is the transactional process between two or more parties. It is, therefore, a complex process and its role is to convey information. Communication is the process, the mechanism, the action and the information is the product that will be reported.

Davis (1967), argues that the interdependence between the communication and the information is so strong that some people set the communication as a, *process of passing information and understanding from one person to another*. Drucker (2002), says that communication has become a major concern for managers, students and professionals of any organization.

The promotion makes use of communication to attract new consumers, sell ideas and strengthen brands, making it necessary that organizations have increasingly skills and competences in the development of their activities of promotion and communication of their products and services and values to conquer customers/consumers well informed and demanding in this global market.

Krieg-Sigman (1995), considers that all the promotion is communication, but all communication is production. The communication of information is addressed in the corporate communication. Corporate communication is a dynamic process that uses various methods and techniques of public relations, journalism, advertising, research, marketing, human resources that interact in both surroundings as internally.

The main goal of the communication in organizations, sometimes persuasively, is winning the sympathy, the credibility and the trust of customers/consumers. The adjective gives a persuasive manipulative communication even treat yourself to an educational action for the benefit of the citizen to whom the message is addressed. That's why organizations must be attentive to the processes of promotion and communication of information, since, *the information explosion requires effective communications* (Drucker, 2002, p. 130).

McCarthy (1950) defined the concept of Marketing-mix into four groups or four P's: product, price, place and promotion. Each variable is considered as an element of this compound. At that time, it was focused on product marketing. Other authors have proposed the inclusion of other P's marketing and various social and technological changes and impacts have been influencing and motivating changes in the 4 P's.

Richards (1980), proposed the 4 A's: analysis, adaptation, activation and avaluation. Lauterbom (1990), suggested the 4 c's: customer, cost, convenience and communication. Without considering the specific differences of each proposal, note that amendment has its origins in the essence of the concept, since in the 4 P's were centered within the organization and the other proposals consider the influences from outside to inside the organization (birth; Lauterborn, 2007, p. 50-74).

To Kotler and Keller (2006, p. 532), the problem is that companies don't communicate, but what to say, how to say it and how often say. Communications become increasingly difficult as, more and more companies try to attract the attention of consumers divided.

Las Casas (2006, p. 377), flip, confusingly the term promotion with sales promotion. *Promote relates to inform, persuade and influence the decision-making of buys from customers/consumers*. In this context the term is more comprehensive than sales promotion. It is considered as promotion activities: advertising, public relations, personal selling, or any other form of communication. A company may have good products, reasonable prices and a good distribution chain, but will only succeed if it has a good communication capacity. Kotler and Armstrong (1998), states that communicate with customers should not be left to chance, namely, promotion and communication inter-relating. Technological advances should be used in the practice of information marketing.

Amaral (1998), states that the information marketing is a process that takes place without communication and proposed the study of the process of promotion/communication of information as the Marketing activity in units of information. The proposal of the model considers the relationship of promotion with communication as something integrated, including the noise element in this process, facilitating the identification of barriers and difficulties faced in promoting the process as a whole, contributing positively to the effectiveness of the promotion in organizations.

The proposed integration of promotion with the communication meets the affirmations of Kotler and Keller (2006), when ensuring that modern marketing demands more than develop a good product, establishes an attractive price and makes it accessible to its consumers. In the process of communicating with customers/consumers and prospects, it is necessary to know what to say, who and when

INTEGRATED MARKETING COMMUNICATION

According to Kotler; Keller, (2006, p. 532), the marketing communication is the means by which organizations seek to inform, persuade and remind customers/consumers on products and services, from the brands they offer. With the technological evolution, propose the replacement of the mix-for promotional marketing communication, including: advertising, sales promotion, events, public relations, personal selling, direct marketing, so as to provide an approximation between the organization and the customers/consumers, i.e. everything communicates something to the customers/consumers and the contact with the brand conveys a message that can strengthen or weaken the vision of the client about the organization.

Note that there is a major change in society, leading to change in focus in relation to the factors of production and development of society. The basis of the transformation is that the information sector is knowledge-intensive, and not in labor. The added value of knowledge, when incorporated in products, causes the industrial processing of raw material information by its value added (Barker, Amaral, & Batista, 1994).

Castells (1999, p. 174-176), states that:

the informational economy is characterized by the development of a new organizational logic, which is related to the current process of technological transformation, but not up to him.

The big change is the transition from mass production to flexible production. The main dilemma of managers is to deal with the uncertainty due to changes in the macroeconomic environment, technological, political, social, cultural, etc.

According to Rowley (2002), in the digital environment, where operating digital libraries, electronic commerce and electronic business (*e-business*) the digital information has become increasingly important to play double role in marketing.

E-commerce relates the marketing with delivery of information products, such as electronic newspapers and magazines, *newsletters*, *e-books*, digital music, video and audio, *software*, images. Moreover, in the digital world, the service providers use the information as a substitute for personal contact (e.g., product catalogs, list of FAQ (Frequently Asked Questions), promotion of products and services to support customer relationships (Rowley, 2002).

Rowley (2006), highlights the marketing in organizations that offer products and/or services of information as a process by which information and knowledge needs of clients are met by means of adequate exchange of information about the products and services of these organizations. This involves the services of these organizations:

- Research on the needs of customers and other market factors;
- Select and define the target clients or groups of clients that seek to achieve information services;
- Set the offer in terms of products and or services associated with the marketing and do this as a reference for the potential value that the bidder can provide to customers;
- Offer products and services in accordance with the expectations and interests of customers;
- Ensure communication with the target customers;

To Amaral (2004, p. 59-60), the study of information marketing refers to systems and organizations involved with the processes of information management, relating to different stages of the cycle of information, such as, the sources, the collection, processing, communication and use of information relating to the supply of products and provision of information services in support of decision-making.

INFORMATION ON THE PRACTICE OF MARKETING

One of the principles of Marketing is the satisfaction of expectations, aspirations and desires of consumers of goods and services to satisfy their needs. The issue of needs and desires requires understanding of these concepts, since there are differences between them: someone needs something and this is manifested in such a way that can be interpreted (Man, 2000, p. 3).

For the economy, the need is *the sense of a certain lack the desire to be eliminated.* The need is *the desire to have a means to prevent to end an unpleasant feeling or to provoke, maintain or increase a nice feeling* (Sabaté, & f. Tarragó, 1989).

The media that serve people to meet their needs are designated by economists like goods. A *"well"* is what we recognize as able to satisfy a need, regardless of any other judgment (for example moral judgment). The goods can be classified in several ways. The first interest is that distinguishes the so-called free goods (ex: the air we breathe, the water) of economic goods. A *well* is an economic one that is characterized by its scarcity, that is, is the one that exists in quantities less than those required, given the need felt (Sabaté, & f. Tarragó, 1989).

The usefulness of a good refers to the quality and capacity of this well in meeting the needs of its own that well. The usefulness of an asset is the ability of this well to meet the needs, and the needs are subjective sensations. It follows that the same well can have different use for different people, depending on the intensity with which these may experience or feel the needs that can be satisfied with this well (Sabaté, & f. Tarragó, 1989).

Closely associated with the concept of utility, is the concept of "*economic value*", of which we can say that it is the quality of everything that gives importance, because it is considered that deserves esteem. So, the value of a given well also depends on each person, since each can feel distinct appreciation for a really good (Sabaté, & f. Tarragó, 1989).

Organizations are seeking the economic return through the sale of their products or services on the market. In a competitive environment, an organization can achieve better performance than their competitors, because it has better features, or because the organization makes better use of its distinctive skills (Penrose, 1959). A distinctive competency is defined as a distinguished qualification, complementary assets and the organization's routines that together enable the company to coordinate a specific set of activities that provide the basis for competitive advantage (Dosi & Teece, in: Williamson 1999, p. 1094).

Competitive advantage is referred to as an economic profitability (Porter 1980). Subsequently, the resource base sees the strategy as a continuous research of profitability. Profitability is then defined as the excess return of a resource that you are the owner of an opportunity cost (Mahoney & Pandian, 1992). In other words, profitability can be measured as a normal financial return.

In order to differentiate the possible sources of profitability, some types are distinguished. For example, the return can be performed by (Mahoney & Pandian, 1992):

- Superior ability to manage to coordinate the resources (Penrose, 1959);
- Possession of a valuable resource which is rare;
- Government Protection when barriers to entry of new competitors are high;
- Take risks and business acumen in uncertain and complex immersive media;
- Make better use of the resources of the organization, of any of the physical assets and human capital.

In short, the resource base is an attempt to explain why the organizations differ in terms of features and capabilities, and how these differences can lead to positions of sustained profitability, producing superior financial return. Base resources, therefore, serve the purpose of focusing explicitly the function of resources and the capacity of organizations, such as the origin of strategy and organizational performance. This exploration of the relationship between resources, competition and profit includes, among other issues, the role of imperfect information in the creation of differences in profitability between competitors (Grant 1991; Barney, 1986; Itami, 1987).

CONCLUSION

The adoption of marketing techniques in organizations, the concept of information market, essentially depends on the knowledge that marketers have about the marketing concepts in the field of information science and the true understanding of the role social, political, educational, cultural and informational professional of information that goes beyond the rules of technical treatment without the belittling.

Recognizing the need for effective treatment of information, regardless of the physical medium and format of presentation, its focus on user/decision maker, considering it target audience to be served by your offer as long as you are facing the business and the information market in which it operates. So, the organization must be concerned with maintaining a two-way relationship with its target audience (customers and/or consumers) to meet their interest profiles for information products and services to be offered.

REFERENCES

Amaral, S., & Silveira, A. (1993). *Marketing em unidades de informação: estudos brasileiros. Organizado por Amélia Silveira e Sueli Angelica do Amaral*. Brasília: IBICT /SENAI.

Amaral, S. A. (1998). *Marketing: abordagem em unidades de informação*. Brasília: Thesaurus.

Amaral, S. A. (2001). *Promoção; o marketing visível da informação*. Brasília: Brasília Jurídica.

Amaral, S. A. (2004). *Marketing da informação na Internet; ações de promoção. Colaboração de S.P. Santos e W.J.Araújo*. Campo Grande: Editora da UNIDERP.

Amaral, S. A. (2008). Marketing da informação: Entre a promoção e a comunicação integrada de marketing. *Informação e Sociedade Estudos, João Pessoa, 18*(1), 31–44.

American Marketing Association. (2007). *Dictionary of Marketing Terms*. Disponível em: http://www.marketingpower.com/mgdictionary- view1862.php

Ansoff, H. I. (1978). *Strategic Management*. Macmillan.

Barney, J. B. (1986). Strategic Factor Markets: Expectations, Luck and Business Strategy. *Management Science, 32*(10), 1231–1241. doi:10.1287/mnsc.32.10.1231

Barreto, A. de A. (1996). A eficiência técnica e econômica e a viabilidade de produtos e serviços de informação. *Ciência da Informação, 25*(3).

Barreto, A. de A. (1999). Oferta e a demanda da informação: condições técnicas, econômicas e políticas. Ciência da Informação, 28(2), 168-173.

Barreto, A. (2000). O mercado de informação no Brasil. *Informação & Informação, 5*(1), 25–34. doi:10.5433/1981-8920.2000v5n1p25

Barreto, A. de A. (1994). *A transferência de informação, o desenvolvimento tecnológico e a produção do conhecimento*. Disponível na Internet no endereço: http://www.alternex.com.br/_aldoibct/infor/informa.html

Barreto, A. de A. (2002). Transferência da informação para o conhecimento. In O campo da ciência da informação: gênese, conexões e especificidades. João Pessoa: Editora Universitária, UFPB.

Bateson, G. (1972). *Steps to an Ecology of Mind*. Nova Iorque, Balentine Books.

Bateson, G. (1972). *Steps to an Ecology of Mind*. Nova Iorque, Balentine Books.

Beijerse, R. P. (2000). Knowledge management in small and medium-sized companies: Knowledge management for entrepreneurs. *Journal of Knowledge Management, Bingley, 4*(2), 162–179. doi:10.1108/13673270010372297

Borko, H., (1968), Information science: what is it? *American Documentation, 19*(1), 3-5.

Botelho, T. M., Amaral, S. A., & Baptista, S. G. (1994). Information and society: An inteligent society in transformation? *International Journal of Information and Library Research, Londres, 6*(1/2), 64–76.

Bronowski, J. (1977). O senso comum da ciência. In *Belo Horizonte*. Editora da USP.

Brum, M. A. C., & Barbosa, R. R. (2009). *Comportamento de busca e uso da informação: um estudo com alunos*. Heidelberg, Germany: Springer.

Castells, M. (1999). *A sociedade em rede*. São Paulo: Paz e Terra.

Checkland, P., & Holwell, S. (1998). *Information, systems and information systems: making sense of the field*. John Wiley & Sons.

Choo, C. W. (1994). Perception and use of information sources by chief executives in environmental scanning. *Library and Information Science Research. Boston*, *16*(1), 23–40. doi:10.1016/0740-8188(94)90040-X

Choo, C. W. (2006), Research in information and Knowledge Management. Brasília: PUC/DF, 29 abr. Palestra.

Cunha, M. B. (2001). *Para saber mais: fontes de informação em ciência e tecnologia*. Brasília, DF: Briquet de Lemos.

Davenport, T., & Prusak, L. (1998). *Working Knowledge*. Cambridge, MA: Harvard University Press.

Davis, K. (1967). Human relations at work: The dynamics of organizational behavior (3rd ed.). Nova Iorque: International Student Edition.

Devlin, K. (1999). *Info Sense: turning information into knowledge*. W. H. Freeman and Company.

Dosi, G. (1996). The Contribution of Economic Theory in the Understanding of a Knowledge-Based Economy. In Foray & Lundvall (Eds.), Employment and Grow in the Knowledge-Based Economy. OCDE.

Drucker, P. F. (2002). *O melhor de Peter Drucker: obra completa. Tradução de Maria L. Leite Rosa, Arlete Simille Marques e Edite Sciulli*. São Paulo: Nobel.

Eppler, M. J. (2006). Managing information quality: increasing the value of information in knowledge-intensive products and processes (2nd ed.). Heidelberg, Germany: Springer.

Ferreira, A. B. H. (2002). *Novo Aurélio século XXI*. São Paulo: Nova Fronteira.

Fuld, L. M. (1994). *The new competitor intelligence: the complete resource for finding, analyzing and using information about your competitors*. John Wiley & Sons.

Goffman, W. (1970). Information Science: Discipline or disappearance. *Aslib Proceedings*, *22*(12), 589–596. doi:10.1108/eb050268

Grant, R. (1991). The Resource-based Theory of Competitive Advantage: Implications for Strategy Formulation. *California Management Review*, *33*(3), 114–135. doi:10.2307/41166664

Grant. (1991). The Resource-Based Theory of Competitive Advantage: Implications for Strategy Formulation. *Californian Management Review, 33*(3), 114-125.

Guinchat, C., & Menou, M. (1994). Introdução geral às ciências e técnicas da informação e documentação (2nd ed.). Tradução de Miriam Vieira da Cunha. Brasília: IBICT.

Hax & Wilde. (2001). *The Delta Project. Discovering New Sources of Profitability in a Network Economy*. Palgrave.

Homem, M. L. (2000). *Necessidade e desejo*. Disponível na Internet no endereço: www.psico.org.br

Itami, H. (1987). *Mobilizing Invisible Assets. Massachusetts*. Cambridge, MA: Harvard University Press.

Keller, K. L. (2006). *Administração de marketing. 12* (p. 750). São Paulo: Pearson Prentice Hall.

Kotler, P. (1978). *Marketing para organizações que não visam o lucro*. São Paulo: Atlas.

Kotler, P. (1998). *Princípios de marketing. Tradução Vera Whately* (p. 527). Rio de Janeiro: Prentice Hall do Brasil.

Kotler, P. (2000). *Administração de marketing* (p. 764). São Paulo: Prentice Hall.

Kotler, P. (2003). *Marketing de A a Z: 80 conceitos que todo profissional precisa saber* (p. 95). Rio de Janeiro: Campus.

Kotler, P., & Keller, K. (2006). *Administração de marketing. 12*. São Paulo: Prentice Hall.

Krieg-Sigman, K. (1995). Kissing in the dark: Promoting and communicating in a public library setting. *Library Trends*, *43*(3).

Las Casas, A. L. (2006). *Administração de marketing: conceitos, planejamento e aplicações à realidade brasileira*. São Paulo: Atlas.

Las Casas, A. L. (2010). *Marketing de varejo. 4*. São Paulo: Atlas.

Lautenborn, R. (1990). New marketing litany: 4P's passé; C-words take over. *Advertising Age*, 26.

Le Coadic, Y. (1996). *A ciência da informação*. Brasília: Briquet de Lemos.

Le Moigne, J. L. (1979). Informer la décision ou décider de l'information. *E.S./ Sciences de Gestion*, (1), 863-888.

Mahoney, J., & Pandian, J. R. (1992). The Resource-Based View Within The Conservation of Strategic Management. *Strategic Management Journal*, *13*(5), 363–380. doi:10.1002mj.4250130505

Mattar, F. N. (1996). *Pesquisa de marketing: metodologia, planejamento*. São Paulo: Atlas.

Mc Gee, J. V., & Prusak, L. (1994). *Gestão estratégica da informação*. Rio de Janeiro: Campus.

Mccarthy, E. J. (1978). Marketing básico: uma visão gerencial. Rio de Janeiro: Zahar, 2v.

Moody, D., & Walsh, P. (1999). *Measuring the value of information: an asset evaluation approach*. European Conference on Information Systems. Disponível: http://wwwinfo.deis.unical.it/~zumpano/2004-2005/PSI/lezione2/ValueOfInformation.pdf

Moody, D., & Walsh, P. (2009). *Measuring the value of information: an asset evaluation approach*. European Conference on Information Systems, 1999. Disponível: http://wwwinfo.deis.unical.it/~zumpano/2004-2005/PSI/lezione2/ValueOfInformation.pdf

Morin, E. (1991). Da necessidade de um pensamento complexo. In *Para navegar no Século XXI*. Rio de Janeiro: Vozes.

Pacheco, C. G., & Valentim, M. L. P. (2010). Informação e conhecimento como alicerces para a gestão estratégica empresarial: um enfoque nos fluxos e fontes de informação. In M. Valentim (Ed.), *Gestão, mediação e uso da informação* (pp. 319–341). São Paulo: Cultura Academica.

Penrose, E. (1959). *The Theory of the Growth of the Firm*. Londres: Basil Blakwell.

Petró, B. (2008). Análise do fluxo informacional dos gestores turísticos da unidade de conservação Parque Natural Municipal Mata Atlantica de Atalanta – SC, p. 184. Dissertação (Mestrado em Ciência da Informação). Universidade Federal de Santa Catarina, Florianópolis.

Porter, M. (1980). *Competitive Strategy: Techniques for Analysing Industries and Competitors*. Free Press.

Porter & Millar. (1985, July). How Information Gives You Competitive Advantage. *Harvard Business Review*.

Richards, G. (1996). Production and Consumption of European Cultural Tourism. *Annals of Tourism Research*, *23*(2), 261–283. doi:10.1016/0160-7383(95)00063-1

Richardson, R. J. (1999). *Pesquisa social: métodos e técnicas*. São Paulo: Atlas.

Robredo, J. (2003), Da Ciência da Informação revisitada aos sistemas humanos de informação. Brasília: Thesaurus; SSRR Informações.

Rosenfeld & Morville. (2002). *Information Architecture for the World Wide Web*. O'Reilly.

Rowley, J. (2002). Information marketing in a digital word. *Library Hi Tech*, *20*(3), 352–358. doi:10.1108/07378830210444540

Rowley, J. (2006). Information marketing (2nd ed.). Ashgate Publishing Limited.

Sabaté, F. T. (1989). *Fundamentos de Economia de la Empresa (2nd ed.)*. Barcelona: Librería Hispano Americana.

Sampaio, L. A. C. (2009). *Mapeamento das competências gerenciais necessárias aos gerentes das unidades de informação dos Tribunais Superiores do Poder Judiciário Brasileiro*. Brasília: Universidade de Brasília. Faculdade de Ciência da Informação. Dissertação de Mestrado.

Saracevic, T. (1991). Information Science: origin, evolution and relations. In *Conceptions of library and information science. Proceedings of the international conference for the celebration of 20th anniversary of the Departament os Information Studies*. University of Tampere.

Sugahara, C. R., & Januzzi, P. M. (2005). Estudo do uso de fontes de informação para inovação tecnológica na industria brasileira. *Ciência da Informação, Brasília*, *34*(1), 45–56.

Tarapanoff, K. (2006). Informação, conhecimento e inteligência em corporações: relações e complementariedade. In Inteligência, informação e conhecimento. Brasília: IBICT: UNESCO.

Tomaél, M. I. (2004). Critérios de qualidade para avaliar fontes de informação na Internet. In M. I. Tomaél & M. L. P. Valentim (Eds.), *Avaliação de fontes de informação na Internet* (pp. 19–40). Londrina: Eduel.

Valentim, M. P. (2000). O mercado de informação brasileiro. *Informação & Informação*, *5*(1), 35–42. doi:10.5433/1981-8920.2000v5n1p35

Weingand, D. E. (1987). *Marketing/planning library and information services*. Littleton, CO: Libraries Unlimited.

Wurman, R. S. (1995). *Ansiedade de informação: como transformar informação em compreensão*. São Paulo: Cultura.

Zikmund, W. G. (2000). Business Research Methods (6th ed.). Dryden Press.

Zorrinho, C., Serrano, A., & Lacerda, P. (2003). *Gerir em Complexidade*. Edições Sílabo.

Chapter 6
Patent Information Quality to Stimulate Innovations

Sérgio Maravilhas
Universidade Salvador, Brazil

Sérgio Goes Oliveira
Universidade Salvador, Brazil

Paulo Melo
Universidade Salvador, Brazil

ABSTRACT

Information, a tool to reduce uncertainty and develop knowledge in organizations, is an important aid in the decision-making process. There are several characteristics that describe the quality of information that will allow the analysis of its value. Quality information warrants us best results when competing with other organizations. Its value is related to the results that it will allow us to obtain, and it depends on the context. Patent information must be of high quality to permit the search and retrieval of the documents needed to solve a problem or stimulate new ideas and solutions. Old inventions can generate new ideas; technologies for one application can be introduced for a new domain and can be applied in ecologically sustainable solutions. The current high number of patent applications reduces the quality of patent information due to the time needed to filter and search for all the prior art available. Some standards, together with machine translation, have been set up to avoid this situation and improve the quality of the patent information retrieved by the interested public. A comprehensive survey of the relevant literature available made us aware that commercial databases supply some value-added information to help the researchers and improve the efficiency of the search queries. Some of these features could be applied by the national and international intellectual property offices.

INTRODUCTION

We are experiencing some important changes in our era, and the truth is that information plays an important role in it. Information has become the basis of the production system instead of the materials produced in factories and this has changed our way of life at all levels. We witnessed the beginning

DOI: 10.4018/978-1-5225-6225-2.ch006

of the production of information and knowledge in massive proportions with financial and economic interests (Drucker, 2000).

Some authors (Brown & Duguid, 2000; Castells, 2000; 2004; Webster, 2000) often talk about the knowledge and information based transformation of the world economy we are living in, with flux and flows of information gaining advantage to the exchange of goods (Godeluck, 2000).

Information, as a tool to reduce uncertainty and to develop knowledge in organizations (Best, 1996b; Kahaner, 1997; Porter & Millar, 1985) is an important aid in the decision-making process and must be of quality to improve its value (Best, 1996b; Beuren, 1998; Choo, 2003; Davenport, Marchand, & Dickson, 2004; Marchand & Horton Jr., 1986; Tapscott, 1999; Wilson, 1985; Wilson, 1987).

There are several characteristics that describe the quality of information that will allow the analysis of the value of the information used.

Quality information warrants us better results when competing with other organizations (Brophy & Coulling, 1996; Redman, 1996; Wormell, 1990) enabling the chance to get a competitive advantage.

Its value is related to the results that it will allow to obtain and it depends on the context (Best, 1996b; Davenport, 1997; Marchand & Horton Jr., 1986; Orna, 1999; Penzias, 1989; Tapscott, 1995).

Patent information, contained in patent documents, must be of high quality to permit the search and retrieval of the documents needed to solve a problem or stimulate new ideas and solutions (Albrecht, Bosma, van Dinter, Ernst, van Ginkel, & Versloot-Spoelstra, 2010; Brünger-Weilandt, Geiß, Herlan, & Stuike-Prill, 2011; Chakroun, 2012; Mueller & Nyfeler, 2011; Philipp, 2005; 2006; Scott, 2010).

Old inventions can generate new ideas (Adair, 2011; Jolly, 2003; Michalko, 1991; Petroski, 2008), inventions never marketed can become real products and satisfy needs and desires (Maia, 1996), inventions for one application can be introduced in a new domain (Buchanan, 2008; Ernst, 2003; Haberman, 2001; Koch, 1991) and technologies can be applied in ecological sustainable solutions to develop 'green' products (Dresner, 2008; Esty & Winston, 2008; Krupp & Horn, 2009).

Patent offices are acting to improve the quality of the patent applications and the patent information retrieved by the search queries performed by examiners, attorneys, inventors, scientists and entrepreneurs will have higher value-added.

Patent information value will be evaluated by the level of acceptance of those innovations by the customers in a given market where they are introduced (Kotler, Armstrong, Saunders, & Wong, 1999; Mohr, Sengupta, & Slater, 2010; Rogers, 1995; Trott, 2008; Utterback, 1994).

INFORMATION QUALITY

Quality is a difficult concept to define because what is quality for one person may be diferente for another. Wagner reminds us that, *the term 'quality of information' is scarce in the literature* (1990, p. 69).

As Ginman puts it there is no *generally accepted definition of quality information. (…)* Both

quality and information are in themselves difficult concepts to grasp and to evaluate, and the whole process is further coloured by the subjective views of the person making the evaluation (1990, p. 18).

Therefore, several different definitions of quality arise, *however, they all accept the notion that quality is defined by the customer* (Cortada, 1996, p. 6) and Redman suggests the need to *understand what customers want in their terms* (1996, p. 141). Other authors state that *quality is achieved when customer needs are met; quality is central to all organizations, not an optional extra; quality is not dependent on high price or high levels of resourcing* (Brophy & Coulling, 1996, p. 7).

Some managers state that they will recognize quality when they see it. That, although very subjective and dependent on individual perceptions and varying emotional states, may be possible if we are evaluating a tangible product. Applying this to immaterial and intangible goods – like data, information and knowledge - makes it more difficult.

Companies:

must define, measure, analyze, and improve the quality of information, treating information as a product because creating quality information and organizational knowledge is the prerequisite for any firm to gain competitive advantage (Huang, Lee, & Wang, 1998, pp. 5-6).

Relating the subject of information quality with business management and the need of quality information for good decision-making, Marchand describes the differences and common points that connect the information quality and the product quality. He concludes that *understanding the interrelationships among the dimensions of information quality is what is critical to strategic information management* (1990, p. 10). His information quality list comprises eight dimensions:

1. The actual value an information product or service may have for the information user;
2. The 'features' associated with an information product or service;
3. The reliability of the information product or service;
4. Its meaning over time;
5. Relevance;
6. Validity;
7. 'Aesthetics' and;
8. Perceived value (1990, pp. 11-12).

For most of the evolution of the information management role in business management function, information quality has not been a major strategic asset, but, *has been perceived and thought about as simply one aspect of decisionmaking (…) only one of the many dimensions of decision-making and often not a very importante one* (1990, p. 8). Thinking about the business value of information quality and its relation with profitability, some authors (Brophy & Coulling, 1996; Cortada, 1996; Garvin, 1988) mention that quality, in general, increases profits by gaining the consumers preference and obtaining a bigger market share. In information, the same occurs because quality lowers the costs of production and the adequacy of the information product or service to the user's needs makes it preferable to the ones who use it (Marchand, 1990, pp. 15-16).

Therefore, cost plays a significant role to lower the price charged to the user of that information.

Information Quality Criteria

To be considered of quality, information must meet several criteria such as: completeness, accessibility, accuracy, precision, objectivity, consistency, relevancy, timeliness, and understandability (Brünger-Weilandt, Geiß, Herlan, & Stuike-Prill, 2011).

Burk and Horton state that the quality of information itself includes attributes and examples, such as Accuracy–Precision; Comprehensiveness–Relevance; Credibility–Reliability; Currency–Simplicity; Pertinence–Validity (Burk Jr. & Horton Jr., 1988).

Other authors mention that information, to have quality, must be:

1. **Precise:** Correct and true;
2. **Timely:** Available when and where needed;
3. **Complete:** Ensuring the presence of its several elements;
4. **Concise:** Easy to handle;
5. **Authoritative:** Quality depends on the provider (Brown & Duguid, 2000; Cleveland, 1985; Cooke, 2001; Davenport, 1997; Davenport, Marchand, & Dickson, 2004; Hinton, 2006; Marchand & Horton Jr., 1986; McGee & Prusak, 1995; Penzias, 1989, 1995; Shapiro & Varian, 1999).

For Marchand the quality of information products depends on their attributes such as coverage, reliability and validity, which makes it highly measurable and quantifiable. Also, when looking at the production level, the quality of information is associated with *meeting requirements and with doing the job right the first time within budget, and on time* (1990, p. 9).

Huang, Lee and Wang divide Information Quality (IQ) in four categories, each one with several dimensions, as follows:

1. **Intrinsic IQ:** Accuracy, objectivity, believability, reputation;
2. **Contextual IQ:** Relevancy, value-added, timeliness, completeness, amount of information;
3. **Representational IQ:** Interpretability, ease of understanding, concise representation, consistent representation;
4. **Accessibility IQ:** Access, security. These authors consider that, defined by the perspective of the consumer, information quality *can be used by researchers and practitioners to direct their efforts for information consumers instead of the IS professionals* (Huang, Lee, & Wang, 1998, pp. 42-43).

If information is not delivered with the right quality, proportion and measure it can originate "information overload" (Dearlove, 1998), also called "infoxication" (Barbosa & O'Reilly, 2011), which occurs when excess of information suffocates businesses and causes employees to suffer mental anguish and physical illness, exacerbated nowadays by the huge amount of information from Web 2.0 and Social Media (Tapscott & Williams, 2008). The effects of the:

information glut are procrastination and time wasting, leading to the delaying of important decisions, distraction from main job responsibilities, tension between colleagues and loss of job satisfaction (Dearlove, 1998, p. 111).

Information overload causes high levels of stress that can result in illness and the breakdown of individuals' personal relationships.

Remember that *information is a bit like money: the more you've got, the more headaches it causes, but you still can't get enough* (Suhr, 2004, p. 41).

So, it's important to understand that customers will evaluate the quality of information according to its adequacy to the satisfaction of their needs.

After all, that's the reason why information is stored and disseminated, to satisfy someone's information needs. That is why it must be of high quality in order to be used and aggregate value to the users' tasks being performed.

INFORMATION VALUE

Burk and Horton define information value as *the value attributed to information produced or acquired by organizations, entities and persons, and delivered in the form of an information product or service.* For them, this value can be perceived immediately but, sometimes, this valuation occurs only several years later like in the case of *the values attached to information created by scientific research* that *are often realized long after it is created* (Burk Jr. & Horton Jr., 1988, p. 79).

For Orna, the functional part of valuing can be defined as *the process of determining and applying appropriate criteria for estimating the value of things* (1996, p. 19).

Wagner states that the *terms value and quality share a common meaning: the degree of excellence, but the former also denotes an economic exchange worth* (1990, p. 69), being the value of information and not the quality of information the preferred concept.

According to Best it is not easy to assign a value to information and there is no *commonly accepted and universally applicable way of valuing the information resource* (1996a, p. 14).

What happens is that an indirect valuation occurs, and the measures used are the value of the investment in information technologies used to hold the information, the costs associated with staff that collects and maintains the information resources, or the value the information has when it is put to a specific utilization. Orna also points out the fact that information has no implicit value in itself, since it depends on its use, purpose and context (1996, p. 20).

To explain why the process of valuing is so difficult, we must remember that,

fixing a value is always an indirect process that involves finding appropriate equivalents and standards, not necessarily or always in money terms, and the estimation of those who use it has to be taken into account as well (Orna, 1996, p. 19).

McPherson (1994, pp. 207-208) proposes three models for information valuing. The first one is related with the cognition process, in which the value is dependent on the contribution of information to the achievement of organizational goals. The second relates to the value of information when it enhances the knowledge not only in the user's brain but also in external, operational, and recognizable actions that are observable. The third deals with traditional accounting, highly cost oriented that does not recognize the value of information unless it can be sold externally as a good or product. Orna, to make this clearer, says that *to have value, information has to be transformed by human minds into knowledge, without which no products of tangible value can be created or exchanged* (1999, p. 141).

That happens because the valuation has three parts, the value triangle:

1. Object;
2. Human judgement;
3. Use to which the person judging puts the object.

It is a three-way relationship, involving human judgment, based on the relationship between the object valued and the person that judges and evaluates, and the relationship between that object and the uses that the human will give to the object. It also implies criteria, which involves thinking and feeling (1999, pp. 143-144).

What we can observe is that it is not easy to apply to information the traditional economic and accounting measures applied to tangible products, the cost benefit analysis. Nevertheless, it is important to do so to avoid,

failure to spot potential threats in time because of lack of intelligence gathering and correlations; failure in attempts to innovate; (…) and failure to recognize opportunities for using information resources more productively (Orna, 1996, p. 25).

Information is called the *glue* that holds the organizational structure together, and there are four ways of using information to create value for a business:

1. Minimize risks;
2. Reduce costs;
3. Add value, guiding the output to the Market and customers;
4. Innovation, through the creation of new realities (Davenport, 1997; Davenport, Marchand, & Dickson, 2004).

Information Value Characteristics

Information has some characteristics that differentiate it from tangible products, such as:

If information is exchanged and traded, the value from using it can increase for all parties to the transaction. The value of information is not diminished by being used; it can be (…) used many times by many users for adding value to many activities and outputs (Orna, 1999, p. 141).

Cleveland highlights the value of information and its characteristics, saying that information is:

1. **Expandable:** The more we have, the more we use and more valuable it gets;
2. **Compressible:** Can be concentrated, integrated, summarized, miniaturized for easier handling;
3. **Substitutable:** Replaces land, labour and capital;
4. **Transportable:** At the speed of light, using e-mail or video-conference we can be anywhere as if we were there;
5. **Diffusive:** It tends to leak and the more it leaks the more we have and the more of us have it, making it available to a growing number of people;
6. **Shareable:** Allows exchange and sharing transactions because giving or selling a information lets the seller to keep it anyway, unlike physical goods (Cleveland, 1985).

For Grant, information is the medium through which an organization relates to its environment and allows the individuals to know how to react and adapt to external changes. For him, the value of information is related to what he designates *imperfect availability of information* and *imperfection of information* (2002, pp. 242, 245-246, 516). If, in a given market, not all the players have access to all the information available, or it's difficult to identify all the information needed to decide effectively in accordance to other competitor's moves, then the ones who possess or know the sources where the information needed exists, have a competitive advantage in that market.

This is going to be very important because even if all the players have access to the same information, again it will be the use that they make of it that is going to distinguish the success that they can obtain. It's only after the information gets inside someone's mind and is applied and used in something useful, like a product or a process that has the preference of the consumers in that market, that it becomes valuable and originates a big return on investment to his owner, the "value for money" valuation (Best, 1996a; Grant, 2002; Kotler, Armstrong, Saunders, & Wong, 1999; Mohr, Sengupta, & Slater, 2010; Orna, 1996; 1999).

Even if everybody can access quality patente information, not everyone will innovate thanks to that. Furthermore, not everyone that does innovate will be successful because the results will be dependente on the output obtained and the accession of the consumers to the solution found, that is to say, it will depend on the product's characteristics, on its adequacy to the market and its consumers' needs, and on consumer satisfaction.

Again, a method to value information can be *through seeking indirect evidence of its value to businesses in promoting competitiveness, productivity, or innovation and successful marketing of the results* (Orna, 1996, p. 26), *and in avoiding risk and reducing uncertainty* (Orna, 1999, p. 142).

Although information can be valuable for a long period of time, certain types of information are only valuable within a given time frame limit (Huang, Lee, & Wang, 1998). Information may lose its value because its timing has passed, like if you could know the lottery numbers in advance for a certain week, after that week those numbers would be useless, or *a tip on the fourth race at Belmont might be valuable at lunchtime and valueless by dinnertime. Yesterday's weather forecast is of merely historical interest tomorrow* (Cleveland, 1985, p. 29). Nevertheless, some information retains its value and can be applied in the same field or in complementary fields, enhancing the final result obtained with this strategy. That is the case of patent information.

PATENT INFORMATION

There is an important information resource that is freely available, the patent information, which is easily accessible across different digital platforms based on the Internet, enabling a stimulus to creativity that can motivate new innovations (Maravilhas, 2009; Maravilhas & Borges, 2009).

Repositories of patent information, in the form of databases and digital libraries, are a major source of scientific and technical information[1]. There are about 100 million published patent documents worldwide, most of them containing information not available anywhere else (Albrecht, Bosma, van Dinter, Ernst, van Ginkel, & Versloot-Spoelstra, 2010; Bregonje, 2005; Greif, 1987; Marcovitch, 1983).

Even the information that can also be found in other documents, such as scientific papers, technical reports, conference proceedings, and dissertations, it is not described with the same degree of detail and they take longer to reach the public.

In addition, approximately 2.2 million new documents are created every year (Mueller & Nyfeler, 2011, p. 384) and their publication allows public consultation even before protection is granted.

Over 30% of patents are in the public domain (they have reached the time limit for protection, or for non-payment of annual fees), or are not being exploited due to lack of funding or technical incapacity of the holder (Godinho, 2003; Idris, 2003; Maia, 1996).

Regarding the technical information contained in patent documents, through their consultation, researchers, scientists or entrepreneurs, can withdraw precious ideas about the state of the art in any field of science and technology (Maravilhas & Borges, 2011a; 2011b).

For Idris (2003), the main reason for patent information analysis is related to the updated information contained in patent documents that can help avoid erroneous investments concerning the possibility of duplicating research work already done by others.

The insufficient use of patent information has caused a significant waste of funds invested in R&D programs whose existence is threatened by the return on investment of patented technology. According to estimates, European industry wastes between 20 to 32 billion Euros annually (Brünger-Weilandt, Geiß, Herlan, & Stuike-Prill, 2011; Ribeiro, 2007), primarily motivated by the lack of patent information, resulting in a duplication with the reinvention of existing inventions, and redeveloping products already available on the market that could easily have been identified if patent information was consulted.

In certain circumstances, it is possible to use patent information to develop new products and processes, if the resulting new invention does not infringe the patents that already exist.

This is a perfectly legitimate and one of the most important justifications for the existence of the patent system (Idris, 2003, p. 88).

The advantage of consulting the technical information is that it encourages creativity and innovation by turning inventions into innovations (Maravilhas, 2009). Either by using patented technologies that are not being exploited; by finding ways to integrate these inventions in the business plan of a company for further development; encouraging the creation of companies to exploit these inventions; identifying Technologies and finding those who want to produce them; or selling ready-made solutions or compounds to be included in finished products (Maravilhas & Borges, 2009).

Patent information is not the only source of information that can stimulate innovation but it is one of the most important ones. This is mainly due to its detail, depth, timeliness and coverage, constituents (description, schematics, drawings, charts), analysis of the state of the art, external inputs of experts that complement and enrich the document (patent examiners and their research reports), among others.

This source of information covers all technical and scientific activities of human knowledge and is encoded to allow its easy recovery and utilization. It nurtures insights not only from the scientific area in which we are investigating, but also of complementary areas that can increase the value of the solution developed. It can also be used for other applications not initially considered and solve other problems from intermediate crossover areas between scientific disciplines.

It also improves the time necessary to introduce new products to markets, lowers costs associated with R&D and monetize the installed capacity in some industries or scientific activities (Maravilhas, 2009).

Searching this information will allow the finding of certain inventions for free exploration, without the obligation to pay any license if the patent is in the public domain and free to use. Such is the case of generic drugs which are the free use of active substances that have reached their limit of protection and are free to be explored, what has been done successfully by several national and international companies.

Origin and Sources

To obtain patent protection, the applicant must formalize an application with the Patent and Trademark Office (PTO) of his country or other supranational office such as the European Patent Office (EPO) or the World Intellectual Property Organization (WIPO). It is also necessary to reveal a full description of the details of the invention, together with a series of claims that are the core of the invention itself and the matter to which legal protection is required (Ulrich & Eppinger, 2012).

During the process of registration and grant of the patent, the official entity like the United States Patent and Trademark Office (USPTO), EPO or WIPO (if protection via Patent Cooperation Treaty—PCT—is required), will generate one or more legal documents that are called patente literature. These documents contain information that is referred to as patent information.

After the publication of the patent application, typically 12 to 18 months after it has been entered in the respective office, this information becomes publicly available to those who wish to consult it.

In exchange for the protection by a patent granted for your invention for 20 years, the information concerning the invention will be in the public domain and these inventions may be made by third parties for research purposes. The purpose of this disclosure is intended to catalyze inventive activity, making possible the advancement of technology that would otherwise still be kept secret.

One of the conditions allowing a patent to be assigned is that the information in the patent application is so detailed that a person skilled in the art, the area in question, is capable of reproduce her own invention (product or process) being able to duplicate it. The patent document not only describes the invention necessarily capable of industrial application, but also the scope of protection intended, if the respective patent is obtained.

Scientific and Technical State-of-the-Art

Much of the information contained in patent documents is not published anywhere else, making it a source of information essential to meet the needs for new technical information (Bregonje, 2005; Greif, 1987; Marcovitch, 1983).

Patent literature is the major source of technological information available worldwide, thus constituting the largest repository of technical knowledge possessing a huge monetary value.

The revelation of the secrets contained in the technical documentation resulting from a patente application discloses valuable information to the public about the state of the art in a given area, thus promoting, through this knowledge, technological development (Maravilhas & Borges, 2009).

Every time a research program is initiated it is advisable to research all the scientific and technological work already done in the area (Naetebusch, Schoeppel, & Fichtner, 1994; Schoeppel & Naetebusch, 1995). Moreover, it is always worth to go back some years in the search to get to know many of the ideas contained in patent documents whose legal validity has expired. They could contain valuable inventions or technical information which, at the time they were made, were quite ahead of their time and, therefore, had no means of effectively being carried out, or for which consumers were not yet ready to appreciate and give due value (Marcovitch, 1983, p. 492). Currently, they can be regarded as an opportunity worth exploiting, taking advantage of recent technical developments and current knowledge and, hopefully, become extremely profitable.

Problem Solver

According to Marcus (1995), the patent information, in addition to providing an excellent source of information for generating ideas, also has the advantage of being a source of inspiration when it is necessary to find a solution to assist in solving problems of technological nature.

Drucker (1987) refers to several representative examples of using an invention for applications other than that which was initially intended. Drugs developed to solve human problems eventually can be adapted to solve problems for veterinary; the creation of DDT during World War II to protect American soldiers against tropical parasites was later widely used in agriculture to protect crops and cattle; 3M® developed several products for industry and only much later realized that they could be used to solve domestic problems with minor changes and improvements, as the case of Scotch Tape® and Post it®.

The Teflon®, from DuPont®, invented in 1938 and marketed since 1946, is waterproof and is the material with the lowest known friction coefficient.

Widely used in aircraft wings to improve aerodynamics and prevent waste to accumulate (Drucker, 1987; Lattès, 1992), its advantages can be applied in another area of activity that requires these no sticking qualities: the kitchen.

DuPont® started putting this coating into pans, preventing food cling during making. When the patent expired, the technology was free to be used, leveraging competition with this knowledge and this extra advantage without any costs.

Business Ideas and Opportunities

Organizations need information to enable them to innovate and gain competitive advantages in the markets in which they operate. Getting the latest and greatest technologies has motivated the search for information that would maintain the productivity, competitiveness, superiority and status towards adversaries and competitors (Burgelman, Christensen, & Wheelwright, 2009; Christensen, Anthony, & Roth, 2004).

Based on the knowledge of the existence of some invention, interested parties may initiate licensing contracts for the exploitation of these technologies in certain geographic areas or markets where they have better penetration capability than the patent holder (Maravilhas, 2009). In terms of market analysis and competition, if we develop technological surveillance and competitive intelligence, patent information can provide many surprising clues that might share a light on the competition strategies, new trends, new products or technological improvements made to existing products (Ashton & Klavans, 1997; Ojala, 1989; Wilson, 1987a; 1987b).

Marcus (1995, p. 70) says that beyond the obvious parameters that can be found in the patent information - such as the name of the inventor, patent holder, priority dates, patent family, among other aspects - the researcher can use his imagination to find a huge variety of commercially valuable information, such as:

1. Ability to find or identify potential customers or business partners;
2. Provide supporting information to a business meeting;
3. Identify trends in R&D, new technologies and new products;
4. Identify trends and movements between companies;
5. Facilitate technology transfer or licensing;
6. Prevent duplication of research projects, avoiding the unintentional copy;
7. Identify experts in a specific field of technology or scientific area;

8. Establish new applications and uses for existing technologies and products[2];
9. Find solutions to technical problems;
10. Support the generation of ideas for new products or processes;
11. Identify trends in Marketing;
12. Establish the expiration date of a patent that will allow the free use of the invention;
13. Identify potential competitors;
14. Monitor the activities of competitors;
15. Establish the state of the art in a certain scientific or technological area.

All these possibilities may trigger a business opportunity profitable for a company.

FROM INVENTION TO INNOVATION

Innovation is defined as the application of new knowledge, resulting in new products, processes or services or significant improvements in some of its attributes. It became an instrument of competition between businesses that search the Market for the acquisition of increasingly sophisticated and developed products.

This development has lead to new innovations in shorter and shorter time periods. The companies were forced to find new solutions and became aware that new technology was the only solution (Carneiro, 1995; Dantas, 2001; Gupta, 2008; Laranja, Simões, & Fontes, 1997; Oliveira, 1999; Sarkar, 2009).

Innovations can be of three types:

1. **Incremental:** Which are technological improvements that occur continuously, resulting from activities of R&D, the work of corporations' design and engineering departments and suggestions from users;
2. **Radical:** Resulting from discontinuous events and the product of R&D work;
3. **Systemic:** Whose amplitude affects the economy as a whole.

Following this classification from the point of view of its nature, innovation can be seen as being on:

1. **Process:** Adding new operations necessary for the company to produce a product, aimed at increasing productivity, reducing costs and improve quality;
2. **Product:** Introducing new products or alterations by aggregating knowledge into existing products, especially in industries that compete for product differentiation;
3. **Organizational:** Changing the way we organize and distribute the work in a particular department or service (Castro, 1999, p. 22).

We should make a distinction between the concepts of innovation and invention[3]. By invention we mean the creation or discovery of a new idea, including the concept, design, model creation or improvement of a piece, product or system. Even though an invention may allow a patent application, in most cases it will not give rise to an innovation. The innovation refers to something new, for example a new invention, which is placed on the market to the consumers benefit (Trott, 2008).

We can therefore say that an invention that does not give way to something that is made available to people to satisfy a particular need, be it a product or service, will never translate into an innovation.

Creativity Stimulation

Speaking of innovation necessarily requires talking about creativity. Although not synonymous, both concepts are inseparable.

Creativity is part of the wider process of innovation. The main difference between both lies in the fact that innovation can be the result of the adaptation of something existing outside the organization (Open innovation), not implying a creative behaviour by the organization. In any case, most innovations result from creative processes, individual and collective, spontaneous or intentional (Baxter, 2000). Creativity[4] is based on reasoning that produces imaginative new ideas and new ways of looking at reality (Birch & Clegg, 1999).

Innovation implies change, requires a combination of creativity, reasoning and ability to act (Tidd, Bessant, & Pavitt, 2003)[5].

The assiduous consultation of information sources, which include patent information together with all that is new and relevant in the scientific areas in question, and even complementary scientific areas, may allow a bank of knowledge that awakens a response when confronted with a technical or scientific problem to overcome. Also, if others before have had the same problem and succeeded, people can benefit from this knowledge and find a better way to overcome difficulties (Maravilhas & Borges, 2011a). It may happen that from the existing knowledge we can find a better solution.

Serendipity and Creative Imitation

There are also two concepts related to innovation that should be considered: serendipity and creative imitation. Serendipity[6] is the ability to make importante discoveries by accident. Not all the ideas for new products or processes appear voluntarily and intentionally, thus generating innovations.

Sometimes a mixture of luck and preparation provides valuable discoveries. A serendipitous discovery results from the combination of a happy coincidence with perspicacity. When developing a strong glue, in 1968, Spencer Silver from 3M® obtained a weak glue that apparently was useless.

A co-worker, Arthur Fry, had the genius to associate that weak glue to paper to mark pages allowing them to be removed and mark new ones (Krols, 2010; Utterback, 1994). As to the notion of creative imitation[7], it refers to the strategy followed by some companies, that imitate something already existing but adding value. Often the imitator can foresee—better than the original creator—how the product or service can be best suited to meet the needs of consumers, changing it to match this observation. The imitator can compete in the market without having the initial costs of R&D to develop the product, presenting something that surpasses the original by better adapting its function and preventing defects. Companies from countries like South Korea, Japan, Taiwan, and China use this strategy and they got excellent results, allowing them to be innovators nowadays.

Old Inventions, New Innovative Solutions

Another advantage provided by patent information is the use of technology to solve problems in areas other than those for which the invention was patented[8]. The Cosworth® factory of high performance engines adapted an invention of na electromagnetic pump, developed in the area of nuclear power, to force the molten metal to enter the moulds used in their traditional mould castings, eliminating the air that, normally, makes the metal parts porous (Tidd, Bessant, & Pavitt, 2003, p. 266).

Another known example is the case of Nylon®, a synthetic polymer of wide application in textile fibres which is characterized by being extremely tough, highly elastic and able to be also processed in the form of moulded articles. This product was developed by a group of scientists led by Wallace Carothers from DuPont® in USA. Even today it is used in varied products such as underwear, shirts, sweaters, raincoats, parachutes, surgical suture line, fishing line and nets, rope, tennis nets and racket lines and, if moulded, in kitchen appliances and industrial machinery parts. However, its technology is about 80 years old and its patent is expired (Maravilhas & Borges, 2011a).

The same happens with Lycra®, Terylene®, PVC®, Polyester® and other synthetic products such as Kevlar®. Invented in 1965 by Dupont®, Kevlar® is still used in bullet proof vests, helmets and protective masks for policemen and soldiers (and Paintball practitioners), protective equipment to high and low temperatures, sports equipment, building materials, cables and sails for boats.

At the time they were invented all these applications were not foresaw but they continue to be widely used in many different and extremely useful products.

Ecological, Sustainable, "Green" Products

With the start of oil scarcity and environmental costs unaffordable for both fossil materials, oil and coal, new solutions must be found to meet the energy needs of humanity (Dresner, 2008), by which hydrogen[9] is a candidate to consider (Yeomans, 2006).

We have the example of the solar oven developed by Prof. Manuel Colares Pereira (university Professor and researcher) based on an expired patent (the *Pirelióforo* from a Portuguese priest, Manuel António Gomes, known as MAG Himalaya) from an invention used for other functions (create fertilizers and melt metals). This invention, which in 1904 won the first prize in the Universal Science Fair in St. Louis, USA (Rodrigues, 1999), motivated the development of an ecological oven meant for preparing food, using the sun's energy[10].

In an Indonesian province, a remote island, with no major economic resources had plenty of coconuts. They ate the fruit and drank the coconut juice, but the shells resulted in a serious environmental problem because they were not given any use and constituted piles of debris around the island, sending tourists away. Training was given to people who could use the computers in the public library that had Internet access to conduct research on Espacenet® to find patented technologies involving the use of coconuts. They discovered they could use the fruit and its juice in several different applications such as jams, sweets, cakes, liqueurs, and the bark or shell could also be treated profitably. They found out that shells could be used as fertilizer; to build furniture; decorative articles; toys and craft products; transformed into construction materials and insulation; and, if processed, could be used as filling material for pillows and mattresses. Such knowledge triggered many successful home industries, transforming a small island without economic resources in a more prosperous and equal society, and several small entrepreneurs and business owners have expanded their businesses across borders. Most of the technologies and patents discovered were of Brazilian origin, not protected in Indonesia, and could be used without any payment rights (Dou, 2004; Dou, Leveillé, Manullang, & Dou Jr., 2005).

Other examples can be found on the use of natural compounds such as clay, shells, and algae as filters for industrial pollution, cleaning the water with more efficiency than the highly expensive activated coal filters (Maravilhas, 2012).

PATENT INFORMATION QUALITY

For Eppler (2006) information quality, applied to patents, can be defined as information that is in accordance with the cognitive, functional and technical requirements of patent owners, patente attorneys, patent offices, patent information providers and patent information searchers.

According to Scott, *good quality patent information should be accurate data stored to be (easy) retrievable* (2010, p. 233).

To make it possible, the quality of the application submitted will be of high importance because the result is highly dependent on the quality of the supplied original text and drawings (Kovac, 2010), or as Philipp puts it "quality in, quality out" (Philipp, 2006).

Since patent offices like the European Patent Office (EPO) and the United States Patent Office (USPTO), amongst others, cannot control the quality of the content of the applications, to improve the quality of patent documents to be properly recovered and used by the interested public—researchers, entrepreneurs, innovators—some amendments must be made to increase the quality. These improvements can take the form of correction, improvement, or adding further data to the title and abstracts, indicating what the invention is about.

More value is added by the search report performed by the patent examiner (Philipp, 2005), citing the best prior art available and proper keywords, and by the classification data that indexes the invention in a category for easy retrieval.

Internally, patent offices like EPO perform routine checks on applications, both by humans and by software, provide extensive training to the examiners, and execute quality audits to several random complete applications (Albrecht, Bosma, van Dinter, Ernst, van Ginkel, & Versloot-Spoelstra, 2010; Brünger-Weilandt, Geiß, Herlan, & Stuike-Prill, 2011; Kovac, 2010; Mueller & Nyfeler, 2011; Scott, 2010).

As an example, to illustrate the task and its difficulty, in the EPO *18.9 million documents were added to the collection in the period 2004-2008, and 5.4 million corrections performed on them* (Albrecht, Bosma, van Dinter, Ernst, van Ginkel, & Versloot-Spoelstra, 2010, p. 280).

We can be certain that *database assurance processes at the EPO contribute significantly to the quality of the database* (Brünger-Weilandt, Geiß, Herlan, & Stuike-Prill, 2011, p. 232). The quality of the data and information at the EPO helps in the classification and search of the documents in a time-efficient and resource-friendly manner, allowing patente families and citations becoming extraordinary sources of information for patent examiners (Albrecht, Bosma, van Dinter, Ernst, van Ginkel, & Versloot-Spoelstra, 2010) and, in the last instance, for all the interested public.

Problems and Opportunities

Problems in patent information quality relates to each year's increasing number and size of the documents augmenting the 'noise' in search results (Philipp, 2006; Scott, 2010), originating that *the ever-increasing number of applications entails backlogs in many offices* (Kovac, 2010, p. 230).

The classification schemes, good titles, abstracts, and keywords allow searchers to easily find relevant information amongst the 'noise' generated by applications created by companies and patent attorneys with the objective of making it harder for competitors to extract the core value of the invention. For Scott, *a renewed fee structure may enhance the quality of the incoming applications and hence the quality of patent information coming out of the EPO* (2010, p. 236).

In the USPTO, where everybody files for patent protection from around the world, the amount of applications to examine is so big that in 2008 they contracted about 1.200 extra examiners to make each month's legal deadlines attainable, and they kept contracting the same number in 2009. Nevertheless, the examination process took between 41 and 44 months instead of the 36 in Europe (Caraher, 2008, p. 151).

Reduced time to examine an application means lower quality of the patent document. Due to this *flood of patent information in which we are already drowning* (Philipp, 2006, p. 118), indexing and coding the patent documents accurately will prevent errors and omissions to occur. Lack of complete indexing may originate that many of the relevant documents will not be found, leading to deflation in the quality of patents.

Abbott (2004) states that the number of patent applications processed by each examiner have doubled since 1995 and the number of hours available for examination of each patent has halved since 1992, decreasing from 23.8 hours to 11.8 hours in 2001.

Some value-added database providers of worldwide patents like Derwent World Patents Index (DWPI) and STN International rewrite original titles and abstracts to highlight the invention and the advantages and principal utilization of the patented technology (Butler, 1995). They use manual codes for patents in chemical and engineering allowing the classification of new technical features of the inventions and, in some cases, its commercial and market application.

They include English human translations of the claims and provide chemical coding to compounds and access to the structure-base of the patented chemicals (Brünger-Weilandt, Geiß, Herlan, & Stuike-Prill, 2011, p. 232).

These characteristics increase the benefits for information professionals with search results more precise and complete, assigning more value to the information retrieved.

Another problem is related with the language barrier that can only be minimized by machine translation. The increasing volume of Chinese, Indian, Russian, Korean and several Asian countries makes it very difficult for American and European examiners to perform an accurate search report because most of them can't understand those languages to ascertain the state of the art in that technical subject. The solutions for the language problem are *improved machine translation tools* (Mueller & Nyfeler, 2011, p. 384) and *semantic search and ranking of documents by relevance* (Brünger-Weilandt, Geiß, Herlan, & Stuike-Prill, 2011, p. 234).

If we add to all these difficulties the complexity of interdisciplinary inventions, like the ones in the field of nanotechnology, bio-engineering or bio-informatics, we can easily acknowledge that it is very hard for a researcher with only one scientific specialization to be able to find relevant patent documents in several scientific and technical fields. For Philipp all these constraints have a huge cost for the public, resulting in *reduced rates of innovation, decreased patent-based economics, and higher prices for goods and services to the detriment of all* (2006, p. 118).

To avoid this situation, several measures are being developed and put in practice by the offices in the form of standards. The EPO drafted quality standards for the process, including quality control, and these standards have been in use since 2008. These apply to: classification, enhanced search report, official action and patent documents. This envisages the search report performed by the examiner in the following criteria: full classification of the entire claimed invention; search in the Patent Cooperation Treaty (PCT) minimum documentation (including the biggest Asian offices); search in all relevant classes and databases; the search report must list all the relevant retrieved documents from patent and non-patent literature; and the report shall be available on time, six months after filing or in time for the 18th month public disclosure (Kovac, 2010). All these measures only apply after the application enters the office. So, to improve the quality of the application Philipp suggests a bigger investment in educating the public, mainly the individual applicant with no professional support from a company legal department, stating that the EPO should focus in offering more patent information advisory services and supplying a 'quick scan' pre-filing search of the state of the art to improve the chances of an application become a granted patent (2006), thus reducing the 'noise'.

FUTURE RESEARCH DIRECTIONS

Future projects consider not only the translation of the essential information of the patent document for English, French, Spanish and German, but also making the search process more user friendly by allowing to search within human indexed keywords, by means of a thesaurus, avoiding the need to know chemical compounds or substances' name, which is not so easy for beginners (Cavalier, 2001; Dulken, 2000; Kovac, 2010; Oda, 2009; Wang, 2009).

All these measures aim to improve the number of people that effectively search patent information with benefits for the whole society.

CONCLUSION

In today's globalized world, quality information warrants best results when competing with other organizations, enabling a competitive advantage (Porter, 1985; Porter & Millar, 1985).

Patent information, contained in patent documents, must be of high quality to permit the search and retrieval of the documents needed to solve a problem or stimulate new ideas and solutions (Albrecht, Bosma, van Dinter, Ernst, van Ginkel, & Versloot-Spoelstra, 2010; Brünger-Weilandt, Geiß, Herlan, & Stuike-Prill, 2011; Chakroun, 2012; Ginman, 1990). Old inventions can generate new ideas, inventions never marketed can be brought to market and satisfy needs and desires; inventions for one application can be introduced for a new domain and Technologies can be applied in ecological sustainable solutions with 'green' products.

The information value will be evaluated by the level of acceptance of those innovations by the customers in a given market where they are introduced (Brophy & Coulling, 1996; Huang, Lee, & Wang, 1998).

It also allows to identify the technical fields where competitors are more active; the leading countries and companies for each technology and the patenting patterns followed in a certain period; the technological trends and the state of the art of a given technology; the number of citations a patent document has; the inventor and the owner of the invention, amongst other relevant information (Idris, 2003; Macedo & Barbosa, 2000; Maia, 1996; Maravilhas, 2009).

Several measures are being taken by the patent offices to improve the quality of the information retrieved by the public. Standards are applied to warrant the quality and validity of the documents and its content, and audits are performed to correct possible gaps. Some value-added information is supplied by commercial databases like DWPI and STN (Albrecht, Bosma, van Dinter, Ernst, van Ginkel, & Versloot-Spoelstra, 2010). That procedure could be followed by the intellectual property offices to facilitate and improve the use of this information and the results it promotes in the form of new patent applications and the introduction of new innovations.

Due to the huge number of documents from China, Russia, India, Korea and other Asian countries, machine translation and semantic ranks of documents are being used (Cavalier, 2001; Dulken, 2000; Kovac, 2010; Oda, 2009).

Further research should allow to measure if all those strategies implemented are improving the value and quality of the information available to the public, measurable by the number of citations in the new patent filings.

In the end, the quality and value of this information will be dependent on the user and related to the results that it will allow to obtain, solving a problem or improving a solution, always integrated on its context and present needs.

REFERENCES

Abbott, A. (2004). Pressured staff 'lose faith' in patent quality. *Nature*, *429*(6991), 423. doi:10.1038/429493a PMID:15175718

Adair, J. (2011). *The art of creative thinking: How to be innovative and develop great ideas*. Norfolk, VA: Kogan Page.

Albrecht, M. A., Bosma, R., van Dinter, T., Ernst, J.-L., van Ginkel, K., & Versloot-Spoelstra, F. (2010). Quality assurance in the EPO patente information resource. *World Patent Information*, *32*(4), 279–286. doi:10.1016/j.wpi.2009.09.001

Ashton, W. B., & Klavans, R. A. (1997). *Keeping abreast of science and technology: Technical intelligence for business*. Columbus, OH: Battelle Press.

Barbosa, P., & O'Reilly, A. (2011). *Harvard trends: Tendências de gestão*. Porto, Portugal: Vida Económica.

Baxter, M. (2000). Projeto de produto: Guia prático para o design de novos produtos (2ª ed.). S. Paulo: Edgard Blücher.

Best, D. (1996a). Business process and information management. In D. Best (Ed.), *The Fourth Resource: Information and its Management*. Hampshire, UK: Aslib/Gower.

Best, D. (1996b). *The fourth resource: Information and its management*. Hampshire, UK: Aslib/Gower.

Beuren, I. M. (1998). *Gerenciamento da informação: Um recurso estratégico no processo de gestão empresarial*. São Paulo: Atlas.

Birch, P., & Clegg, B. (1999). *Criatividade em negócios: Um guia para gestores*. Viseu: Pergaminho.

Branscomb, L. (2004). *Where do tech comercial innovations come from?* Paper presented at the Meredith and Kip Frey Lecture in Intellectual Property at Duke Law School. Retrieved 06-09-2010, from http://www.law.duke.edu/conference/namedlectures.html

Bregonje, M. (2005). Patents: A unique source for scientific technical information in chemistry related industry? *World Patent Information*, *27*(4), 309–315. doi:10.1016/j.wpi.2005.05.003

Brophy, P., & Coulling, K. (1996). *Quality management for information and library managers*. Guildford: Aslib/Gower.

Brown, J., & Duguid, P. (2000). *The social life of information*. Boston: Harvard Business School Press.

Brünger-Weilandt, S., Geiß, D., Herlan, G., & Stuike-Prill, R. (2011). Quality: Key factor for high value in professional patent, technical and scientific information. *World Patent Information*, *33*(3), 230–234. doi:10.1016/j.wpi.2011.04.007

Buchanan, B. (2008). Unlocking the value of patent data: Patent informatics services at the UK intellectual property office (UK-IPO). *World Patent Information*, *30*(4), 335–337. doi:10.1016/j.wpi.2008.07.002

Burgelman, R., Christensen, C., & Wheelwright, S. (2009). *Strategic management of technology and innovation* (5th ed.). Singapore: McGraw-Hill.

Burk, C. Jr, & Horton, F. Jr. (1988). *InfoMap: A complete guide to discovering corporate information resources*. Upper Saddle River, NJ: Prentice Hall.

Butler, J. T. (1995). Patent searching using comercial databases. In Lechter (Ed.), Successful patents and patenting for engineers and scientists. New York: The Institute of Electrical and Electronics Engineers (IEEE) Press.

Caraher, V. (2008). The evolution of the patent information world over the next 10 years: A Thomson scientific perspective. *World Patent Information*, *30*(2), 150–152. doi:10.1016/j.wpi.2007.07.008

Carneiro, A. (1995). *Inovação: estratégia e competitividade*. Lisboa: Texto Editora.

Castells, M. (2000). *The information age: The rise of the network society* (2nd ed.). London: Blackwell.

Castells, M. (2004). *A galáxia internet: Reflexões sobre internet, negócios e sociedade*. Lisboa: Fundação Calouste Gulbenkian.

Castro, J. (1999). *Invento & inovação tecnológica: Produtos & patentes na construção*. Annablume Editora Comunicacao.

Cavalier, T. (2001). Perspectives on machine translation of patent information. *World Patent Information*, *23*(4), 367–371. doi:10.1016/S0172-2190(01)00045-X

Chakroun, N. (2012). Improving patent information quality: Development and the disclosure requirements. *The Journal of World Intellectual Property*, *15*(3), 199–220. doi:10.1111/j.1747-1796.2012.00438.x

Choo, C. W. (2003). *Gestão de informação para a organização inteligente: A arte de explorar o meio ambiente*. Lisboa: Editorial Caminho.

Christensen, C., Anthony, S., & Roth, E. (2004). *Seeing what's next: Using the theories of innovation to predict industry change*. Boston: Harvard Business School Press.

Cleveland, H. (1985). *The knowledge executive: Leadership in an information society*. New York: Dutton.

Cooke, A. (2001). *A guide to finding quality information on the internet: Selection and evaluation strategies* (2nd ed.). Cornwall, UK: Library Association Publishing.

Cortada, J. (1996). *TQM for information systems management: Quality practices for continuous improvement*. Singapore: McGraw-Hill.

Dantas, J. (2001). *Gestão da inovação*. Porto: Vida Económica.

Davenport, T. (1997). *Information ecology: Mastering the information and knowledge environment*. New York: Oxford University Press.

Davenport, T., Marchand, D., & Dickson, T. (2004). *Dominando a gestão da informação*. Porto Alegre: Bookmann.

Dearlove, D. (1998). *Key management decisions: Tools and techniques of the executive decisionmaker*. Wiltshire, UK: Financial Times/Pitman.

Dou, H. (2004). Benchmarking R&D and companies through patent analysis using free databases and special software: A tool to improve innovative thinking. *World Patent Information*, *26*(4), 297–309. doi:10.1016/j.wpi.2004.03.001

Dou, H., Leveillé, V., Manullang, S., & Dou, J. Jr. (2005). Patent analysis for competitive technical intelligence and innovative thinking. *Data Science Journal*, *4*(31), 209–236. doi:10.2481/dsj.4.209

Dresner, S. (2008). *The principles of sustainability* (2nd ed.). Chippenham: Earthscan.

Drucker, P. (1987). *Inovação e gestão* (2nd ed.). Lisboa: Presença.

Drucker, P. (2000). *Desafios da gestão para o século 21: Civilização*. Academic Press.

Dulken, S. (2000). The future of patent information. *World Patent Information*, *22*(4), 273–275. doi:10.1016/S0172-2190(00)00082-X

Eppler, M. (2006). *Managing information quality: Increasing the value of knowledge-intensive products and processes* (2nd ed.). New York: Springer.

Ernst, H. (2003). Patent information for strategic technology management. *World Patent Information*, *25*(3), 233–242. doi:10.1016/S0172-2190(03)00077-2

Esty, D., & Winston, A. (2008). *Do verde ao ouro: Como empresas inteligentes usam a estratégia ambiental para inovar, criar valor e construir uma vantagem competitiva*. Cruz Quebrada: Casa das Letras.

Garvin, D. (1988). *Managing quality*. New York: Free Press.

Ginman, M. (1990). Quality information, and information for quality. In Wormell (Ed.), Information quality: Definitions and dimensions. London: Taylor Graham.

Godeluck, S. (2000). *A explosão da economia na internet*. Lisboa: Livros do Brasil.

Godinho, M. (2003). *Estudo sobre a utilização da propriedade industrial em Portugal* (Vol. 1). Lisboa: Instituto Nacional da Propriedade Industrial.

Grant, R. M. (2002). *Contemporary strategy analysis: Concepts, techniques, applications*. Oxford, UK: Wiley-Blackwell.

Greif, S. (1987). Patents and economic growth. *International Review of Industrial Property and Copyright Law, 18*(2), 191–213.

Gupta, P. (2008). *Inovação empresarial no século 21*. Porto: Vida Económica.

Haberman, M. (2001). The role of intelectual property and patent information in successful innovation, production and marketing: Case study 1: The non-spill drinking vessel. *World Patent Information, 23*(1), 71–73. doi:10.1016/S0172-2190(00)00105-8

Hess, P., & Siciliano, J. (1996). *Management: Responsibility for performance*. New York: McGraw-Hill.

Hinton, M. (2006). *Introducing information management: The business approach*. Burlington, MA: Elsevier Butterworth-Heinemann.

Huang, K., Lee, Y., & Wang, R. (1998). *Quality information and knowledge*. Upper Saddle River, NJ: Prentice Hall.

Idris, K. (2003). *Intellectual property: A power tool for economic growth*. Geneva: World Intellectual Property Organization.

Jolly, A. (2003). *Innovation: Harnessing creativity for business growth*. New York: Kogan Page.

Kahaner, L. (1997). *Competitive intelligence: How to gather, analyze, and use information to move your business to the top*. New York: Simon & Schuster.

Koch, A. (1991). Patent information to stimulate innovation in small and medium sized companies. *World Patent Information, 13*(4), 201–205. doi:10.1016/0172-2190(91)90193-9

Kotler, P., Armstrong, G., Saunders, J., & Wong, V. (1999). *Principles of marketing* (2nd ed.). Cambridge, UK: Prentice Hall.

Kovac, W. (2010). Documentation and quality: Quo vadis patent system? A patent attorney's view. *World Patent Information, 32*(3), 229–232. doi:10.1016/j.wpi.2009.07.008

Krols, B. (2010). *Invenções acidentais que mudaram as nossas vidas*. Booksmile.

Krupp, F., & Horn, M. (2009). *Reinventar a energia: Estratégias para o futuro energético do planeta*. Alfragide: Estrela Polar.

Laranja, M., Simões, V., & Fontes, M. (1997). *Inovação tecnológica: Experiências das empresas portuguesas*. Lisboa: Texto Editora.

Lattès, R. (1992). *O Risco e a fortuna: A grande aventura da Inovação*. Lisboa: Difusão Cultural.

Macedo, M., & Barbosa, A. (2000). *Patentes, pesquisa & desenvolvimento: Um manual de propriedade intelectual*. Rio de Janeiro: Fiocruz. doi:10.7476/9788575412725

Maia, J. M. (1996). Propriedade industrial: Comunicações e artigos do presidente do INPI. Lisboa: Instituto Nacional da Propriedade Industrial (INPI).

Maravilhas, S. (2009). A informação de patentes: Vantagens da sua utilização como estímulo à criatividade, I&D, inovação e competitividade das empresas portuguesas. In IAPMEI (Ed.), Parcerias Científicas para a Inovação (pp. 91-110). Lisboa: IAPMEI.

Maravilhas, S. (2012). *A informação de patentes como fonte de ideias para produtos e negócios sustentáveis*. Paper presented at the Vth GECAMB, Conferência CSEAR Portuguesa de Gestão e Contabilidade Ambiental, Porto, Portugal.

Maravilhas, S., & Borges, M. (2009). O impacto das bibliotecas digitais de patentes no processo de inovação em Portugal. In Borges & Sanz-Casado (Eds.), A ciência da informação criadora de conhecimento (Vol. 2, pp. 47-63). Coimbra: Actas do IV Encontro Ibérico EDIBCIC 2009.

Maravilhas, S., & Borges, M. (2011a). Os recursos de informação usados na I&D em Portugal: Caracterização dos centros de investigação do ensino superior público das áreas de ciência, tecnologia e medicina. In P. Guerrero & V. Moreno (Eds.), *Límites, fronteras y espacios comunes: Encuentros y desencuentros en las Ciencias de la Informacíon* (pp. 321–333). Badajoz: Actas do V Encontro Ibérico EDICIC.

Maravilhas, S., & Borges, M. (2011b). A utilização da informação de patentes pelos centros de investigação do ensino superior público: o seu impacto no processo de inovação em Portugal. In P. Guerrero & V. Moreno (Eds.), *Límites, fronteras y espacios comunes: Encuentros y desencuentros en las ciencias de la informacíon* (pp. 364–376). Badajoz: Actas do V Encontro Ibérico EDICIC.

Marchand, D. (1990). Managing information quality. In Wormell (Ed.), Information quality: Definitions and dimensions. London: Taylor Graham.

Marchand, D., & Horton, F. Jr. (1986). *Infotrends: Profiting from your information resources*. New York: Wiley.

Marcovitch, J. (1983). *Administração em ciência e tecnologia*. São Paulo: Edgard Blücher.

Marcus, D. (1995). Benefits of using patent databases as a source of information. In Lechter (Ed.), Successful Patents and Patenting for Engineers and Scientists. New York: The Institute of Electrical and Electronics Engineers (IEEE) Press.

McGee, J. V., & Prusak, L. (1995). *Gerenciamento estratégico da informação: Aumente a competitividade e a eficiência de sua empresa utilizando a informação como uma ferramenta estratégica*. Rio de Janeiro: Campus.

McPherson, P. (1994). Accounting for the value of information. *Aslib Proceedings*, *46*(9), 203–215. doi:10.1108/eb051366

Michalko, M. (1991). *Thinkertoys: A handbook of business creativity for the 90's*. Berkeley, CA: Ten Speed Press.

Mohr, J., Sengupta, S., & Slater, S. (2010). *Marketing of high-technology products and innovations* (3rd ed.). Upper Saddle River, NJ: Pearson.

Mueller, H., & Nyfelcr, T. (2011). Quality in patente information retrieval: Communication as the key factor. *World Patent Information*, *33*(4), 383–388. doi:10.1016/j.wpi.2011.06.012

Naetebusch, R., Schoeppel, H. R., & Fichtner, H. (1994). Patent information in a large electrical company, exemplified by the situation at Siemens. *World Patent Information*, *16*(4), 198–206. doi:10.1016/0172-2190(94)90003-5

Oda, S. (2009). JP-NETe - A English-language search tool for Japanese unexamined patents. *World Patent Information*, *31*(2), 131–134. doi:10.1016/j.wpi.2008.09.002

Ojala, M. (1989). A patently obvious source for competitor intelligence: The patent literature. *Database*, *12*(4), 43–49.

Oliveira, J. (1999). *Gestão da inovação*. S. João do Estoril: Sociedade Portuguesa de Inovação |Principia.

Orna, E. (1996). Valuing information: Problems and opportunities. In D. Best (Ed.), *The Fourth Resource: Information and its Management*. Hampshire, UK: Aslib/Gower.

Orna, E. (1999). *Practical information policies* (2nd ed.). Cambridge, UK: Gower.

Penzias, A. (1989). *Ideas and information: Managing in a high-tech world*. New York: W. W. Norton.

Penzias, A. (1995). *Harmony: Business, technology & life after paperwork*. New York: Harper Collins.

Petroski, H. (2008). *Inovação: Da idéia ao produto*. São Paulo: Edgard Blücher.

Philipp, M. (2005). Why pay for value-added information? *World Patent Information*, *27*(1), 7–11. doi:10.1016/j.wpi.2004.06.001

Philipp, M. (2006). Patent filing and searching: Is deflation in quality the inevitable consequence of hyperinflation in quantity? *World Patent Information*, *28*(2), 117–121. doi:.wpi.2006.01.00610.1016/j

Porter, M. E. (1985). *Competitive advantage: Creating and sustaining superior performance*. New York: Free Press.

Porter, M. E., & Millar, V. E. (1985). How information gives you competitive advantage. *Harvard Business Review*, *63*(4), 149–160.

Redman, T. (1996). *Data quality for the information age*. Norwood: Artech House.

Ribeiro, D. (2007). Propriedade intelectual: Mais de 30% da investigação em Portugal é redundante. *Jornal de Negócios, Quinta-feira, 34*.

Rodrigues, J. (1999). *A conspiração solar do padre Himalaya*. Porto: Árvore - Cooperativa de Actividades Artísticas.

Rogers, E. (1995). *Diffusion of innovations* (4th ed.). New York: Free Press.

Roman, K., Maas, J., & Nisenholtz, M. (2009). *How to advertise: What works, what doesn't – and why* (3rd ed.). London: Kogan Page.

Sarkar, S. (2009). Empreendedorismo e inovação (2ª ed.). Lousã: Escolar Editora.

Schoeppel, H. R., & Naetebusch, R. (1995). Patent searching in a large electrical company, as exemplified by the situation at Siemens. *World Patent Information, 17*(3), 165–172. doi:10.1016/0172-2190(95)00016-S

Scott, J. (2010). Does the EPO really care about the quality of patent information? *World Patent Information, 32*(3), 233–236. doi:10.1016/j.wpi.2009.08.004

Seymore, S. B. (2009). Serendipity. *North Carolina Law Review, 88*, 185.

Shapiro, C., & Varian, H. (1999). *Information rules: A strategic guide to the network economy*. Boston: Harvard Business School Press.

Slater, P., Twyman, P., & Blackman, M. (2000). The smart way for patent information to help small firms. *World Patent Information, 22*(4), 337–341. doi:10.1016/S0172-2190(00)00071-5

Suhr, C. (2004). A change of paradigms: Looking back to the pioneer years of patent information management (1960-1990). *World Patent Information, 26*(1), 41–43. doi:10.1016/j.wpi.2003.10.017

Tapscott, D. (1995). *The digital economy: Promise and peril in the age of networked intelligence*. New York: McGraw-Hill.

Tapscott, D. (1999). *Creating value in the network economy*. Boston: Harvard Business School Press.

Tapscott, D., & Williams, A. (2008). *Wikinomics: A nova economia das multidões inteligentes*. Matosinhos: Quidnovi.

Tidd, J., Bessant, J., & Pavitt, K. (2003). *Gestão da inovação: Integração das mudanças tecnológicas, de mercado e organizacionais*. Lisboa: Monitor.

Trott, P. (2008). *Innovation management and new product development* (4ª ed.). Essex, UK: Prentice Hall I Financial Times.

Ulrich, K., & Eppinger, S. (2012). *Product design and development* (5th ed.). Singapore: McGraw-Hill.

Utterback, J. (1994). *Mastering the dynamics of innovation: How companies can seize opportunities in the face of technological change*. Boston: Harvard Business School Press.

Wagner, G. (1990). The value and the quality of information: the need for a theoretical synthesis. In Wormell (Ed.), Information Quality: Definitions and Dimensions. London: Taylor Graham.

Wang, D. (2009). Chinese to English automatic patent machine translation at SIPO. *World Patent Information, 31*(2), 137–139. doi:10.1016/j.wpi.2008.10.003

Webster, F. (2000). *Theories of the information society*. Cornwall, UK: Routledge.

Wilson, R. (1987a). Patent analysis using online databases—I: Technological trend analysis. *World Patent Information, 9*(1), 18–26. doi:10.1016/0172-2190(87)90189-X

Wilson, R. (1987b). Patent analysis using online databases—II: Competitor activity monitoring. *World Patent Information, 9*(2), 73–78. doi:10.1016/0172-2190(87)90131-1

Wilson, T. (1985). Information management. *The Electronic Library*, *3*(1), 62–66. doi:10.1108/eb044644 PMID:2498741

Wilson, T. (1987). Information for business: The business of information. *Aslib Proceedings*, *39*(10), 275–279. doi:10.1108/eb051066

Wormell, I. (1990). *Information quality: Definitions and dimensions*. London: Taylor Graham.

Yeomans, M. (2006). *Oil - Petróleo: Guia conciso para o produto mais importante do mundo*. Lisboa: D. Quixote.

KEY TERMS AND DEFINITIONS

Creativity: Reasoning that produces imaginative new ideas and new ways of looking at reality. Creativity is an individual process, arises from the idea that popped into someone's head. Relates facts or ideas without previous relationship and is discontinuous and divergent. No creative process exists if there is no intention or purpose. The essence of the creative process is to seek new combinations.

Information: A set of data arranged in a certain order and form, useful to people to whom it is addressed. Reduces uncertainty and supports decision-making. Information is considered to support human knowledge and communication in the technical, economic and social domains. Results from the structuring of data in a given context and particular purpose.

Innovation: The application of new knowledge, resulting in new products, processes or services or significant improvements in some of its attributes. A new solution brought to the market to solve a problem in a new or better way than the existent solutions.

Invention: The creation or discovery of a new idea, including the concept, design, model creation or improvement of a particular piece, product or system. Even though an invention may allow a patent application, in most cases it will not give rise to an innovation.

Knowledge: Is a fluid composed of experiences, values, context information and apprehension about their own field of action that provides a cognitive apparatus for evaluating and incorporating new experiences and information. It originates from data and information and allows acting upon it.

ENDNOTES

1. "Patents provide a constraint on the uses of commercial knowledge, while providing knowledge diffusion. It was intended to promote innovation (…), by making sure that if people have access to patent protection they would file the patent, and through the patent disclose the invention. And from that disclosure, inventors could get other ideas and move ahead" (Branscomb, 2004, parág. 48).
2. "Remember that everything new is just na addition or modification to something that already existed. Much creative thinking involves synthesis, the process of combining previously unrelated ideas, goods, or services to create something new. Many creative individuals have made their fortunes by combining the ideas of others" (Michalko, 1991, pp. 71, 80-82).

3 "Most writers distinguish innovation from invention by suggesting that innovation is concerned with the *commercial and practical application* of ideas or inventions. Invention, then, is the conception of the idea, whereas innovation is the subsequent translation of the invention into the economy (Trott, 2008, p. 14).

4 "These illustrate the first principle–creativity doesn't create something out of nothing. It uncovers, selects, reshuffles, combines, synthesizes already existing facts, skills, and ideas. So, the object is not something 'new' but a new combination of existing elements" (Roman, Maas, & Nisenholtz, 2009, pp. 4-5).

5 "Creativity in the sense we are using here might be described as the ability to discover relations between things that were formerly considered unrelated" (Hess & Siciliano, 1996, p. 107).

6 "Serendipity, finding something of value initially unsought, has played a prominent role in science and technology. These "happy accidents" have spawned new fields of science, broken technological barriers, and furnished countless products that have altered the course of human history" (Seymore, 2009, pp. 185, 188).

7 According to Drucker (1987, pp. 231 – footnote 1) "Describes a strategy that, in essence, is 'imitation'. But it's "creative" because the entrepreneur who implements the strategy of 'creative imitation' understands better what the innovation represents than the people that made the innovation."

8 "Patent information to locate existing technology used in a new context: Industrial Copolymers won a award for a product for waterproofing a roof. The additive used was spotted in a 20-year old patent as a potential candidate to achieve this" (Slater, Twyman, & Blackman, 2000, p. 340).

9 "The idea of using hydrogen as fuel was born in 1839. Only in the 60s, more than a hundred years after its invention, General Electric has used this technology in space missions, not as fuel but to produce drinking water" (Yeomans, 2006, pp. 198-205).

10 Sun Co. Retrieved December 15, 2012, from http://solarcooking.wikia.com/wiki/ Sun_Co.

Chapter 7
Reflecting on Analytics Impacts on Information Architecture Contexts as a Source of Business Modelling for Healthcare Services

Liliane Carvalho Jamil
Independent Researcher, Brazil

Augusto Alves Pinho Vieira
Independent Researcher, Brazil

Antônio José Daniel Xavier
Federal University of Minas Gerais, Brazil

ABSTRACT

This chapter approaches the context of healthcare, observing how information system conceptual background can help to comprehend the evolution and adoption of emerging technologies on the applicable solutions. As a critical sector, both from economic and life quality points of view, healthcare is an excellent environment to observe how these tools and its associated infrastructure make it possible for new business organizations to be proposed. The chapter is aimed to develop a theoretical and practical comprehension around concepts through isolated and integrated analysis.

INTRODUCTION

Information architecture can be regarded as a rich conceptual background that complements information management principles and studies, allowing best opportunities for practical applications. Its intrinsic fundamentals offer possibilities to propose efficient business models, permitting them to be simple, effective and optimized. In this chapter, it is our objective to understand how analytics brings innovation

DOI: 10.4018/978-1-5225-6225-2.ch007

for a sector where information and knowledge are intensively produced and consumed – Healthcare services – specifically on how it can promote new business models to be strategic positioned, as a competitive advantage.

Information architecture is defined as a set of concepts that orient the design for planning, proposition and management of informational processes in organizations (IA Institute, 2014). Its comprehension impacts and relate to several scientific and practical areas, involving technology, design, interface projects, data and information sciences approaches, among others. Its "mapping" and "design" capabilities, which will deserve more deep attention in the theoretical development in the following sections, result in versatile and flexible specifications that can help any organization to understand information impacts on its activities, may it be programmed, planned or even risk-oriented, sudden reactions. As for SAS (2017), analytics is a way to produce knowledge from continuous analysis over information coherently and homogeneously produced and treated. These concepts, in the purpose of this chapter based on a previous research, perform a new context to understand how information management can produce organizational alternatives, both for actions and structures.

The relationship of the main concepts of information and architecture is challenging to study, if any researcher does not focus on the specific objectives. It is now more relevant when we face the situation of a massive data generation age with the introduction of "app-based" service, mobile services, internet of things and the immense diffusion of social media platforms. Cited by Richard Saul Wurman, IA sets a perspective for the "series of systems", "systemic design" and "performance criteria" definitions which will allow treating information, and its complex inter-related context, producing design artifacts for this purpose. This way, analytics can present a new systematic path to logically connect ideas on how information can be produced from data, adding content for this potential knowledge provision.

It is our purpose, in this chapter review, to introduce the analytics background as a contributive theme like a valid update on the original approach. This chapter develops a contextual reflection around IA concept adding the impacts of analytics trends over information architecture, again evaluating on how these new definitions potentialize benefits or challenges for healthcare business models.

This chapter intends to develop this multifaceted conceptual relationship, adding the consideration about analytics, observing some of its potential aspects, and then study it on the arena of business models for healthcare services. This area always offered perspectives for Information Science and other multidisciplinary researchers and professionals, especially because it is a major producer and consumer of data, information and knowledge for various tasks and levels of decision – reaching from merely operational to critical ones.

The first section presents a discussion of theoretical background, relating information architecture, analytics and business model concepts, also approaching some of the supporting conceptual definitions which allow the understanding of this new research context. After, a perspective of conceptual integration is worked, with the study of the continuity of analytics, as an information management phenomenon, producing a potential definition for new business models. As the last section, a discussion on how analytics promote an information architecture which possibly defines a business model for healthcare is approached, resulting in a practical, oriented result of this study.

BACKGROUND

The main concepts to be approached in this review are, undoubtfully, analytics, information architecture and business models. However, a previous, brief study around information and its related context – supporting concepts, like data, knowledge and information management – must be addressed to produce a coherent approach for the desired development. This way, we start leveling those definitions, and then studying the main topics of this chapter.

Data, Information, Knowledge and Its Associated Managerial Aspects

The elementary base of several of Information Science is adopted here as the main origin of discussion for these concepts. Nevertheless, other disciplines and scientific areas, such as Information Technology, Information Science, Management Information Systems, and emerging topics, like internet of things and big data can contribute expressively to clarify its formation and relationships. Information, the main focal point of this study, is a critical concept, which must be observed for the continuity of the study. Aligned to our methodological strategy, information is the concept which relates to the others, enabling a sufficient comprehension around its various implications and impacts over the theoretical background.

The following discussion is totally based on the previous work, Jamil *et al.* (2013), which served as the research platform for this chapter development. Authors defined data, information and knowledge not only with isolated definitions, but also in terms of their relationships and functional interactions (for example, in automated and computerized information systems for tasks such as collection, storage and distribution). Tuomi (2000) adopted an apparent "reversal", contradictory way of thinking, when it was defined that data could be produced from knowledge when developing models and database schemas for application in software development. It is interesting to note, however, that this study reinforces each concept characteristics, along with their potential relationships. From sources like Davenport and Prusak (2000), Lucas Jr. (2005), Turban, Mc Lean, and Wetherbee, 2002, Turban, Rainer and Potter, 2007 and Stair and Reynolds (2009), it is possible to perceive how Information Technology (IT), and its associated critical view, approached such concepts in themes like technological infrastructure, information systems design and web-based applications development.

Jamil (2005) conceptualized data as absolute value that, for example, can be obtained directly from a measurement, observation or collected from an automated source and is expressed in formal, simple and standardized measurement units (as currency, metering system units, etc.). Although showing flexibility and easiness to generate, store and communicate, data does not have much of context, or immediate support for more complex decisions, being an understandable building block for more consistent concepts. This definition recalls several of others, as from those cited authors, like Davenport and Prusak (2000) and Stair and Reynolds (2009), considered remarkable works of their areas.

It is affirmed that information is a collection of homogeneous, correlated data, added with notions from the process environment where it was generated. Information develops conditions for organizational decisions, but it demands several additional, more expensive resources and works for it to be generated and applied like when one information system successful and planned execution, along with IS infrastructure (Turban, Rainer & Potter, 2007; Stair & Reynolds, 2008). Allen (1995) enunciated a more provocative definition, which added to this first a perception of information as a process that also defines a perspective opportune for this chapter study, as an "architectural" contribution. Considering information as a process, it is possible to relate it as an indisputable component to plans, plots, diagrams, specifications, methods, processes definitions, objectives, risk-management tasks, and other infrastructural requirements that will implicate in organizational design – i. e. a business model – among several other unfolding aspects for the organizational conformation and its relation, pressures and interactions with the competitive environment in which it competes. This way, we can relate information to architecture, in an initial approach of this final concept.

Buckland (1995) approach, explored after by Marchand and Davenport (2000) and other authors, reached the explanation of "informational services". These units were understood as corporative areas, service units or even companies that perform processing tasks with information to produce predicted results, required by end-users (Marchand & Davenport, 2000; Marchand, Kettinger & Rollins, 2001). These "informational services" also promotes the comprehension of information architecture and this set of disciplines could help to define its structures, forms of communication (both to be contracted, required, bought by a final user and to deliver its "informational products").

Based on these concepts, knowledge is defined from a gathering of information, including conceptual and practical definitions of the processes that generated that set of information. Knowledge is a complex concept, more difficult to obtain, manage (control its production, storage, sharing, communication, application, etc.) organizationally. It represents, on the other hand, a higher level of detail and comprehension about the organization environment (Ackoff, 1989) and its competitive behavior, producing comprehension on actions, reactions, planning scenarios and several other essential aspects and factors, enabling even competitive predictions (Davenport, 2000; Akbar, 2003; Jamil, 2005; Nonaka, 2008), as it supplies decision elements and propositions. Its related processes are designed, implemented, coordinated and comprehended when we reach the level of a process usually identified as knowledge management (Choo, 2003; Kearns & Lederer, 2003; Jamil, 2005; El-Bashir, Collier & Sutton, 2011).

It is possible to think about knowledge generation from data. This idea, quite provocative, is augmented, as we face big data and analytics events (Johnson, 2012; Jamil, 2013; Linke, 2017), in sectors such healthcare services (Srinivasan & Arunasalam, 2013). The conceptualization above illustrates how information plays a critical role in this transformation, as it can be comprehended as present in the middle of the conceptual relationship, resulting in a practical evidence, leading to understand, in a first level, how it is possible to relate information architecture to organizations building and, therefore, for business models.

Previously, Jamil (2013) defined that it is possible to perceive cycles of data and information gathering, processing and communication, which can lead to knowledge formation in each round. Processing these cycles, in an evolutionary maturity grading, it is possible to expect its evolution towards a better application of knowledge for strategic decisions and competitive comprehension, directly influencing its managerial capabilities, turning possible to understand knowledge management process evolution in such cases (Jamil, 2001; Jamil, 2013; Dimitrios, Sakas & Vlachos, 2013).

Organizations try to manage information and knowledge for several years with many objectives (Lang, 2001; Gal, 2004; Jacobson & Prusak, 2006). These processes, however, are not intuitive, standardized (as it is possible to adapt from a high-level modelling), easy to implement and usually do not return immediately (pondering from quantitative answers) or consistently without including several aspects of their strategic, long-term management (Davenport & Prusak, 2000; Silva, 2002; Choo, 2003; Jamil, 2005).

In this attempt to express information (IM) and knowledge management (KM), several authors proposed models – as it can be seen from studies like Cristea and Capatina (2009) and Haslinda and Sarinah (2009), which aimed to consolidate its achievements – allowing first to comprehend IM and KM sub processes and elements, after being guided for managerial actions regarding those IM and KM processes itself.

Here, a concise description of one model proposed by Jamil (2005) follows. It is affirmed that this modelling is not restrictive, a study on being an integrative platform, that could receive additional contributions, revisions and other updates. This approach was correct and we understand that analytics could promote one productive review of its main propositions. This model is structured based on the following sub processes:

- **Generation:** Observes the production, collection or capture of information and knowledge, both from external and internal organizational contexts. Events and facts like information technology integration, systematic collection about customers' interaction, laboratory tests, library composition or human resources hiring processes are typical for this sub-process. Here, we face an undoubtful positive pressure from big data, social media and analytics events, which proposes to produce data, information and knowledge.
- **Retention:** It is a subprocess that details actions, processes and methods to register, store and keep information and, if possible, knowledge treated by the "generation" sub-process. Information technology infrastructure, knowledge codification, "big data" classification, among many others, are aspects of study.
- **Sharing:** This sub-process evaluates how information and knowledge obtained are put available, in practical, reliable and optimal conditions for those who possibly interest on their application, whatever these end-users intend to do. Social media environments, communication methods, formal and informal teaching are some of the various possibilities for this sharing. A specific attention is recommended to strategic application of information and knowledge because of the critical results that depend on this availability. Social media contexts, platforms, processing big data elements and, mainly, analytics answers are among the new structures and methods that will produce content-sharing objectives.
- **Valuation:** Approaches organizational propositions to evaluate costs, prices, financial results, production indicators and other quantitative factors and aspects that can expose some of the gains obtained for information and knowledge management. This sub process potentially relates information and knowledge to other managerial processes and strategy for one organization.
- **Internal Cultural, Social and Organizational Forces, Motivations and Constraints:** It is essential to understand how forces like habits, hierarchy, market standards, organizational structure and other specific factors, typical for the competitive scenario of one organization, can impact on the integration for these sub processes. Some are adopted or simply exist for all organizations in a market sector (as legal demands for publishing formal account reports, for example), spontaneously originated throughout sometime of actions and competition. Others are developed by one business or social organization specifically, from these general conditions, as a training program.

Information and knowledge management can be defined, proposed and implemented in terms of these sub processes, integrated as operational and methodological contexts. For this chapter objective, it is possible to understand than IM and KM are conceptions that strictly relate to information architecture and both describe managerial contexts where information (and knowledge) flow in an organization, being treated, negotiated, processed, generated and used by organizational actors can propose organizational design and living interaction with its environment. Information architecture, as it is detailed in the following sub-section, promotes the analysis for infrastructural and managerial base for processes, such as IM and KM, allowing to compose a better design aspect of the information architecture. It is a possible "two-way" relation, where processes of IM and KM dynamically correspond with organizations´ information architecture.

Analytics

Processing data for further business purposes is a task which is becoming strategically critical as time evolves. A view, interesting for this chapter text, will be devoted to Healthcare services, but with expressive similarities to what can be found in other sectors and services. Not a long time ago, data was collected and manipulated, through application of basic statistics, to address simple problem, as to plan, predict and follow resource allocations, like rooms, surgical devices and facilities, even personnel to answer patients´ needs – seasonal, peak – for healthcare services.

Significant facts, however, changed this situation dramatically. Aligned to what Yonce *et al.* (2017), it was possible to understand crescent offer of information technology implements – devices, computer networks structures, configuration and implementation of software, software development platforms, etc. – competitive emerging technologies – data base, data mining, data warehouses (Obenshain, 2004; El-Gayar & Timsina, 2014) – and now, new strong market offers, still undefined from commercial, market and technologic points of view, as big data resources, internet of things applied for human monitoring, free software platforms, among other significant factors and signals (SAS, 2014; Jamil, 2015; SAS, 2017). Interestingly, as Medicine and associated disciplines and supportive topics evolve, complexity also reaches Hospitals, research institutions and other healthcare services excellency centers, producing more provocations to understand phenomena, processing data, information trying to obtain knowledge in demanding and intricate processes, adding additional requirement for more IT and information system resources and services. All of that with the characteristic growing demand for services, by all mankind, as healthcare perspectives also results in more demands for professional answers (Gillespie, 2014; Drell & Davis, 2014).

Analytics is one of the strongest trends with potential to be applied in healthcare services (Jamil, 2013; Gillespie, 2014; SAS, 2017). Analytics is presented by specific definitions, as in SAS (2017), where it was conceptualized like:

an encompassing and multidimensional field that uses mathematics, statistics, predictive modeling and machine-learning techniques to find meaningful patterns and knowledge in recorded data.

This definition is also supported by other authors, such as Jain (2016), who approached both as:

Big Data are the data sets that are so huge they cannot be captured, stored, processed, and analyzed with traditional tools. Predictive Analytics, on the other hand, is used to extract information from data to make better decisions.

Dodge (2010) and Diakantonis (2016) addressed, in opportune ways, how these informational services and processes are necessary components for answers to regulatory demands, being examples on how analytics could promote new business models answers to customers and organization's needs.

Some typical services of analytics are commercially exemplified in Mehmod (2017), an objective and ludic reference which shows how these tools can be immediately applied. It is possible to identify how initial level data, collected from usual processes in services (as in healthcare units) can produce solid, formal evidences which can result in stronger analysis, which, by its turn, will support further decisions and planning abilities.

Analytics usage, together with big data perspectives, result in a proposition of processing data collected from several sources to produce effective, application-designed knowledge. In general, this arrangement supports the knowledge management process definition and implementation desired for many organizations (Jamil, 2005). But, for healthcare units, it answers for life support dynamics, quality of life improvements and medicine knowledge development, producing scientific, commercial and market opportunities that impact expressively how this essential field will evolve (El-Gayar & Timsina, 2014; Gillespie, 2014; Raghupathi & Raghupathi, 2014; Diakantonis, 2016; Oliveira, 2017).

The apparent simplicity of analytics, when addressing the massive production and collection of data, with the huge number of organizational implications, the most relevant as to dynamically generate knowledge for decision-making, hides an evolutionary array of software, hardware, connections infrastructure, logical and operational design, implementation techniques and process definitions, which are consolidated in simple, routinely tasks and products, like mobile applications, interface-oriented systems, internet of the things devices, and so forth (Jamil, 2015; Siegel, 2016; SAS, 2017). This is an evolutionary field that will allow the implementation of intelligent devices and its associative connections, together with implementation of solutions which will answer more sophisticated problems in the forthcoming period, also providing knowledge to be implemented in alternatives for business models which will encompass these technologies to change how healthcare services can be offered to customers. Along with operational promises, like "telemedicine", it is possible to find optimization for healthcare plans, governmental and private investments and business model developments (Dodge, 2010; Diakantonis, 2016; Oliveira, 2017).

The interest in analytics on the healthcare sector is evolving and it is perceived in the comprehensive development from the studies of information system applications (such as LeRouge, Mantzana & Wilson (2007) to updated researches, conceived by Siegel (2016). These dynamics find, nowadays, the fertility of new techniques and process implemented, producing the unquestionable dilemma with cultural and regulatory standards which will be accessed in the last section of this chapter.

Information Architecture

Information architecture is a multidisciplinary approach, defined by authors as concerning information, both as a concept and with some related aspects, such as systems, technology, management, with a proposal unlimited perception of its interaction, which makes it possible to understand the "architectural" relationship.

Architecture is an actual, vivid, wide area of applied knowledge whose most known results are those of remarkable, artistic, useful, cultural and social propositions over buildings and infrastructural projects. Bringing representative institutes conceptualizations, it is a scientific field, which works over design principles, considering esthetical and functional aspects of constructions, taking into considerations user interests, social and environmental implications, cultural and historical influences, along with constructive technology and processes (RIBA, 2014; RAIC, 2014; IAB, 2014). These fundamental aspects of Architecture form a first alignment with information, calling attention to a possible context for information architecture, where the intention, need and plan to produce optimal, esthetical, rationalized and efficient blueprints of information processes and artifacts in one organization life cycle are demanded by users.

A simple research around IA concept produces the evidences of a new, innovative subject, even considering studies some years or decades ago, and notions about the continuity of theme´s evolution, mainly because of the second symptom. Also, it is possible to define that information architecture is conceived not only as a theoretical concept, but frequently related to practical, application-oriented implementations, like several functions and concepts of "Design" (here presented as a concept with spread discussion in modern times) and other related aspects for example usability, accessibility, technology, applications, systems and organizational environments and process.

One of the original points for information architecture concept is the work of Richard Saul Wurman (Wurman, 2014). His studies around the topic, as an experienced architect and designer who observed information-related phenomena, resulted, since the mid-1970s, in considering that information was massively generated when organizations and practitioners faced tough demand of its application in critical, networked environments. It is a situation already approached before, in this review when analytics theme was discussed.

Wurman questioned the need for structures that allowed the overall perception and productive control of information, reaching a way where it will be possible a minimal level of management. His interest over information sharing persists, as information continues to show easiness to produce, difficulty on being organized stored and very complex to control. As we present a reality where data and information have an uncontrolled way of production, and somewhat close to a chaotic fact, Wurman´s studies and conclusions, which based information architecture initial conceptualizations, are completely valid nowadays and it is supported by an increasing number of citations frequently of his works. Information architecture is a discipline that helps to promote this arrangement, providing methods for various information-related functions, like identification, classification, mapping, storage organization, systematic access, among many others, improving also the domain of the interaction of information contents and process with the environment of their usage.

This way, it is correct to understand that information architecture conceptualization was approached by different scientific areas and practical contexts. From Davenport and Prusak (2000) or Davenport and Marchand (2000), it is possible to identify how IA is expected to serve as part of the informational infrastructure for information technology projects, providing a qualitative level, resulting in benefits for its implementations. Information architecture relationships and motivations to information and knowledge management were studied by Choo (1996), Jamil (2001), Choo (2005) and Nonaka (2008), presenting concept-based works that will serve to map, identify and allow to better address practical issues and perspectives. Information architecture also appears in developments of ideas on the information systems arena, covered by Stair and Reynolds (2008) or information technology researchers, as Turban, Mc Lean and Wetherbe (2002), Lucas Jr. (2005) or Turban, Rainer Jr. and Potter (2007), when they usually observe IA as a field oriented to provide business and usability detailed specifications for information technology projects, enabling to reach best results for external user satisfaction. Other authors worked on the concept specifically or others, related to it, leading to appreciation of IA, like a base for practical projects as Leidner and Elam (1995); Marchand, Kettinger, and Rollins (2001); Inmon, Strauss and Neushloss (2008).

Bringing other contributions to information architecture definition, aiming to understand its relationship from one side to data, information and knowledge management fueled by analytics, and, on the other side, as a contribution for business modelling techniques and principles, other works are cited in the following. Dillon (2002) proposed a *small information architecture*, from a dual view of the basic conceptualization, identifying it mainly from tactical user interactions with information for decision-making in strategic planning executive processes. For the same author, *big* IA relates to a wider view of information architecture, where it relates to its strategic management for complex, high-level decision-making, planning and overall structuration of a complete organization point, in a macro vision of architectural principles, inspiring the evaluation of its perspectives for effective business models design. Bailey (2002) also advanced his studies this way, addressing functional aspects for information like finding, navigating, storing and updating were considered elements, which compose the needed architectural plot, serving for information usage, again producing some hints on organizational structure and associations with external processes.

These relationships compose a critical part of business models proposition, and information flow studies, which was proposed from Morville and Rosenfeld (2009), make it possible to understand how the architectural fundamentals produce implications for optimal ways to accomplish strategic goals. These authors pointed out four principles that reinforce our considerations of IA – Business model relationships:

- Initially, it is a "structural design of shared information environments" which defines information architecture design purpose and possibilities;
- Information architecture has a focused practical view about information management and its implications.
- IA is the "art and science of shaping information products and experiences to support usability and findability". This way, these authors relate information architecture to its perspectives of producing results which, for this chapter purposes, design business models.
- Is an *emerging discipline and community of practice focused on bringing principles of design to the digital landscape*. Aligned to the definition of Information Architect Institute (IA Institute, 2014), IA it is an opportune field which aligns for information technology innovations, new business models and other provocative and new offers by the market and propositions from the scientific context.

Francke (2009) studied information architecture as a series of associated disciplines, regarding both to information mapping, systems design, retrieval and usage, which will present a focused information management paradigm. This citation reinforces information architecture principles, a topic which allows to understand functional aspects and design for organization models.

Covert (2014) defined that *IA a set of concepts that can help anyone make anything to make sense of messes caused by misinformation, disinformation, not enough, too much information*. From this point of view, it is possible to comprehend Wurman´s motivations over the need for information architecture are still present, and facts presented in the big data approach (Mc Kinsey, 2011; Ohata & Kumar, 2012; Park, Huh, Oh & Han, 2012; McAfee & Brynjolfsson, 2012; Budford, 2014), suggesting, in support of our findings in this text, that information is easier and faster to be produced these days, generating even the perception of an "information overload".

Jamil *et al.* (2013) composed these references, producing a conceptualization for information architecture as *a set of disciplines and knowledge that is applied to allow information life cycle in organizations*. This way, IA involves methods, procedures, communication symbol standards, professional profile definitions, specifications, regulatory demands (legal, ownership, etc.) and other aspects and items that:

will enable the optimal information processing – generation, collection, storage, sharing, communication, processing – but also its continuous application for end-users planning and development purposes.

From these works, with the conceptual consolidation presented, a detailed view of information architecture principles can be announced, affirming that, from Jamil *et al.* (2013):

- IA is a function-related concept, application-oriented. It is strictly related to real-time information functions, especially its usage – retrieval, application, etc. – by the final user.
- Information must be precisely identified. It must be classified, correlated and documented, supported by mapping and other techniques.
- Information architecture can reach goals not addressed by the original fundamentals and propositions of Architecture main field, eventually proposed as a wider concept, although still not presenting a closed, terminative definition.
- It is needed by an organization like a conceptual fundamental (tools, methods) for its own planning, as an information management support, represented by standardized symbols, methods, professional profiles and many other resources.

Approaching its results, it is possible to define usual objectives of information architecture, as:

- IA helps one organization implement knowledge and information management;
- IA contributes precisely for information systems and technology tools implementation, methods and processes development, aligned to strategic organizational needs;
- IA also enables to evaluate human resources management for information services and works, both for operational and managerial tasks.
- IA defines standards of informational specifications for an organization and its related business environment, contributing to define business rules and demands for an aggregated-value chain.

Because of all definitions and reflections, information architecture has a strong potential to organizational design, going beyond the intended expectations of just a "drawing scheme", but to apply information management principles and impacts to the best way for organizational perspectives proposing new business models, which will be studied in the next section.

Business Models

Management research history shows how difficult is to know when business models (BM) became a specific investigation theme or objective. Several studies from academic centers, like Harvard or London Business School, cited terms such as "organization", "business methods", "business rules", "structures", "business process". So, it is fair to affirm that this theoretical front was proposed several years ago, although contributions and developments of this topic are among the most relevant topics. There is no doubt that several new companies and innovative markets called special attention, some of big successes, other of big failures.

From Deshler and Smith (2011), citing works from Alex Osterwalder, define business models:

define how one organization meets the unique needs of its customers, delivers a set of differentiated set of products and services, highlights required capabilities (…) and generates value and profit.

Business model innovation is at the realm of most of strategic intentions by competitive companies nowadays (Khanagha, Volberda & Oshri, 2014) when they can define competitive advantage as a resource *per se*, producing a differentiated value for final customers, also supporting new ways to reach and compete for markets and customers.

It is possible, for instance, to perceive, from the seminal studies from Edith Penrose, conceptions about the operations and its related strategic definitions for firms, with the fundamental idea of a company being a "set of controlled resources" (Penrose, 1959). In this work, it is possible to find conceptions for various attributes for processes, methods, strategic planning and other critical aspects of an organization, guiding how it can define and apply its main resources for competitiveness. This is a genuine affirmation regarding definition of business model towards market orientations.

For definitions and suggestions, we start with Malone *et al.* (2006), who discussed some definitions for business models, from information technology and digital markets typical companies. In their study, they approached authors that defined BMs as a set of processes, rules, systems and methods to offer their services or products through a predictable way of doing business. They also focused on factors, like customer relationships, internal and external technological connections, tools and methods for performance control and definitions of commercial and marketing value. To contribute for a formal definition, US Federal Government (US Government, 2007) developed and adopts reference business models – Business reference model (BRM) and Performance reference model (PRM) – in the proposition of "FEA – Federal Enterprise Architecture", which constitutes in a documentation designed mainly to *facilitate cross-agency analysis and the identification of duplicative investments, gaps and opportunities for collaboration within and across agencies*. These business models – not applied to commercial companies – considered frameworks, composed by diagnostics, auditing specifications, structural definitions and communications details that will serve for any public manager interested in implement new functions or use some available service.

From Mintzberg, Ahlstrand and Lampel (2009), we understand how business models can be created, as a strategic proposition, answering with "structures and power" for competition. Through the exposition of seven alternatives, the authors build, adapting from an earlier publication of Henry Mintzberg, conceptions for an organization to perform, considering its internal forces – cohesion, systems, process, communication, power negotiations, etc. – and external aspects – such as market, regulatory, entrepreneurial abilities, politics, etc. It is possible to enumerate this factors as some of the most affected when designing a competitive and value-oriented business model.

Chesbrough (2010), approaching open innovation theories, evaluated how a technology, by itself, does not constitute *per se* a way of doing business, but supported by the definition of a business model, can deliver expressive value potential. It is possible, from this affirmation, to relate these fundamentals to information architecture, as it can be conceived as a design and managerial configuration that can support an effective technological implementation. This way, it is possible to foresee how a potential BM – IA interaction implements a business model, like flows, processes and goals set for information and knowledge-supported value offer, can result in a differentiated positioning. Chesbrough had been discussing the BM relating it to a field where his contributions are remarkable, the proposition of "Open innovation", which can be understood also as an openness of a business model by one organization to allow a more effective innovation generation process, created a remarkable area of applied research and entrepreneurship.

Choudary (2013) presented two main conceptual frameworks, propositions of business models structures, allowing a modern organization to compete: the pipes and platforms. "Pipes" refer to serial, almost one-way process of production or services offering, which is adopted in its classical or adapted fashion. It is a referential proposition, as various markets and companies still adhere to those traditional definitions. Alternatively, "Platforms", in a newer way, are assigned to organizations like Internet-based companies (social media platforms, digital markets, electronic commerce, etc.), providing layers of doing business, offering "connections" and "experiences" for final users, like value transactions, associated to functions, resources and processes, to build a system for "create and consume value".

Finally, Mullins (2014) studied and researched modern business alternatives, reaching five business model architectures:

- **Signature Model:** Where users pay periodically to receive a set of products and / or services focused on a same theme, topic or sector (like magazines, it is possible to identify clothing, beverages, tourism packages, etc.)
- **Scarcity Model:** Where the providing organization controls the offer and associate it to some quantitative parameter for value offer, imposing negotiations with customers – like fashion companies which extinguish one seasonal offer with a previous, rigid schedule, announced for customers.
- **Pay in Advance Model:** Where some users pay anticipated for a value, a benefit, which they can receive with a discount. For example, one software user can anticipate 20% of the value of a new version, paying (financing) for its development, and then receiving the new version before their competitors, or with a financial award, a possible discount that compensates the amount of 20% offered for developers.
- **"Servicing" for Products:** When one product is embedded in a service, allowing an expansion of value offer. For example, when sportive material is sold not as the object material itself, but as part of an inscriptional kit for a marathon. So, if the customer wants to join this competition, he or she can buy the kit, composed by sneakers, clothing and additional services.
- **Matchmaking Model:** Where one company provides a platform to proportionate negotiations between resource owners – cars, rooms, programming code, installations, facilities, shop spaces, etc. – to potential users. This is recognized as the "Uber" model, like that applied by the successful, global competitor for mobility using application supported interactions of drivers and potential users.

For this chapter study, a business model can be conceptualized as a definition reached at Jamil *et al.* (2013), as an artifact which exposes the way an organization perform its functions towards its customers, through a documentation that includes diagrams, specification of business rules and relations, along with definition of systems and processes, structural definition and, most important, how an organization conceives its functioning. With this definition, we have a higher-level conception that will serve as a framework for any flow, integrative function, process or system implementation or adaptation, working as a "roadmap" on how the organization must use all managerial tools to produce.

With this BM definition in mind, it will be possible do develop the analysis about its relationship to information architecture aligned to the main objective for this chapter.

Supporting Concepts

Some concepts are opportune to define, to compose the final picture of this theoretical background. Here, briefly, information technology, information systems, strategy and strategic planning are addressed and defined.

Information technology (IT) can be consolidate from several sources, as Lucas Jr. (2005), Turban, Mc Lean, and Wetherbee (2002) and Turban, Rainer and Potter (2007). From these, also perceiving the citation in some additional references, it is possible to understand IT as the set of computational resources and its related materials, artifacts, process definitions and structures to answer business's needs. IT has been evolving continuously from an operational support provided by "computers" and "software", typically a connected solution, which became the place for informational platforms and mobile communications. Information technology provides integrated support for network-model companies to a business model support, nowadays even representing new forms of entrepreneurship (Henderson & Venkatraman, 1999; Jamil, 2001; Turban, Rainer & Potter, 2007). It is an active proposition for new organizational models, for example, associating providers to consumers in marketplace-shaped business models (Mullins, 2014).

Information systems can be understood as an organizational definition to encompass a set of resources, like information technology tools – software, hardware, associated processes – methods for their usage, skilled personnel and structural design, as communication specification, processes and infrastructure and other coordination details. This set of integrated components provides information and knowledge to answer user´s informational needs (O´Brien & Marakas, 2008; Stair & Reynolds, 2009; Laudon & Laudon, 2013). This area is expressively related to information architecture, depicting the overall way on how information can be processed to produce knowledge, from the conceptual point of view of its elementary components.

Finally, strategy is defined as the long-term coordination of the positioning for one organization, proposed from organizational perception about external, internal environments and associated alignment of resources to be applied to improve this positioning (Penrose, 1959; Porter, 2008; Mintzberg, Ahlstrand & Lampel, 2009). With this objective, organizations aim to produce idealized visions for this future, design it into plans which coordinate goals and actions to reach these goals, execution processes and resources definition and, finally, the monitoring of this execution, as criteria to understand its potential for success. This process is defined as strategic planning (Jamil, 2005; Porter, 2008).

With this conceptual background developed, it is possible now conduct the desired level of reflections on how information architecture, analytics proposition and business models can be related to propose strategic differentiation, studying it for healthcare organizations.

APPLIED VIEW FOR IA AND BM RELATIONSHIPS

In this conclusive section, we have the main objective to relate information architecture, business modelling and analytics, producing a potential correlation which will serve for new business models design. Those concepts are still being debated, challenging authors, researchers and practitioners, their relationship is also difficult, leading us to propose, in a first level observation, a reflection on how they can be discussed in an integrated way.

One organization has many implicit and explicit conceptualizations, as statements, plans, directions, decision-making processes, structures, systems, profiles, processes, methods and several other items that can determine and describe how it works internally and, mainly, how it produces, offers its results and interacts with the external people and other organizations. Generally, it is not a simple plot, it is not a mere design principle, but complexes relationships, formal and informal agreements, goals possible and impossible to achieve, human resources implications and relations, among many other facts that impact these specifications. Analyzing this way, a business model can be considered an overall guidance, a plan specification, a more global, high-level design, which will orient an organization for some specific action (as to attend to a customer), or to a more permanent way of performing – as a process structuration for one information technology implementation. All these aspects can constitute a business model.

In a healthcare unit, this comprehension must consider the relevance of two strong external forces: culture and regulations. The first one comprehends the behavioral standards adopted, defined and developed by healthcare players, organizations by themselves and in relationship with each other and with external stakeholders. Additionally, regulatory systems exist to promote the safety and access to healthcare services and products, guaranteeing its governance, transparence and risk management conditions, as this sector deals with the most critical existing asset: human lives. This way, some announcements of innovations, new business models, for example, simply can not be implemented (because of regulations) or face though difficulties (originated from cultural factors and determinants). This is a strong delimitation for the intended association of information architecture and business models, studied in this chapter.

At a first glance, we can observe the relationship where IA serves to design a configuration for a BM, which will analyze information (through its architectural aspects) to build new organizational propositions – structures, processes, methods, command requirements, etc. – that will imply on how the healthcare provider performs. In another point of view, we also can understand that one business model can be designed just to make the definitions of an information architecture guidance result in a practical structure, implemented for the final services for customers. In this first appreciation, we did not check the role played by analytics in this situation, in the sake of simplicity.

It is possible to compose a simple picture were a healthcare private service provider, such as a "health insurance plan" company, can, for example, review how one hospital will check-in urgent patients, implementing internet of things resources, when still monitoring these customers at home. So, hypothetically, if a patient arrives at hospital, after being monitored with these online resources, his admission will be immediately processed by the plan, associated with the hospital, in a faster way, simpler than the usual, implemented a "fast-track" admission subsystem which was especially designed for this information architecture – based internet of the things implementation, as innovative business model. This monitoring can, for example, include data that will help answer faster to avoid some critical events, as, also, to avoid conducting the patient to a wrong way of medical service, also contributing for operational optimization. Along with a strategic plan, this can become a competitive advantage for the healthcare unit, performing with additional, valuable benefits, to better profit conditions and management of human resources.

As Drell and Davis (2014) approached, some simple analysis could be defined and implemented to produce studies of rudimental predictions, like those done by commercial marketers, to understand the evolution of a symptom or a patient´s behavior. This way, pharmaceutical industries, marketers and, mainly, healthcare institutions, could try to predict this happenings in the future, expanding or adjusting technical and commercial offers to customers and future patients. This could allow logistics, human resources and financial management to be held in better conditions. Thinking this way, it is possible to identify an "entry point" for analytics, when these resources will contribute to produce this periodical analysis of evolution, providing a better planning condition.

Srinivasan and Arunasalam (2013) pointed also the processing of structured data – collected from standardized documents – and unstructured data – filled in textually, by healthcare center professionals in a non-standard way, when they freely write texts (for example with the patient´s service story) – to automatically understand how several healthcare procedures´ schedules and events, and, this way, detect possible errors or even frauds, when opening new service processes and preventing managerial mistakes and risks. This way, it can be expected that this units´ management could be optimized, as a potential evolution of big data (structured + unstructured) analytics are applied. A new business model, which will provide each professional with a combination of more autonomy and responsibility, generating a potential matrix-shaped process design for business operation, can be implemented. Information architecture principles can be addressed to guarantee that all produced and collected information will be managed according to stakeholder´s principles and regulatory demands.

It could be analyzed, from Gillespie (2014), how data collection, processing and filtering, together with the consolidated results exposed through accessible dashboards, could support even basic hygiene methods to be reinforced, trained and shared. This way, this leads to a provocation regarding to what Oliveira (2017) called "data governance models" for healthcare, with close demands for structural organization issues, or, eventually, new business models design and implementation in healthcare. It is an evidence of relationship of information architecture and business models' definition, based on data and information management. For Diakantonis (2016), this is an opportune way on how these informational services and processes are necessary components for answers to regulatory demands, being examples on how analytics could promote new business models answers to customers and organization's needs, according to information architecture principles.

In the original work, which is the base for this study, two ways of influence about these concepts were addressed: business model relating to final information architecture adoption and, on the other hand, how information architecture emphasizes characteristics for new business models demands. Let us review those findings, aiming to introduce analytics.

When thinking how a business model influences the information architecture adoption, it is possible to observe this relationship as a potential inference, a context of high level definition when those concepts – BM and IA – are, indeed, sets of principles. A proposition of a new business model, which, for example, will take advantage of analytics, happening in a "marketplace" hypothetical offer, where specialists of one healthcare area will compose a set of providers to be contracted by the marketplace registered users, can result in a preview to be tempered by information architecture principles. The final organizational design will have possibilities to define an optimal design of analytics processing, identifying relationships needed by customers (for example, when the customer did not feel its needs supplied by the offers in the marketplace), adapting its design, its operational conditions and ways to reach new customers, based on information architecture principles – for example, when the platform will be programmed to acquire information from clinics and other healthcare centers, filling in the final needs for that type of customer.

On the other hand, extremely flexible organizational with dynamic structures, theoretically offer optimal conditions to implement business models which facilitate definitions for information architecture (Kettinger & Marchand, 2005; Kettinger, Zhang & Marchand, 2011). This can happen, for example, analyzing the signature-based business model. If one healthcare insurance provider, for example, structures its services dynamically, planning its processes information adopting analytics resources, allowing customers to pay in advance for potential treatments (monthly installments) and adopting a compensation system, which will return to these customers some benefits if they do not use the complete set of healthcare services. This way, the adaptive business model will respond to the patient needs potentially in advance, resulting in a perspective of information architecture processes to integrate other organizations.

Aiming a final reflection, it is useful to recall how business models can affect information architecture because of strategic planning. When studying and proposing its future competitive developments, through strategic planning, an organization adopts, for instance, generic business models, trying to implement standardized and integrated processes. One case is the "low cost" strategic positioning (Porter, 2008), which determines that one company will propose all its efforts for an optimal cost management approach, enabling this company to offer its products or services with the lower competitive cost in its sector, in a strictly managed way, to be perceived by a final customer (Hitt, Ireland & Hoskisson, 2012). With this market approach, this hypothetical company can modify its information management principles, for instance, implementing analytics through customer-related transactions to reduce costs of commercial interaction, changing also its model for information architecture, and the automatic implementation, will both reduce costs and produce a more agile company, with the risk of less flexibility, that must be strategically addressed.

As exposed in the previous discussion, it is a provocative, endless and challenging association when rethinking the complementarity of business models and information architecture conceptual arenas, applying a new wave of technological assets, as it happens with analytics. From the previous work, Jamil *et al.* (2013), is useful to remark the citations of some reflections around healthcare services, such as:

- Healthcare services can be considered as one of the most focused services for human living, receiving demographical, technological, social and political impacts for its services offer. So, several analytical models, from different areas, can be applied for any study, and this happens with this association, which can be analyzed from computing science, information science, knowledge management, business operation, marketing, logistics and several others.
- Taking HS services as its more frequent, practical operative units, such as hospitals, clinics, medical staff offices, it is possible to understand how information architecture is a fundamental conceptual base to understand how the value-aggregated chain produces knowledge, improving business models' application and management.
- Healthcare units can be observed as informational units, or associations, with immense potential to produce knowledge for its own managerial purposes, or, through a conception provided by information architecture definition, with relation to other markets, scenarios and sectors (as, for example, governmental services, pharmaceutical industries, etc.).

It is the final intention of this development, to understand how it is possible to apply information architecture principles as one of the topics, still not completely understood by the market, associated with technology assets and resources, also in a comprehension process of its perspectives of application, to propose, design, develop, implement and adapt business models. In this study, a focus on healthcare units and services called reader´s attention on how one essential, critical sector can be affected. This can invoke additional perception on how our lives can become differently associated to organizational actions, as this association – IA and BM through analytics – becomes more practiced, eventually without ideal planning conditions and requisites.

CONCLUSION

In this chapter, we aimed to study how analytics, as an emerging and affirming technologic trend, can be applied to promote the outcomes of the high-level relationship of business models design and information architecture. One business model can be proposed, analyzing findings from several sources. One can compose a business model and then implement it strictly following a plan, then promoting adjustments on it over time. It is just the scientific way to think about an organizational model, as a general guidance to build a real implementation. Alternatively, a business model can be designed from the operational conditions, emerging as an evidence, as those were perceived by a company in its operational areas. From theory, several ways of conception for a business model can be defined, as it can be created from one aspect – exactly planned – to the philosophical opposite –intuitively generated for design and implementation.

A business models will relate company´s main objectives to its tactical planning and, in the final, to a complete operational definition, making it possible to understand how the organization strategic planning will result in final position and deliver of value for its customers. As information architecture signals how information will flow internally and externally in one organization, exposing and setting continuity for its works, it is interesting and advantageous on appreciate IA fundamentals as those plans are composed. The association of IA and BM will result in a more robust proposition, an additional coherency to commercial, tactical plans, leading to a potential new source of competitive advantage.

In this scenario, analytics emerge as a strong force to allow more comprehension for information architecture principles, enabling the design of optimal and or adaptive, flexible business models. It is a competitive advantage that can be promoted from one innovation, needed by healthcare systems in all world, and there are several debates around its perspectives, changes and challenges facing its cultural factors, strategic demands and life evolution.

NOTE

This chapter presents a review of the former publication Jamil, Jamil, Vieira, and Xavier (2015), keeping its basic structure and parts of its original content, enhancing it with additional references and theoretical and practical work around analytics and related themes, as a valid context application for healthcare business models design and implementation.

REFERENCES

Ackoff, R. L. (1989). From Data to Wisdom. *Journal of Applied Systems Analysis*, *16*(1), 3–9.

Akbar, H. (2003). Knowledge Levels and their Transformation: Towards the Integration of Knowledge Creation and Individual Learning. *Journal of Management Studies*, *8*(40), 1999–2020.

Allen, B. (1996). *Information tasks: Toward a user-centered approach to information systems*. San Diego: Academic Press.

Bailey, S. (2002). Do you need a taxonomy strategy? *Inside Knowledge, 5*(5). Accessed at http://www.ikmagazine.com/

BPM Institute – Business process management Institute. (n.d.). Access at http://www.bpminstitute.org

Buckland, M. (1995). *Information and Information Systems*. New York: Ed. Praeger.

Budford, S. (2014, October). A grounded theory of the practice of web information architecture in large organizations. *Journal of the Association for Information Science and Technology*, *65*(10), 2017–2034. doi:10.1002/asi.23098

Chesbrough, H. (2010). Business model innovation: Opportunities and barriers. *Long Range Planning*, *43*(2-3), 354–363. doi:10.1016/j.lrp.2009.07.010

Choo, C. W. (1996). The knowing organization: How organizations use information to construct meaning, create knowledge and make decisions. *International Journal of Information Management*, *16*(5), 329–340. doi:10.1016/0268-4012(96)00020-5

Choo, C. W. (2005). *The knowing organization: how organizations use information to construct meaning, create knowledge and make decisions.* Oxford, UK: Oxford University Press. doi:10.1093/acprof:oso/9780195176780.001.0001

Choudary, S. P. (2013). Why business models fail: pipes and platforms. *Wired Magazine*. Accessed at http://www.wired.com/2013/10/why-business-models-fail-pipes-vs-platforms/

Covert, A. (2014). *How to make sense for any mess: Information architecture for everybody*. Charleston, SC: CreateSpace Independent Publishing. Retrieved from www.createspace.com

Cristea, D. S., & Capatina, A. (2009). Perspectives on knowledge management models. *Annals of "Duranea de Jos". Galati University*, *15*(2), 355–366.

Davenport, T. H., & Marchand, D. (2000). *Mastering information management.* Boston: Financial Times Prentice Hall.

Davenport, T. H., & Prusak, L. (2000). *Working knowledge: how organizations manage what they know* (2nd ed.). Harvard Business Press.

Deloitte. (2017). *Five questions about applying analytics to risk management*. Available at https://www2.deloitte.com/content/dam/Deloitte/au/Documents/risk/deloitte-au-risk-risk-angles-applying-analytics-risk-management-250215.pdf

Diakantonis, D. (2016, February). Data Matters in Healthcare. *Mergers and Acquisitions*, 27-28.

Dillon, A. (2002). Information architecture in JASIST: Just where did we come from? *Journal of the American Society for Information Science and Technology*, *53*(10), 821–823. doi:10.1002/asi.10090

Dimitrios, N. K., Sakas, D. P., & Vlachos, D. S. (2013). The role of information systems in creating strategic leadership model. *Procedia: Social and Behavioral Sciences*, *73*(1), 285–293. doi:10.1016/j.sbspro.2013.02.054

Dodge, J. (2010). Mining for Gold. *Information Management*, *10*(5), 38-40.

Drell, L., & Davis, J. (2014, Fall). Getting started with predictive analytics. Marketing Healthcare Services, 23-27.

Dremel, C., Herterich, M. M., Wulf, J., Waizmann, J., & Brenner, W. (2017, June). Digital transformation often requires big data capabilities. *MIS Quarterly Executive*, *16*(2), 81–100.

Dreshler, R., & Smith, K. (2011). Making business model innovation stick. People and Strategy, 34(4), 19-23.

El-Bashir, M. Z., Collier, P., & Sutton, S. G. (2011). The Role of Organizational Absorptive Capacity in Strategic Use of Business Intelligence to Support Integrated Management Control Systems. *The Accounting Review*, *86*(1), 155–184. doi:10.2308/accr.00000010

El-Gayar, O., & Timsina, P. (2014). Opportunities for Business Intelligence and Big Data Analytics In Evidence Based Medicine. In *Annals of 47th Hawaii International Conference on System Science*. IEEE. 10.1109/HICSS.2014.100

Francke, H. (2009). Towards an architectural document analysis. *Journal of Information Architecture*, *1*(1). Accessed at http://journalofia.org/volume1/issue1/03-francke/jofia-0101-03-francke.pdf

Gal, Y. (2004). The reward effect: A case study of failing to manage knowledge. *Journal of Knowledge Management*, *8*(2), 73–83. doi:10.1108/13673270410529127

Gillespie, G. (2014). How reporting and analytics and metrics affect outcomes and save lives. *Information & Management*, A2–A5.

Haslinda, A. & Sarinah, A. (2009). A review of knowledge management models. *Journal of International Social Research*, *2*(9), 187-198.

Henderson, B., & Venkatraman, N. (1999). Strategic alignment: Leveraging information technology for transforming organizations. *IBM Systems Journal*, *32*(1), 472–484. doi:10.1147j.382.0472

Hitt, M., Ireland, R. D., & Hoskisson, R. E. (2012). *Strategic management: concepts, competitiveness and globalization*. Boston: Cencage Learning.

IA Institute – Information Architecture Institute. (2014). *Recommended reading*. Accessed at http://iainstitute.org/en/learn/education/recommended_reading.php

IAB – Instituto de Arquitetos do Brasil. (2014). Accessed at http://www.iab.org

Inmon, B., Strauss, D., & Neushloss, G. (2008). *DW 2.0: The Architecture for the Next Generation of Data Warehousing*. Morgan Kaufmann.

Jacobson, A., & Prusak, L. (2006). The cost of knowledge. *Harvard Business Review*. Accessed at https://hbr.org/2006/11/the-cost-of-knowledge

Jain, C. L. (2016). How to use big data and predictive analytics to improve success of new products. *Review of Business*, *37*(1), 48–55.

Jamil, G. L. (2001). *Repensando a Tecnologia da informação na empresa moderna*. Rio de Janeiro: Axcel Books do Brasil.

Jamil, G. L. (2005). *Gestão da Informação e do conhecimento em empresas brasileiras: estudo de múltiplos casos*. Belo Horizonte: Ed. Con / Art.

Jamil, G. L. (2013). Approaching Market Intelligence concept through a case analysis: Continuous knowledge for marketing strategic management and its complementarity to competitive intelligence. *Annals of Centeris 2013 – Conference on Technology and Information Systems*.

Jamil, G. L., Jamil, L. C., Vieira, A. A. P., & Xavier, A. J. D. (2015). Challenges in modelling Healthcare services: a study case of information architecture perspectives. In G. L. Jamil (Ed.), *Handbook of research on information architecture and management in modern organizations*. Hershey, PA: IGI-Global.

Johnson, J. E. (2012, July). Big data + Big Analytics + Big opportunity. *Financial & Executive*, 51-53.

Kearns, G. S., & Lederer, A. L. (2003). A resource based view of IT alignment: How knowledge sharing creates a competitive advantage. *Decision Sciences*, *34*(1), 1–29. doi:10.1111/1540-5915.02289

Kettinger, W., & Marchand, D. (2005). Leveraging information locally and globally: the right mix for flexibility and standardization. *IMD Bulletin*. Available at https://www.imd.org/research/publications/upload/Marchand_Kettinger_WP_2005_02_Level_1.pdf

Kettinger, W., Zhang, C., & Marchand, D. (2011). CIO and Business executive leadership approaching to establishing company-wide information orientation. *MIS Quarterly Executive*, *10*(4), 157–174.

Khanagha, S., Volberda, H., & Oshri, I. (2014) Business model renewal and ambidexterity: structural alteration and strategy formation process during transition to a Cloud business model. Research and Development Management, 44(3), 322-240. doi:10.1111/radm.12070

Lang, J. C. (2001). Managerial concerns in knowledge management. *Journal of Knowledge Management*, *5*(1), 43–59. doi:10.1108/13673270110384392

Laudon, K., & Laudon, J. P. (2013). *Management information systems: Managing the digital firm* (13th ed.). New York: Prentice Hall.

Leidner, D., & Elam, J. J. (1995). The impact of executive information systems on organizational design, intelligence and decision making. *Organization Science*, *6*(6), 645–664. doi:10.1287/orsc.6.6.645

LeRouge, C., Mantzana, V., & Wilson, E. V. (2007). Healthcare information systems research: Revelations and visions. *European Journal of Information Systems*, *16*(1), 669–671. doi:10.1057/palgrave.ejis.3000712

Linke, R. (2017). Applying analytics – Ideas and examples from leading companies and researches. *MIT Management Newsroom*. Available at http://mitsloan.mit.edu/newsroom/articles/reading-list-applying-analytics/

Lucas, H. C. Jr. (2005). *Information technology: strategic decision making for managers*. Hoboken, NJ: John Wiley and Sons.

Malone, T. W., Weill, P., Kai, R. K., Urso, V., Herman, G., Apel, T. G., & Woerner, S. (2006). *Do some business models perform better than others?* MIT Sloan Research Paper no. 4615-06. Accessed at http://papers.ssrn.com/sol3/papers.cfm?abstract_id=920667##

Marchand, D. (n.d.). Value Add through effective use of information. *IMD Bulletin*. Available at http://www.imd.org/research/challenges/tc00110.cfm

Marchand, D., & Davenport, T. (Eds.). (2000). *Mastering Information Management*. New York: Financial Times Prentice Hall.

Marchand, D., Kettinger, W., & Rollins, J. (2001). *Making the invisible visible: how companies win the right information, people and IT*. Wiley.

Mc Kinsey. (2011). *Big Data: the next frontier for innovation, competition and productivity*. Accessed at http://www.mckinsey.com/insights/business_technology/big_data_the_next_frontier_for_innovation

McAfee, A., & Brynjolfsson, E. (2012). Big data: The management revolution. *Harvard Business Review*, *90*(10), 60–68. PMID:23074865

Mehmod, H. (2017). *A big list of what Google Analytics can and can not do*. Available at http://marketlytics.com/blog/list-of-things-google-analytics-can-and-cannot-do

Mintzberg, H., Alhstrand, B., & Lampel, J. (2009). *Strategy Safari: the complete guide through the wilds of strategic management* (2nd ed.). Pearson Education Limited.

Morville, P., & Rosenfeld, L. (2009). *Information architecture for the world-wide-web: Designing large scale web-sites* (3rd ed.). Sebastopol, CA: O'Reilly Media.

Mullins, J. (2014). *The customer funded business*. Hoboken, NJ: Wiley.

Nonaka, I. (2008). *The knowledge creating company*. Harvard Business Review Classics.

O'Brien, J., & Marakas, G. (2008). *Management Information Systems*. Irwin: Mc Graw Hill.

Obenshaim, M. K. (2004). Application of data mining techniques to healthcare data. Statistics for Hospital Epidemiology, 25(8), 690 – 695.

Ohata, M. & Kumar, A. (2012, September). Big Data: A Boom for Business Intelligence. *Financial Executive*.

Oliveira, J. (2017). Data "government" models for Healthcare. Business Intelligence Journal, 17(4), 34-39.

Park, S. H., Huh, S. Y., Oh, W., & Han, S. P. (2012). A social network-based inference model for validating customer profile data. MIS Quarterly, 36(4), 1217-1237.

Penrose, E. (1959). *The theory of the growth of the firm*. New York: John Wiley and Sons.

Porter, M. (2008). *On Competition*. Harvard Business School Press.

Pressman, R. (2006). *Software Engineering* (6th ed.). Boston: Makron Books.

Raghupathi, W., & Raghupathi, V. (2014). Big data analytics in healthcare: promise and potential. Health Information Science and Systems, 2(3), 1-10. doi:10.1186/2047-2501-2-3

RAIC – Raic Canada. (2014). *Raic / Irac Architecture Canada*. Accessed at https://www.raic.org/

RIBA – Royal Institute of British Architects. (2014). Accessed at http://www.architecture.com/Explore/Home.aspx

SAS. (2017). *What is Analytics?* Available at https://www.sas.com/en_us/insights/analytics/what-is-analytics.html

SAS. (n.d.). *SAS Enterprise Miner – SEMMA*. Available at http://www.sas.com/technologies/analytics/datamining/miner/semma.html

Siegel, S. (2016). Applying analytics to clinical data. *Tech Crunch*. Available at https://techcrunch.com/2016/08/31/applying-analytics-to-clinical-data/

Silva, A. M. (2008). Modelos e modelizações em Ciência da Informação: o modelo e-lit e a investigação em literacia informacional. *Prisma.Com, 13*(1). Accessed at http://revistas.ua.pt/index.php/prismacom/article/viewFile/785/710

Silva, S. L. (2002). Informação e competitividade: A contextualização da gestão do conhecimento em processos organizacionais. *Revista Ciência da Informação., 31*(2), 142–151.

Srinivasan, U., & Arunasalam, B. (2013). Leveraging Big data analytics to reduce healthcare costs. *IEEE Computer.org*. doi:10.1109/MITP.2013.55

Stair, R., & Reynolds, G. (2009). *Principles of information systems*. Course Technology.

Tuomi, I. (2000). Data is more than knowledge: Implications of the reversed knowledge hierarchy for knowledge management and organizational memory. *Journal of Management Systems, 16*(3), 103-117.

Turban, E., Mc Lean, E., & Wetherbe, J. (2002). *Information technology for management: transforming business in the digital economy* (3rd ed.). Hoboken, NJ: John Wiley and Sons.

Turban, E., Rainer, R. K. Jr, & Potter, R. E. (2007). *Introduction to information systems*. Hoboken, NJ: John Wiley and Sons.

US Government. (2007). *FEA – Consolidate reference model document* (Version 2.3). Accessed at http://www.whitehouse.gov/sites/default/files/omb/assets/fea_docs/FEA_CRM_v23_Final_Oct_2007_Revised.pdf

Wurman, R. S. (2014). *Richard Saul Wurman's website*. Accessed at http://www.wurman.com/

Chapter 8
Methodology for Identifying Drivers of Innovative Sectorial Nuclei

Brenner Lopes
Nous SenseMaking, Brazil

Luander Falcão
Nous SenseMaking, Brazil

ABSTRACT

In order to subsidize the public policy planners, identifying innovative sectorial nuclei that support the design of public policies to support innovation in companies, the chapter proposes and constructs an indicator of innovation potential with characteristics that fill an important gap, where it is possible to reach the groups of innovative companies in each sector (and their territories in which they are inserted), with a process of data collection focused on secondary sources and yet of high effectiveness and assertiveness. In this way, it is possible to make a detailed analysis of what is most directly relevant to decision makers, with the need for a low investment and the possibility of constant updating of the information. Therefore, this chapter will discuss the construction and demonstration of practical examples of the innovation potential indicator and their respective subindexes.

INTRODUCTION

It is unanimous on the part of researchers, teachers, experts, entrepreneurs and policy makers that innovation is the factor that has generated the greatest impact on the competitive advantage of companies and nations in recent decades. The prospect is that its importance will become increasingly blunt and its impacts deeper in the coming decades.

On the other hand, it is possible to perceive that not only the definition, but also the delimitation of what will become the innovation, has undergone changes and changes of scope. But even so, a final position of general recognition that was accepted by all stakeholders was not yet achieved. In the same

DOI: 10.4018/978-1-5225-6225-2.ch008

way, there are also several approaches, methods and visions on how innovation could be identified and measured.

The definition of innovation that will be used in this work is the same as that found in the Oslo Manual (3rd edition, 1997), that is,

innovation is the implementation of a new or significantly improved product or services, and yet or a new marketing method, or a new organizational method in business practices, workplace organization, or external relations.

In general, thinking about the policymakers' perspective, these issues are of fundamental importance. Firstly, it is necessary that they have a priori an overview of where the nuclei of the most innovative companies are located in their respective territories. So, that they can take proactive (and not only reactive) initiatives for the design of policies to support and develop competitiveness of the respective companies and territories. According to the Oslo Manual (3rd edition, 1997, p. 21), the forces driving innovation within the enterprise and the successful innovations to improve firm performance are central to policy making "which require better information" (also according to the Oslo Manual itself).

Secondly, it is crucial that these formulators have an approach to identify innovative business nuclei in all sectors of a given territory that is capable of being updated as often as necessary, not only to the current government strategy, but also to possible customization of strategies democratically elected by future governments representatives. It is worth emphasizing two questions: 1) that in this chapter, the term "government" is understood in a broader sense, where not only the central government itself but also its local authorities, foundations and other institutions are considered in one way or another maintained by government resources, and; 2) it is assumed from the Oslo Manual (3rd edition, 1997) that innovation can occur in any sector of the economy.

As a third point, it is critical that these approaches are based on a compatible budget that allows for at least annual updates as well as based on a conceptual tool that is sufficiently broad to be able to capture the necessary nuances from official secondary data, while maintaining the objectivity and rigor needed to obtain robust results capable of subsidizing policies and proposals, support, development and improvement in the innovation environment.

For all this, the locus object of the methodology proposed here is necessarily the companies with business activities (commercial), as is the Oslo Manual (3rd edition, 1997), in which innovation is associated with uncertainty and investments. This is characterized as an overflow substrate that requires the use of new knowledge or combination of existing knowledge, aiming to improve the performance of a company with the gain of a competitive advantage.

It should be noted that the methodology proposed here is the identification of territories with innovative business nucleuses in their multiple component sectors to support the design of development policies. This methodology is based on the knowledge acquired by the authors in the execution of consulting projects that had as general objective the context described here, being, therefore, a consolidation of knowledge improved from the practice of these projects, and not only a theoreticcal proposition. This approach holds immense value, as it can be replicated for any reality, in different territories and countries, since its structure is based on official data with international alignment.

As a way of demonstrating the applicability of the methodology discussed in this chapter, an economic sector will be chosen within the Brazilian territory, and on it will detailed the whole approach to identify the poles of innovation in this sector, emphasizing that this same approach could be applied to multiple sectors or to one or more productive chains with the realization of small adjustments that could be customized for each reality.

Despite its comprehensive application to various analytical extracts, the sectoral clipping chosen based on the logic indicated by the Oslo Manual (3rd edition, 1997, page 45), which states that *innovation processes differ greatly from sector to sector in terms of development, rate of technological change, interactions and access to knowledge.*

Thus, in order to fulfill the objectives proposed here, the present work will be succinctly structured as follows: an introduction, which will establish the general logic of the chapter; a general overview on the theme and its contextualization based on other works and authors of reference; the description of the methodology, i.e., a delineation of the methodological approach of the indicator and data collection format proposed; it will be demonstrated the application of the proposed approach in a segment / groups of companies of a selected territory, and finalizing, we will come to the conclusions.

BACKGROUND

Innovation: Concepts and Characteristics

In the last decades, this may have been the most discussed and sought-after question in the sphere of companies, states and nations. When one examines, among other transformations, the evolution of the business world, the ever new perceptions and desires and new social configurations, it is clear that striving for the forefront of innovative processes is a sound strategy for companies, states and nations. According to Davila and Epstein (2007, p. 23), *quality innovation gives a company the opportunity to grow faster, better, and smarter than its competitors and as a dictating the direction of its industry.*

In order to understand the concept of innovation, it is necessary to return to its first definitions, proposed by Joseph Alois Schumpeter and his challenging idea about the concept of "creative destruction". From a lato sensu understanding, the idea of the concept of creative destruction as the logic of replacing products and old habits and/or in force with new products and / or consumption habits could be put forward.

In Schumpeter's view, in his work "Theory of Economic Development", economic development would be based on innovation or, as the author precepted in his work, from the so-called "creative destruction", which would be characterized by a logic in that new technologies, processes, products and business models have necessarily replaced the old alternatives. Within this logic, technologies, processes, products and business models that no longer meet the new conditions, aspirations and demands that were established should necessarily "die", thus giving space to the "new".

The Oslo Manual (1997, p. 55) has been bringing together an evolving conceptualization that seems to bring together most of the views on the subject. This manual defines an innovation as the:

implementation of a new or significantly improved product (good or service) or process, or a new marketing method, or a new organizational method in business practices, workplace organization, or in external relations.

The Oslo Manual (1997, p. 24) also incorporates an important change in the scope of the conceptualizations involving the issue of innovation:

One change is the removal of the word "technology" from the definitions, since the word evokes the possibility that many service companies interpret "technology" as "user of high technology plants and equipment", and thus does not apply to many of its product and process innovations."

When talking about innovation, it is important to understand that innovations do not occur only in the technological field, as stated in the Oslo Manual. Davila and Epstein Handbook (2007, p. 48) argued that *one of the most common misunderstandings is to understand that innovation mainly, if not exclusively, of transitional Technologies*, and complement the reasoning adding that in the *truth, however, is that innovation is not just about technological change*.

This fact is reinforced by Utterback (1996), where the author works the technological issue in his work linked to other aspects of the company, emphasizing that it would be isolated, and much less would be seen as a presupposition.

A further deepening of the context and concepts that surround the innovation theme also allows us to understand that, as Tidd, Bessant and Pavitt (2008, p.12) affirm,

innovation does not consist only in the opening of new markets and can, therefore, also signify new ways of serving already established and mature markets.

Arbix (2007, p. 29) points out that innovation is *all processes capable of transforming an idea into a product or process with a market differential, whether in industry, services, commerce or agriculture*.

Because it is such a complex and dynamic topic, an analysis of the literature on the subject made it possible to identify that several authors, with different typologies and biases, point to several types of innovation, which, even in very different contexts, seem to be part of a same general logic that tends towards the same direction. In this sense, by way of example, the definitions proposed by Schumpeter (1982), from which the various other definitions derived, from the Oslo Manual (1997) and from Tidd, Besson and Pavitt (2008) will be punctuated.

In Schumpeter's (1982) view, the types of innovation would be five:

1. Introduction of new products;
2. Introduction of new production methods;
3. Opening of new markets;
4. Development of new sources providing raw materials and other inputs;
5. Creation of new market structures in an industry

In the view of Tidd, Besson and Pavitt (2008, p.30), when talking about innovation, the context of analysis is the process of change, and this process can take several forms, which, in the view of the authors, are centered in four categories:

1. Product innovation: changes in the things (products/services) that a company offers;
2. In-process innovation: changes in the way products/services are created and delivered;
3. Innovation of position: changes in the context in which products/services are introduced;
4. Paradigm innovation: changes in the underlying mental models that guide what the company does.

On the other hand, four typologies of innovation are prescribed in the Oslo Manual (1997, p.23):

1. Product innovation;
2. Process innovations;
3. Organizational innovations;
4. Marketing innovations.

Finally, it should be stressed that much more than concepts or types, innovation is inserted in a context where several other issues influence it, positively or negatively. One of them, of course, is the issue of multiculturalism, that is, the degree of diversity to which an organization is exposed to new ideas, different professional backgrounds and different backgrounds (nations) of professionals and entrepreneurs, as well as different worldviews, colors and races.

To have or to build a multicultural environment is to have an environment with different perceptions, visions of the world and of things and processes, expectations and values, which supports a rich and potentiating environment of the innovation process (Cox, 1993; Maznevski, 1994).

There is no doubt that the lack of a favorable environment for these issues is one of the major obstacles, which is not usually covered by traditional innovation indicators.

Another key issue is the role of the professionals involved in the innovation process and their tacit and explicit skills and knowledge. This refers in large part to the educational process to which these individuals have been subjected and, therefore, also to their degree of specialization. As quoted in the Oslo Handbook (1997, 53), a great deal of knowledge about innovation is embodied in people and their abilities, therefore, *the role of human capital in innovation is important both to the company and to the aggregate level.*

Research and Indicators on Innovation

A first, more inattentive look at the practice discussed in this chapter may lead the reader to feel that he is accessing a content and a theme that is already widely studied and discussed, which is no more than a mistake. In the literature and practical initiatives in general, we find several publications on the issue of innovation and the possibilities of identification and measurement of this, which almost always is based on the companies or the territory. In this case, the basic would be information and indicators of a broader nature, capable of portraying the dynamics of a given territorial cut, from the innovative business nuclei present in that territory.

Throughout the years, several approaches to measuring innovation initiatives and potentials have been developed and applied.

Bernardes (2003) makes a chronological survey of several methodologies used over time to construct science, technology and innovation indicators, as shown in the Table 1.

Although they are sound and highly relevant approaches, public policy makers have to make an effort and incur a greater cost (of financial and personnel resources, especially among others), by raising the level of innovation from each individual company, and then tracing the consolidation of data into the image of a group of companies or sectors. Another situation that those responsible for public policies may incur is to try to measure them from indicators linked to a given territory. On the one hand, this makes it easier and less expensive to collect these data, but on the other hand, it tends to bring about an inflated or even more augmented reality. This increased reality brings with it smaller possibilities for a deeper understanding of certain nuances and greater chances of constituting more general public policies

Table 1. Measurement methodologies for construction of science, techonology and innovation indicators

Year	Methodological Proposition
1963	Frascati manual
1978	Unesco. Recommendation concerning the International Standarization of Statistics on Science and Technology. Paris, 27 November
1984	Unesco. Manual for statistics on scientific and technological activities (Unesco Division of Statistics on Science and Technology, Office of Statistics ST – 84/WS/12)
1992	TEP – The Technology-Economy-Productivity Program – The Key Relationsships
1994	Using Patent Data as Science and Technology Indicators – Patent Manual
1994	Oslo Handbook (5th edition)
1994	Canberra Manual
1995	The Mensurement of Scientific and Technological Activities Manual on the Mesarument of Human Resources Devoted to S&T "Canberra Manual"
1997	Manual of Oslo (2nd Version). Proposed Guidelines fo Collecting and Interpreting Technological Innovation
1997	Revision of the high-technology sector and product classification
1997	Committee for information, computer and comunication policy: measuring eletronic commerce. Paris, OECD/GD (97), 185.
1998	Mesuring intangible investiment. Intangible investment in the statistical frameworks for the collection and comparison of Science and technology statistics.
2000	Stuz, J. "Latin American innovation surveys: a comparative analysis of the forms of inquiry". Work prepared for the Project Standardization of Indicators of Technological Innovation in Latin America. OAS, June.
2000	Manual of Bogotá - Standardization of Indicators of Technological Innovation in Latin America and the Caribbean - OAS Organization of American States - Ricyt, Colciencias, Cyted, Ocy T.
2000	Brisolla, S. and Quadros, R. Innovation in innovation indicators. A study of the methodologies adopted in the developing countries. Work prepared for the Project Standardization of Indicators of Technological Innovation in Latin America. OAS, June.

Source: Bernardes, 2003. Prepared by the authors.

that would require a new layer of efforts, in order to identify, in a clearer way, the possible beneficiaries of the actions to be undertaken.

This, in turn, could lead to a dissipation of the impact of the planned actions, causing greater resources to be invested to try to promote a certain dynamic. It could be said that, in this case, we would have a loss of productivity, a cost-benefit ratio not optimal, and a lack of optimization of available resources.

With regard to the most commonly used approaches to innovation research, Bernardes (2003) schematically condenses the characteristics of approaches to innovation surveys, comparing those based on the object (whose focus is centered on the results of innovation), and these based on the subject (whose focus is centered on the company), according to the Table 2.

The main focus of the innovation measurement processes has been the generation of indicators that allow not only to understand the magnitude and depth of innovation in companies, sectors and countries, but also to enable the monitoring of this process, as well as the conditions for the construction of initiatives to support the development of innovation by public and private initiative.

In this context, Farias et al. (2013, p.6) point to a set of more accepted traditional indicators when it comes to innovation. A quick review of them makes it clear that there is scope for improvement in this area. They are:

Table 2. Forms of innovation survey approaches: Subject-based approach

Characteristics of the Approach	Object-Based Approach	Subject-Based Approach
Unit of analysis	Tecnologic innovation	Company
Method of collecting information	Collection of information from different public and private production sources, specialized and detailed surveys. Examples are patent information, R & D, bibliometrics, etc.	Collection of information within the company through the application of questionnaires or direct interviews.
Form of innovation capture	Registration of information about the output of the innovation process	Information record on input of the innovation process
Periodicity	Casual surveys	Periodic and timeless Surveys for the construction of statistical series through the CIS (Community Innovation Survey)
Coverage	Sample of successful innovations	Sample of good and unsuccessful innovations. Innovative and non-innovative businesses, covering the industry and services sectors
Classification criteria	• Technological Area • Product • Main economic activity of the company	• Company size • Types of innovation • Forms of cooperation and interaction • Main economic activity of the company
Typical examples	Small business administration, EUA (Acs & Audretsch, 1991)	Innovation Surveys (CIS – Communty Innovation Surveys) (Guellec and Pattinson, 2002)

Source: Bernardes, 2003. Prepared by the authors.

- Research and development statistics which measurement focus is on the volume of expenses and labor involved in the process that could be understood more as a component of effort than an improvement or degree of innovation, at least when viewed individually. Within the logic of a composite indicator, this information would enter as one of its components, composing a more robust and complex structure;
- Patents which, according to the authors, can be considered more effort measures than effective results;
- Direct innovation monitoring, where they point to the fact that the count/enumeration of identified innovations does not necessarily have a direct link with the market success of the products or services linked to these "innovations";
- Bibliometric indicators, which make a count of articles or quotes on articles. The authors emphasize the fact that innovation focuses on the development of solutions, products and services, and not basic research, which would make this indicator adequate for some situations and not others, depending on the scope of what is intended to be measured;
- Semi-quantitative techniques, which would address performance and productivity issues in R & D units and/or organizations, which serve as an indicator of individual organization effort rather than as an indicator of the degree, improvement or impact of innovation.

In Brazil, specifically, as pointed out by Cavalcanti and De Negri (2011, p.9),

Table 3. Deposits of type PI patents, considering 1st depositor: Major depositors

Rank	Nome	Total 2000-2012	Participação Residentes	Participação Total Depósitos
1	Petróleo Brasileiro S.A. - Petrobras	730	1,4%	0,3%
2	Whirlpool S.A.	659	1,3%	0,2%
3	Universidade Estadual de Campinas - Unicamp	620	1,2%	0,2%
4	Universidade de São Paulo - USP	468	0,9%	0,2%
5	Universidade Federal de Minas Gerais - UFMG	425	0,8%	0,2%
6	Universidade Federal do Rio de Janeiro - UFRJ	235	0,4%	0,1%
7	Universidade Federal do Paraná - UFPR	208	0,4%	0,1%
8	Vale S.A.	173	0,3%	0,1%
9	Universidade Federal do Rio Grande do Sul - UFRGS	163	0,3%	0,1%
10	Empresa Brasileira de Pesquisa Agropecuária - Embrapa	133	0,3%	0,0%
Top 10 - Total		3.814	7,3%	1,4%
Total de depósitos de residentes		52.450	100,0%	19,1%
Total de depósitos		274.728		100,0%

Source: INPI, Advisory for Economic Affairs, BADEPI v2.0, data extracted in December / 2013.

innovation surveys have been carried out frequently since the 1990s and, despite the initiatives of the National Association for Research and Development of Innovative Companies (ANPEI), in the 1990s, innovation surveys began to be produced from 2000, when the first edition of PINTEC (Innovation's Survey from IBGE) was launched.

Still on the issue of patents, it should be emphasized, especially in the case of Brazil, that institutions, as public universities, are the ones that register the most patents rather than private companies.

While in the US, for example, schools like Harvard, Columbia, MIT and others do not account for 5% of total patents. In Brazil, data collected by INPI (National Institute of Industrial Porperty), from 2000 to 2012, show six public universities among the ten institutions with the highest number of patent registrations in the period. In addition to these schools, we also have two public companies and one private sector company, but with a strong decision component linked to the public sector, as shown in the Table 3.

In order to give more depth to the indicator of "research and development statistics", particularly in Brazil, there are two key issues that make it difficult to use this indicator as an adequate parameter to measure the innovation situation in the country. Cavalcante and De Negri (2011) argued that *while high technology companies in Germany invest, on average, 6.85% of their R & D revenues, in Brazil the percentage invested only 1.89%*. Still according to the authors, only in sectors of medium-low technological intensity *investments in Brazil are proportionally higher than those observed in Germany.*

The second question refers to the fact that recent studies indicate that, in recent years, there has been a greater availability of public resources destined to support the innovation processes of the industrial sector in Brazil, which grew much more than proportionally to the number of innovative companies. This demonstrates that much more than resources is necessary to understand where and how to apply them to have the expected effect, maximizing the comparison between the costs of these initiatives and their expected benefits, especially when this equation involves public resources.

It also demonstrates that, while being a powerful driving force, innovation isn't possible to be achieved only with public resources, which leads us to the concrete fact of the paternalistic view still valid in

Brazilian business. Obviously, recognizing that it's not the reality of a small group of companies and entrepreneurs, who have already fully understood the dynamics of the markets and which are the pillars to compete and survive in that market.

METHODOLOGY

The methodology that will be dealt with in this chapter was initially applied to the Brazilian context, since it is the most direct business territory of researchers and its logic and method are replicable for any territory/reality.

For this purpose, the Innovation Potential Indicator was created, based on the analysis of 1,267 businesses, distributed in the Brazilian municipalities, in the 27 states of the federation. The Innovation Potential Indicator aims to identify which businesses and in which regions there is greater or smaller potential for innovation in relation to a business in the same or different region. The business concept worked out here are economic activities, based on the precepts of the CNAE (National Classification of Economic Activities), and grouped by activities purposes. The indicator also has the ability to compare different businesses, such as commerce and service, and to indicate which ones have the potential to have more innovations in relation to the other.

For the elaboration of the Indicator of Innovation Potential, five sub-indices were constructed to make up the main indicator. These sub indexes were constructed from the literature presented so far, and named as follows: Potential Supply, Gazelle Business, R & D Activities, Masters and Doctors and finally, Multiculturalism. Data on business and municipalities were obtained from the RAIS (Annual Social Information Report) of the Brazilian Ministry of Labor, between 2009 and 2015. Only data on the number of employees, and the salary mass generated were considered. The Indicator of Innovation Potential can be synthesized in the following formula. The weights of each sub-indicator were given according to the relevance of each one to the innovation debate.

$$IIP = \beta_1 PS + \beta_2 GB + \beta_3 RDA + \beta_4 MD + \beta_5 MC$$

At where:

$$\beta 1 + \beta 2 + \beta 3 + \beta 4 + \beta 5 = 1$$

IIP = Indicator of Innovation Potential
PS = Potential Supply
GB = Gazelle Business
RDA = R & D Activities
MD = Masters and Doctors
MC = Multiculturalism
$\beta 1$ = Potential Supply sub-indicator weight
$\beta 2$ = Gazelle Business sub-indicator weight
$\beta 3$ = R & D Activities sub-indicator weight
$\beta 4$ = Masters and Doctors sub-indicator weight
$\beta 5$ = Multiculturalism sub-indicator weight

The first sub index built was Potential Supply, with the objective of capturing how much a business in a given municipality has the conditions to offer goods and services, both for the region and for others. For this purpose, this sub index captured the size of this business in the municipality, in absolute terms, the number of employees and the salary mass generated. The size of the business indicates the economic movement that the business companies generate in the municipality. The sub index also captured the business specialization, which indicates the importance of a business within a municipality in relation to its participation in the country. For this, the concept of QL - Locational Quotient was used, in which it is understood that there will be specialization of some business, if this result is greater than 1. This criterion was calculated in relation to the country, based on the number of employees of 2009 and 2015, and in the salary mass generated in these two periods, to consider the effect of specialization in two moments in time. This sub-index also captured Growth, or Business Growth Dynamics, which indicates the increase in the volume of business generated by the companies of a given business, in a period of 7 years. As a premise if it was assumed as a proposition that a business in process of growth hires more employes, being a tangible and measurable reflection for the sector as a whole, and not only for large companies. For this calculation, the Compound Annual Growth Rate (CAGR) was used, between the years of 2009 and 2015, for both the business in the municipality, and the number of employees and the mass generated.

The second sub index built was Gazelles Enterprises, which are high growth businesses, according to the methodology of the IBGE, Business Demography study that classified as "high growth" companies that had an average growth of people employed equal to, or greater than 20% per year in three years. The premise adopted is that businesses that grow at this pace, by that time, have a strong innovation approach.

The third sub index built was "R & D Activities", based on activities of the Brazilian Classification of Occupations - CBO. Activities with strong adherence to the innovation process were identified, such as employees with a position or function of Director of Research and Development (R & D), Bioengineer, all engineering activities, researchers, physicians and other knowledge intensive activities. All were selected 273 activities. To understand if a business in a given municipality possessed latent potential in R & D activities, the same concepts of calculation, Size, Specialization and Growth were used for this group of occupational activities selected by the researchers.

The fourth sub index built was "Masters and Doctors", based on the number of masters and doctors in business in municipalities, as this is one of the variables used to measure the innovation capacity of a company or segment. The same concepts of calculation, Size, Specialization and Growth were also used, similarly to what was done for the Sub-index of Potential Supply and R & D Activities.

Finally, the last sub index, Multiculturalism, in which it captured the diversity of the business of the municipality, in terms of race and nationality. Nationality was only captured in 2015, in terms of the number of non-Brazilian employees. The concept of race took into account indigenous, white, black, yellow and brown, in terms of number of employees and salary mass, in 2009 and 2015.

In order to obtain the Indicator of Innovation Potential, some methodological procedures were adopted. The first was the exclusion of some sectors, such as Public Administration and Education, as they did not fit into the concept of companies with business activities (commercial). The second methodological procedure was to avoid the effect of QL when there are few employees in the business in the municipality and, consequently, few employees in the business in the country. When you have this situation, the result is a very high number, which distorts the final result of the indicator, not reflecting reality. Therefore, it was parameterized that every QL result above 100 would be defined as 100. The third procedure was to standardize the results of the variables, to bring them to the same basis of comparison and, finally, weights were given for each variable within each sub index and consequently for each sub index, according to their importance to the context of innovation.

This methodology also allows a greater degree of deepening in the analyses, since it contemplates the segmentation by sector and, within that, a specific business. An example of this is the possibility of analyzing only the Transformation Industry Sector, or, within that, analyzing the dairy business in one or more states or municipalities.

After the adoption of this methodological procedure for the creation of the indicator, the results showed that some businesses in the municipalities contributed negatively to the promotion of innovation in the region, as the Indicator of Innovation Potential was negative. This points to the fact that, in a given municipality, there are not only businesses that are more innovative than others, but also there are businesses that contribute very little or do not contribute to the logic of innovation, since they do not have they grow at high speed, activities related to research and development, masters and doctors and / or do not have collaborators from other countries, thus presenting a low or nonexistent degree of multiculturalism.

To identify which businesses and in which municipalities these businesses had a greater capacity for innovation, all businesses with the Indicator of Innovation Potential above the 3rd quartile, in this case above 0.00912, were selected. Of the 379,408 businesses of the original base, about 94,852 businesses, were selected, which can be understood as those with effective capacity for innovation.

The analysis carried out with this new framework as a parameter has raised the question about the correlation between the sub-indices, as well as with the main index (Indicator of Innovation Potential), with the objective of *quantifying the intensity and direction of the association between two variables* (MAROCO, 2007, p. 42), based on the difference between the sectors. To make this comparison between the pigs of each sector, the Pearson Correlation Coefficient measure was used. According to Maroco (2007, p. 42), *a correlation simply measures the association between variables without any implication of cause and effect between them.* Pearson's coefficient is defined by Maroco (2007: 43) as a coefficient that *measures the intensity and direction of linear association between two quantitative variables.*

Based on these methodological considerations, the case study was carried out in which the correlation between the indicator and the subscripts by sector was compared.

CASE STUDY

For the indexation of the indicator and the subscripts, the following abbreviations were maintained: IIP - Indicator of Innovation Potential; PS - Potential Supply; GB - Gazelle Business; RDA - R & D Activities; MD - Masters and Doctors; MC - Multiculturalism.

The first sector to be analyzed was the Manufacturing Industry. The data show that there is no strong linear relationship between the variables because no correlation exceeds 0.75 or -0.75. When analyzing the Indicator of Innovation Potential (IIP), it is observed that the strongest correlations are with the R & D Activities (RDA) and Masters and Doctors (MD) subscripts, and these are positive, indicating both subindices tend to increase with the Indicator of Innovation Potential. Hence, increasing the subindices will also raise the indicator.

When analyzing the subscript Multiculturalism, the data show that there is a weak linear correlation between the variables, and this correlation is negative. As the IIP grows, the MC tends to regress, which raises the question: how much multiculturalism in the Transformation Industry contributes to innovation? This questioning is corroborated by the negative correlation values with the other subscripts.

The second sector analyzed was Services. With a wide range of business, the data show that there is a moderate to strong correlation between the indicator and Business Gazelles. The correlation value of this subscript is higher than the RDA and MD subindices. This indicates that, in the services sector, the sub-index Gazelle Business (GB) and IIP tend to grow together. One of the justifications for this fact can be found in a characteristic of the sector, which is low entry barriers, and companies can open with little capital and structure. An example of this is business based on intellectual activities, such as IT and Consulting. It also seems to make sense that in the Services sector, perhaps more than in others, it is necessary to print an accelerated growth in order to guarantee the maintenance of the existence of the business. However, this subscript does not keep a linear correlation with the other subscripts.

The data also show a moderate and positive linear correlation between Potential Supply (PS) and R & D Activities (RDA). This indicates that both variables tend to grow together. The justification for this fact may be that the sector's supply capacity comes from R & D activities, but this cause and effect relationships were not evaluated in this work. Also in this sector, Multiculturalism (MC) showed a weak and negative correlation with the other variables, however, it is more homogeneous than in the Transformation Industry sector.

The third sector analyzed was Commerce, and also shows a correlation between the Indicator of Innovation Potential (IIP) and Gazelle Business (GB), and this linear correlation is strong and positive. This indicates that both variables grow together, which shows a strong association between the variables. In

Figure 1. Industry of transformation
Source: Prepared by the authors.

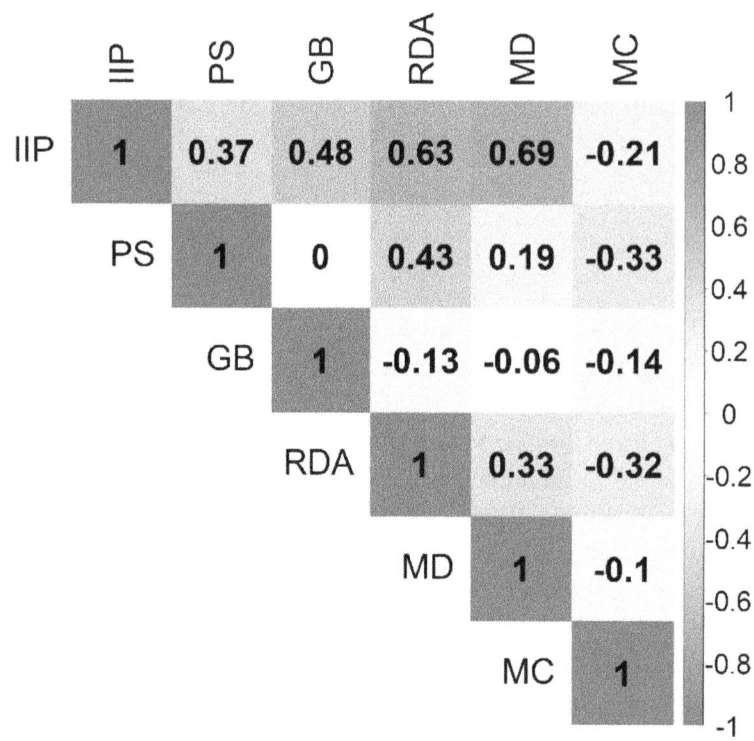

Figure 2. Service sector
Source: Prepared by the authors.

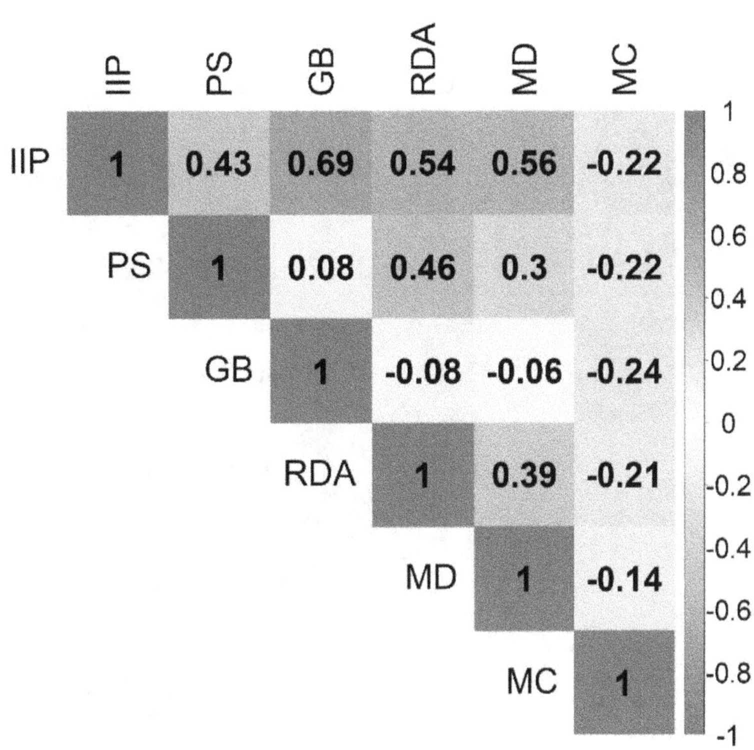

this case, an assessment of how much the Gazelle Business (GB) variable impacts the Indicator of Innovation Potential (IIP) is necessary, since in the Commerce sector innovation can come from fast growing businesses that may, or may not, be related to the scope of innovation. This fact can be corroborated by the low correlation values of the Indicator of Innovation Potential (IIP) with R & D Activities (RDA) and Masters and Doctors (MD). In the case of Services, there is a greater correlation between Potential Supply (PS) and R & D Activities (RDA) than in the Commerce sector, so that it can be inferred that there is a weak correlation between supply capacity and R & D activities.

The fourth sector analyzed was that of Agropecuária. In this sector, there is a moderate and positive correlation between the Indicator of Innovation Potential (IIP) and the sub-indices Gazelle Business (GB), R & D Activities (RDA) and Masters and Doctors (MD), which evidence that these fourth variables grow together in this sector. It is important to note that the agricultural sector has grown a lot in recent years in Brazil, being one of the main pillars that has sustained the country's economy. This impacts on the accelerated growth of some segments, and, for example, hiring more specialized employees for the operation of sophisticated machines in terms of technologies. In addition, there is investment in research and development in this sector, encouraging the training of masters and doctors.

Of all the correlations to which it has the lowest result is in this sector among the variables Potential Supply (PS) and Multiculturalism (MC), evidencing that when the supply of goods and services in this sector increases, multiculturalism decreases. Some factors arise from this result, such as a low rate of foreigners and people of various races. This may also show that the supply of goods and services grows based on local labor.

Figure 3. Trade sector
Source: Prepared by the authors.

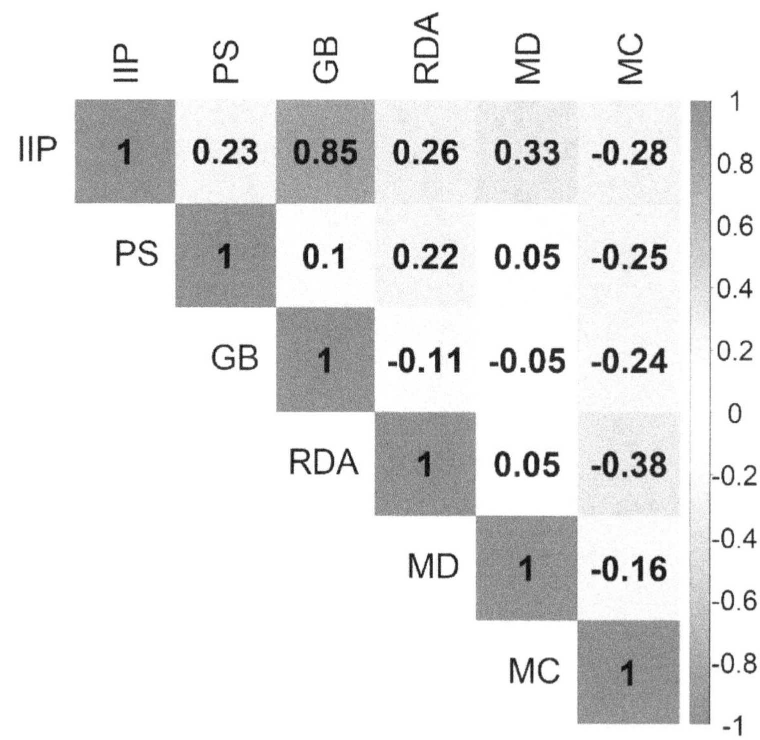

When analyzing the correlation between the sub-indexes, it is observed that there is no linear correlation between the variables, and tend to be negative.

The fifth sector analyzed was Civil Construction, and also shows a correlation between the Indicator of Innovation Potential (IIP) and the Gazelle Business (GB), being this linear correlation strong and positive. This indicates that both variables grow together, which shows a strong association between the variables. The sector has experienced strong growth in recent years in Brazil, but has undergone a period of low growth. It also obtained, as a result of the analyzes, low correlation values of the Indicator of Innovation Potential (IIP) with R & D Activities (RDA) and Masters and Doctors (MD). There is a weak and positive correlation between Potential Supply (PS) and R & D Activities (RDA), indicating that there is a weak correlation between supply capacity and R & D activities.

The sixth sector analyzed was Mineral Extractive and presents a strong and positive correlation of the R & D Activities (RDA) and Masters and Doctors (MD) sub-indexes with the Indicator of Innovation Potential (IIP). This indicates that the variables grow together. It can be speculated that in this sector the effect of activities related to R & D and masters and doctors impact the innovation indicator more than other activities. It is important to note that in this sector there are companies called World Class, such as Petrobrás and Vale (AGTMAEL, 2009). The companies in this sector work in competitive and high rivalry businesses, because they are pressured by suppliers, buyers, and the economic, technological and regulatory impacts are high. Due to this, the innovation factor is also here, in a prominent way, to be a factor of competition, because the competitors will use it to become better and more competitive or, at least, to maintain their market shares.

Figure 4. Agricultural sector
Source: Prepared by the authors.

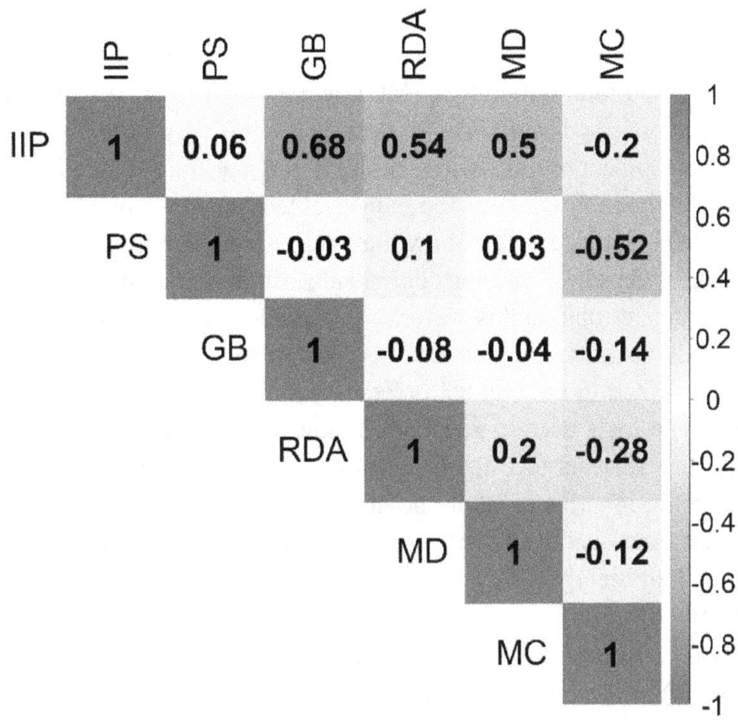

Figure 5. Construction sector
Source: Prepared by the authors.

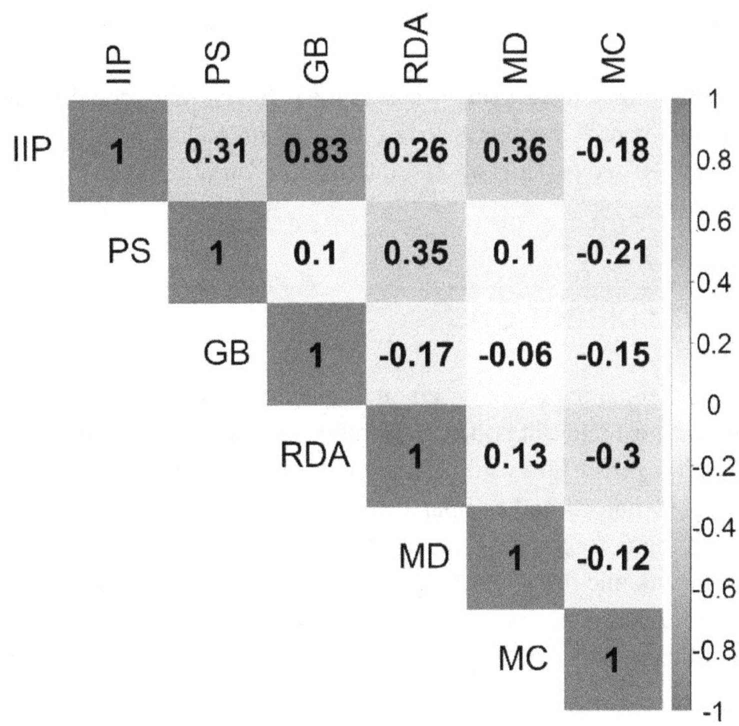

It is also observed that there is a moderate and positive correlation between R & D Activities (RDA) and Masters and Doctors (MD) variables. This correlation may be due to the fact that masters and doctors occupy R & D-related jobs, which should not be surprising, since they show how much innovation is linked to the activities of companies in this sector.

Potential Supply (PS) and Multiculturalism (MC), which may indicate that multiculturalism is not a factor linked to the supply of goods and services, and that in this sector the impact of multiculturalism does not impact the development of innovation, contradicting part of the literature on the subject.

The seventh and last sector analyzed was the Industrial Services of Public Utility, such as energy and water supply. With little business, the data show that there is a moderate to strong correlation between the indicator and Business Gazelles. The correlation value of this subscript is higher than the RDA and MD subindices. This indicates that, in this sector, the sub-index Gazelle Business (GB) and IIP tend to grow together. One of the justifications for this fact may be the increase in the number of employees and the salary mass generated in the sector in recent years, which in a broader perspective shows an opposite effect, that is, the logic used to gauge the innovation potential in other sectors, it should not be used as a ruler for it. A relevant fact that bases this conclusion can be seen in the swelling of most of the sectors in this sector, in terms of labor and high salaries, almost always without correspondence with the competitiveness and quality of services rendered and marked by strong political influence, if this segment is analyzed differently.

The data also show a moderate and positive linear correlation between Potential Supply (PS) and R & D Activities (RDA). This indicates that both variables tend to grow together. Also, in this sector, Multiculturalism (MC) showed a weak and negative correlation with the other variables.

When comparing the sectors, it is observed that the sub-index Multiculturalism has weak and negative correlation with all other variables and in all sectors, only being moderate in the sector of Agriculture and with the Variable Potential Supply (PS). This subindex has a positive and moderate correlation (value above 0.4) only with the R & D Activities (RDA) variable in the Transformation and Services Industry sectors. This shows that, in these sectors, these variables grow together, being necessary to estimate the impact of one variable on the other.

This comparison between the sectors shows how is the behavior of association between the variables related to the innovation potential among those businesses with greater potential and, when comparing these businesses within their sectors, it is observed that some variables have the same level of development, while others do not.

FUTURE RESEARCH DIRECTIONS

Having the innovation potential indicator, based on a selection and prioritization of variables (subindices) from a survey of the most relevant issues to the innovation approach present in the world literature, when analyzing its set, it is possible to verify the existence of a logical and consistent order, and its use in future strategies of governments and companies is highly recommended.

But when analyzing the correlation between this index, its subindices and the macro sectors that make up the Brazilian economic fabric, it was possible to verify that the strength demonstrated by the set of subindices in the formation of the innovation pointer indicator does not hold when we analyze its correlations with the others, with the potential of innovation indicator and from the standpoint of the individual sectors.

Figure 6. Mineral extractive sector
Source: Prepared by the authors.

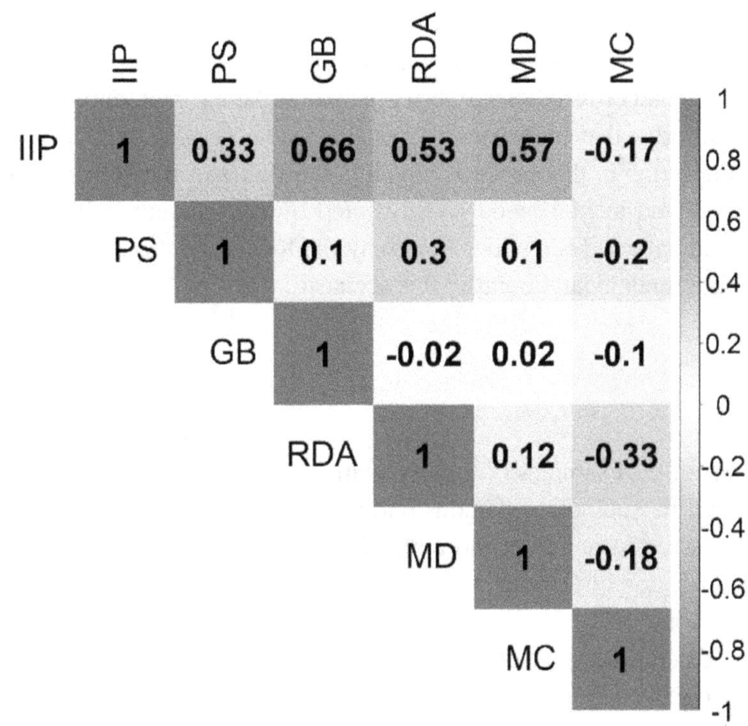

Figure 7. Industrial utilities sector
Source: Prepared by the authors.

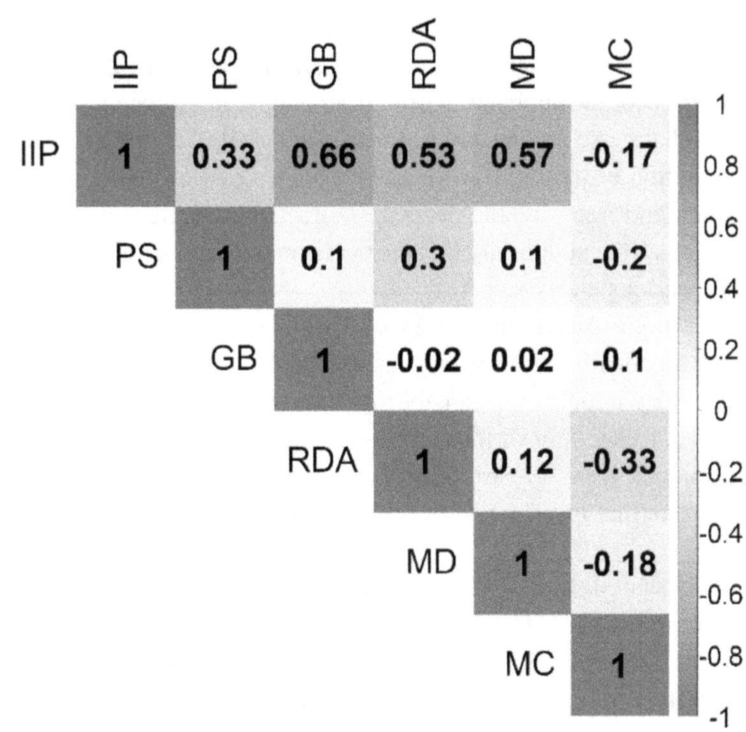

In this sense, in a specific way regarding the subindividual multiculturalism, based on its behavior in the analysis of correlation in almost all sectors analyzed, it is understood as deserving and necessary an in-depth and comprehensive analysis of this subindice, for a better understanding of its role in the potential of innovation in the sectors inserted in their respective territories.

It seems, for all logic and criteria already established on the basis of studies and practical applications, that this is a sub-index that represents one of the fundamental points of the innovation process and the fundamental question in this process, which is the figure of the "contradictory" that is, of the different views, cultures and mental models, constituted in environments with diverse ethical, moral, cultural and educational values. Then understand why the low correlation with the main indicator and the other subindices is a fundamental point in this scenario.

CONCLUSION

The results obtained with the examples of the application of the methodology used in the "Innovation Potential Indicator" and its subindices demonstrate the applicability of this approach and explain an alternative to support the construction of more effective innovation incentive policies, at least, in relation to the logic Identification of territories with cutting-edge sectoral nuclei of innovative companies and their respective sectors.

The focus on sector analysis has enabled a much more assertive and fruitful view of what "pointers" should be activated and at what intensity to deepen innovative dynamics. From this approach, it is then possible for each sector to know exactly where to invest, that is, which subindice (pointer), which question, should be addressed and developed to impact innovation more forcefully. To understand and use individually for each sector a set of forces that knows if it impacts directly on its degree of innovation, allows the focus and prioritization of resources (material, economic, financial, talents, etc.) necessary for the maximum socialization of this (innovative) capacity.

The good conformation of the construct worked in the applications of this study points to the existence of a good consistency for the application of the same logic in an even more granular form, that is, within the subsegments that make up a macro segment, such as those analyzed in this study. As all the basic concern of constitution of the logic outlined here was based on the possibility of working with data available in an open and free way, the deepening of the application of this methodology is quite extensive, being able to reach the granularity that the decision maker considers necessary for the analysis and decision support at various levels.

It is worth noting the importance of the sub-index "business gazelles", in most of the analyzed sectors, demonstrating that the identification of businesses that grow in an accelerated manner and distinct from others, bring innovative initiatives with capacity to transform local and sectoral realities. Another key factor, with a strong correlation in various sectors, is that of qualified professionals and investments in research and development. Here, in particular, there seems to be a certain symbiosis, since several practical cases demonstrate that investments in research and development demand professionals with high qualification.

As has been shown, it is possible to drill down using several criteria, to specify with less granularity, the information of a certain segment; This indicator can therefore be used by public policy managers at various levels and spheres (national, state and municipal), by providing practical tools that can be used to identify innovative sectors of innovative companies to support the design of development policies.

REFERENCES

AGTMAEL. (2009). *Antoine van. O Século dos Mercados Emergentes*. São Paulo: Cultrix.

Arbix, G. (2007). *Inovar ou Inovar. A indústria brasileira entre o passado e o futuro*. São Paulo: Editora Papagaio.

Bernardes, R. (2003). Produção de Estatísticas e Inovação Tecnológica: PAEP 1996 – 2001. *São Paulo em Perspectiva*, *17*(3-4), 151–167. doi:10.1590/S0102-88392003000300016

Cavalcante & De Negri. (2011). *Trajetória recente de indicadores de inovação no Brasil*. IPEA; Texto para discussão, 1659.

Classificação Nacional De Atividades Econômicas – CNAE. (n.d.). Retrieved April 18, 2016 from http://cnae.ibge.gov.br/?view=secao&tipo=cnae&versaosubclasse=9&versaoclasse=7&secao=C

Cox, T. H. (1993). *Cultural Diversity in Organizations*. San Francisco: Berrett Koehler.

Davila & Epstein. (2007). *As regras da inovação*. Porto Alegre: *The Bookman*.

Farias, Andrade, Freitas, & Junior. (2013). *Metodologias para mensurar o impacto da inovação nas atividades empresariais*. Anprotec, XXIII Seminário Nacional de Parques Tecnológicos e Incubadora de Empresas. Recife, Pernambuco.

Manual de Frascati, OCDE. (2013). *Metodologia proposta para definição da pesquisa e desenvolvimento experimental*. F. Iniciativas, Assessoria em P&D.

Manual de Oslo. (1997). *Diretrizes para coleta e interpretação de dados sobre inovação. Terceira Edição. OCDE*. FINEP.

Maroco, J. (2007). *Análise Estatística: com utilização do SPSS*. Lisboa: Edições Sílabo.

Maznevski, M. L. (1994). Understanding our diferences: Performance in decision-making groups with diverse members. *Human Relations*, *47*(5), 531–552. doi:10.1177/001872679404700504

Neeley, T. (2015). Equipes globais que funcionam: Uma estrutura para diminuir a distância social. *Harvard Business Review*.

Schumpeter, J. A. (1982). *Teoria do desenvolvimento econômico*. São Paulo: Editora Abril Cultural.

Tidd, J., Bessant, J., & Pavitt, K. (2008). *Gestão da Inovação*. Porto Alegre: Bookman.

Tigre, P. B. (2006). *Gestão da inovação: a economia da tecnologia no Brasil*. Rio de Janeiro: Elsevier.

Utterback, J. M. (1996). *Dominando a dinâmica da inovação*. Rio de Janeiro: Editora Qualitymark.

Chapter 9
Internet of Things and Internet of All Things

Cláudio Roberto Malhães Pessoa
Universidade Fumec, Brazil

Cássio Luis Batista
Universidade Fumec, Brazil

Marco Elisio Marques
Universidade Fumec, Brazil

ABSTRACT

Interaction between devices, systems, and services is provided in internet through a new concept: internet of everything (IoE). This new concept should present numerous opportunities for the creation of new devices and applications in the years that follow and can dramatically affect the day-to-day life of all of us. New possibilities of information access, transmission, analysis, and interaction are practically infinity. With so many benefits, it is expected to come with several challenges. System structure and architecture and information confidentiality, integrity, and availability are aspects that someone should keep in mind to design an IoE application. This chapter offers to the reader an overview of the evolution of devices connectivity until it reaches the IoE. Some IoE concepts and applications, together with some challenges, are presented.

INTRODUCTION

The Internet of Everything (IoE) represents a significant paradigm shift in Internet applications.

The evolution of the Internet can be briefly represented in five phases, as can be seen in Figure 1, according to Perera et al (2014). In an initial phase, this apllication was created to connect two computers, using a network interconnection protocol, the IP. In the next phase, the World Wide Web (WWW) was created, allowing the connection of a large number of computers. In a third phase, the Mobile Internet appears, connecting mobile devices to the Internet. In a fourth phase, Identities were connected on the

DOI: 10.4018/978-1-5225-6225-2.ch009

Figure 1. Development of the Internet until the inclusion of the Internet of Things
Source: Batista adapted from Perera, 2017.

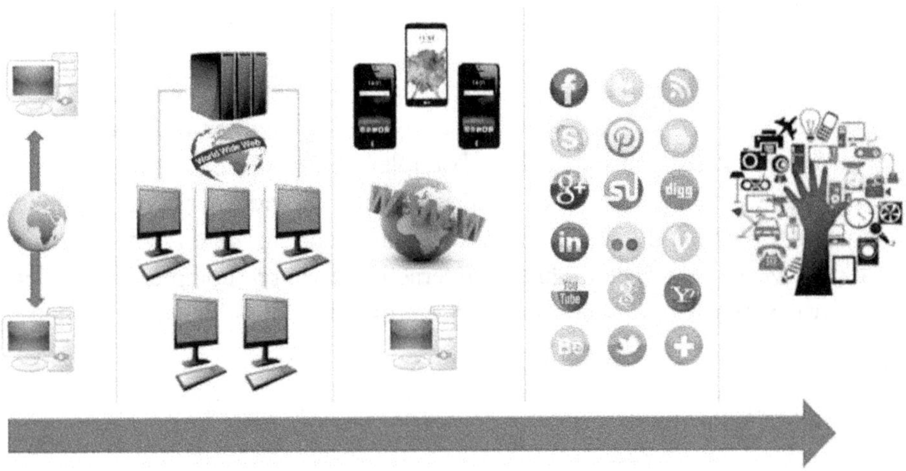

Internet through social networks. Finally, in the last phase, the Internet now includes objects that connect through the Internet, thus, forming the Internet of Things (IoT), as can be seen in Figure 1.

The expression "Internet of Things" was initially used by Kevin Ashton and, in 2001, the MIT Auto-ID center presented its vision of IoT (Perera, Zaslavsky, Christen, & Georgakopoulos, 2014). IoT was formally introduced by ITU (International Telecommunications Union), in Workshop Report (2005).

According to Chui (2010), Internet of Things is the network of physical objects or "things" embedded in electronic products, software, sensors and connectivity to enable objects to exchange data with the manufacturer, operator and / or other connected devices. Typically, IoT provides advanced device, systems and service connectivity, and goes beyond machine-to-machine (M2M) communication and covers a wide variety of protocols, domains and applications.

Esteves (2015) refers to Internet of Things (IoT) as the creation of a network of objects that have built-in technology, usually sensors and microprocessors, and that can interact with each other by sending or receiving information in an internal or external way. The Internet of Things connects people, processes, data and things to make connections more relevant and valuable than ever: converts information into action that creates new skills, richer experiences and unprecedented economic opportunities for businesses, individuals and countries.

Pessoa et al. (2016) have presented the concepts, applications, challenges and future trends of the Internet of Things (IoT) in their overview. The authors have reported good perspectives in the creation of solutions (products and services) that meet the "market needs" and / or needs of individuals. An example shown, is the concept established by Pandikumar et al. (2014), which states that the IoT architecture is a convergence of several technologies such as pervasive / ubiquitous computing, sensors / actuators, Information and Communication Technologies (ICT) and embedded systems, which gives the technology the necessary flexibility to adapt the real demands of the market.

The IoT paradigm has its own associated characteristics and concepts, and it also brings with it concepts from other areas of computing. IoT focuses on several technologies, such as: sensor hardware and firmware, WEB services, cloud, data modeling and storage, signal processing and information and telecommunications.

Due to the offered potential by IoT technologies, it has made possible to develop a large number of applications. According to Calvo et al (2016), typical Internet Applications of Things (IoT) involve information automatic collection from several geographically distributed intelligent sensors and information concentration on powerful computers.

Cisco defines Internet of Everything (IoE) as a comprehensive variation of IoT. While IoT is based on sensors/physical actuators, IoE generalizes to include people inputs and instrumentation of business processes (Etzion, Fournier, & Arcushin, 2014). Examples of people's input are the apliccations Waze, Facebook and LinkedIn, whose main source of information is provided by people's input.

IoT support requires machine-to-machine (M2M) communication. M2M is defined as the communication between devices without the need for human interaction. It can be communication between devices and a server or between devices ("device-to-device"), either directly or over a network.

POTENTIAL MARKET

There is a general expectation that IoE should present numerous opportunities for the creation of new devices and applications in the years that follow. The Internet of Things (IoT) interconnects the "Things" and allows the autonomous exchange of data between them. "Things" can be machine components, intelligent sensors / actuators, or even objects, such as wearable or retail articles. This capability can bring significant advancement in efficiency and user experience in the system.

Wired and short-range communication should be used for most connections. However, there is an emerging long-range technology and integration with the cellular mobile network.

IoT covers a large number of categories of business opportunities. There are some interpretations of how one could segment applications into key verticals. Generally, there are five key vertical highlights: Connected Wearables, Connected Cars, Connected Homes, Smart Cities, Connected Cities, and IoT Industrial. Figure 2 illustrates the five key verticals.

Figure 2. Key vertical segments for IoT
Source: 5G Americas, 2016.

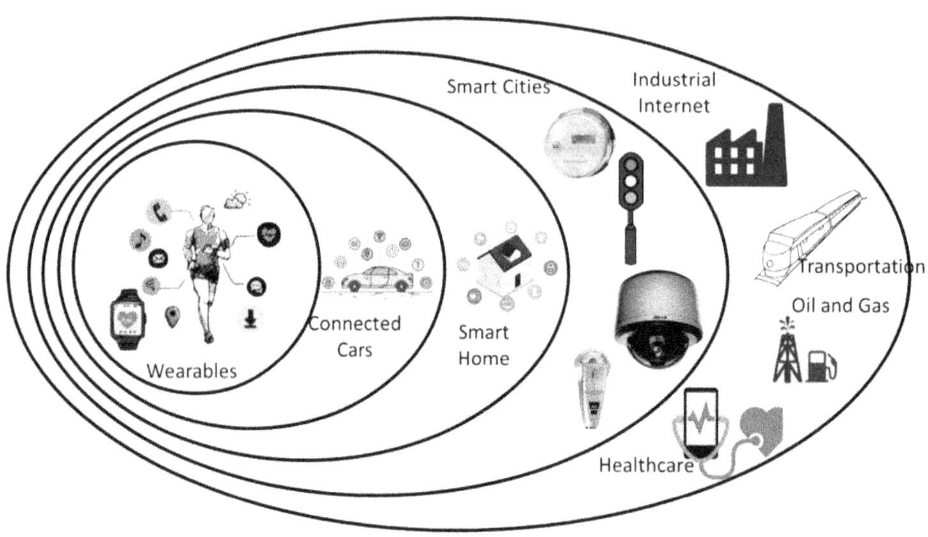

As described by 5G Americas (2016), the four key factors to boost IoT are:

- The costs of internet-connected sensors are coming down;
- There is more money poured in the IoT;
- High adoption of "remote" devices, especially smartphones, phablets, tablets and systems on a chip (SoC) that can support the growth of IoT market penetration;
- Expanded Internet connectivity: The ITU (International Telecommunication Union) estimated that, in 2015, 40% of the global population was connected to the Internet and by 2019, 57% would be connected.

Currently, mobile phones remain the main category of connected devices and, by 2018, they are expected to be overtaken by IoT. Specifically, a Compound Annual Growth Rate (CAGR) of 21% IoT devices is expected in the period from 2016 to 2022. In total, around 29 billion connected devices are estimated to be in place 2022, of which 18 billion are related to IoT, according to the Ericsson Mobility Report (2016). Figure 3 illustrates this growth.

IoT services provision requires an object with a mobile, flexible and ubiquitous connectivity to the network. The first option would be the connection to expensive devices, such as cars and industrial machinery in which the business model makes sense because of the availability of electric power and use of low-cost modems, compared to the cost of the objects.

Due to the large number of possible IoT services, the variety of uses with different requirements is enormous. Table 1 illustrates some of the most relevant uses, in terms of a business plan and their need to be exploited in the coming years.

Depending on the specific service and the values achieved with the four IoT challenges, different requirements can be considered:

- Traffic patterns (throughput and activity cycles);
- Identity / Security needs;
- Ease of installation;
- Mobility;
- Service Level Agreement (SLA);
- Reliability;
- Possible sector regulation.

Figure 3. Estimated growth of connected devices up to 2022
Source: 5G Americas, 2016

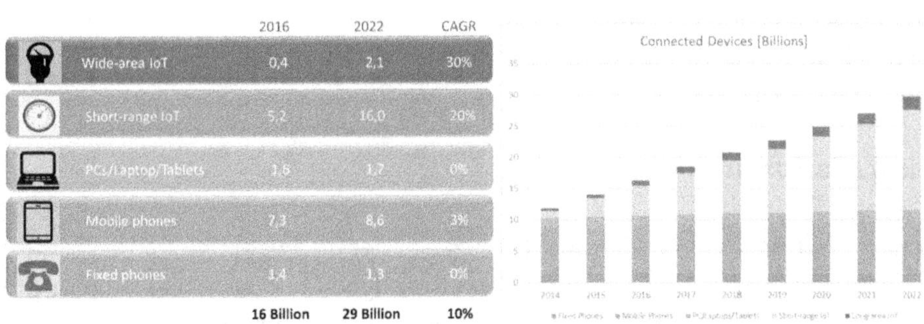

Table 1. Use cases and main requirements by sector

Setor	Caso de Uso	Principais Requisitos
Industry	High Volume (eg, mining)	Range, Coverage, Reliability, Cost
Agriculture	Dynamic Agriculture (eg Animal Tracking)	Battery, Range, Coverage, Reliability
	Static (eg irrigation fields)	Battery, Range, Coverage, Reliability
Public utilities	Utilities Energized, for example, Electricity	Indoor Coverage, SLA, Reliability
	Utilities not energized, eg Water / Gas	Indoor Coverage, SLA, Reliability
Logistics	Management and Localization (eg Fleet location)	Easy installation, Mobility, Coverage, Cost
	Basic Monitoring (eg delivery condition, storage)	Battery, Easy Installation, Mobility, Coverage, Cost
Smart Cities	Dynamic Systems (eg, traffic lights)	SLA, Coverage, Reliability
	Basic Sensing (eg pollution monitoring)	SLA, Coverage, Reliability
Payments	Portable sales terminal	Indoor Coverage, Interoperability, SLA, Reliability
	Fraud detection	Indoor Coverage, Interoperability, SLA, Reliability
Wearable (including e-Health)	Continuous Tracking (eg diabetes)	Indoor Coverage, Battery, Mobility, SLA, Coverage, Reliability
	Localized Tracking (SPOTS) (eg step tracking)	Battery, Easy Installation, Mobility
Security	Large Volume (e.g. video)	Indoor Coverage, Throughput, Security, SLA, Reliability
	Small volume (eg motion sensor)	Indoor Coverage, Throughput, Security, SLA, Reliability
Connected Cars	Integrated Solution (eg, traffic tracking))	Easy Installation, Mobility, Coverage, Cost
	Basic Monitoring (eg location)	Easy Installation, Mobility, Coverage, Cost
Buildings (including House)	Complex Solution (eg energy management)	Indoor Coverage, Safety, SLA, Reliability
	Basic Solution (eg presence, air pollution)	Indoor Coverage, Safety, SLA, Reliability
IoT Complex systems	Autonomous Cars, Drones and Ecosystems	Battery, Security, Reach, SLA, Coverage, Reliability

Source: 5G Americas, 2016.

IOT SOLUTION COMPONENTS

IoT solutions are typically composed of three basic components - devices, networks and software platforms. The devices are the intelligent sensors, machines, products or gateways that contain the communication modules. Devices often include CPUs and embedded software platforms that enable distributed intelligence. Multiple wireless or wired networks, used in various possible combinations, enable remote communication. Several protocols are used to communicate and enable functionality in the solution. A standardization effort is being made by different standardization forums. Software platforms include various IT environments and systems that provide functionality, such as device management, data management, business logic and integration with other enterprise IT systems. Figure 4 illustrates the various components of the IoT solution.

Software and Platforms for IoE

IoE solutions have platforms and middleware that make easier the integration of devices, networks and applications. There are a variety of software platforms designed to enable and support IoE solutions. These platforms provide components, tools and Application Programming Interfaces (APIs) that can be

Figure 4. IoT solution components

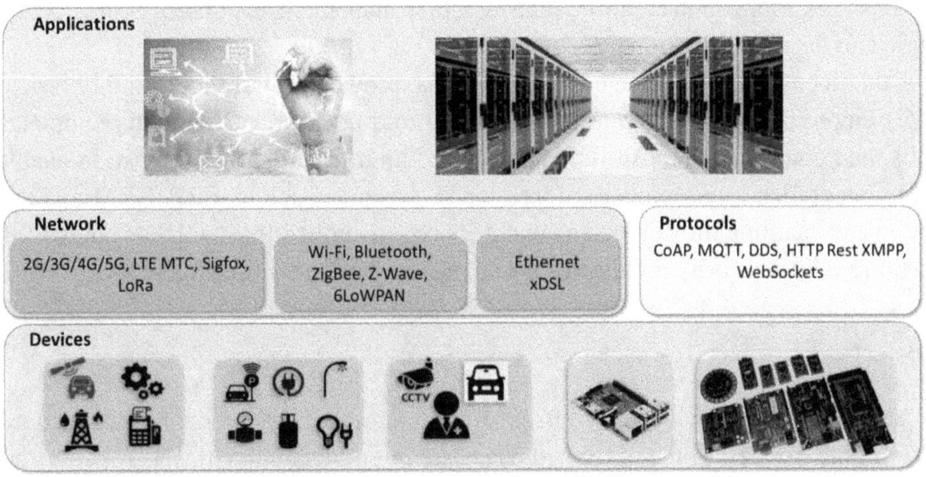

shared across multiple applications and across multiple vertical segments. For developers, the platforms come with a great quantity of built-in modules that enable faster integration into a growing community.

For network operators, the platforms give access to external developers to business-related processes information, such as pricing and access to user provisioning, as well as solution performance optimization capabilities. IoT platforms are relatively new to the market and offer a wide range of applications and functionalities. Fundamentally, three types of IoT platforms can be highlighted: connectivity management platform, device management platform and application development platforms.

Connectivity Management Platforms

Connectivity management platforms provide a wide range of functionality, in order to facilitate information exchange through communication services over wireless or wired networks.

Device Management Platforms

They enable remote management of IoT devices, such as activation / deactivation, configuration and software updates. Some IoT hardware providers provide device platforms that allow integration with the network operator platforms. In addition, some application-enablement platforms include management functionalities for multiple devices from multiple vendors.

APPLICATION ENABLEMENT PLATFORMS

They are developed to accelerate and simplify the development of IoT solutions and applications, providing horizontal components that can be reused across industry and market segments. These platforms can provide standardized interfaces for connectivity and integration, device management features, device data collection and other functionality for developers.

However, the core value proposition of application enablement platforms are: the tools, environment and APIs for business application creation, such as data management, event processing, automated tasks, business logic and data visualization.

Platforms that focus on applicative development also provide prebuilt tools and libraries, and a user interface (UI) environment to facilitate the modeling and creation of interactive applications, work environments, and control panel with little need for coding. In addition, application enablement platforms provide integration environment tailored for IT systems, such as ERP, CRM and analytics. For data protection and enabling information exchange between applications and data sources, the enabling platforms require robust security architectures and user authorization management.

Devices

Connected "things" involve a wide array of sensors, actuators and devices for embedded systems, machines and wearable devices. Some of these devices are integrated with chipsets for connectivity, while others connect to external modems, or gateways, including smartphones.

Typically, the IoT developments are based on standardized modules built into circuit boards for development. In many cases, the module is packaged as a general purpose terminal or gateway that is integrated into the final solution.

Modules, Chipsets and Mobile Terminals

IoT modules are communication devices that can be embedded into "things", enabling them to communicate in wireless networks. Today, wireless modules are available in different formats for easy integration into a myriad of applications. IoT modules are currently available for major standard wireless networks: cell phones, WPANs and WLANs, as well as emerging LPWA technologies.

Due to the development and complexity, these devices depend on the market scale, in order to the costs be compatible with the expected cost for this type of application

The increasing communication modules processing capacities are handled by a central processing unit. Module builders support software platforms with languages, such as Python and Java, Open AT

Figure 5. IoT communication modules

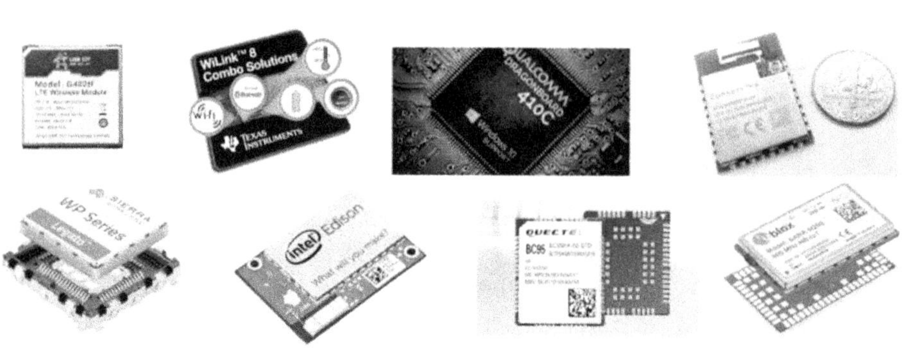

interfaces and operating systems, mostly Linux, allowing developers to create specific applications. Several manufacturers have software upgrades service supports over-the-air (OTA), enabling the stack of protocols, operating systems and applications loaded in the module to be upgraded to reduce maintenance costs and field device upgrades. This is especially important for long lifecycle equipment.

Embedded Software

Embedded software for embedded computers are often an integral part of devices and systems, such as consumer electronics, household appliances, machinery and industrial automation equipment. Embedded operating systems and network stacks enable remote control, monitoring and operation of connected devices. A growing number of device developers now use off-the-shelf operating systems, that are available from a large number of vendors and development communities. These operating systems enable the development of applications, using standard programming languages. Because the functionalities of the various operating systems are significantly different, application portability often involves extensive modification or adaptation of the source code. Software platforms, such as Java, have been developed to allow cross-platform software portability.

Raspberry Pi

Raspberry Pi is a microcomputer compatible with operating systems based on the ARMv6[1] architecture and, therefore, any language that can be compiled in this architecture can be used for the software development. Thus, Raspberry Pi can be applied to countless purposes: electronic projects, high-definition video reproduction and various activities that the conventional computer can perform, such as spreadsheets operations, word processing and games (Raspberry Pi Foundation, 2014).

This credit card-sized computer was developed by the UK-based Raspberry Pi Foundation, in order to teach and encourage children to program. For that purpose, such device has low acquisition costs. Figure 4 shows a Raspberry Pi 3 model B (Raspberry Pi Foundation, 2014).

Raspiberry Pi has been used in several IoT applications. As an example, Vujović and Maksimović (2016) propose to use Raspberry Pi as wireless sensors network node and Web sensor network node, highlighting the great advantages that the device has for this application, such as size, cost, operating system and other features.

Awasthi et al. (2016), also proposes Raspberry Pi as hardware for a Home Surveillance system. Xu et al. (2016), have used Raspberry as the smart wireless sensor node of a wireless sensor network architecture for rain monitoring.

All this demonstrates the numerous possibilities of technology use, thus, giving credit to the application presented in the context of this work.

IoT Networks and Technology

Virtually, any network can support M2M applications. Wired networks, such as Ethernet or xDSL, are appropriated for monitoring fixed things, while wireless networks can be used under any circumstances. In wireless communications, there is a wide area of application with technologies and standards suitable to several use. The standards of the mobile cellular network, in its latest versions, have been adopting

Figure 6. Raspberry Pi 3 model B
Souce: Raspberry Pi Foundation, 2017.

several modifications in order to allow IoT applications. Low Power Wide Area (LPWA) systems and various Wireless Local Area Network (WLAN) technologies and several Personal Area Networks (PANs) can be used for applications of IoT. Figure 7 shows various wireless technologies and their relation to the range and occupation of spectrum. The higher the data transmission rate, the greater the radio frequency spectrum width is used.

Figure 7. Personal, local, and long-distance wireless systems

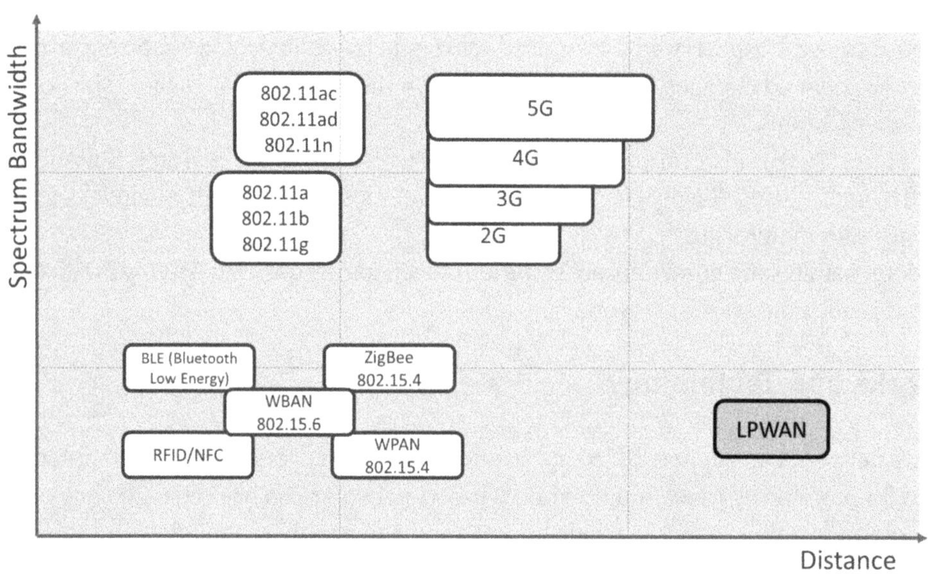

Mobile Cellular Technology and LPWA

According to Cisco (2014), the number of connected devices in the world will grow from 15 billion to 50 billion by 2020. While this forecast represents all forms of wireless connectivity, with the growth of the reach of Mobile Internet, cellular connectivity becomes more valuable as an important form of access for IoT.

Figure 8 shows the growth estimate of cellular technology, as a IoT access technology, and the percentage predicted for the IoT through the mobile network, using Machine Type Communication (MTC), with changes in the network core for this type of communications, specified as CIoT, or, Cellular Internet of Things.

The first evolution of MTC technology was aimed at increasing possible scenarios, focusing on cost reduction, since Release 10 of 3GPP (Mobile Systems Standardization Entity).

The same low-cost requirements were addressed more than a decade by other standardization forums and entities, such as: ZigBee, Z-wave, all using ultra-low power devices and focused on providing mesh networks, addressing four challenges to massive deployment of IoT services:

- Low cost devices, needed to be integrated not only into a single chip modem, but also in sensors and actuators on all objects, including wearables;
- Power efficient system, which allows the maximum possible autonomous operation of the IoT devices with reduction of the battery size;
- Ubiquitous coverage in the implanted scenario. Network deployment should ensure a high degree of indoor and outdoor coverage. The mesh networks should use auto-tuned routing, to enable node deployment without any manual configuration;
- Scalability, considering the large number of devices that can be deployed in ultra-dense scenarios such as the city center, and with expectation of exponential growth of the connected devices.

These basic requirements were initially addressed with the combination of short range devices and mesh network.

However, the use of multihop mesh networks presents a degree of complexity and performance that imposes limits to its application.

To get around this bottleneck, thanks to the improvement of microelectronics technology, allowing the integration of high-power modems (including power amplifiers (PA) and other radiofrequency parts),

Figure 8. Growth of IoT devices and percentage of CIoT devices
Source: 5G Americas, 2016.

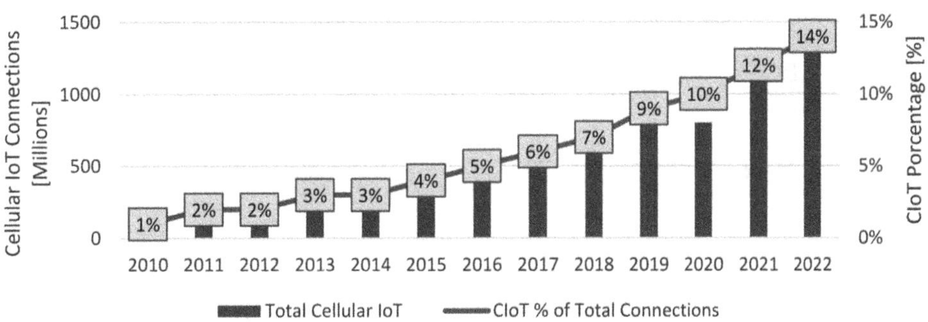

a new approach to meet the four requirements, mentioned as challenges of IoT was adopted a few years ago, it is known as Low Power Wide Area (LPWA). The basic idea is to move to a star connection by increasing cell coverage, through modulation technology, which complexity and power requirements are centered as much as possible on a base station. This star approach provides a low operation cost (OPEX) since no coordination node is needed near the end devices, with simple and highly desirable deployment for infrastructure companies and carriers.

In order to meet the challenges listed with an LPWA approach, several radio parameters must be degraded, as they are not required in some predicted IoT services:

- Throughput, usual scenarios require only a few kbps or less;
- Packet length, messages will be very short with only a few bytes;
- Cycle of activity, very short periods of activity, in all, less than 10% of the time;
- Latency, values will depend on services, but several of them have loose requirements, with only a small number of connections every day;
- Handover sessions will not be required in most scenarios.

Figure 9 shows the topology used in short-range networks with mesh topology and star topology using a gateway node and IP connectivity with the short-range application and the long-range star topology for an LPWA network.

Initially, mobile cellular systems focus was the high data rate broadband service. Currently, a series of improvements have been made in network access and in network core of, as well as in users provisioning to make it suitable for the various uses of IoT applications.

The radio access network is optimized to enable low-cost, low-power transmission devices. Broadband protocol transmissions are designed for minimum power for a battery life of more than 10 years and provide long distance operation range, including rural area and achieving significant indoor coverage. New categories of User Equipment (UE) were categorized as LTE-M, or category M1 and NB-IOT, or category NB1, based on the adopted technology in LTE.

LTE-M provides requirements for a low-complexity terminal category that supports reduced bandwidth, ultralong battery lifespan utilizing consumption reduction and coverage expansion techniques.

Figure 9. Topology for short-range networks and LPWA networks

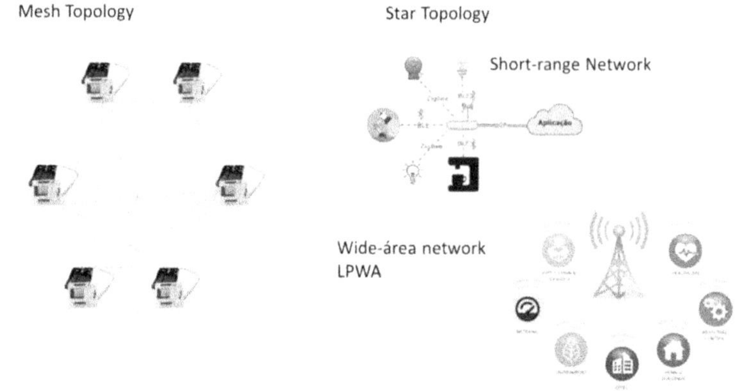

NB-IoT improves indoor coverage, "things" mass-scale support with low transmission rates, ultra-low cost devices, lower power consumption and optimized network architecture. The technology was developed to allow deployment in the LTE (in-band), or LTE carrier guard band, or in stand-alone mode. Figure 10 illustrates the use of NB-IoT.

NB-IOT can be seen as a standardized LPWA technology with the use of a regulated range in the frequency spectrum.

Most of the new LPWA systems are based on the use of the unlicensed radio frequency band in the ISM band, and use low power radio transmitters to create optimized network solutions for long battery life, low cost hardware and ubiquitous coverage. There is always a trade-off between cost and energy consumption, in one side, against network performance, on the other, which need to be weighted according to the specific needs of the application. LPWA networks were initially developed for early telemetry applications and evolved into technologies that use the most advanced wireless networking techniques.

There are currently LPWA systems that use proprietary or semi-proprietary platforms for specific applications, such as intelligent meters or Supervisory Control and Data Acquisition (SCADA). Larger deployments connect millions of nodes through point-multipoint or mesh configurations. Recently, there have been some initiatives to create LPWA standardizations with open standards. So far, LoRa and Sigfox have gained more support, with long-range networks planned or under implementation in several countries.

WLAN/WPAN Technologies

The major wireless local area network (WLAN) and wireless personal area network (WPAN) technologies include Wi-Fi and Bluetooth. A family of WPAN technology based on the IEEE 802.51.4 family of standards, is designed to support a wide range of predominantly industrial use cases. They include ZigBee, developed for control and sensor reading applications and WirelessHART, developed for wireless networks for field devices.

WiFi

In recent decades, Wi-Fi based on the IEEE 802.11 standard family has emerged as the universal platform for homes and businesses wireless LANs. WiFi is the world leader in local wireless that enables electronic devices to participate in a computer network, using unlicensed frequency bands in the 2.4 GHz and 5.8 GHz frequency bands. The term Wi-Fi has been adopted by WLAN products based on 802.11

Figure 10. NB-IoT options of and coexistence with LTE
Fonte: 5G Americas, 2016.

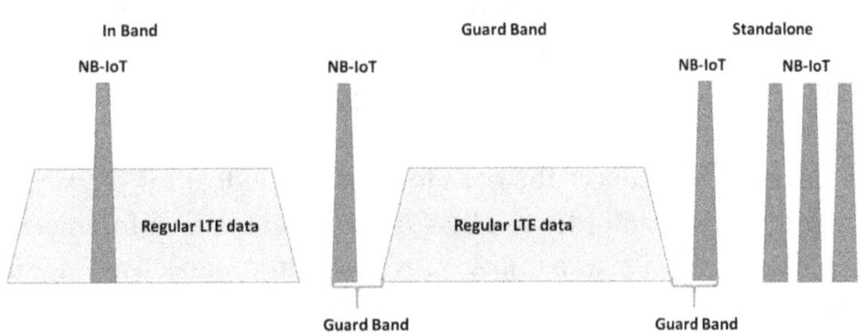

standards, by the association of Wi-Fi Alliance industries in 1999. Today, more than 2 billion Wi-Fi certified devices are sold annually, and there are around 50 million Wi-Fi hotspots deployed worldwide. Consecutive technology generations have increased rates, from 11 Mbps, to 1 Gbps, and added new features, such as Wi-Fi direct to peer-to-peer networks and Wi-Fi Aware for standalone network discovery. The 802.11ah standards extend the scope of Wi-Fi technology to an extended range, with lower power consumption. The 802.11ah will use the frequency range below GHz to reduce power consumption, by allowing large groups of devices to share a common signal.

Bluetooth

Bluetooth was created by Ericsson Company, in 1994, and actually is the standard technology for short-distance data exchange using the 2.4 GHz unlicensed frequency band. Since 1998, the standardization is managed by the Bluetooth Special Interest Group (SIG), which has more than 30,000 companies members in the areas of telecommunications, computing and consumer electronics. If we compare to Wi-Fi, we would conclude that Bluetooth technology is optimized for low data rates and low power consumption. The Bluetooth 4.0 specifications, adopted in 2010, have introduced three subsets of standardizations to better support the wide range of applications. Classic Bluetooth consists of a Bluetooth protocols legacy, Bluetooth High Speed , that is based on Wi-Fi, and Bluetooth Smart (formally, Bluetooth Low Energy) optimized for a considerably reduced power consumption, low cost and reduced latency. The key features of Bluetooth Smart, making it desirable for a wide range of IoT applications, include long operating times for small battery powered devices (months or years), small size and compatibility with a huge installed base of mobile phones, tablets, phablets and notebooks. The application data rate is limited to 0.3 Kbps, with a latency of 6 ms and an energy consumption of 0.01 - 0.5 mW, depending on the configuration. The recent specifications of Bluetooth 4.1 and 4.2 have added full support to the IPv6 protocol, making it possible to have an infrastructure to manage Bluetooth Smart Edge devices. Bluetooth Smart is a key technology for the emerging market of wearables, and is becoming widely adopted for connected healthcare, gymnastics, sports, signage, safety, and consumer devices.

The new specification of Bluetooth 5 brings improvements, such as range, speed and broadcast messaging capability, as well as support mesh networks, which address numerous IoT use cases.

Network Architecture

A generic architecture for IoE is presented in Figure 11. The number of hierarchical layers varies from solution to solution.

According to Gluhak et al. (2011), architecture is a key dimension of systems for IoT, as it inherently determines its properties. The structure of a system refers to the organization of the hardware components present in it, and most follow a two-layer structure: a server layer and a directly attached IoE device layer. There is also a 3 layers structure that introduces a Intermediate Gateway (GW) device layer.

According to Gluhak et al. (2011), larger systems typically exhibit a three-layer architecture, in which, one or even dozens of IoT devices, are connected by means of wired or wireless connections to an intermediate GW device. This reduces the need for each individual IoT device to provide a possibly expensive network interface card, in order to allow communications with remotely located test servers. The three layers system uses GW as a utility, providing only connectivity aggregation and passive experimentation support.

Figure 11. Generic architecture for IoE solution
Source: Batista, 2017.

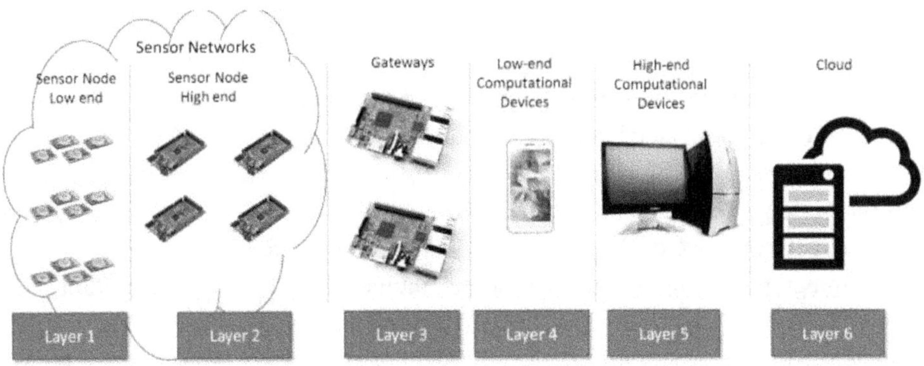

As established by Toma e Popa (2014), the role of the IoT gateway is to process the data collected from the IoT Nodes, establish local feedback behavior and send them to the Datacenter/Cloud system to process them with dedicated solutions.

The valuable exploited information needs to be stored and processed somewhere on the IoE platforms.

According to Hausenblas (2015), a data platform, that needs to process data from IoE devices reliably and in scale, must meet the following requirements:

- **Support Native Raw Data:** In terms of both data reception and processing, the platform must be able to deal natively with IoE data. Some platforms make it possible to receive input data in its raw format (JSON, XML, log files, etc.), and - for optimization purposes - convert upstream data to more sophisticated formats.
- **Support for a Variety of Workload Types:** IoE applications typically require the platform to natively support stream processing and handle low latency queries against semistructured, scaled data items.
- **Business Continuity:** Commercial IoE applications generally come with Service Level Agreements (SLAs) in terms of availability, latency and disaster recovery metrics (Recovery Point Objective / Recovery Time Objective). Thus, the platform must be able to fully guarantee these SLAs. This is especially critical in the IoE applications context in domains, such as health care, where people's lives are at risk
- **Security and Privacy:** The platform must ensure secure end-to-end operation, including integration with existing authentication and authorization systems within the company. Last, but not least, user privacy must be guaranteed by the platform, from data source support to encryption and data masking. This is the most contradictory point in the evolution of IoE and Big Data.

CLOUD DEVELOPMENT

Infrastructure development cost is never low. Nevertheless, IoE systems are driven by low sensor and microprocessor costs and computing requirements by exploiting the new generation of native cloud technologies, avoiding significant infrastructure costs and datacenter support costs. Many cases involve event responses and event-driven computing services, such as AWS Lambda and IBM OpenWhisk, that provide low-cost computing processing without the need for investment in servers.

Modern application architecture, like microservices, matches perfectly with event-driven use cases.

The massive view of IoE can generate Big data, while analytical view, based on learning machine, is the ideal one for data mining. However, the growth of intelligent programming, with deep learning, opens up opportunities for automated intelligence. Remote systems can be better managed and controlled by intelligent systems, and have been applied in IoE projects. In this regard, Silva et al. (2016) point out that, as systems are integrated, they also become an important source of input for public policy decision making, which characterizes one of the objectives of state intelligence activity.

IoE solutions integrate with numerous IT systems, including ERP, CRM and database management systems, as well as analytics and business intelligence applications.

Derhamy et al. (2015) present cloud IoE development environments. As presented by Malm (2016), most M2M and IoE platforms are cloud-based, using Infrastructure as a Service (IaaS) or Service Platform (PaaS) solutions from vendors, such as Amazon, Google, IBM, Microsoft, Cisco, Oracle and SAP and cloud systems management software companies, such as VMWare, BMC, CA Technologies. Offers of IaaS typically include virtual machines, storage, load balancers and virtual local networks, available as on-demand resources. PaaS offerings also include operating systems, runtime environments, databases and development tools.

REFERENCES

Awasthi, S., Singh, S., Soni, R., & Jaidka, P. (2016). Internet of Things Using Raspberry pi 2. *IJEECS, 5*(5).

Blackstock, M. (2015). Toward a Distributed Data Flow Platform for the Web of Things. *IEEE IoT conference: Developing IoT Applications in the Fog: a Distributed Dataflow Approach. 2015 International Conference on the.*

Brown, C. (1994). Unix Distributed Programming. In *Building IoT Applications with Raspberry Pi and Low Power IQRF Communication Modules*. Academic Press.

Carboni, D., Serra, A., Pintus, A., Kro, S., Nati, M., & Gurgen, L. (n.d.). *Specific Targeted Research Projects (STReP) SocIoTal - Creating a socially aware citizen-centric Internet of Things*. Available at: http://cordis.europa.eu/docs/projects/cnect/2/609112/080/deliverables/001-SOCIOTALD41V1.pdf

Chauhan, P., & Kumar, T. (2015). Power Optimization in Wireless Sensor Network: A Perspective. *International Journal of Engineering and Technical Research, 3*(5).

Chui, M., Loffler, M., & Roberts, R. (2010). *The Internet of Things*. McKinsey Quaterly.

Coulouris, G., Dollimore, J., Kindberg, T., & Gordon, B. (2013). *Sistemas Distribuídos - Conceitos e Projetos*. Bookman Editora.

Dahlman. (1998, September). UMTS/IMT-2000 Based on Wideband CDMA. *IEEE Communications Magazine*, 70–80.

DoD 5000. (1995). *Modeling and Simulation Master Plan*. Author.

Ênfase em Eficiência Energética. (n.d.). Disponível em: http://www.inf.ufpr.br/aldri/disc/artigos/Relatorio_Tecnico_Mauricio_Barros.pdf/

Esteves. (2015). *Tese Mestrado A Internet das Coisas: Avaliação do grau de aceitação da tecnologia RFID pelo cidadão comum UCP*. Available at http://repositorio.ucp.pt/bitstream/10400.14/19429/1/Andr%C3%A9%20Gil%20Capela%20Esteves_355413001_Tese%20Mestrado%20Gest%C3%A3o_A%20Internet%20das%20Coisas-avalia%C3%A7%C3%A3o%20do%20Grau%20de%20Aceita%C3%A7%C3%A3o%20da%20tecnologi~1.pdf

Evaluation of Restful Web Services in Raspberry Pi. (n.d.). Available at https://www.researchgate.net/publication/283865130_Performance_and_energy_evaluation_of_RESTful_web_services_in_Raspberry_Pi

Fielding. (2000). *Architectural Styles and the Design of Network-based Software Architectures* (Tese de Doutorado). University of California, Irvine, CA. Available at: http://www.ics.uci.edu/~fielding/pubs/dissertation/rest_arch_style.htm

Folkens, J. (n.d.). *Building a gateway to the Internet of Things*. Available at: http://www.ti.com/lit/wp/spmy013/spmy013.pdf

Fonseca, J. J. S. (2002). *Metodologia da pesquisa científica*. Fortaleza: UEC.

Gammu. (n.d.). *SMSD*. Available at: https://pt-br.wammu.eu/smsd/

Gluhak, A., Krco, S., Nati, M., & Pfsterer, D. (2011). A Survey on Facilities for Experimental Internet of Things Research. *IEEE Communications Magazine*, *49*(11), 58–67.

Gubbi, J., Buyya, R., & Marusic, S. (2013). Internet of Things (IoT): A vision, architectural elements, and future. *Future Generation Computer Systems, 29*(7). Available at: http://ac.els-cdn.com/S0167739X13000241/1-s2.0-S0167739X13000241-main.pdf?_tid=68615f94-c2f6-11e6-856400000aacb361&acdnat=1481827605_39fd450812ee030ddd19162300d8d3f

Gubbi, J., Buyya, R., Marusic, S., & Palaniswami, M. (2013). Internet of Things (IoT): A vision, architectural elements, and future directions. *Future Generation Computer Systems*, *29*(7), 1645–1660. doi:10.1016/j.future.2013.01.010

Hausenblas, M. (n.d.). *Key Requirements for an IoT Data Platform*. Available at: https://www.mapr.com/blog/key-requirements-iot-data-platform

Hill, J. (2003). *System Architecture for Wireless Sensor Networks. Spring 2003* (Vol. 1). Berkeley, CA: University of California.

IBM. (n.d.). *Node-RED*. Available at: https://developer.ibm.com/open/openprojects/node-red/

International Telecommunication Union. (2005). *ITU internet reports 2005: The internet of things*. Available at: https://www.itu.int/osg/spu/publications/internetofthings/

Junior. (n.d.). *Instalando o Raspbian no Raspberry Pi*. Available at: http://www.jairjunior.eng.br/artigos/instalando-o-raspbian-no-raspberry-pi/

Kleinfeld, R., Doukas, C., & Radziwonowicz, L. (2014). *Glue.things – a Mashup Platform for wiring the Internet of Things with the Internet of Services*. Research Gate.

Mantovani. (2012). *Estudo sobre computação ubíqua*. Available at: http://www.unifil.br/portal/arquivos/publicacoes/paginas/2012/11/516_867_publipg.pdf

Marini. (n.d.). *Comparativo da comunicação de dados em dispositivos móveis: WEB Servicese e Sockets*. Available at: http://painel.passofundo.ifsul.edu.br/uploads/arq/20150522102340173874 1499.pdf

Marion. (2002). *Monografia para cursos de administração, contabilidade e economia*. São Paulo: Atlas.

Mobberley, C. (n.d.). *What is Node-Red?* Available at: https://learn.adafruit.com/raspberry-pi-hosting-node-red/what-is-node-red

Oracle. (n.d.). *The Java EE 6 Tutorial*. Available at: http://docs.oracle.com/javaee/6/tutorial/doc/gijqy.html

Pandikumar, S., & Vetrive, R. S. (n.d.). *Internet of Things Based Architecture of Web and Smart Home Interface Using GSM*. Available at: https://ijirset.com/upload/2014/iciet/it/8_611.pdf

Perera, C., Zaslavsky, A., Christen, P., & Georgakopoulos, D. (2014). *Context Aware Computing for The Internet of Things: A Survey. IEEE Communications Surveys & Tutorials, 16(1).*

Pessoa, C., Silva, T., Rosa, M., & Augusto, T. (2016). Internet das Coisas: Estudo de mercado para aplicação em Pet Shops. *XXIII SIMPEP Proceedings*.

Pessoa, C., Silva, T., Rosa, M., & Jamil, G. (2016). A Internet Das Coisas: Conceitos aplicações, desafios e tendências. *XIII CONTECSI Proceedings*.

Pilon. (2009). *Estudo Para Aplicação De Redes Sem Fio No Ambiente Industrial*. Universidade Tecnológica Federal Do Paraná.

Postscapes. (n.d.). *IoT Gateways*. Available at: http://www.postscapes.com/iot-gateways/

Prehofer, C. (n.d.). *From IoT Mashups to Model-based IoT*. Available at: https://www.w3.org/2014/02/wot/papers/prehofer.pdf

Raspberry Pi Foundation. (n.d.). *Raspberry Pi*. Available at: https://www.raspberrypi.org/help/faqs/#introWhatIs

Rouse. (n.d.a). *Gateway*. Available at http://internetofthingsagenda.techtarget.com/definition/gateway

Rouse. (n.d.b). *Ethernet*. Available at http://searchnetworking.techtarget.com/definition/Ethernet

Rozas, N. (2004, November). O que é telemetria? *Revista Gás Brasil, São Paulo, 1*(15), 13–15.

Santos, R. A. S. (n.d.). *Domótica via dispositivos móveis*. Universidade Federal De Ouro Preto.

Silva, B. L. R. (2012). *Sistema De Controle Do Trio Automotivo Por Meio De SMS*. Centro Universitário De Brasília –UniCEUB.

Tanenbaum & Van Steen. (n.d.). *Sistemas Distribuídos: Princípios e Paradigmas* (2nd ed.). Pearson.

Toma, C., & Popa, M. (2014). IoT – Internet of Things Architecture for Context Aware Sensors Data Processing in Waste Management Solution. *Journal of Mobile, Embedded and Distributed Systems, 6*(4).

Valente. (n.d.). *Um middleware para Internet Das Coisas*. Available at: http://repositorio.ul.pt/bitstream/10451/9211/1/ulfc104490_tm_Bruno_Valente.pdf

Vujovic, V. (2016). *Raspberry Pi as a Wireless Sensor Node: Performances and Constraints*. ResearchGate.

Xu, Z., Pu, F., & Fang, X. (2016). Raspberry Pi Based Intelligent Wireless Sensor Node for Localized Torrential Rain Monitoring. Journal of Sensors.

Yunbiao, S. (n.d.). *Internet of Things: Wireless Sensor Networks*. Available at: http://www.iec.ch/whitepaper/pdf/iecWP-internetofthings-LR-en.pdf

ENDNOTE

1 Version 6 of the 32-bit Processor Architecture used primarily in embedded systems.

Chapter 10
Customer Experiential Knowledge's Contribution to Innovation Management:
Toward the Definition of a New Organizational Competence

Dhouha Jaziri
Sousse University, Tunisia

ABSTRACT

Many researchers have explored the knowledge management theory. However, to the author's knowledge few were interested in the tacit knowledge construct, whether it is gained inside or outside the organization. This chapter has a challenge to analyze in-depth the embedded knowledge gained from the customer, especially as it sheds light on the role of customer experiential knowledge by defining the customer experiential knowledge. It follows an emphasis on the customer experience and its close relationships with innovation management. Hence, a thorough theoretical background is presented progressively in order to define a new organizational competence labeled CEKMC. The first part presents an overview of the knowledge status, fundamental knowledge views, the evolutionary theory to the tacit knowledge construct. The second part stresses the definition of customer tacit knowledge related to customer experience. Finally, the conclusion defines a new organizational competence relative to this knowledge while discussing its contribution, especially to the experiential innovation type.

INTRODUCTION

As researches which are interested in the tacit knowledge are scarce, it is of paramount importance to revisit the tacit knowledge. This will be done as follows: a theoretical analysis will be presented through an extensive literature review, concerning the knowledge management in general as well as the tacit knowledge in particular. The first part treats the knowledge status and the related fundamental knowledge views as mainly *the Resource Based view* (RBV) and its extension *the Knowledge based view*. This first

DOI: 10.4018/978-1-5225-6225-2.ch010

section concludes by regarding the evolutionary theory of knowledge creation of Nonaka and Takeuchi (1995), while clarifying the tacit/explicit types of knowledge. The second part will be devoted to the construct of tacit knowledge (TK), while highlighting the importance of the customer tacit knowledge related to customer experience. In this case, the objective is to present the tacit knowledge related definitions and roots, by defining especially the customer experiential knowledge (Jaziri-Bouagina, 2017). The third and last section underlines the relationship of tacit knowledge with the innovation management. In this case, it is the objective to ascertain that tacit knowledge is a key input of innovation. In particular, to show that the customer experiential knowledge nourishes the innovation. On this issue, the proposed chapter concludes by the proposal of a new organizational competence relative to the customer experiential knowledge, while discussing its contribution to the experiential innovation in the case of well-being tourism.

BACKGROUND

The Knowledge Underpinnings

Knowledge Status

A knowledge research synthesis, established by Shin *et al.* (2001), presents a conceptualization of "Knowledge" according to three views on the status of knowledge made from information (Shin, Holden, & Schmidt, 2001).

The first view considers knowledge as an object that is stored and manipulated. In this case, MacQueen (1998) considers the Knowledge as the pathway to information (p.609) and as a set of rules (p.610). Respectively, McQueen (1998) supports that *The Knowledge can be retrieved through access to documents and databases containing data and information...* (p. 610), and that *the set of rules is extracted by a knowledge engineer, example are diagnostic procedures, from a domain expert who has extensive experience* (p. 610). Once information is ascertained to be useful in a context, it becomes Knowledge and this it is deposited (Shin *et al.*, 2001).The second view defines knowledge as a process related to the application. Regarding this second view, McQueen (1998) states that *Knowledge can be stored in repositories of electronic communication* (p.610). He discusses this view in the context of consulting firms. The repository is an archive of comments and opinions of experts useful for other less expert consultants.

As reported by Shin *et al.* (2001), the implication of considering Knowledge as "process" for Knowledge management is that organization focuses on the conception of the process of Knowledge creation and dissemination. Also, considering *Knowledge as process* requires a system/technology in order to connect the source and recipient of knowledge (Shin, Holden, & Schmidt, 2001, p. 339).

The third view conceptualizes Knowledge as belief *in mind*, personalized information or the cognitive status of knowing (Shin, Holden, & Schmidt, 2001). According to Von Krogh (1998):

The individual justifies the truthfulness of his or her beliefs from observations of the world, which depend on the individual's unique viewpoint, sense-making processes and individual experience (Von krogh, 1998 in Sara, 2008).

Belief in mind coincides with the view given by McQueen (1998), regarding *Knowledge as Understanding*. This corresponds to a philosophical orientation (McQueen, 1998). It follows that knowledge exists only in humans; therefore, some researchers agree that it is not possible to mechanize it. In this case, McQueen (1998) stresses that:

The role of information technology and knowledge management is to provide sources for searching (information repositories) and stimulation (information streams) so that individuals can expand their personal knowledge and apply that knowledge to assist the organization in meeting its goals. (McQueen, 1998, p. 610)

In relation to the above, Bennet & Bennet (2008) highlighted the idea that knowledge is a creation of the human mind (the result of associative patterning in the brain). These researchers acknowledge that knowledge is composed of two parts: Knowledge (Informing) (or KnI) and Knowledge (Proceeding) or (KnP). This division corresponds, respectively, to "knowing that" and "knowing how".

- "Knowing that" corresponds to '*What something means*' (Dinur, 2011), in other terms insights, meaning, expectations, theories, symbols, facts that lead to useful action (Bennet & Bennet, 2008; Dinur, 2011).
- As to "Knowing How", it represents *the process of selecting and associating the relevant information (KnI) from which specific actions can be identified and implemented, that is, actions that result in the desired outcome*, according to Bennet and Bennet, 2008 (p. 74). In addition to this, Dinur (2011) denotes that KnP is the accumulated practical skill or expertise that leads someone to act in an efficient way.

Moreover, through the literature review, it is concluded that knowledge is primarly considered as a process rather than a result (Bennet & Bennet, 2008). In this sense, Eisenhardt and Santos (2002) also conclude that the consideration of knowledge as knowing, which is related to processing information, overcomes the view of knowledge as a resource. In other words, it reflects a new epistemological view that is ascribed to *The process phenomenon of knowing, which is clearly influenced by the social and cultural settings in which it occurs* (Eisenhardt & Santos, 2002, p.4).

FUNDAMENTAL KNOWLEDGE VIEWS

The Resource Based View (RBV): A Determinant of Firm's Innovation Capacity

According to Rollines & Halinen (2005), the resource-based view (RBV) considers knowledge as a resource embedded in the firm. Three years earlier, Kostopoulos, Spanos and Prastacos (2002) indicated that the firm's strategic resources heterogeneity is the basis of RBV. Kostopoulos *et al.* (2002) presented the firm as a set of tangible and intangible resources and capabilities. The former set is expressed in terms of financial or physical resources. The second corresponds to Knowledge, skills of employees, or the organizational procedures. The notion of capabilities is defined by Kostopoulos *et al.* (2002) as *the firm's capacity to deploy and coordinate different resources usually in combination, using organizational processes, to affect a desired end* (p.4). Moreover, it is important to distinguish between capability and resource according to two aspects. The first aspect is linked to *organization and its processes*; hence, capability is related, while "the resource" is not.

The second aspect depends on the targeted purpose. Hence, capability tends to improve the effectiveness of the resources that a firm possesses, in order to attain its purposes. Therefore, the resource-based view attempts to gain a competitive advantage which is the outcome of organizational capabilities reflected by the processes of selection, accumulation and deployment of a set of tangible and intangible resources (Kostopoulos, Spanos, & Prastacos, 2002).

The resource-based view represents an interesting theoretical view to approach the innovation research in terms of resources and capabilities (Kostopoulos, Spanos, & Prastacos, 2002). The later impacts positively the outcome of the innovation processes. From here, the highlighted assumption is that

organizational resources and capabilities are those that underlie and determine a firm's capacity to innovate. They are taken to provide the input that, in turn, is combined and transformed by capabilities to produce innovative forms of competitive advantage (Kostopoulos, Spanos, & Prastacos, 2002 p. 8).

The Kostopoulos *et al.* (2002)'s theoretical paper sets an interesting illustration. On the one hand, it shows the relationship between the intangible resources as Knowledge and the firm's capacity to innovate. Specifically, the intangible resources (human or knowledge) have a decisive anchor point since they are valuable. On the other hand, the model indicates the importance of capabilities as determinants of the firm's capacity to innovate. Particularly, the focus will be on the dynamic capabilities which refer to the firm's ability to integrate, build, and reconfigure internal and external competencies to address rapidly changing environments, while showing that the absorptive capacity is embedded in the debate on dynamic capabilities.

It's worth mentioning that the relationship between RBV and innovation is a mutual beneficial relationship. In this case, the RBV allows to concentrate on the determinants of the firm's capacity to innovate. In return, the implementation of innovation implies the renewal of its assets.

The Knowledge Based View (KBV): An Extension of Resource Based View

The importance given to intangible assets in the resource-based view has led to the emergence knowledge based view (Kostopoulos, Spanos, & Prastacos, 2002; Eisenhardt & Santos, 2002). The KBV conceptualizes knowledge as a resource that can be acquired, transferred or integrated to achieve sustained competitive advantage (Eisenhardt & Santos, 2002). Knowledge, both explicit and tacit, represents a strategic resource.

However, Eisenhardt and Santos (2002) revealed some challenges regarding the knowledge based view, while underscoring their concern about its nature. Is Knowledge based view a theory of strategy? (a theory that links independent variables to a specific conception of firm performance). The authors explain that the Knowledge based thinking is essentially associated with the conceptualization of knowledge as the most important resource. This fact restricts the approach to a particular case of RBV, and constitutes a constraint to the evolution of KBV toward a theory of strategy. In order to gain the status of theory, it's important to see the firm as a system of knowing activity rather than as a system of applied knowledge bundles (Eisenhardt & Santos, 2002). As a matter of fact, Eisenhardt and Santos (2002) support the conception of knowledge as a process of knowing, since the organization is a complex system of knowledge creation and application.

The perfect position, according to Eisenhardt and Santos (2002), is to adopt a combined view in order to provide a global comprehension of the role of knowledge. Consequently, it is important to bridge the two theoretical orientations *Knowledge as a Resource* and *Knowledge as Knowing*. This is the case for Cook and Brown (1999), who argue that *it is through the interplay of knowledge and knowing that innovation takes place in organizations* (Cook & Brown, 1999 in Eisenhardt & Santos, 2002). It follows that we must concentrate on the knowledge integration processes. The latter is explained by Eisenhardt and Santos (2002) as the capacity to understand *how specialized knowledge is integrated from different sources to generate new knowledge, or to apply that knowledge to the creation of new products and services.* (p.34)

The Evolutionary Theory of Knowledge Creation (Nonaka & Takeushi, 1995)

It is through the dynamic theory of organizational knowledge creation of Ikujiro Nonaka (1994), that a philosophical conception of knowledge adopts a more practical direction (Stillwell, 2003). Hence, *what matters is the embodiment of knowledge.* As a conceptualization, Nonaka (1994) proposed a "spiral" model of knowledge creation. Thus, he agrees the idea that organizational knowledge is understood as a process that arises principally through informal networks (Einshardt & Santos, 2002). This model shows the creation of a new concept in terms of a continuous dialogue, or interplay between tacit and explicit knowledge at individual and organizational levels (Einshardt & Santos, 2002). The model presents an important structural distinction between the explicit and tacit knowledge. This latter theoretical and practical contribution is very significant for the organizational field, although it has been confronted with some critics. The main critic is addressed by Ray and Clegg (2007), who emphasized the fact that the explicit knowledge is assimilated to be objective one in the knowledge creation theory. This approximation is misleading since explicit knowledge is globally subjective in nature, and the objective knowledge requires scientific methods.

Many distinctions have been addressed in the literature regarding the tacit and explicit knowledge. It is important to retain the explanation of Einshardt and Santos (2002), who stated that:

Tacit knowledge is linked to the individual and is very difficult, or even impossible, to articulate. As knowledge is explored, put into action and is socially justified, some part of it may be codified (made more explicit), by being converted into messages that can then be processed as information and transmitted. (Einshardt & Santos, 2002, p.3)

At the organizational level, Nonaka (1994) exposes both components of tacit knowledge: "cognitive" and "technical". The former consists of:

mental models, or working models that include schemata, paradigms, beliefs and viewpoints that provide "perspectives" helping individuals to define the world, this dimension refers to an individual's images of reality and visions for the future. (p.16)

As for the technical component, it includes the know-how, crafts, skills that are implemented in particular contexts. Generally, the tacit knowledge equals the experience obtained from the organizational context (Chalkiti & Sigala, 2007). In this case, the individuals' tacit knowledge is developed by permitting them to deeply understand the knowledge received (StillWell, 2003).

Boisot (1998) highlighted that it is fundamental to determine, acquire, use and convert tacit knowledge into an explicit one in order, to transform the firms' knowledge assets into competitive capabilities. Nonaka (2001) noted that the central concern of Spiral model (also called by SECI processes) is to create a knowledge in and for the organization in which members engage in practices which first dis-embody and then re-embody tacit knowledge (Stillwell, 2003).

The SECI process is an abbreviation of the four consistent processes. The first stage is the socialization where the tacit knowledge is shared among individuals (Nonaka, 2001). This means that each person's tacit knowledge is converted into another person's tacit knowledge. This conversion is realized by individual interactions or by what Nonaka (2001) called "the physical proximity" with colleagues, suppliers or customers (Nonaka & Konno, 1998). This implies the empathy between them (Nonaka, 2001). Tacit knowledge can be obtained without language through shared experience. For example, Nonaka (1994) presents on-the-job training (OJT) as a learning technique based on observation, imitation and practice. In other terms, a long period of apprenticeship confers to organization's members the ability to comprehend the feelings and thoughts of others (Nonaka & Konno, 1998).

In this case, the socialization process implies a knowledge acquisition and dissemination. The latter consists in the sharing of personal knowledge which imposes the use of such means as formal or social meetings, team discussions and projects (Sigala & Chalkiti, 2007). Furthermore, Nonaka (1991) emphasized that socialization has no intention of transferring such tacit knowledge into an explicit one.

The second stage is externalization. Tacit knowledge is converted into an explicit one. This means that the conversion and translation of the tacit knowledge will be made in comprehensible forms (Nonaka,

Figure 1.

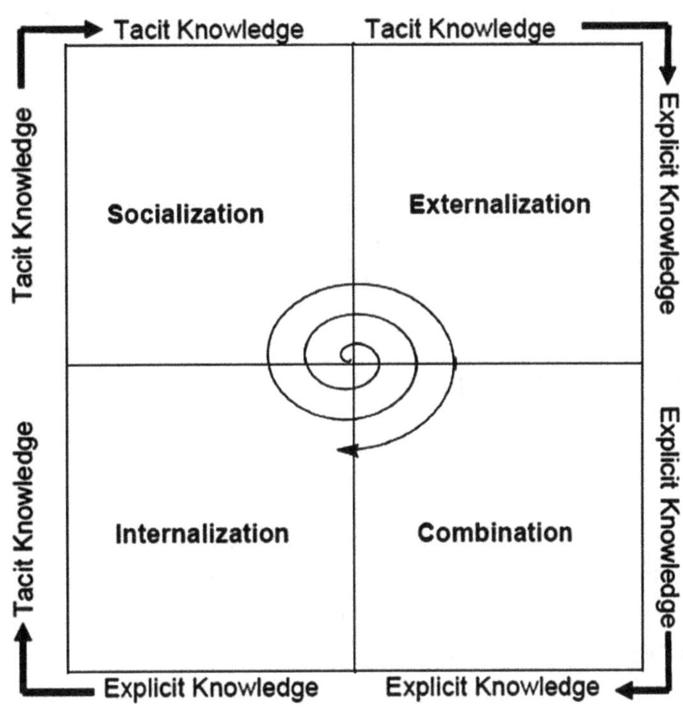

Note: SECI => Socialization-Externalization-Combination & Internalization

2001; StillWell, 2003). Later on, at the Knowledge Advantage Conference, Nonaka (1997) reported that there are two cases.

The first is the articulation of one's own tacit knowledge - ideas or images in words, metaphors, analogies. The second is eliciting and translating the tacit knowledge of others - customer, experts, for example - into a readily understandable form, e.g., explicit knowledge (Nonaka, 1997).

The researcher stresses the importance of dialogue (during face-to-face communication, people share beliefs and articulate their thinking). Dialogue strongly supports the externalization process (Nonaka & Konno, 1998). Regarding the externalization of customer knowledge, Nonaka and Konno (1998) stressed the need for *a deductive/inductive reasoning or creative inference (abduction).* This stage can take place through other forms of interactions, such as brainstorming and experts' interviews (Sigala & Chalkiti, 2007).

The third stage is combination. In this stage, newly explicit knowledge becomes widely discussed, redesigned and modified (StillWell, 2003). The combination process relies on three processes (Nonaka & Konno, 1998), and its objective is to re-arranges existing explicit knowledge into a more structured form for an organizational use (Asiamah, 2009). The first process is the capture and integration of the new explicit knowledge. The second consists in the dissemination of the explicit knowledge. In this order, the organization gets more composite assortments of explicit knowledge, which are communicated and diffused among the organizational members through presentations or meetings. To that end, the information technology becomes useful and it conducts the explicit knowledge by emails, data bases, etc. (Nonaka, 1997). Additionally, the more advanced the technologies are, the more the outsourcing of the knowledge is enhanced (Yakhlef, 2005). Finally, the third process is to make the explicit knowledge more usable (e.g. market data, reports...).

Finally, the last stage is internalization. This stage is represented by the process where the explicit knowledge is absorbed and transformed, again, into a new tacit organizational knowledge form via real world experiences (e.g. learning by doing) or simulated world experiences (e.g. trial and errors, Sigala & Chalkiti, 2007). The individual identifies the knowledge pertinent to his work objective in order to internalize it. In other words, it is a question of learning through codified knowledge and its attributions (Asiamah, 2009). Likewise, it is worth mentioning that each knowledge stage is particular in itself as the experience of members changes from one process to another as well as the leadership involved (Nonaka, 2001). In this case, the internalization involves the accomplishment of concepts, innovation or improvement and the use of simulations to learn new concepts in virtual situations (Nonaka & Konno, 1998).

THE TACIT KNOWLEDGE CONSTRUCT

Definition and Roots

As shown is previous discussion of this chapter, tacit knowledge in knowledge management represents the fundamental concept of the theory of knowledge spiral of Nonaka and Takeuchi (1995). It involves a continuous interplay between explicit or codified knowledge and tacit knowledge at individual and organizational level (Einshardt & Santos, 2002). Many definitions have been advanced by management researchers regarding this concept. On the one hand, some researches unveiled the philosophical roots of this concept as: Nonaka, 1994; Nonaka and Takeushi, 1995; Eisenhardt and Santos, 2002; Polanyi, 1966; Castillo, 2002; Küpers, 2005; Dinur, 2011; Stillwell, 2003; Mascitelli, 2000; Mooradian, 2005 and many others. On the other hand, according to different research fields, such as neuroscience and psychology, the concept's taxonomies were defined differently (Bennet & Bennet, 2008).

Two categories of human knowledge have been classified by Polanyi (1966), explicit and tacit knowledge. The explicit knowledge is articulated and shared through formal and systematic language (Nonaka & Takeuchi 1991), whereas the *tacit/implicit* knowledge has a personal quality, profoundly attached to people. This makes it hard to formalize and communicate (Nonaka & Konno, 1998; Dinur, 2011). This Knowledge is "characterized by "analogue" qualities – parallel processing of continuous complex variables – while explicit knowledge shows the discrete discontinuities characteristic of "digital" processing" (Nonaka, 1994). Consequently, subjective insights, intuitions are situated at the domain of tacit knowledge. Similarly, Chalkiti and Sigala (2007) indicate that the type of knowledge that cannot be expressed in written forms is ingrained in our minds.

In other words, Nonaka and Konno (1998) and Mascitelli (2000) acknowledge that the tacit knowledge is intensely embedded in an individual's experience, values and emotions. In this way, the tacit dimensions are extracted from experience (Polanyi, 1966) and are deeply rooted in action and context. In the same way, Nonaka (1994) expressed that tacit knowledge takes up residence in a comprehensive cognition of the human mind and body.

However, according to Mooradian (2005) and Castillo (2002), generally the literature has presented a hazy concept of tacit knowledge caused by the multiplicity of interpretations. The haziness involves confusion to implement a knowledge management strategy and requires the comprehension of the philosophical roots of the concept (Mooradian, 2005).

Starting from the beginning, Michael Polanyi, a scientist turned-philosopher (1966), was the first developer of tacit knowledge conception. Thus, the conception is dominated by the philosophical aspect. According to Polanyi (1958), the implicit knowledge is considered basic to all human knowing.

In his book entitled *The Tacit Dimension*, Polanyi argues that the knowledge that it can be expressed in speech, writing, and other explicit forms is only the tip of the intellectual iceberg. "Beneath the surface of conscious thought lies a vast sea of tacit knowledge, derived from a lifetime of experience, practice, perception and learning. Therefore, Polanyi (1966) states that *We can know more than we can tell* in order to capture better the tacit side (Polanyi, 1966 in Mascitelli, 2000).

Consequently, Polanyi proposed a subjective side in contrast to the positivism epistemology. Science is not only the objective aspect (Mooradian, 2005). The distinction between tacit/implicit and explicit knowledge reflects, respectively, the distinction between subsidiary knowledge and focal Knowledge.

On that account, the subsidiary knowledge (implicit knowledge) was characterized as follows:

(1) It is active in the mind, but not consciously accessed at the moment of knowing; and (2) it somehow brings about the focal knowing (Mooradian, 2005). Hence, the focal knowing corresponds to the act of knowing. For example, the tactile sensations and visual perceptions, as explains Mooradian (2005): *they are the subsidiary or implicit acts of knowing, through repetition, experience and integration with memory, they are transformed into their subsidiary role as an enabler of an expanded knowledge* (Polanyi, 1969). Therefore, and in line with Mooradian (2005), the tacit knowledge is a *structural/relational concept* which describes a relation between different kinds of knowledge. Mooradian (2005) concludes that the tacit knowledge shouldn't be treated as having a value in itself because this fact is contrary to the relational explicit/tacit distinction. In other words, *for any explicit Knowledge (Ke) there is some tacit knowledge, (Kt)* (Mooradian, 2005).

Furthermore, it is interesting to note that the tacit knowledge can differ in difficulty along a scale from 'easy' to 'impossible expressible'. In this regard, key ideas were expressed by Mooradian (2005, p.110):

Some tacit knowledge is more easily expressible in natural or formal language than other kinds of tacit knowledge. It may be the sort of knowledge expressed by saying what it is like to be in a certain situation or have a certain experience. Similarly, some knowledge can be described as a bodily skill and involves our sensory awareness of our own bodies as well as brain signals of which we are not fully conscious. This sort of knowledge is hard, if not impossible to articulate.

Tacit Knowledge: The Phenomenology Underpinning

According to Ladkin (2012), the phenomenology found its roots in the work of the phenomenologist Maurice Merleau-Ponty (1908-1961). Merleau-Ponty (1962) advanced that reflection and volition, which are high functions of consciousness, are grounded in *pre-reflective, bodily existence*. Hence, Ponty (1962) described the human bodies as "immanent" and "transcendent".

- **Immanence:** The material corporeal flesh and bone aspect of the human body. It is through the immanent body that the sensations are experienced and physically presented in the world (Ladkin, 2012).
- **Transcendence:** Refers to non-material aspects like intellectual, imaginative and cognitive processes (Ladkin, 2012).

The principle of Ponty (1962) is to announce that the interaction with the world is made through the cognitive processes and bodies responses. Kupers (2005) proposed a phenomenology underpinning the embodied implicit and narrative knowing in organizations while showing, on the one hand, the significance of experiential dimensions of implicit and narrative knowing; and on the other hand, their mutual interrelations in organization.

According to Kupers (2005), it is needed a processual and relational understanding of knowledge in contrast with the resource-based view. In this order, Kupers (2005) provides a phenomenological based processual perspective on implicit and narrative knowing.

Kupers (2005) explains the embodied implicit knowing in organizations through the concept of experience. Hence, the embodiment is a way of knowing through "lived situations" and its encounters. *The living body mediates between "internal" and "external" or "subjective" and "objective" experience and meaning* (Merleau-Ponty, 1962 in Kupers, 2005). In this case, the knowledge about encounters dependent on the level of lived experience as a bodily – engaged beings (Kupers, 2005). Hence, in line with Kupers (2005) *the embodiment doesn't simply means* physical manifestation. Rather, it means being grounded in mundane experience and being inherently connected to our environment in an ongoing interrelation (p.115).

In this sense, the individual thought, feeling or action is dependent on his exposition to an environment of inter-related senses (tactile, visual, olfactory or auditory way). In accordance with Kupers (2005), taking a phenomenological understanding means taking into account the body- and sense- related contact, and embodied correlations as the constitutive base of knowledge and also narrative processes as medium of expressing and sharing tacit knowledge. The attempt of Kupers (2005) is to apply a phenomenological framework in order to understand the knowledge in organizations. The principle emphasized is that people are immersed in an embodied world of experience in which the lived is always greater than the known (Merleau-Ponty, 1962 in Kupers, 2005). Hence, tacit knowledge is not 'a resource', but it always represents a process of knowing and acting, which is called by "an embodied knowledge in –use" (Kupers, 2005).

The Narrative Knowing in organizations corresponds to living narratives adopted in organization. For example, in the case of organization, Kupers (2005) referred to social exchanges like conversations, face to face discussion translating the accounts of events. To demonstrate the relationship between embodied implicit and narrative knowing, Kupers (2005) referred to "stories".

On the one side, the embodied self is situated in a world of signs, symbols and communication. (Kupers, 2005). In this way, Kupers (2005) noted that:

Embodied practices of knowing are enmeshed through the mediation of artefacts – objects and symbols, including language – through which knowers extend themselves in the world and which, in turn, re-form their bodily movements and perceptions. (p.119)

On the other side, language represents an event of expression, and is an establishment of an embodied sign system. Consequently, the embodied thought and gesture expressed in narratives are living from what has already been experienced (Merleau-Ponty, 1962).

Therefore, it is possible to link between embodied implicit and narrative knowing because the latter can be considered as expressive medium to transfer implicit knowledge, indirectly. According to Kupers (2005), the tacit knowledge and implicit knowing are shared and externalized by means of storytelling and shared experiences, which will be discussed fully in the next section.

CONCEPTUALIZATION

The Tacit Knowledge Role Into the Innovation Management

Literature recognizes that tacit knowledge is a key input of innovation (Yakhlef, 2005). However, it follows that its capture is a challenge (Mooradian, 2005). Thus, in this case, organizations which are willing to enhance the effectiveness of the innovation process, should engage customers to have tacit knowledge (Yakhlef, 2005).

Yakhlef (2005) supports that the locus of innovation is displaced from a hierarchical model to a distributed environment including customers, (lead) users, intermediaries and other external stakeholders, which is expressed in terms of the immobility of tacit knowledge. As found by Yakhlef (2005): *Innovation-related activities will tend to be allocated between companies and other external sources (customers and users, etc.) depending on the location of tacit knowledge underlying them* (p. 227). Hence, the R&D function is disintegrated. The innovation should be nourished not only from the knowledge in organization, but also from the outside while integrating the internal skills and expertise.

The use of technology helps largely in the acquisition of customer knowledge but, often, the tools neglect the feelings and emotions of customers. Taking all of the above into consideration, Mascitelli (2000) wrote that:

If we accept the concept that tacit knowledge is fundamentally based on bodily experiences and emotional involvement, however, it is hard to imagine that something so personal can be digitized and downloaded. (p.189)

Hence, Yakhlef (2005) stressed the idea that the face-to-face interactions and the sharing of context and perspectives are still the preferred ways to capture tacit knowledge (Yakhlef, 2005). Furthermore, according to Yakhlef (2005), it is important that the organization knows how to achieve the essential competencies to effectively exploit underline external knowledge flow, in order to transform them into innovation.

Still with the case of research papers establishing the link between tacit knowledge and innovation, Mascitelli's paper (2000), which connects, specifically, the breakthrough innovation type regarding "product" and the tacit knowledge of employees organization 'and' that of customers is referred. According to Mascitelli (2000), to evince the creative power of tacit knowledge, managers should induce the deep emotional commitment and involvement to the innovation process. The establishment of commitment and involvement is an essential condition, in order to move from the externalization to innovation.

Researchers advanced then the intimate physical interaction by encouraging the face to face interaction with customers, a co-location of teams. In this case, the use of metaphor and analogy plays a fundamental role in proposing the best product design. Mascitelli (2000) concludes that innovations are often an outgrowth of a highly personal form derived from a lifetime of experience and learning. On the other hand, Seidler *et al.* (2004) are interested in the role and impact of tacit knowledge in innovation management, while aiming at the success of innovation. Following Seidler *et al.* (2004), this impact can be anticipated through three characterizations of tacit knowledge:

1. The evolution of TK in organization i.e., the importance of making the personal knowledge available to others for an innovator firm which involves a knowledge process, as supported by the theory of knowledge creation of Nonaka and Takeushi, (1995). Referring to Senker (1993), Seidler *et al.* (2004) annotate that the tacit element of innovation evolve through practical experience in/or outside the organization.
2. The vitalization of TK which consists in the activation of TK, for example, through the learning by using, once the good, service or process was used internally inside or outside the organization as by external consumers (Senker, 1993), and
3. The transfer of TK in organizations. Regarding this transfer dimension, it's important to *reveal that much tacit knowledge is transferred through body language* (Seidler, Hartmann, & Gemünden, 2004, p.9).

Seidler *et al.* (2004) present a classification model through which an assumption was adopted: *that tacit knowledge management needs to be adapted quite closely to the innovation management of a firm for being able to achieve maximum innovation success* (p.18). Concerning, the tacit knowledge management dimension, authors represented it through the three key levels discussed above: Evolution, vitalization and transfer of tacit knowledge. Hence, the interpretation is:

The more 'relevant tacit knowledge' is involved in the different phases of the innovation process, the more effective and efficient innovation management is carried out, which leads to an increase in innovation success (Seidler, Hartmann, & Gemünden, 2004, p.18).

Moreover, an interesting theory, as well as a practice case, presents this closely relationship between the tacit knowledge and the innovation; it is the knowledge marketing theory developed by Curbatov (2003). This theory applies the spiral model of knowledge of Nonaka and Takeushi (1995) within an

Figure 2.

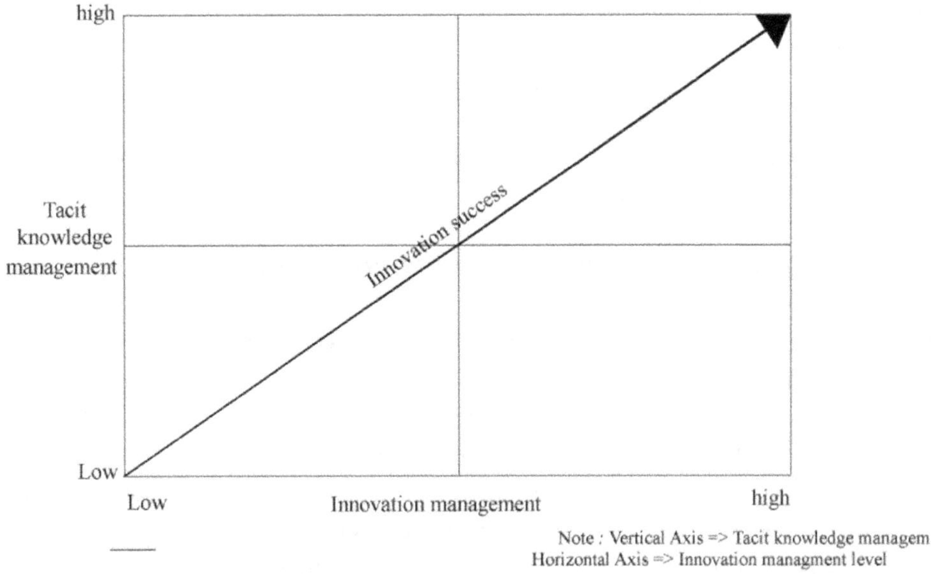

experiential context. Hence, Curbatov (2003) advocates the idea that experience is a source of knowledge. The experience described by Curbatov (2003) was concerned with *the experience of creation of personal product (fragrance)* with the organization and the related knowledge creation.

Moving to the field of customer knowledge, as supported by Yakhlef (2005), knowledge marketing reviews the customer through his competencies which corresponds to the philosophy of the co-creation marketing. Hence, the principle of the knowledge marketing conceived as a new conceptual approach of marketing is integrated in the current thinking movement of the customer empowerment. The knowledge marketing proposed in this case, activated both the customer and organization competencies that are integrated in the process of knowledge co-creation (Curbatov, 2003). The customer will regain some control on his experience of consumption, on the creation of his personalized product. The organization will have the chance to innovate with its competent customers.

The approach founded its main idea in the marketing of "procuration" (Cova & Cova, 2001), which recognized the importance of company's resources and collective competencies of consumers. The idea is to connect between the experience of consumption to the desire of appropriation expressed by consumers. Hence, it follows the use of their competencies in order to produce the lived experience.

Moreover, Curbatov (2003) and Curbatov *et al.,* (2006), recognized the importance of experience consumption as an important part of the tacit knowledge field. They describe it as a procedural knowledge. According to this orientation, hence, they recognized the importance of the experiential knowledge as a pivotal part of the tacit knowledge. It is related to activities, and is unique to each person and to the context of use. This is a knowledge related to a personal experience. They described it as a practical knowledge that exists at individual and collective level.

The co-creation experience of personal fragrance product conception stressed the importance of experiential knowledge gained from the interaction and the share of experiences related to the consumption of perfumes between customers and enterprise employees, on the one hand. But on the other hand, Curbatov (2003) emphasized the importance of prior customers' experiences He also pointed up the capture of experiences riches in knowledge as a source of innovation for companies. In this case, the knowledge marketing was limited to a method of the olfactory knowledge marketing developed for olfactory web. Curbatov (2003) promoted the idea that the fragrance is the support of tacit knowledge, and its externalization is achieved by the means of dialogue. The latter is integrated in the frame of experience of personal product creation in presence of company's members.

This amassed knowledge was transferred through the creation of fragrance concept. Hence, the innovation reflects the creation of new knowledge.

Defining the Customer Experiential Knowledge (CEK)

A depth theory review showed that customer experiential knowledge can result from customer experience through three dimensions: physical or sensorial, praxiological, and rhetorical, (Jaziri-Bouagina, 2017) (See Table 1). All are integrated in a dynamic interaction between the person (customer), object or stimuli (good or service) and the situation (Roederer, 2008).

In parallel, a deeper understanding of tacit knowledge aspects in literature leads us to identify some taxonomies of tacit knowledge, and hence, to express the customer experiential knowledge in terms of knowledge aspects. The physical dimension corresponds to the embodied tacit knowledge as defined by Bennet and Bennet, (2008). The rhetorical dimension is connected to the semantic tacit knowledge as defined by Castillo (2002). Moreover, we think it is necessary to integrate the affective tacit knowledge as it represents unexpressed feelings (Bennet & Bennet 2008; Dinur, 2011) that may be important for the customer service experience, and may have a close relationship with other knowledge aspects (See Figure 3).

METHODOLOGY

An exploratory documentary research was followed by targeting specifically the thalassotherapy and spa tourism sector. To this regard, an exploration of different online and offline resources was made. Two methods of analysis were carried out. The first concerned Tunisian thalasso guidebooks and especially sites specializing in well-being tourism. In this case, a lexical analysis of 40 thalasso-spa centres' eadvertising presentations on a French tour operator's website was made. The identification of major repeated segments led us to summarize it into two main categories: experiential and technical arguments. This

Table 1. Summary of consumption experience dimensions

Dimension	Definition
Physical dimension	It covers all the sensory aspects involved by an interaction between a subject and an object of consumption in a given context.
Praxeological dimension	The different categories of actions starting from the individual to its environment (exp interactions between persons, actions in the experience place...)
Rhetoric dimension	it reflects the production of meaning associated with consumer experiences, real or fictional. The individual interprets objects as signs.

Roederer 2008 in Jaziri-Bouagina, 2017, p.10.

Figure 3. The customer experience and the customer tacit knowledge correspondence
Jaziri-Bouagina, 2017, p.11.

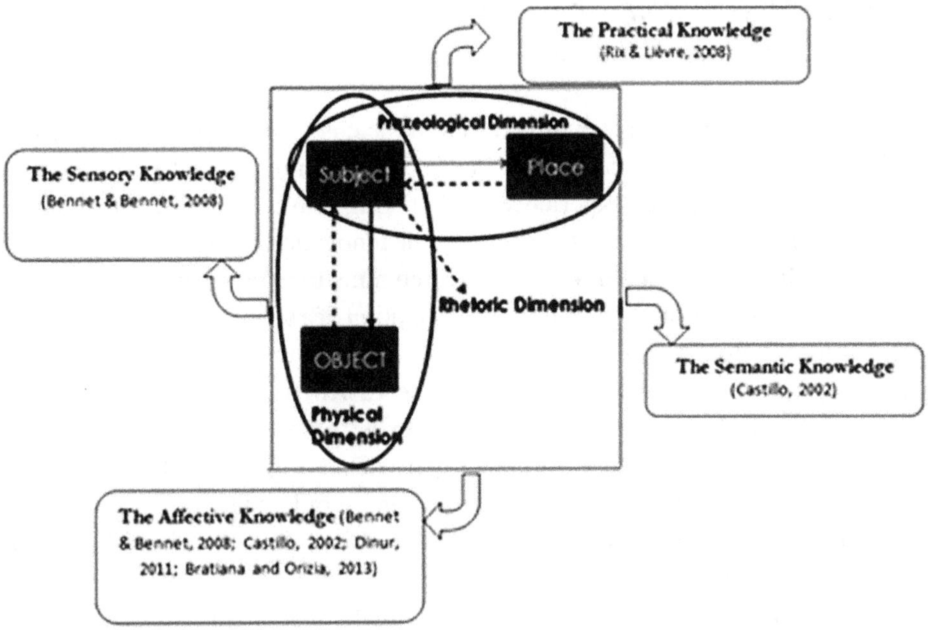

first analysis concluded that the use of experiential aspects was very limited. Instead, the emphasis was on the technical arguments.

The concept of experience was restricted only to sensorial aspects. The second analysis is thematic and it concerned directly the websites content of 36 thalasso and spa centres. This analysis included texts and images. The results supported the earlier conclusions and showed there were only six centres trying to present themed cures. In this case, we can connect the aspects of this offer to the definition of Roederer (2008, p.38), concerning the experience offer creation, namely:

the ability to create experience offers, that is to say, the ability to deliberately assemble products and services in order to stage experiences,which are valued by the consumer, can be a significant competitive advantage.

Thus, we noticed the lack of innovation in terms of experience offer. The thalassotherapy and spa tourism is a core-experience industry. However, there is a lack of recognition related to the concept of experience in this industry.

In this framework, a key question is essential to ask as follows:

What managerial competence can help wellbeing organizations succeed in producing innovation based on the concept of experience?

DISCUSSION

Toward the Definition of a New Organizational Competence: A Focus on the Customer Tacit Knowledge

On theoretical level, it was noticed a lack of research in the customer knowledge management (CKM) field that highlights the tacit knowledge of customer. Hence, it is imperative to conceptualize the CKM on the basis of the customer service experience while taking into account the innovation research. In this case, the research is interested in a specific type of innovation, which is the innovation through experience creation. On the empirical level, the research aims to explore the use of knowledge-based customer experience and its contribution to the implementation of a successful experiential offer –themed cures. The previously mentioned gaps represent a basis to propose the customer experiential knowledge management approach, which is defined as follows: the CEKM is the association of the knowledge management process with the customer service experience in order to enhance the future customer service experience, or to create an offer of service experience. As suggested by the competence-based management theory, we have noted a perceivable co-evolution of knowledge and competence management in recent research and practice (Hong & Stahle 2005). Day (1994) uses interchangeably the words 'capabilities' and 'competencies' and defines them as "complex bundles of skills and accumulated knowledge exercised through the organizational process that enable firms to coordinate activities and make use of their assets". He concludes that there is a direct connection between distinctive capabilities and superior performance (Day 1994). Based on the knowledge-based competencies literature, critical competencies are summarized as follows:

On the one hand, it is deduced that the researches are mostly conceptual; few of them propose a measurement scale. On the other hand, the main conclusion is that the competencies proposed deal with market knowledge, or more specifically, with the customer knowledge in general terms. However, the CEKM requires a specific organizational competence to make full use of the customer's experiential knowledge in order to enhance performance. Hence, we define the CEKM competence (CEKMC) as follows: it is the degree to which an organization demonstrates competence to generate and to integrate the knowledge-based customer experience in order to obtain a successful experience innovation. In other words, CEKMC is the competence of the organization to propose a new experience offer on the basis of effective management of customer experiential knowledge. It follows that main research questions regarding CEKMC are issued:

1. How can customer experiential knowledge be generated, integrated and utilized effectively within the well-being organization?
2. What methods and platforms are used by managers in order to externalize the customer experiential knowledge CEK?
3. What are the internal and external factors that may influence the CEKM in the context of an innovation through experience offer?

Table 2. Knowledge based competencies synthesis

Authors	Competence	Definition	Competence Dimensions
Li & Calantone (1998) (c) & (e)	Market Knowledge competence	Is the process that generates and integrates market knowledge	(1).Customer knowledge process, (2).Competitor knowledge process (3). Marketing research and development interface
Campbell (2003) ©	Customer Knowledge Competence	Based on internal firms processes that generate and integrate specific customer information which enables firms to develop customer specific strategies	(1).Customer information process (2).Marketing-it (inf. technology) interface (3).Senior management involvement (4).Employee evaluation and reward system
Rollins& Halinen (2005) ©	Customer Knowledge management competence	The management of 5 aspects that the generation, dissemination and use of customer knowledge are enhanced	(1).Interfunctional cooperation (2).Supportive organizational systems (3).Cooperation with customers (4).Supportive (IT) systems (5).Organizational culture that support customer orientation
Chen & Huang (2009) (e)	Knowledge management capacity	Is the extent to which the firm is able to acquire, share and apply new or improved Knowledge.	---------
Hou & Chien (2010) © & (e)	Market Knowledge management competence	The process capability to acquire, convert, apply to use and to protect the market knowledge	(1).Acquisition (2).Conversion (3).Application (4).Use (5).protection
Sun (2010) © & (e)	Customer knowledge management competence	Competitive organizational resources for implementing CKM in an organization which composes the capability to exploit, integrate and utilize CK.	-Customer knowledge process capability -Customer Knowledge infrastructure capability

Jaziri & Triki, 2013.
Notes: © Conceptual study, (e) Empirical study. The presentation of the knowledge based competencies follows a chronological order.

CONCLUSION

This chapter treated the various aspects of literature regarding the knowledge management and specifically focused on the tacit knowledge definitions and also related taxonomies and the externalization tools presented in knowledge theory. Recognizing the difficulty of approaching the relationship between tacit knowledge and innovation, this chapter follows a progressive theoretical analysis in order to show the existence of a close relationship while targeting the customer as an important contributor of tacit knowledge.

In this case and as shown by Vargo and Lush (2008), it is believed that the value is phenomenologically determined by the beneficiary. Hence, making the experience of use, not only important, but also primordial to externalize the experiential knowledge. The experiential knowledge was addressed by research in organizations' management and was explicitly recognized through the conceptualization of marketing in terms of knowledge perspective. This is done by the knowledge marketing theory (Curbatov, 2003). However, the latter did not deal in depth with the essence of experiential knowledge rather it had an interest in the creation of knowledge through the interaction between tacit/explicit knowledge in the case of customers creation experiences of their personal fragrance. It follows that this chapter goes beyond and conceptualizes thoroughly, on the one hand, the hidden part of customer experiential tacit knowledge by determining its facets. This is done by presenting in depth different types of customer experiential knowledge other than sensory knowledge. On the other hand, this chapter defines a new organizational competence labeled CEKMC which addresses the knowledge-based customer experience. In this case, CEKMC focused on the relationship between the customer tacit knowledge and the innovation management, particularly the research targeted the experiential innovation into the thalassotherapy and spa centers.

REFERENCES

Arnould, E. J., & Wallendorf, M. (1994). Market-oriented ethnography: Interpretation building and marketing strategy formulation. *JMR, Journal of Marketing Research*, *31*(4), 484–504. doi:10.2307/3151878

Asiamah, F. N. (2009). *Exploiting tacit knowledge through knowledge management technologies.* Paper presented at the Learning Forum London Conference (ELFEL). Retrieved: http://bura.brunel.ac.uk/bitstream/2438/3613/4/EXploiting%20Tacit%20Knowledge%20-%20Frank%20Nyame-Asiamah.pdf?origin=publication_detail

Beijerse, R. (1999). Questions in KM: Defining and conceptualizing a phenomenon. *Journal of Knowledge Management*, *3*(3), 94–109. doi:10.1108/13673279910275512

Beijerse, R. (1999). Questions in knowledge management: Defining and conceptualizing a phenomenon. *Journal of Knowledge Management*, *3*(2), 94–110. doi:10.1108/13673279910275512

Bennet, D., & Bennet, A. (2008). Engaging tacit knowledge in support of organizational learning. *The Journal of Information and Knowledge Management Systems*, *38*(1), 72-94.

Boisot, M. H. (1998). *Knowledge Assets: Securing Competitive Advantage in the Information Economy.* New York, NY: Oxford University Press.

Bratianu, C., & Orezea, I. (2013). Emotional Knowledge: The Hidden Part of the Knowledge Iceberg. In *Proceedings Book of the 14th European Conference on Knowledge management* (pp. 82-90). Kaunas, Lithuania: Academic Conferences and Publishing International Limited.

Byosiere, P., & Luethge, J. D. (2008). Knowledge domains and knowledge conversion: An empirical investigation. *Journal of Knowledge Management*, *12*(2), 67–78. doi:10.1108/13673270810859523

Cardenas, J. M., & Spinola, M. (2011). The roles of tacit knowledge and knowledge management systems for writing academic papers: a research case. *Proceedings of the 12th European Conference on Knowledge management.*

Caru, A., & Cova, B. (2008). Small versus big stories in framing consumption experiences. *Qualitative Market Research*, *11*(2), 166–176. doi:10.1108/13522750810864422

Castillo, J. (2002). A note on the concept of Tacit knowledge. *Journal of Management Inquiry*, *11*(1), 46–57. doi:10.1177/1056492602111018

Churchman, C. W. (1971). *The design of inquiring systems: basic concepts of systems and organizations*. New York: Bencis Books.

Cova, V., & Cova, B. (2001). *Alternatives Marketing: réponses marketing aux évolutions récentes des Consommateurs*. Paris: Dunod.

Curbatov, O. (2003). *L'intégration du consommateur par le knowledge Marketing: conception, production et consommation d'un produit personnel* (doctoral dissertation). Université de Nice-Sophia Antipolis en cotutelle avec l'Institut International de Management, Nice, France.

Curbatov, O., Pavlidis, P., & Gay, M. (2006). Le knowledge marketing ou le détournement du marketing (pourquoi le web perfume détourne le marketing vers la "competence clients-entreprises"? (The Knowledge Marketing or Marketing Reloaded: smell web matrix reloaded). *5th International Congress Marketing Trends Venice*.

Davenport, T. (1998). Managing customer knowledge. *CIO Magazine*. Retrieved from http://www.cio.com/archive/060198/think.html,10.5.2004

Davenport, T. H., & Prusak, L. (1998). *Working Knowledge: How Organizations Manage What They Know*. Boston: Harvard Business School Press.

Denning, S. (2000). *The Springboard: How Storytelling Ignites Action in Knowledge-era Organizations*. KMCI Press.

Dinur, A. (2011). *Tacit Knowledge Taxonomy and Transfer: Case Based Research*. Working paper. Institute of behavioral and Applied Management.

Dubet, F. (1994). *La sociologie de l'expérience*. Paris: Seuil.

Eisenhardt, K. M., & Santos, F. M. (2002). Knowledge-Based View: A New Theory of Strategy? In A. Pettigrew, H. Thomas, & R. Whittington (Eds.), *Handbook of Strategy and Management*. Sage Publications.

Heavin, C., & Adam, F. (2011). Optimizing customers as knowledge resources and recipients: Cases in Small to Medium Sized Software Enterprises. In M. M. Al Shammari (Ed.), *Customer-centric Knowledge management concepts and applications*. Information Science Reference.

Jaziri-Bouagina, D. (2017). The Tacit Knowledge Through the Customer Experience: Conceptualization, Externalization Methods, and Use Application to Tunisian Thalassotherapy Centers. In Handbook of Research on Tacit Knowledge Management for Organizational Success. Academic Press.

Jaziri-Bouagina, D., & Jamil, G. L. (2017). *Handbook of Research on Tacit Knowledge Management for Organizational Success*. Hershey, PA: IGI Global. doi:10.4018/978-1-5225-2394-9

Jaziri-Bouagina, D., & Triki, A. (2013). The customer experiential knowledge management (CEKM) - concept proposition and research framework development. *Proceedings of the 14th European Conference on Knowledge Management ECKM 2013, 1*, 307-314.

Kane, H., Ragsdell, G., & Oppenheim, C. (2006). Knowledge Management Methodologies. *Electronic Journal of Knowledge Management*, *4*(2), 141–152.

Kostopoulos, C. K., Spanos, Y. E., & Prastacos, G. P. (2002). The Resource – Based View of the Firm and Innovation: Identification of Critical Linkages. *The 2nd European Academy of Management Conference*. Retrieved from http:// ecsocman.hse.ru/data/165/663/1219/rb_view.pdf

Kozinets, R. V. (2002). The field behind the screen: Using netnography for marketing research in online communities. *JMR, Journal of Marketing Research, 39*(1), 61–72. doi:10.1509/jmkr.39.1.61.18935

Küpers, W. (2005). Phenomenology of embodied implicit and narrative knowing. *Journal of Knowledge Management, 9*(6), 114–133. doi:10.1108/13673270510630006

Ladkin, D. (2012). *Perception, Reversibility, "Flesh": Merleau-Ponty's Phenomenology and Leadership as Embodied Practice*. Retrieved from http://integralleadershipreview.com/6280-perception-reversibility-flesh-merleau-pontys-phenomenology-and-leadership-as-embodied-practice

Mascitelli, R. (2000). From Experience: Harnessing Tacit Knowledge to Achieve Breakthrough Innovation. *Journal of Product Innovation Management, 17*(3), 179–193. doi:10.1016/S0737-6782(00)00038-2

Matthews, C., & Sternberg, R. (2009). Developing experience-based (tacit) knowledge through reflection. *Learning and Individual Differences, 19*(4), 530–540. doi:10.1016/j.lindif.2009.07.001

McQueen, R. J. (1998). Four views of knowledge and knowledge management. *Proceedings of Fourth Americas Conference on Information Systems*, 609-611. Retrieved from http://www.mngt.waikato.ac.nz/departments/staff/bob/WEBDOCS/pubs/aiskm98.pdf

Mooradian, N. (2005). Tacit knowledge: Philosophic roots and role in KM. *Journal of Knowledge Management, 9*(6), 104–113. doi:10.1108/13673270510629990

Nonaka, I. (1991). The knowledge creating company. *Harvard Business Review, 69*, 96–104.

Nonaka, I. (1994). A Dynamic Theory of Organizational Knowledge Creation. *Organization Science, 5*(1), 14–37. doi:10.1287/orsc.5.1.14

Nonaka, I. (1997). *Organizational Knowledge Creation*. Presentation summary at the Knowledge Advantage Conference.

Nonaka, I., & Konno, N. (1998). The concept of 'Ba': Building a foundation for knowledge creation. *California Management Review, 40*(3), 40–54. doi:10.2307/41165942

Nonaka, I., & Takeuchi, H. (1995). *The Knowledge Creating Company*. New York, NY: Oxford University Press.

Polanyi, M. (1966). *The Tacit Dimension*. London: Routledge & Kegan Paul.

Ray, T., & Clegg, S. (2007). Can We Make Sense of Knowledge Management's Tangible Rainbow? A Radical Constructivist Alternative. *Prometheus, 25*(2), 161-185. Retrieved from http://opus.kobv.de/tuberlin/volltexte/2010/2789/pdf/CKM_embedded_Innovaiton_Marketing.pdf

Roederer, C. (2008). *L'expérience de Consommation: Exploration Conceptuelle, Méthodologique et Stratégique* (doctoral dissertation). Centre de recherche en marketing de Bourgogne IAE, Dijon, France.

Rollins, M., & Halinen, A. (2005). Customer Knowledge Management Competence: Towards a theoretical Framework. *Proceedings of the 38th Hawaii International Conference on System Sciences*. 10.1109/HICSS.2005.180

Sara, V. (2008). The Interplay of Reificative and Participative Processes of Customer Knowledge Creation: An exploratory study of commercial lending. *Journal of Financial Services Marketing*, *12*(4), 287–298. doi:10.1057/palgrave.fsm.4760082

Seidler, R., Hartmann, E., & Gemünden, H. G. (2004). *The role of tacit knowledge in innovation management*. Competitive Paper submitted to the 20th Annual IMP Conference in Copenhagen.

Shin, M., Holden, T., & Schmidt, R. A. (2001). From knowledge theory to management practice: Towards an integrated approach. *Information Processing & Management*, *37*(2), 335–355. doi:10.1016/S0306-4573(00)00031-5

Sigala, M., & Chalkiti, K. (2007). Improving performance through tacit knowledge externalization and utilization: Preliminary findings from Greek hotels. *International Journal of Productivity and Performance Management*, *56*(5), 456–483. doi:10.1108/17410400710757141

Speigler, I. (2003). Technology and Knowledge: Bridging and generating gap. *Information & Management*, *40*(6), 533–539. doi:10.1016/S0378-7206(02)00069-1

Spender, J. C. (2003). Exploring uncertainty and emotion in the knowledge-based theory of the firm. *Information Technology & People*, *16*(3), 266–288.

Stillwell, W. D. (2003). *Tacit Knowledge and the Work of Ikujiro Nonaka: Adaptations of Polanyi in a business Context*. Retrieved: www.missouriwestern.edu/orgs/polanyi/tad%20web%20archive/tad30-1/tad30-1-pg19-23-pdf.pdf

Sun, H. (2010). *CKM-embedded Innovation Marketing as Success Driver for Product Innovation, Theoretical Framework and Empirical Research* (doctoral dissertation). Von der Fakultät Wirtschaft & Management der Technichen Universität Berlinzur Erlangung des akademischen Grades, Berlin, Germany.

Thusy, A., & Morris, L. (2004, January). From CRM to Customer Experience: a new Realm for Innovation. Business Digest.

Vance, D. (1997). Information knowledge and wisdom: the epistemic hierarchy and computer-based information systems. *Proceedings of the Third America's Conference on Information Systems*, 15-17. Retrieved from http://hsb.baylor.edu/ramsower/ais.ac.97/papers/vance2.htm

Vargo, S. L., & Lusch, R. F. (2008). Service-dominant logic: Continuing the evolution. *Journal of the Academy of Marketing Science*, *36*(1), 1–10. doi:10.100711747-007-0069-6

Yakhlef, A. (2005). Immobility of tacit knowledge and the displacement of the locus of innovation. *European Journal of Innovation Management*, *8*(2), 227–239. doi:10.1108/14601060510594684

KEY TERMS AND DEFINITIONS

CEKMC: Customer experiential knowledge management competence; the degree to which an organization demonstrates competence to generate and to integrate the knowledge-based customer experience in order to obtain a successful experience innovation.

Customer Experiential Knowledge: It is a customer tacit knowledge based on his/her lived experience of consumption.

Customer Tacit Knowledge: It is in-depth embedded knowledge that is related to the customer entity.

Knowledge-Based View (KBV): Integrated under the resource-based view, KBV stresses the strategic role of knowledge into the organization that should be processed in order to reach a competitive advantage.

Resource-Based View (RBV): It is an organizational perspective/view that structures the company through a set of physical and intangible resources.

Tacit Knowledge: It is a deeply embedded and subjective knowledge. It exists both on the individual and the organizational level and its acquisition depends widely on its level of hardness. Example of tacit domain: intuitions, action, values, cognition, emotions.

Chapter 11
Information Strategy:
Implementing and Managing a Digital Strategy in a Portuguese Company

Sérgio Maravilhas
Universidade Salvador, Brazil

Paulo Melo
Universidade Salvador, Brazil

Sérgio Goes Oliveira
Universidade Salvador, Brazil

ABSTRACT

In this chapter, the implementation of an information strategy project in a Portuguese real estate company is analyzed. This involves its framing—historical and socio-economic—and a brief description of the activity sector in which the company operates. Several well managed projects have been developed to improve the competitive position of the company, but without focus all the activities lose strength because they might not reach their proposed targets. Some tools to identify the information needs of business activity developed are described, as well as the role of information as a promoter of competitive advantages. Social media tools were utilized and proved to be a great strategic decision. To conclude, a few reminders of the factors to consider in developing the information strategy to implement and that information management without a strategy could result in several diversified decisions without any positive consequence for the organization.

INTRODUCTION

The implementation of an information strategy is vital for any organization in the historical era in which we live, because we came from an industrial society to the information age (Gleick, 2011), where the efficient use of information as an economic resource and a production sector became a factor of strategic, economic, social and political (Best, 1996a) importance.

DOI: 10.4018/978-1-5225-6225-2.ch011

Information, along with natural and economic resources, proves to be an unprecedented social and strategic expedient (McGee & Prusak, 1995; Beuren, 1998). It has therefore a potential strategic value in identifying new market opportunities, and in identifying potential threats to the company (Gomes & Braga, 2001; Liautaud, & Hammond, 2001), two aspects to consider for any company that wants to achieve its goals.

Therefore, the importance of information for organizations is now universally accepted, being, if not the most important, at least one of the resources whose management influences the success of organizations (Ward & Griffiths, 1996). Information is also considered and used in many organizations as a structural factor and a tool for managing the organization (Zorrinho, 1991), as well as an essential strategic weapon for competitive advantage (Porter, 1985).

This information is called strategic information (Earl, 1998; Hinton, 2006), because it's indispensable for the full functioning of the organization and needed to sustain its competitive position. The systems that identify, store, organize and provide this information, are called strategic information systems (Alturas, 2013; Amaral & Varajão, 2000; Davenport; Marchand & Dickson, 2004; Rascão, 2001; 2004; Varajão, 1998; Ward & Griffiths, 1996).

Information management relates to the organizational ability to make the right information available for use in decision making (Davenport, 1997; Rascão, 2008).

Hence, it is advantageous to any decision maker to hire professionals with these skills, since this function is not restricted to technological knowledge, but with the ability to properly organize and interpret the information obtained, allowing it to be used by those who need it, when they need it (Castells, 2001; 2004; Earl, 1998; Hinton, 2006; Ward & Griffiths, 1996).

Only someone with communication and language skills can more easily find and use the information available, making it profitable (Tapscott, 1999). Professionals in the area of information management have a multidisciplinary training that involves information sciences, information technology and communication, management and several of its sub-disciplines (Hinton, 2006; Wilson, 1989a; Wilson, 2002).

This multidisciplinary approach allows a comprehensive and overall vision that enables not only the collection of relevant information, but also its analysis and, especially, their synthesis to support the decision making of managers and organizational leaders, transforming the informational chaos into useful and practical knowledge application, leading to benefits for the organizations.

The situation of an existing Real Estate company, BV[1], will be analyzed. Some advantages of bringing a properly defined information strategy, with the participation of all employees of the organization, will be discussed, leaving its coordination to the care of an information manager. The risks of leaving that task only to the care of the information technology (IT) department and the danger of information overload will be alerted. The conclusion will try to clarify some of these points stressing the benefits of conducting such activity.

All the information about the company and the activity described herein was obtained through participant observation and subsequent personal interviews with the three managing partners, five of the current ten branch directors and the head of the information systems department and his two collaborators.

Please remember that it's never easy to obtain confidential strategic information from an organization due to competitive reasons.

The information revealed here is just the one that had clearance and was authorized to be disclosed, given the need to keep certain confidential information outside the scope of competition, extremely aggressive in this sector in the current economic environment, and to maintain the image and reputation of the company among its customers, partners and collaborators.

BACKGROUND

For Edgar Morin (1996) *the whole universe is based on only three constituents and the relationships between them: matter, energy and information.*

Generally, energy is defined, operationally, as the ability to produce work, not the work itself. Similarly, the information is defined, operationally, as the ability to store and transmit meaning or knowledge, not knowledge or meaning in themselves (Gatlin, 1972, p. 23). The ability to store and transmit knowledge should not be confused with the knowledge that it can produce (Choo, 2003; Gomes & Braga, 2001; McGee & Prusak, 1995).

Therefore, it is concluded that the very nature of information is mainly relational, organizational (Choo, 2003; Gleick, 2011).

That the human individual needs information is a common-sense observation and a factor generally accepted by experts, who attribute to cognitive aspects an important role in the genesis of behavior.

Gordon Pask (1969) supports the idea that the human being is characteristically an organization that processes information and learns. The amount of effort expended in the search for information differs, therefore, with its connection to diverse human needs and the intensity of the need in question.

Thus, it is understood that the energy expended in the search for information, such as a need to orient ourselves in the environment in which we operate, is higher in periods of growth of the individual, without direct connection with the satisfaction of other needs.

In a society in continuous technological and scientific development it is also understood that the need to know is constantly activated and represents a predominant role in the life of the individual (Jorge, 1995).

Regarding information theory, what had fascinated me was to discover that the information could be defined physically. In reality it was a partial truth. To define the information is necessary to give a biophysical-anthropological definition: information, unquestionably, has something physical, but only appears with the living, and moreover, we are the ones who have named and discovered it (Morin, 1996, p. 107).

The importance of information stands out for its ability to allow management strategies that translate into competitive advantages of high importance (Porter & Millar, 1985), such as price leadership, differentiation and focus on a particular product, market area or specific consumer, which may involve a differentiation of costs, or the publication of differentiated services.

The Real Estate Brokerage Firm BV: Historical and Socio – Economic Background

BV is the second largest real estate company with Portuguese owners and capital in the Northern region of Portugal, and was one of the largest in the country, both in number of departments, stores and employees, as well as in sales and net income.

With over forty years of existence, its headquarters are based in Porto and the majority of its current ten stores are located in the city or its surroundings. In 2001 it had fifty-one stores, which were distributed from Valença do Minho, in the North, to Leiria, in the lower center. The distribution of its stores was as can be seen in Table 1.

Table 1. Number of BV stores and their location in 2001

City/Location	Number of Stores
Valença do Minho	1
Viana do Castelo	1
Ponte de Lima	1
Braga	1
Trofa	1
Vila do Conde	1
Póvoa do Varzim	1
Vila Nova de Famalicão	1
Guimarães	1
Stº Tirso	1
Paços de Ferreira	1
Maia	2
Porto	5 (one was also the Head office)
Matosinhos	1
Srª da Hora	1
Gondomar	2
Rio Tinto	2
Paredes	1
Penafiel	1
Felgueiras	1
Lousada	1
Valongo	2
Vila Nova de Gaia	3
Stª Maria da Feira	1
São João da Madeira	1
Estarreja	1
Espinho	1
Ovar	1
Aveiro	3
Ílhavo	1
Vagos	1
Águeda	1
Oliveira de Azeméis	1
Mealhada	1
Coimbra	2
Lousã	1
Figueira da Foz	1
Leiria	1

In 1999, the company recorded a turnover of around 100 million Euros[2]. This value has been declining every year, due to the socio-economic situation the country is facing, being currently barely sufficient to meet the expenses of maintaining the company. It consists primarily of sales and rent departments, supported by law and legislation departments, insurance, documentation, marketing and advertising, information technology, administrative support, accounting and also training (currently disabled due to the drop-in turnover and number of new employees to train).

The company can be very flexible in its decision making process because it's very flat and horizontal in its levels of decision and command, having only four levels: 1) Three owners and directors who operate as CEO (chief executive officer), CFO (chief financial officer) and COO (chief operations officer); 2) Eight partners, who act as Heads of the commercial departments of each of the stores; 3) Supporting staff like accountants, lawyers, IT engineer and technicians, marketers, etc., and; 4) Several sellers/brokers who work as independent professionals, allocated at one of the departments but selling where the customer wants to buy, even if it is in another department commercial area.

It had over six hundred employees, being the most brokers (realtors) that developed their activity in the area of sales. This number currently is less than one hundred employees.

Nowadays, the distribution of its stores is as shown in Table 2.

Since the company operates in an extremely aggressive and highly competitive industry, with a rotation of products for sale and human resources (HR) of the highest in the trade, services and related areas, it's essential for its survival to define the development and implementation of a strategy for organizational information to preserve and maintain its activity in an increasingly demanding market such as that of real estate brokerage.

INFORMATION

The efficient exploitation of information[3] as an economic resource and one of the production sectors has become a factor of strategic, economic, social and political importance (Best, 1996b).

A new vision must be considered in relation to information and the role it plays in the organization.

If, on the one hand, an organization cannot function without information, on the other it is important to know how to use this resource to improve its functioning.

Table 2. Number of BV stores and their location in 2014

City/Location	Number of Stores
Viana do Castelo	1
Braga	1
Maia	1
Porto	2 (one is also the Head office)
Matosinhos	1
Rio Tinto	1
Vila Nova de Gaia	1
Aveiro	1
Coimbra	1

Thus, the faster the identification of the relevant information to the organizations and the quicker the access to this information more easily their goals will be achieved.

Information is an intuitive, indefinable principle, such as energy, whose precise definition always seems to escape through the fingers like a shadow (Gleick, 2011; Jorge, 1995; Morin, 1996).

But, most important is that information plays a key role in business since this affects the competition at three levels:

1. Modifies the industrial structure and, therefore, changes the rules of the competition;
2. Creates competitive advantage for organizations offering new ways to overcome their rivals;
3. Create new business opportunities, most often from the organization's internal processes (Porter, 1985; Porter & Millar, 1985).

It is important to remember that the primary function of information is to try to eliminate uncertainty (Wilson, 1985; 2001), a problem that plagues any area of activity.

Organizations should follow some recommendations to successfully achieve their goals, such as:

1. Value information as a resource so or more important than any other that it needs to function (Maravilhas, 2013b);
2. Give the employees the relevant and necessary information to the excellent performance of their function minimizing, where possible, the overhead with unnecessary information (information overload);
3. Pay attention not only to the internal information generated within the organization, necessary to carry out the organizational tasks it undertakes, but also external information, from various points of interest to the sector to maintain their activity profitable.

Effective managers and decision-makers should not ask about the cost of obtaining the information needed. They should ask instead how much will be the loss if they don't have it.

Information Management

Information management is usually defined as a comprehensive organizational capacity to create, maintain, retrieve and make available the right information, at the right place, at the right time and in the hands of the right person, at the lowest cost, in the best support for its use in decision making (Choo, 2003; Davenport, 1997; Earl, 1998; Hinton, 2006).

Information management must be based on developing a strategy that involves the entire organization, considering mainly the users, information resources and appropriate technology available (Wilson, 1985, p. 65).

Consequently, information management (IM) should be seen as the conscious process by which information is gathered and used to assist in decision making at all levels of an organization.

(...) A final point about the definition is that it makes no reference to computers or information technology. Information management is as much about paper-based systems, or even human voice-based systems, as it is about technology-based systems.

It is a popular misconception that information management is only concerned with information technology management (Hinton, 2006, pp. 2, 3).

It is also an economic coordination of the efficient and effective production, control, storage, retrieval and dissemination of information from internal and external sources, to improve the performance of a given organization (Best, 1996a; Beuren, 1998; Choo, 2003; Davenport, Marchand, & Dickson, 2004; Prusak & McGee, 1995).

The important role of information management and its integration into the organizational strategy must be emphasized, revealing itself as a key factor in creating added value for the company, as it helps to detect new opportunities and create competitive advantages and enables to defend them from the competitors.

Information management is an activity that aims to regulate the information, the information and communication technologies (ICT) and their respective users, through the application of management techniques to process and provide updated and relevant information, using electronic means or not, depending on the user's needs.

Information management is much more than the ability to obtain information from computer form (the computer will be just a tool here, as a catalog of a library can also be). Managing information means information processing so that, whoever needs it, can then get some help to achieve their goals.

Armed with relevant information, at the appropriate time, at the lowest possible cost, the organization will be better prepared to face the adversities of the industry.

The Information Manager Role: Importance and Attributions

Everybody associated with the business world knows that some function that saves time and allows to easily access information within the area of interest of the organization will be positively valued.

The functions assigned to an information manager are many and varied (Choo, 2003; Wilson, 1989a) and he may be asked to:

1. Develop and disseminate web sites that aim to optimize the service provided to customers (ex. bringing together all existing information published in any format) (Marchand & Horton Jr., 1986; McGee & Prusak, 1995);
2. Perform information audits, internal and external (Best, 1996b; Orna, 1999);
3. Use the internet as a base to search databases of companies and organizations, conducting competitive/market intelligence and technology watch (Gomes & Braga, 2001; Kahaner, 1997; Liautaud, & Hammond, 2001);
4. Develop skills that enable him to act as a consultant to his customers, deciding what is the best search engine for use in a research, what sites can be used to support specific information needs, or where are the best information sources for decision making support, etc. (Choo, 2003);
5. Implement systems where services could go from the traditional way of transmitting information to a form of information analysis, i.e., to filter, monitor and evaluate sources of information, and/or prepare and gather summaries of executive decisions (Beuren, 1998; Brown & Duguid, 2000; Varian & Shapiro, 1999);
6. To train professionals that could build new paths to make available web site information to end-users (Adeoti-Adekeye, 1997).

In sum, the role of the information management professional should be to focus on optimization, as well as in service personalization, and not be just a mere unqualified worker, carrying only simple tasks like the collection of e-mails and other types of unfiltered information.

For any organization, in any industry, will be vital to its success to have the capabilities of a information management professional that will give a quick and efficient access to information in a particular subject, saving time and resources, providing the company with innumerous benefits, including financial, increasing the competitiveness of its activity (Hildebrandt, 1997) with the quality of the information collected and filtered to allow managers a more coherent and efficient decision making (Wilson, 1989a).

With regard to support for the management and administration of organizations, and recruiting professionals who have this function, it's positive for any decision maker to hire professionals with training in information management, since this function does not only refer to technical knowledge and technology, but the ability to properly interpret the information obtained, as well as its organization, so it can be used by those who need it.

The information manager, by its extensive network of multidisciplinary knowledge, will be the key element in the collection, analysis, and dissemination of relevant information to those responsible for decision making in organizations.

At this point, it should be noted that it is essential for the information manager, who should be responsible for the definition of the information strategy to implement, to know what are the aims of the organization, including its goals and strategic plan (Ward & Griffiths, 1996), or in other words, what is the business strategy of the organization, what are their goals for next year and what kind of problems the company is currently facing (Cleveland, 1985; Wilson, 2002).

If such issues are not available to the information manager, it will be extremely difficult to plan the obtainment and delivery of information services related to the strategic planning of the company or organization (Orna, 1999).

The information manager must ensure that this does not happen, as long as the organization gives him the necessary resources for the proper performance of its duties.

THE STRATEGIC MANAGEMENT OF INFORMATION

An information strategy defines the organization's information needs, ensuring that existing information services to meet those needs are properly organized and managed (Wilson 1987; 1989b; 1994b), and has access to ICT for storage, search, distribution, communication, and information security.

Information must be managed like any other vital resource of the organization. In fact, sometimes, all other resources are dependent on this one and it is imperative for the organization to safeguard and take advantage of the information resources gathered internally with the same or more relevance of the ones gathered externally. Research and development (R&D) results, clients lists and respective contacts, suppliers of sensitive or specific materials used in production, business models and marketing strategies, distribution and operations logistics, trade secrets, among other invaluable information's are vital for the success and survival of every organization (Liautaud, & Hammond, 2001).

The lack of information might induce mistakes or originate wrong and costly decisions. It is necessary to adapt and adequate the information content to the needs of the users. In every organization exists huge volumes of information but, frequently, the vital information arrives after it was needed.

Vital information exists but, occasionally, it is dispersed and it's very hard to find the answer for simple needs and questions. There is a lot of information without any value to the organization and the right one is not available in an accessible format. Sometimes it is kept by someone in a department without being shared with all the organization and others who need it. So, it's necessary to develop three tasks to avoid or minimize those situations:

- In all the organization information sources available should be identified and centralized;
- These activities should be regarded as parts of a whole;
- These activities should be managed by someone capable of understanding the information needs of the company and each actor involved with its specific tasks to perform (Donnelly, Gibson, & Ivancevich, 2000).

The organization should start the strategic planning by preparing an industry profile (Markides, 2000; 2008). This is an essential management function, because it allows a more effective assessment of the situation of the company in the sector in which it operates or that the company depends on for survival and involves:

1. Knowledge of the environment where the company operates (STEEP factors)[4];
2. Statistics and forecasts, starting from the current state of the industry and related areas;
3. Identification and analysis of the biggest companies in the industry, their main competitors, and business practices developed by them.

This profile will allow building a coherent map of the position of the company in its industry sector and the situations that they will have to deal with and should be included in the strategy.

Instruments and Tools to Be Used in the Information Strategy

Some instruments for detecting the information needs of the company are (Wilson, 1994a; Romagni, 1999):

1. Comparative analysis of the sector to which the organization belongs (Benchmarking);
2. Value chain analysis and identification of critical areas of activity of the company (Porter, 1985);
3. Analysis of Porter five forces (1980), to understand from which direction comes the most important pressure for the organization;
4. SWOT[5] analysis, to understand the strengths, weaknesses, opportunities and threats that the company is subject to and should seek to prevent and improve;
5. STEEP analysis, to be aware of the risks, chances and benefits that can be exceeded if information is well managed.

According to Porter (1980), it should be borne in mind the five forces that define, regulate and maintain the market in which the organization operates and acts, to avoid unforeseen situations that may endanger its economic and professional integrity.

The implications of the five forces to consider and its relation to the activity undertaken by the organization are:

1. Rival firms or direct competition within the sector;
2. Bargaining power of Suppliers;
3. Bargaining power of Customers;
4. Threat of Substitute products or services;
5. Threat of new entrants, or potential new competitors that may enter the organization's business sector.

This latest threat is undoubtedly one of the most pressing and cumbersome due to the lack of regulation and the effective legislation of the sector. At this conjuncture, information plays a key role, since this affects the competition at three levels:

1. Modifies the industrial structure and, therefore, changes the rules of the competition;
2. Creates competitive advantage by offering companies new ways to overcome their rivals;
3. Create new business opportunities, most often from the internal productive processes of the organization (Porter, 1985).

Analysis of Porter Five Forces Regarding the Organization Industrial Sector

The Real Estate Brokerage is, above all, an activity that provides services since whoever is selling the property is its owner and not the company itself. The company and its professionals operate as agents or brokers.

The core activity is to find the right property that the customer demands, according to the criteria desired for the house, apartment or land, and reach an agreement between the interested buyer and the owner. For that, there will be a monetary reward for the service provided by the company. That is the work done by the agents or brokers, also called sellers, which collaborate with the company.

Thus, for the analysis of Porter's value chain or rather the critical sectors of the same elaborated by Hunsicker (1989), BV situates his practice in the area of sales and services (Porter, 1985; Grant, 2002; Picot, 1989).

For the five forces described by Porter (1980), this activity refers to:

1. **Rival Firms or Direct Competition Within the Sector:** These are all the other real estate companies or other entities that sell or rent properties (ex. construction companies);
2. **Bargaining Power of Suppliers:** The persons or entities who want to sell or rent their properties;
3. **Bargaining Power of Customers:** The people or entities who intend to buy or rent a property;
4. **Threat of Substitute Products or Services:** New ways of looking for a house through the Internet, bank websites or competition websites, fleeing from the traditional activity that the company is dedicated to;
5. **Threat of Potential New Competitors That May Enter the Organization's Business Sector:**
    a. Foreign companies that develop business, using Franchising models, in Portugal (ex. ERA, Remax, Century 21, etc.);
    b. Former employees of the biggest companies who engage on their own in the market;
    c. The banks trying to sell the properties with unpaid loan debts;
    d. Construction companies and builders themselves, who create their own real estate companies to operate in parallel with the construction;

e. The owners trying to sell the property themselves using Social Media Networking (ex. Facebook, Hi5, Pinterest, Google+, Twitter, etc.), trying to avoid the payment of the service to the Real Estate companies.

As can be seen, there are several pitfalls to avoid and various situations that should be known so that the organization can continue to occupy the prominent place that has always occupied.

Following Brandenburger and Nalebuff (1996) advices, the company initiated several Co-opetition deals with construction companies and banks to obtain their products with exclusivity, being it houses or credit loans at the best price and interest rate to surpass their competitors in the market.

Grant (2002) called this the sixth force. It's not present on Porter five forces model but, contrary to the other five that withdraw value from the industry, this is the only one that aggregates value. It is the case of the software industry that aggregates value to the computer business, the ink cartridges that aggregate value to the printers, and the video game industry that aggregates value to certain hardware manufacturers, amongst others.

The company has a website that allows searching for properties available for sale or rent using various criteria such as the type, location, price, purpose and the county where the property is located. Simultaneously, through a partnership with a bank, allows for a simulation of the amount of mortgage loans to be paid with online approval. Social Media has been adopted recently with hardware to support and facilitate its use.

This strategy allows the reduction of the costs in physical facilities and personnel. Being a first point of contact for potential customers, it provides that they can look through photos and short videos all the various options available with all the convenience and may then turn to a physical store for personalized support and make a visit to the properties of their interest and formalize the completion of the deal.

THE IMPORTANCE OF INFORMATION IN THE STRATEGY PROJECT

The importance of information[6] stands out for its ability to allow management strategies that translate into competitive advantages (Faulkner, & Bowman, 1995; Porter, 1985; Porter, Millar, 1985) of high importance, such as:

1. **Price Leadership:** Not the prices to be charged by the property, but possibly the low fees charged for services rendered;
2. **Differentiation:** Which can provide a unique product or service, handling all legal documents and procedures required for the purchase or lease of a building or property;
3. **Focus:** In a particular product, in a particular area or specific consumer market, which may involve the differentiation of fees or the publication and advertisement of differentiated services.

Since October 2012, the Portuguese Government is conceding the so-called Gold Visa to attract foreign investors that can obtain citizenship by the investment of 1 million Euros in a business endeavor, creating more than 10 permanent jobs, or half a million in the purchase of a residence for the purpose of living in the Portuguese territory. Several citizens from China, Russia, Brazil and Angola are adhering to this proposal so they will need a professional service to advise them and take care of all the paperwork involved[7].

The Focus strategy can involve the creation of a special task force, highly educated and knowledgeable of the situation, to take proper care of these segments of customers and their needs. With an adequate price fee, dealing with all the bureaucracy to conclude the business satisfactorily, this can create a differentiation that make this type of customers to search for the company when they need to fulfill these acquisitions or others in the future, passing the good service by word-of-mouth to their associates, creating new customers and new businesses. Social Media will be an important part of this strategy.

The company strategists decided also to entail several negotiations with the Embassies and Consulates of the countries from the Portuguese official language region, like Brazil, Angola, Mozambique and Macau (now Chinese) that were former Portuguese colonies and have an excellent relationship with Portugal, to be certified as a trustable company to do business with in case one of their residents decides to buy a half million house or start a million Euros company. No decision has been made yet from the Embassies side.

But, for this, it is necessary to obtain relevant strategic information, such as:

1. Can the adaptation of the company business model to the consumers' needs be improved? (Debelak, 2006; Osterwalder, & Pigneur, 2010; Rappa, 2010; Watson, 2005).
2. Can the tenacity of the sales force and its location be improved? (Faulkner, & Bowman, 1995).
3. Can the consumer attention for products marketed by the company be improved? (Kumar, 2004).
4. Could a price change improve the market share of the company? (Cruz, 1990).
5. Could the use of new ICT (Smartphone, Tablets and Apps) and Social Media Networks (Facebook, Twitter, Instagram, Tumblr, Flickr, Google+, Hi5, Pinterest, etc.) result in increased consumer demand and loyalty? (Maravilhas, 2013a, 2013b; Ryan, & Jones, 2012, 2013).

These are the issues that the company should pay special attention to, and this is the type of situation that a well-defined information strategy will be able to respond, enabling the company to maintain its market share and, if possible, eventually improve it (Wilson 1987; 1989b, 1994a; 1994b).

Specifications of the Information Strategy to Implement

Information management must be based on developing a strategy that involves the whole organization taking mainly into account:

1. The users;
2. The information resources;
3. The appropriate technology available (Wilson, 1985, p. 65).

According to Wilson (1985), a certain number of questions must be made when we start to develop a strategy, such as:

1. Which type of information is needed so people can be helped in the performance of their tasks?
2. Is that information available to them? If yes, at what cost? If not, at what prejudice?
3. Which priorities are there to direct who should get that information?
4. Which methods of acquisition, storage and distribution of information are used? And, at what cost?
5. Which alternatives exist for each of these processes? How much do they cost?

6. Which systems would be better? How much do they cost?
7. What is the proper balance between manual and computerized systems?

A strategy based on these issues means that a new vision, a new look, should be considered in relation to information and the role it plays in the organization.

The specifications that the information strategy must meet to be effectively implemented and successful are (Back, & Moreau, 2001):

1. The strategy to be implemented should be formally documented;
2. Must be initiated by the top management of the company, with a project sponsor and a project champion[8];
3. Should be monitored by planned revisions;
4. It should be based on the provision of information about key indicators, as well as detailed analysis of the information needs of top management.

It should be stressed that to be successful, the information strategy to develop and implement must include strategies of:

1. Information management;
2. Information systems;
3. Information technology (Wilson 1987; 1989b; 1994b).

The whole process described above should be formalized, since many studies (Kalseth, 1991; O'Connor, 1993; Earl, 1998) show a higher percentage of success if the Management Board is involved and committed, motivating all other actors in the process[9].

Also, if formal documentation does not exist, no one will know what is the information strategy defined, to be able to carry it out effectively and efficiently. With regard to monitoring, if it is not formally performed it's not possible to take into account the defined strategy offered by environmental changes and controlling its effective implementation. Finally, without an effective needs analysis, there is no apparent reason why the strategy exists.

Mintzberg (1987) suggests that outlining a strategy is more of an art than a skill or science, since this involves negotiating various barriers. These barriers may be contrary to the attempt to implement any innovation, such as hostile attitudes by the middle management level, as may be related to the problem of hiring the appropriate employees or even difficulties in quantifying the benefits of such implementation.

It should be noted that it is essential for the Information Manager, the person who will be in charge of the definition of the information strategy to be implemented, to know what are the objectives of the organization, including its goals and strategic plan (Ward & Griffiths, 1996; Rascão, 2001; 2008), being aware of what is the business strategy of the company, including the company's goals for the next five years and what kind of problems the organization is currently facing (Cleveland, 1985; Wilson, 2002).

If such issues are not available to the Information Manager, it will be extremely difficult for him to plan to obtain and delivering the information services related to the strategic planning of the company or organization (Orna, 1999). However, he can only be aware of what is strategic information for the company if faced with the strategic issues for the efficient performance of the organization (Wilson, 1994b, p. 266).

Mistakes to Avoid in Implementing the Information Strategy

For an organization to implement an effective information strategy, two errors must be avoided. The first concerns the use of IT and the people in charge for this sector being responsible for coordinating the strategy. The second relates to the collection and provision of information in excess, or the so-called information overload (Dearlove, 1998). Both are likely to be related, which will be described next. Both assumptions will be analyzed in the light of an overall information strategy, engaging and comprehensive, not compartmentalized to the IT department of the company and the IT strategy.

Thus, it is usual to assign responsibility for implementing an information strategy to the department responsible for computers in the company, constantly confusing information management[10], with information systems management and information technologies (Hinton, 2006).

Even if everybody should cooperate and work together for the benefits it brings to the company, it is not a single issue to be taken into consideration (McGee & Prusak, 1995; Ward & Griffiths, 1996), and this confusion often originates that the investment made in IT does not become effectively profitable (King; Grover & Hufnagel, 1989; Premkumar & King, 1994; Oliveira, 2004), precisely due to the lack of coordination between the various sectors involved.

Although there is an IT department in the company and the person responsible for it must have an extensive knowledge of its industry, typically that person is unaware of all the other factors that could have implications on the real business benefit of these technologies for the company (Cattela, 1981; Hinton, 2006; Wilson, 2002).

Factors such as the degree of knowledge of employees regarding the use of IT, which is usually little or null, activities involving the real estate business such as documentation required, taxes payable and other amounts involved in the process of buying and selling real estate properties, management expertise to respond adequately to the challenges posed by an activity in constant change and constant attacks from inside and foreign competition are essential for the implementation of the information strategy for the proper performance of the organization.

Therefore, unaware of the activity from the 'inside', and thinking that the interests of the company are being satisfied, disseminating and providing information electronically to all the users will occupy many of the employees with issues that are not relevant to their work and will make them lose time for the effective performance of the activity (Edmunds & Morris, 2000). For the same reason stated, it might happen that the information provided is not the one that really matters and it's needed, disseminating unfiltered information from someone not knowledgeable of the intricacies of the profession (Choo, 2003; Gomes & Braga, 2001).

It can also be concluded that the mere purchase of technological equipments alone does not lead to any competitive advantage (McGee & Prusak 1995; Earl, 1998; Davenport; Marchand & Dickson, 2004) unless they are appropriately framed in a wider plan, allowing their potential to be monetized (Oliveira, 2004).

This leads to another problem that should be seriously considered, and with regard to the issue of excess of information, or information overload. If, on the one hand, a company cannot function without information, on the other it is important to know how to use this resource to improve its functioning. Since we live in an era of abundance of information, it is necessary to distinguish the useful, relevant and necessary information from the surplus (Dearlove, 1998; Wilson, 2001). Thus, the faster the identification of the relevant information to the enterprise, and the faster the access to this information, more easily their goals will be achieved.

Since people and organizations, similarly, have a peak in the information processing capacity, there is a saturation point from which an increase of the available information does not correspond to an increase in its use, but rather corresponds to its contraction.

It is considered information overload, in personal terms, when someone has the perception (either the user or an observer) that the information associated with the tasks to perform in the workplace is in greater quantity than can be managed efficiently and effectively (Dearlove, 1998; Edmunds & Morris, 2000; Wilson, 2001). Also involves the perception that this excess creates a degree of stress[11] in the attempt to adapt to the strategies defined that proves itself ineffective.

Excess of information is considered, then, in organizational terms, when the situation in which the extent of perceived individual information is sufficiently dispersed in the organization, thus reducing the effectiveness of all operations management. Although not a new or recent phenomenon, it tends to worsen with the advent of IT, Social Media and the Internet being a risk factor in terms of organizational control, and, also, the health of the employees.

The impact of information overload causes bad time management, delaying decision-making, allowing distractions of the main tasks, causing stress and consequent lack of job satisfaction, reduced social activity, fatigue and, in more severe cases, effective disease (Wilson, 2001).

Finally, it should be kept in mind that the company is not an island (Davenport, 1997), doesn't live closed in on itself, being imperative to monitor everything that goes on the outside and that relates to the sector of activity of the company whether economic, political and social (Gomes & Braga, 2001), both nationally and internationally, without incurring in the error described in the previous point of an eventual excess of irrelevant information that could interfere with the performance of the company.

So, the company must invest heavily in an information strategy to allow the knowledge and analysis of all stakeholders' issues in its activity, both internally and externally.

DECISIONS ADOPTED AND THEIR RESULTS

After defining the information strategy, in conjunction with the overall strategy of the company, some decisions to adopt essential measures were taken. The essential decisions consisted in:

1. Increasing the loyalty of the customers and prospects, in a long-life personal and digital relationship, making them advertise the company to their friends and connections through word-of-mouth and word-of-web (Michael, & Salter, 2006; Ryan, & Jones, 2012; 2013);
2. Improve the work of the sellers to increase the number of deals concluded positively, reducing stress and maximizing profits for them and the company.

To achieve these goals, the company realized that a change in their traditional business model must be produced (Debelak, 2006; Markides, 2008; Osterwalder, & Pigneur, 2010; Rappa, 2010; Watson, 2005).

A project champion and a project sponsor (Boonstra, 2013; Earl, 1998) were chosen to warrant the implementation of the measures defined. The IT manager became the project sponsor for his technology expertise, and one Head of commercial department became the project champion for his leadership qualities and respect gained by his knowledge of the profession.

The Marketing manager applied his knowledge in the digital environment but, in the future, an experimented digital Marketing manager should be integrated to bring more value to the team.

The company Board members realized that they must adapt and change *from selling products to providing solutions* (Kumar, 2004).

Some measures adopted and their results were:

1. Creation of company accounts on major Social Media Networking sites to communicate with prospects and clients, together with the existing Website and e-mail addresses;
2. Training the sales people on the use of the IT resources available on the company, to direct the prospects and customers to the choices available on the company Website and Social Media Networks;
3. The use of smart phones and/or tablets is now mandatory, so the company can keep the communication channels with the sellers and clients always open;
    a. Smart phones and/or tablets allow sellers to take pictures and send them to the company and/or customers to check their interest and make them dream about their new house or answer any question or doubt that they might have;
    b. Smart phones and/or tablets allow communicating via web, using e-mail and/or free Apps, reducing substantially the costs with phone calls and communication;
    c. Smart phones and/or tablets allow the access to Social Media Networks, increasing the number of contacts through their personal connections and the endurance of relationships more permanent and durable with customers and prospects;
    d. When accompanying clients on a visit to a property, the seller can show them in the smart phone and/or tablet other options, documents available on the company website, costs, taxes, and so on, to support the sale, making the time shorter for the client's decision;
    e. Smart phones and/or tablets allow the seller to check competitors Websites, confirming prospects information about the products and prices, analyzing their market strategy, knowing their employees, and avoiding unpleasant surprises;
    f. The GPS available through the smart phones and/or tablets, operating with Google Maps and other geo referencing tools, allow arriving on time to meetings and property visits avoiding the seller to get lost, as sometimes happened in the past (or giving that excuse for their delay);
    g. Geo referencing tools, operating with Google Maps, allow the seller and the clients to check the infrastructures available nearby like: schools, hospitals, restaurants, gyms, ballet classes, green parks, museums, libraries, highways, supermarkets, and other equipments required by the clients as a must to buy the property;
    h. The seller can save time writing reports, checking e-mail from the company or clients, updating the products photos, price and characteristics, asking for documents, and so on, while waiting for a meeting or for a visit with prospects, reducing stress and workload;
4. Cold calls now are not so cold because the seller, almost every time, knows who he/she is talking to, making them more confident and motivated5. The company Intranet has a knowledge database made with a Wiki tool, allowing the seller to check for answers when confronted with a doubt. They can also contact the department responsible, by phone or e-mail, and solve the problem talking with the expert on the subject, being it a lawyer, the insurance expert, the credit/loan expert or another seller.
5. Market segmentation made possible a focus strategy with the constitution of a task force to deal with big accounts, with big numbers involved, and different needs to attend, based in the Porto Head office;

a. According to Markides (2008) a new company or spin-off could be created to address this situation, but BV has a strong brand name and reputation and the Board decided to use all the synergies involved to give a quality assurance to the prospects with the same brand name they already know.

Until now the results are very positive, allowing the company to increase twofold the number of information requests by interested prospects and the number of connections/followers in their networks. Portugal is under International Monetary Fund (IMF) economical rescue so, in this environment, the results are considered good for the company.

Another situation that helped the company in financial terms was that each seller/agent/broker, being independent professionals, had to buy and pay for its smart phone and/or tablet, as a tool of the profession, making the company save several hundred Euros in the process.

For the customers it's also positive because, now they can track the seller and know if there are any complaints about his/her professionalism and ethical conduct.

Some customers have written positive feedback about the company and the service received which can be confirmed by other prospects checking the commentators profiles to confirm that they are real customers and not company employees faking good results. This improves the positive image of the company and its brand.

The company now fulfills a stronger service, even more professional, because the seller has everything needed at his/her fingertips, saving time and money, and increasing sales. With that information the seller can better fulfill its tasks, feeling less stressed and with better results.

To increase the loyalty of the customers, the company helps them with all the legal documents, taxpaying, and even with the annual individual tax deduction, making the customers link and connect long after the sale is done.

The customers receive, by e-mail or a post in their preferred Social Media Network where they have connected with the company, or via Really Simple Syndicate (RSS) tools, information that can be useful to them like deadlines to pay the annual tax for house owners, how to minimize their costs, when to deliver certain documents, and so on and so forth.

Next time they need to sell their house and buy a new one they will, eventually, deal again with whom they know and trust, knowing that everything will be taken care with high professionalism.

FURTHER RESEARCH DIRECTIONS

It would be useful and important to continue to follow this case study on the advantages of implementing an information strategy project, the mistakes to avoid, and essential results that would allow other organizations in the same area of activity or complementary areas to benefit from this experience and can, themselves, implement their own information strategies faster and more consistently, avoiding mistakes and adapting the most effective solutions for their performance and continuity in the market.

The analysis of the future implementation in the same company of a Customer Relationship Management (CRM) solution, based in the Cloud, will also be very useful to allow the measuring of the results that such a solution will bring to this type of company and the benefits obtained with it.

CONCLUSION

For an organization to remain profitable in this new century, it is essential to be well prepared for the challenges that the information-based society entails (Shapiro & Varian, 1999; Brown & Duguid, 2000). To do this, companies should try to develop and implement an information strategy, which involves information management strategy, information systems strategy, and information technology strategy, because only then can involve all the stakeholders needed to create a good, harmonious, understanding of the problem to overcome (Alturas, 2013; Amaral & Varajão, 2000; Rascão, 2004; Varajão, 1998).

For the strategy to prove itself effective, it must be properly documented and formalized as any other company initiative so that all employees may be aware of the measures to adopt, whatever they are after decided by everybody involved (Orna, 1999). It must also have a genuine commitment from the top management assuring its importance and allowing every collaborator to be motivated in their application (Earl, 1998).

Since the real estate sector has been deeply affected by successive external factors such as rising interest rates, the entry of more experienced foreign companies in the business, excess of professionals operating in the sector, oversupply of products for consumption, acute crisis in the political, economic and social life of the country, international recession, among other factors that have serious repercussions on confidence and purchasing power of domestic consumers, it is of utmost importance for the company to develop protection mechanisms, enabling it to anticipate certain events that may influence the sector and make it less stable. This may be possible through the process described above to provide the organization with structures to channel relevant information to be used when needed.

Implementing such a strategy should consider the future business plans of the company, the activity that it develops and related areas, which will only become effective after a preliminary analysis of the sector and its implications for the company's plans (Marchand & Horton Jr., 1986; Wilson, 2002). Such a strategy should under no circumstances be left only to the IT sector of the organization, since it does not have knowledge of all the details involved, although it should also be consulted in relation to its industry and the changes to be made for effective performance of the organization (Ward & Griffiths, 1996; Hinton, 2006; Oliveira, 2004).

With relevant information, at the appropriate time, at the lowest possible cost, the organization will be better prepared to face the rigors of the industry. However, for this reason it should not overload the operation of the business with information in excess, superfluous and redundant, as this will only cause a delay in what could be a competitive advantage transforming it in an organizational nightmare, leading to a situation more unfavorable than previously maintained (Dearlove, 1998; Edmunds & Morris, 2000; Wilson, 2001).

A well-designed information strategy is a step ahead of the closest competitors, which may well serve to avoid certain mistakes and make certain decisions that otherwise, would be totally unknown to the decision makers (Best, 1996b). Always remember that the primary function of information is to try to eliminate uncertainty (Wilson, 1985).

Finally, some recommendations that the company must follow to successfully achieve its goals were described, such as: the company should consider the information as a resource equally or more important than any other that it needs to function; the company should think about investing in the training of its human resources so that they are in a position to be able to respond positively to the changes that the implementation of such a strategy may entail in terms of knowledge of new IT solutions and Social Networking; the management board of the company should set the value of information to the business strategy of the organization; the organization must provide a thorough knowledge of the company plans to its information manager so that he can intervene positively in obtaining valid resources; the mere acquisition of information technology is not sufficient to obtain the competitive advantages that it can provide so, there must be a more comprehensive planning, elaborated previously to the development or acquisition of technological equipments; the organization must give its employees the relevant and necessary information to the excellent performance of its function minimizing, when possible, the unnecessary information to avoid overload; the company must be attentive not only to the internal information generated within the organization, the one it needs to carry out the business tasks it undertakes, but also external information, from various points of interest to the sector, in order to maintain its activity profitable; the company must provide their customers the access to its products and services through various technological solutions so as to meet various market segments that may have interest or need in the business area of the company.

Some measures were put in place, after defining the information strategy to adopt and implement, complementarily to the general strategy of the company (Back, & Moreau, 2001). The results improved twofold the number of prospects that contacted the company to obtain information about properties, augmenting the sales funnel that can increase the number of effective sales. The Social Media solutions also helped the customers to become more loyal, and improve the positive image of the company amongst other potential customers.

Decentralizing the access to the right information and disseminating useful information sources needed by their professionals, allowed the organization to avoid information overload in its sales force.

The limitations of this study are motivated to restrictions imposed by the organization regarding the confidentiality of their strategic and tactic/operational information, understandable situation given its sensitivity and value that it contains.

REFERENCES

Adeoti-Adekeye, W. (1997). The importance of management information systems. *Library Review*, *46*(5), 318–327. doi:10.1108/00242539710178452

Alturas, B. (2013). *Introdução aos sistemas de informação organizacionais* (1st ed.). Lisboa: Sílabo.

Amaral, L., & Varajão, J. (2000). *Planeamento de sistemas de informação* (2nd ed.). Lisboa: FCA.

Back, W., & Moreau, K. (2001, March). Information management strategies for project management. *Project Management Journal*, 10–20.

Best, D. (1996a). Business process and information management. In Best (Ed.), The fourth resource: Information and its management. Hampshire: Aslib/Gower.

Best, D. (1996b). *The fourth resource: information and its management*. Hampshire: Aslib/Gower.

Beuren, I. (1998). *Gerenciamento da informação: um recurso estratégico no processo de gestão empresarial*. São Paulo: Atlas.

Boonstra, A. (2013). How do top managers support strategic information system projects and why do they sometimes withhold this support? *International Journal of Project Management*, *31*(4), 498–512. doi:10.1016/j.ijproman.2012.09.013

Brandenburger, A., & Nalebuff, B. (1996). *Co-opetition: A revolution mindset that combines competition and cooperation: The game theory strategy that's changing the game of business*. New York: Bantam Doubleday.

Brown, J., & Duguid, P. (2000). *The social life of information* (1st ed.). Boston: Harvard Business School Press.

Castells, M. (2001). *A sociedade em rede: A era da informação: Economia, sociedade e cultura* (5th ed.). São Paulo: Paz e Terra.

Castells, M. (2004). *A galáxia internet: Reflexões sobre internet, negócios e sociedade*. Lisboa: Fundação Calouste Gulbenkian.

Cattela, R. (1981). Information as a corporate asset. *Information & Management*, *4*(1), 29–37. doi:10.1016/0378-7206(81)90023-9

Choo, C. (2003). *Gestão de informação para a organização inteligente: A arte de explorar o meio ambiente*. Lisboa: Editorial Caminho.

Cleveland, H. (1985). *The knowledge executive: Leadership in an information society* (1st ed.). New York: Dutton.

Cruz, E. (1990). *Planeamento estratégico: Um guia para a PME* (3rd ed.). Lisboa: Texto Editora.

Davenport, T. (1997). *Information ecology: Mastering the information and knowledge environment*. New York: Oxford University Press.

Davenport, T., Marchand, D., & Dickson, T. (2004). *Dominando a gestão da informação*. Porto Alegre: Bookmann.

Dearlove, D. (1998). *Key management decisions: Tools and techniques of the executive decision-maker* (1st ed.). Wiltshire: Financial Times/Pitman.

Debelak, D. (2006). *Business models made easy*. Entrepreneur Press.

Donnelly, J. H., Gibson, J. L., & Ivancevich, J. M. (2000). *Administração: Princípios de gestão empresarial* (10th ed.). Lisboa: McGraw-Hill.

Earl, M. (1998). *Information management: The organizational dimension*. New York: Oxford University Press.

Edmunds, A., & Morris, A. (2000). The problem of information overload in business organizations: A review of the literature. *International Journal of Information Management*, *20*(1), 17–28. doi:10.1016/S0268-4012(99)00051-1

Faulkner, D., & Bowman, C. (1995). *The essence of competitive strategy* (1st ed.). Exeter: Prentice Hall.

Gatlin, L. (1972). *Information theory and the living system*. Columbia University Press.

Gleick, J. (2011). The information: A history, a theory, a flood (1st ed.). St. Ives: 4th Estate.

Gomes, E., & Braga, F. (2001). *Inteligência competitiva: Como transformar informação em um negócio lucrativo*. Rio de Janeiro: Campus.

Grant, R. (2002). *Contemporary strategy analysis: concepts, techniques, applications*. Wiley-Blackwell.

Hildebrandt, D. (1997). *Internet tools of the profession: a guide for information professionals*. Chicago: Special Libraries Association.

Hinton, M. (2006). *Introducing information management: The business approach* (1st ed.). Burlington: Elsevier Butterworth-Heinemann.

Hunsicker, J. (1989, Summer). Strategies for European survival. *The McKinsey Quarterly*, 37–47.

Jorge, M. (1995). *Biologia, informação e conhecimento*. Lisboa: F. C. Gulbenkian.

Kahaner, L. (1997). *Competitive intelligence: How to gather, analyze, and use information to move your business to the top* (1st ed.). New York: Simon & Schuster.

Kalseth, K. (1991). Business information strategy: The strategic use of information and knowledge. *Information Services & Use*, *11*(3), 155–164. doi:10.3233/ISU-1991-11307

King, W., Grover, V., & Hufnagel, E. (1989). Using information and information technology for sustainable competitive advantage: Some empirical evidence. *Information & Management*, *17*(2), 87–93. doi:10.1016/0378-7206(89)90010-4

Kumar, N. (2004). *Marketing as strategy: Understanding the CEO's agenda for driving growth and innovation*. Boston: Harvard Business School Press.

Liautaud, B., & Hammond, M. (2001). e-Business intelligence: Turning information into knowledge into profit. New York: McGraw-Hill.

Maravilhas, S. (2013a). A web 2.0 como ferramenta de análise de tendências e monitorização do ambiente externo e sua relação com a cultura de convergência dos media. *Perspectivas em Ciência da Informação*, *18*(1), 126–137. doi:10.1590/S1413-99362013000100009

Maravilhas, S. (2013b). Social media tools for quality business information. In Talburt, Yeoh & Zhou (Eds.), Information quality and governance for business intelligence. Hershey, PA: IGI Global.

Marchand, D. (1990). Infotrends: A 1990s outlook on strategic information management. *Industrial Management Review*, *5*(4), 23–32.

Marchand, D., & Horton, F. Jr. (1986). *Infotrends: Profiting from your information resources*. Wiley.

Markides, C. (2000). *All the right moves: A guide to craft a breakthrough strategy*. Boston: Harvard Business School Press.

Markides, C. (2008). *Game-changing strategies: How to create new market space in established industries by breaking the rules* (1st ed.). San Francisco: Wiley.

McGee, J., & Prusak, L. (1995). *Gerenciamento estratégico da informação: Aumente a competitividade e a eficiência de sua empresa utilizando a informação como uma ferramenta estratégica*. Rio de Janeiro: Campus.

Michael, A., & Salter, B. (2006). *Mobile marketing: Achieving competitive advantage through wireless technology* (1st ed.). Oxford: Butterworth-Heinemann.

Mintzberg, H. (1987, July). Crafting strategy. *Harvard Business Review*, 66–75.

Morin, E. (1996). *O Problema epistemológico da complexidade*. Publicações Europa- América.

O'Connor, A. (1993). Successful strategic information systems planning. *Journal of Information Systems*, *3*(2), 71–83. doi:10.1111/j.1365-2575.1993.tb00116.x

Oliveira, A. (2004). *Análise do investimento em sistemas e tecnologias da informação e da comunicação*. Lisboa: Sílabo.

Orna, E. (1999). *Practical information policies* (2nd ed.). Cambridge: Gower.

Osterwalder, A., & Pigneur, Y. (2010). *Business model generation: A handbook for visionaries, game changers, and challengers*. Wiley.

Pask, G. (1969). *Learning strategies and teaching strategies*. Surrey: System Research Ltd.

Picot, A. (1989). Information management: The science of solving problems. *International Journal of Information Management*, *9*(4), 237–243. doi:10.1016/0268-4012(89)90047-9

Porter, M. (1980). *Competitive strategy: Techniques for analyzing industries and competitors*. New York: Free Press.

Porter, M. (1985). *Competitive advantage: creating and sustaining superior performance*. New York: Free Press.

Porter, M., & Millar, V. (1985). How information gives you competitive advantage. *Harvard Business Review*, (July/August): 75–98.

Premkumar, G., & King, W. (1994). The evaluation of strategic information systems planning. *Information & Management*, *26*(6), 327–340. doi:10.1016/0378-7206(94)90030-2

Rappa, M. (2010). Business models on the web. *Managing the Digital Enterprise*. Retrieved 25-05, 2012, from http://digitalenterprise.org/models/models.html

Rascão, J. (2001). *A análise estratégica e o sistema de informação para a tomada de decisão estratégica* (2nd ed.). Lisboa: Sílabo.

Rascão, J. (2004). *Sistemas de informação para as organizações: A informação chave para a tomada de decisão* (2nd ed.). Lisboa: Sílabo.

Rascão, J. (2008). *Novos desafios da gestão da informação* (1st ed.). Lisboa: Sílabo.

Romagni, P., & … . (1999). *10 instrumentos-chave da gestão* (1st ed.). Lisboa: D. Quixote.

Ryan, D., & Jones, C. (2012). *The best digital marketing campaigns in the world: Mastering the art of customer engagement* (1st ed.). Croydon: Kogan Page.

Ryan, D., & Jones, C. (2013). *Understanding digital marketing: Marketing strategies for engaging the digital generation* (2nd ed.). Croydon: Kogan Page.

Shapiro, C., & Varian, H. (1999). *Information rules: A strategic guide to the network economy* (1st ed.). Boston: Harvard Business School Press.

Tapscott, D. (1999). *Creating value in the network economy*. Boston: Harvard Business School Press.

Varajão, J. (1998). *A arquitectura da gestão de sistemas de informação* (2nd ed.). Lisboa: FCA.

Ward, J., & Griffiths, P. (1996). *Strategic planning for information systems* (2nd ed.). Wiley.

Watson, D. (2005). *Business models: Investing in companies and sectors with strong competitive advantage*. Harriman House Publishing.

Wilson, D. (2002). *Managing information: IT for business processes* (3rd ed.). Woburn: Butterworth-Heinemann.

Wilson, T. (1985). Information management. *The Electronic Library*, *3*(1), 62–66. doi:10.1108/eb044644 PMID:2498741

Wilson, T. (1987). Information for business: The business of information. *Aslib Proceedings*, *39*(10), 275–279. doi:10.1108/eb051066

Wilson, T. (1989a). Towards an information management curriculum. *Journal of Information Science*, *15*(4/5), 203–209. doi:10.1177/016555158901500403

Wilson, T. (1989b). The implementation of information system strategies in UK companies: Aims and barriers to success. *International Journal of Information Management*, *9*(4), 245–258. doi:10.1016/0268-4012(89)90048-0

Wilson, T. (1994a). Tools for the analysis of business information needs. *Aslib Proceedings*, *46*(1), 19–23. doi:10.1108/eb051339

Wilson, T. (1994b). The nature of strategic information and its implications for information management. New Worlds in Information and Documentation, 265-269.

Wilson, T. (2001). Information overload: Myth, reality and implications for health care. *ISHIMR - International Symposium on Health Information Management Research.*

Zorrinho, C. (1991). *Gestão da informação*. Lisboa: Editorial Presença.

ADDITIONAL READING

Davenport, T. (1997). *Information ecology: Mastering the information and knowledge environment*. New York: Oxford University Press.

Davenport, T. (2007). *Profissão: Trabalhador do conhecimento: Como ser mais produtivo e eficaz no desempenho das suas funções*. Amadora: Exame.

Davenport, T., & Prusak, L. (1998). *Working knowledge: How organizations manage what they know*. Boston: Harvard Business School Press.

Davenport, T., Prusak, L., & Wilson, J. (2003). *What's the big idea: Creating and capitalizing on the best management thinking*. Boston: Harvard Business School Press.

Dinis, J. (2005). *Guerra de informação: Perspectivas de segurança e competitividade* (1st ed.). Lisboa: Sílabo.

Drucker, P. (1987). *Inovação e gestão* (2nd ed.). Lisboa: Presença.

Drucker, P. (1988). *As fronteiras da gestão*. Lisboa: Editorial Presença.

Drucker, P. (1993). *Gerindo para o futuro*. Lisboa: Difusão Cultural.

Drucker, P. (1993). *Sociedade pós- capitalista*. Lisboa: Difusão Cultural.

Galliers, R. (1993). Towards a flexible information architecture: Integrating business strategies, information systems strategies and business process redesign. *Journal of Information Systems*, *3*(3), 199–213. doi:10.1111/j.1365-2575.1993.tb00125.x

Hampton, D. (1992). *Administração contemporânea* (3rd ed.). São Paulo: Makron Books.

Hess, P., & Siciliano, J. (1996). *Management: Responsibility for performance*. McGraw-Hill.

Instituto de Altos Estudos Militares. (2000). *A gestão da informação e a tomada de decisão*. Sintra: Atena.

Introna, L. (1997). *Management, information and power* (1st ed.). London: MacMillan. doi:10.1007/978-1-349-14549-2

Johannessen, J., & Olaisen, J. (1993). The information intensive organization. In J. Olaisen (Ed.), *Information management: A Scandinavian approach*. Oslo: Scandinavian University Press.

Lacalle, D. (2013). Nós os mercados: O que são, como funcionam e porque são indispensáveis (1ª ed.). Barcarena: Marcador.

Mintzberg, H. (1973). *The nature of managerial work*. New York: Harper and Row.

Mintzberg, H. (1994). *Rise and fall of strategic planning*. New York: Free Press.

Montgomery, C. A., & Porter, M. E. (1995). *Strategy: seeking and securing competitive advantage*. Harvard Business Review Book Series.

Olaisen, J., & Revang, O. (1991). Information management as the main component in the strategy for the 1990s in Scandinavian airline system (SAS). *International Journal of Information Management*, *11*(3), 185–202. doi:10.1016/0268-4012(91)90032-8

Penzias, A. (1989). *Ideas and information: Managing in a high-tech world*. New York: W. W. Norton.

Penzias, A. (1995). *Harmony: Business, technology & life after paperwork*. Harper Collins.

Petit, P. (2001). *Economics and information*. Dordrecht: Kluwer. doi:10.1007/978-1-4757-3367-9

Porter, M. (1986). *Competition in global industries*. Boston: Harvard Business School Press.

Porter, M. (1993). *A vantagem competitiva das nações*. Rio de Janeiro: Editora Campus.

Postman, N. (s/dat.). *Five things we need to know about technological change*. Retrieved 14-10-2008 from http://www.newtech.org/address10_en.htm

Rascão, J. (2012). *Novas realidades na gestão e na gestão da informação* (1st ed.). Lisboa: Sílabo.

Reponen, T. (1993). Information management strategy: An evolutionary process. *Scandinavian Journal of Management*, *9*(3), 189–209. doi:10.1016/0956-5221(93)90016-L

Roberts, N., & Clarke, D. (1989). Organizational information: Concepts and information management. *International Journal of Information Management*, *9*(1), 25–34. doi:10.1016/0268-4012(89)90034-0

Rothschild, W. (1992). *Como ganhar (e manter) a vantagem competitiva nos negócios*. São Paulo: Makron Books.

Rue, L., & Holland, P. (1989). *Strategic management: Concepts and experiences* (2nd ed.). New York: McGraw-Hill.

Vickers, P. (1984). Information management: A practical view. *Aslib Proceedings*, *36*(6), 245–252. doi:10.1108/eb050930

Walsham, G. (1993). *Interpreting information systems in organizations*. Chichester: Wiley.

Ward, J., & Peppard, J. (2002). *Strategic planning for information systems* (3rd ed.). Chichester: Wiley.

Webster, F. (2000). *Theories of the information society*. Cornwall: Routledge.

Wilmot, R. (1989). Change in management and the management of change. *Long Range Planning*, *20*(6), 23–93. doi:10.1016/0024-6301(87)90128-2

KEY TERMS AND DEFINITIONS

Business Management: The activity of conducting all the necessary operations in order to make the organization grow, benefiting all the shareholders and stakeholders and the society in general. It involves several specialized functions like finance, marketing, operations, human resources, production, etc.

Competitiveness: Competence of an organization or country to produce and sell products/services that meet the quality of the markets at the same or lower prices and maximize returns on the resources consumed in producing them.

Competitors: A company in the same business or similar industry which offers a similar product or service. Competitors can lower the prices of products and services and gain a larger market share, reducing clients and lowering profits.

Data: The representation of facts, concepts or instructions. Formalized with a structure suitable for communication, interpretation or processing by humans or by automatic means. The raw material of information. The basic element for the production of new information.

Information: A set of data arranged in a certain order and form, useful to people to whom it is addressed. Reduces uncertainty and supports decision-making. Information is considered to support human knowledge and communication in the technical, economic and social domains. Results from the structuring of data in a given context and particular purpose.

Information and Communication Technologies: Electronic tools based on the Internet, that allow synchronous and asynchronous communication and exchange of information—using sound, text, images, and video—between several points in different locations. Improves and stimulates the economic and societal globalization.

Information Management: Usually defined as a comprehensive organizational capacity to create, maintain, retrieve and make available the right information, at the right place, at the right time and in the hands of the right person, at the lowest cost, in the best support for its use in decision making.

Knowledge: Is a fluid composed of experiences, values, context information and apprehension about their own field of action that provides a cognitive apparatus for evaluating and incorporating new experiences and information.

Market Intelligence: Data and information collected by commercial and industrial organizations about their competitive environment to support good decision making. Makes possible to compare our market share with our competitors and take actions to maintain or improve that share.

Strategy: The way an organization chooses to do his business in order to surpass competition and gain consumers preference. What is going to do, how, for whom, using what resources, and so forth. It's the plan to conduct business, maximize profits, allocate the needed resources, and stay in the market as long as possible. Includes foresee what can happen, using forecasting methodologies and tools, in order to avoid problems or, if they happen, be prepared to minimize their effects.

ENDNOTES

1 For ethical reasons (by request of the management board of the company) the full name of the organization will not be mentioned, but only an acronym (BV) when referring to it in the course of this work.

2 We were not allowed to provide the latest financial information, being the current values a pale reminder of the values between 1994 and 1999. As mentioned above, currently the volume of turnover is practically to cover the fixed monthly expenses.

3 "Information is a process aimed at knowledge, or, more simply, information is everything that reduces uncertainty... A tool for understanding the world and the action on it" (Zorrinho, 1991, p. 21).

4 STEEP - sociological, technological, environmental, economic and political factors.

5 SWOT – Strengths, Weaknesses, Opportunities, Threats.

6 "It is no longer appropriate to consider information as simply a fourth factor in production. Corporate management is nothing else but information work. This means that information has become the prime production factor, and the purpose of information management is to make sure that it is put to good use at both strategic and operational levels" (Picot, 1989, p. 238).

7 Since October 2012 and March 2014 the Portuguese Government granted 772 Gold Visa, being 612 of them to Chinese citizens. That decision has brought to the State about 464 million of Euros. In January and February 2014 the purchase of properties of 500.000€ and above increased about 30% in relation to those months in the previous year. From those 772 Gold Visa granted, only two were for the creation of enterprises with more than 10 employees, one of the conditions for the grant (Cf. http://www.publico.pt/sociedade/noticia/sef-1629438).

8 "Champions are more than leaders and they are different from sponsors. Sponsors have the funds and authority to accomplish their goals. Champions, on the other hand, bring about change in their

organizations in spite of having less than the requisite authority or resources" (Earl, 1998, p. 347. Cf. also pp. 348-380).

9 "If an ISS is to be properly related to the business strategy, it is important that the Board should be closely involved" (Wilson, 1989b, p. 248).

10 "Information Management comprises activities including the acquisition, protection, utilization, accessibility and dissemination of information, and also the promotion and management of thrusts to derive maximum benefit from the resource. It also incorporates the development, management and marketing of an enterprise-wide model, and application of the principles of data management" (Ward & Griffiths, 1996, p. 360).

11 "The machines we have invented to produce, manipulate and disseminate information generate information much faster than we can process it. It is apparent that an abundance of information, instead of better enabling a person to do their job, threatens to engulf and diminish his or her control over the situation. It is now widely recognised that stress can be experienced from a feeling of lack of control" (Edmunds & Morris, 2000, p. 18).

Chapter 12
Organizational Intelligence: A Conceptual Proposal for Value Creation to Economic Organizations

Pedro Fernandes da Anunciação
Instituto Politécnico de Setúbal, Portugal

Antonio Juan Briones Peñalver
Universidad Politécnica de Cartagena, Spain

ABSTRACT

In the present context, the term intelligence has marked the economic and social fields. The increase in computational power of the technologies, associated with the activities of the people and organizations, economic know-how development needs of the market, the imperative of participation in economic dynamics established, among other things, is critical to the survival and for the sustainability of organizations. This is the dimension of competitive intelligence. But the term intelligence is not limited only to the external side of organizations, where this concept has gained broad meaning. This should also encompass the internal domain of organizations, organizational intelligence, and in this sense, refers to the centrality of functioning, flexibility of the articulation of resources, the indispensability of the systems architecture, and to the management insight. The ability to see important issues in the chain of value is crucial to identify the most important factors for success, to anticipate the competition, and to have pro-active management instead of reacting to market pressures.

INTRODUCTION

There are several concepts in the field of management (strategy, competitive advantage, product development, innovation, etc.) that, over time, have demonstrated the centrality of understanding the environment in the development of economic organizations (Liebowitz, 2006). Organizational competitiveness depends on the organizational adaptability to the environment, assuming it as a critical success factor in the sustainability of economic activities. Being sustainable and competing in a market strongly marked by information, technology presupposes owning or developing the 'orchestrate all the leads' capability

DOI: 10.4018/978-1-5225-6225-2.ch012

that the market offers (Berry, Carbone & Haeckel, 2002). Understanding this capability, presupposes that organizations develop capacities to identify and understand the challenges of the market and, simultaneously, appropriately value their capabilities or those that they can develop through partners. If we fit both identified capacities (understanding the environment and organizational knowledge) into a time frame (in a logic of economic functioning strongly conditioned by time-to-market), we understand the importance of the concept of intelligence in the field of management and economic organizations.

At the center of both concepts, we find information and knowledge. Both concepts form the basis of intelligence and, in economic terms, constitute the main merchantable economic resources and the basis of management. It is in the value of information and knowledge that lies the sustainability and competitiveness of organizations (Anunciação, 2014; Anunciação, Esteves & Rocha, 2013). It is on this basis that strategic development must be directed to and influence the market. Thus, the centrality of information systems in the field of intelligence is evident. The functions of collecting, analyzing and making available information depend on the IS and, consequently, on them, the best decisions and the best levels of management effectiveness (Guarda, Santos, Pinto, Augusto & Silva, 2013).

The concept of intelligence comes from the Latin "intelligentĭa", which, in turn, derives from "inteligere". This word includes two expressions: "intus" ("between") and "legere" ("choose"). Therefore, we can consider that, due to the etymological origin of the concept of intelligence, the central reference of the concept is associated with the capacity of choice, that is, intelligence must allow to select/choose the best options when solving a given problem. In this sense, we can affirm that the capacity of choice or decision is conditioned by the defined objectives, the economic context, the available resources and the timings of the decision, because all of them can condition.

The use of the concept of intelligence to the organizational domain should be considered in the scope of this work from the perspective of the analogy. The assumption of intelligence is traditionally associated with human beings. Thus, its association with economic organizations, the subject of research in this work, must be framed through the analogy. On the human side, this concept has been widely discussed and applied. Through the analogy, it will try to emphasize the importance and the potential of this concept in the organizational management, trying to highlight the potential of this proposal for the framework of the complexity of the management and market contingency. In this sense, it will be sought to present or produce knowledge through the design and presentation of a model that allows finding solutions that can be applied by the managers of the organizations and the professionals of the information systems (Costa, 2006). We believe that, through the interconnection of the various dimensions of organizational functioning and the integration of the various subsystems, is possible to adapt, organize and effectively manage the various elements in a context of relational interaction, and to ensure a harmonious and coherent functioning of the economic activities developed, according to the defined objectives and economic requirements of the market.

Only by considering information and knowledge as main differentiating factors in the support of strategies and responses to the market, on the one hand, and basic elements of the internal knowledge of the organizations, on the other hand, it will be possible to think about the dimension of intelligence at the organizational level (Anunciação & Esteves, 2012).

LITERATURE REVIEW

In the permanent technological innovation, companies coexist with the support of management systems for strategic management, where changes in ICT for management and administration are fundamental, together with an intelligence system that is integrated into knowledge management and the business intangibles (López-Cano; Briones-Peñalver & Cegarra-Navarro, 2015; Briones-Peñalver; Roca-Gonzalez & Martinez-Martinez, 2015).

The concept of organizational intelligence is based on two concepts: the concept of intelligence and the concept of organization. We can understand intelligence, as previously mentioned, as the ability to read and evaluate a given reality, as well as identifying the constraints and conditions required to choose the best decision option. Whether in the human or the economic domain, information is at the basis of intelligence, insofar as intelligence provides people with a contact with something they did not know, or confirm anything they suspected. Intelligence presupposes the reception and communication of knowledge and, consequently, intelligence. We can, thus, understand that the concepts of information, knowledge and learning must be part of the organizational vocabulary and must be linked to the intellectual, individual and collective development, in particular as regards the way knowledge is acquired, applied, distributed and even memorized. In short, in aspects related to how to put "intelligence in action".

The focus of the Intelligence allows to analyze the environment, constituting a research in the field of business management where valuable facts and evidences are identified for the competitiveness of the organization, and actions are determined to follow, from the detection of strategic movements that allow to optimize business models (Briones-Peñalver & Poças-Rascão, 2014). The great reflexes of intelligence, in the organizational domain, must be expressed in the capacity of perception of the complex environment, allowing, through solutions of Urbanism, to adapt the organizational performance in the relationship with other stakeholders, in order to improve the competitive capacity in the market. This means that the improvement of the "intelligence of the organization" must necessarily rationalize and make flexible the various organizational systems, in order to adapt the commercial responses to the market.

The dimension of intelligence at the level of economic organizations can be framed in various ways, for example, the emotional intelligence of groups or Task Forces. However, in general, intelligence always presupposes a capacity for commitment and innovation in solving problems and in identifying the best solutions, enhancing the competences and participation of each actor in a wider collective context.

Given that current challenges force corporations to shift the focus from management of change from the anatomical to the physiological (Oxman & Smith, 2003), the organizational flexibility dimension assumes the importance of a governance principle, dethroning the primacy of the structure that, for many years, led the construction and organizational identity strand. Physiology is supplanting anatomy. It is in this sense that we can assume the centrality of the organizational dimension associated with the ability to integrate or "reassemble" the activities, functions and areas. From the various parts/elements/components identified, and in the search for the best solutions recommended, the management should ensure a dynamic and harmonious functioning appropriate to the established objectives.

With traditional hierarchies to disappear and innovative organizational forms, almost always quite complex, to replace them, it is important to perceive organizations differently. Organizations will simultaneously have to approach their customers, anticipate market needs, create opportunities for new products and services, present innovative solutions, engage in the entire value chain to ensure ultimate customer success, develop partnership with customers and suppliers, etc. All these challenges, by analogy with the human dimension, show that organizational intelligence is composed of several more concrete dimensions that must be considered. From the various classifications of intelligence, we can draw three dimensions that emphasize different aspects of this concept: the psychological (understood as the cognitive capacity of learning and relation), the organic (understood as the capacity to adapt to new situations), and the operative (as the capacity to put into operation). From the analogy, we can point out that these three dimensions show, at the level of economic organizations, three important aspects that must integrate a model proposal: intellectual capital, organizational architecture and operational effectiveness.

Organizational intelligence can be understood as an ability to think effectively about an organization from a business point of view (Silber & Kearny, 2010). The adoption of this concept in economic organizations makes sense if we consider the requirements of key-drivers that support the skills required to economic organizations and, according to McConnell and Ward-Perkins (1998), are:

- Accessing to various markets, reaching potential clients (Globalization);
- Having better relationship with customers, anticipating market needs, taking opportunities for new products and innovative solutions (Anticipation);
- Allowing subcontracting, establishing partnerships for various activities, searching for excellence (Specialization);
- Involving the value chain in the final success of the product. Developing a close relationship with clients and suppliers (Relationship);
- Proximity with clients and suppliers (Partnership);
- Developing processes and commercialization before competition (Reactivity);
- Reorganizing the company as to the paperwork and definition of responsibilities, in order to be more quick and flexible (Flexibility);
- Management and communication of quality procedures, according to customer demands (Quality).

When we consider these factors that characterize the market and associate them with the new techno-economical paradigm, based on information and communication technologies, is important to consider the specificity of the required adjustments, namely:

- Adjustment of competitive advantages through information processing, which implies the possibility of creating new operations or improving those already established by applying new knowledge;
- Creating synergies between technological innovation and organizational flexibility, redesigning processes and products;
- Focus on developing a learning organization, being able to increase the value (Bartlett & Goshal, 2000).

The specificity of the key factor of development of the current economic operation corresponds to the knowledge action over knowledge itself. The growing potential processing of technologies, and the support of information technologies to economic activities, create virtuous circles of interaction, between various sources of information, allowing access to opportunities. The effects of virtuous circles of interaction introduce the learning experience and intelligence factor in the course of competences and organizational abilities.

Market development demands intelligent organizational models able to react appropriately to the needs of the environment (Choo, 1999; Quinn, 1992; Senge, 1992). So, two core concepts which are connected emerge, the competitive intelligence (CI) (Zambon, Sousa, Andrade & Anunciação, 2015; Zambon & Anunciação, 2014; Calof & Wright, 2008; Dishman & Calof, 2008; Murphy, 2005) and organizational intelligence (OI) (Silber & Kearny, 2010). Some of the common elements are:

- Integrating disperse data research processes, transformed into relevant information (Tarapanoff, 2001);
- Corresponding to institutional and systematic programs of information collection and analysis (Gomes & Braga, 2002);
- Corresponding to a strategic information management activity, allowing the decision-makers to anticipate the markets and the competition (Gomes & Braga, 2002);
- Implementing a formal and permanently controlled process, to evaluate the capability to aid maintenance or development of a competitive edge.

For some authors, the goal for developing the intelligence processes is to monitor the internal and external environment, researching information with added value, opening doors to new business opportunities, contributing to the development of strategies to reach competitive advantages (Oliveira, Gonçalves & Paula, 2013; Sapiro, 1993). Apart from its value in supporting the development of strategies, it allows the practice of benchmarking with the objective of stimulating the increment of efficiency and competence in organizational development, as the improvement of time-to-market, the agility and efficiency of commercial strategies.

It is true that information systems (IS), supported by information and communication technologies (ICT), constitute a fundamental asset, facilitating the access, use and availability of data (Tsai, Raghu & Shao, 2013). However, 'intelligent organizations' should work from a referential based on the creation of knowledge that aim at the creation of competences, the adequacy of efforts and the balanced use of resources, bearing in mind the defined goals and strategies. It is important that organizations have the ability to make clear decisions, develop actions and introduce stimuli, dominate and control consequences rather than reacting to them, knowing how to play out adequate attitudes or develop actions, as market leadership initiative.

There are many recognized advantages through the implementation of organizational intelligence:

- The reduction of information dispersion;
- Wider interaction space between collaborators;
- Ease of access and opportune availability of information;
- Versatility; and
- Flexibility in the adaptation of the company's reality and usefulness of the process of decision making.

However, detaining information does not equal possessing intelligence. The information constitutes the necessary but insufficient condition to the existence of intelligence which can only be achieved by the structuring of the information (Tarapanoff, Júnior & Cormier, 2000). Also, to know how to capture and detain information and manage and maintain knowledge, bearing in mind its transformation into intelligence. From the quality of the raw material (information) depends the quality of the final product (intelligence) and assuring the quality of the production process (Information System). Tarapanoff, Júnior, and Cormier (2000) refer information as the transformation of interconnected data into strategic intelligence towards organizational decision making and its respective strategic planning. Therefore, some requirements are essential to the viability of quality information, like: exactitude and reliability of data sources, objectivity, usability, relevance, availability and opportunity (Rodrigues, Rechziegel, Esteves & Riscarolli, 2012).

Knowledge is the second of the fundamental attributes of any organization. In the equation of intelligence, other than information, we must consider organizational knowledge as a crucial factor in the quality of the work developed and the strategic decisions. Knowledge constitutes a collection of information considered valid, accepted and related to determine the factors (data, actions, information and, sometimes, hypotheses). However, many organizations consider knowledge management as the activities related to information management (Setia & Patel, 2013). After all that has been mentioned, the central roadmap must contain the following sequence, shown in Figure 1.

The value of knowledge is based on this relationship between the capability to read and understand, precisely, quickly, and reliably, the context and opportune information about facts, events and tendencies about the same. In a competitive environment, it is the relation between them that knowledge is created. It is in the combination of this knowledge with the competences of management and structural flexibility and organizational dynamics that we find intelligence (Zambon, Sousa, Andrade & Anunciação, 2015).

A collection of information does not reveal itself sufficient, if the systems are not appropriate for the analysis of the information, according to organizational needs, and if they are not managed accordingly. Equally, there is little use in the management process of the existence of market information, prices and competitive strategies if there is no capacity to process and withdraw value.

The informational needs must be defined in relation to corporate strategies and supported by actual and real information. If management does not bear in mind the risks, weaknesses and threats, or if it does not extract value from the strengths and opportunities, it will not take advantage of the full potential that competitive intelligence provides, abdicating a pro-active strategic positioning in the markets regarding to the competition (Alves & Falsarella, 2005).

Figure 1. Path to intelligence
Zambon, Sousa, Andrade & Anunciação, 2015;Zambon & Anunciação, 2014.

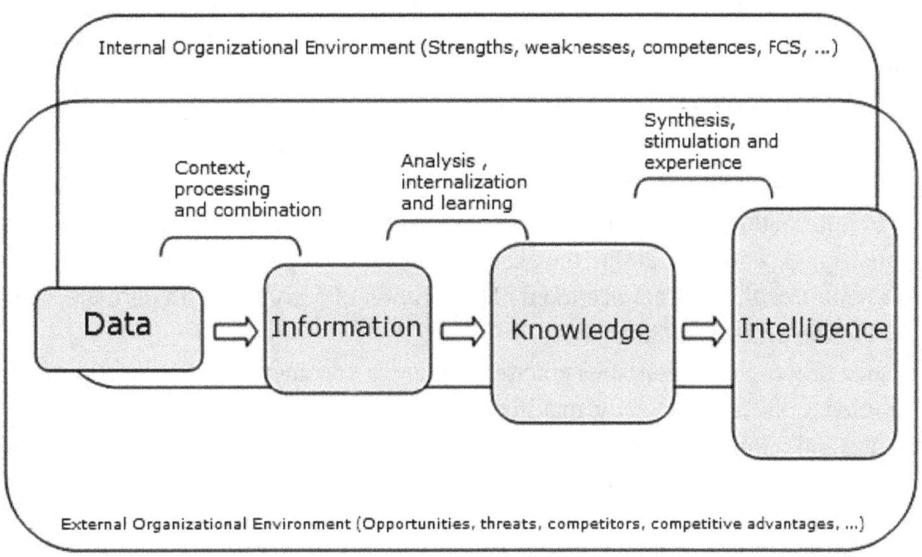

The search for intelligence must constitute a strategic referential of action and development, aiming to reduce uncertainty, reach new opportunities and the optimization of reaction time regarding the competitors. Only the transparence and evidence of the internal processes, production, marketing, design, quality or others, allows a quick identification of the organizational dynamics towards the strategies, facilitating the understanding of the obtained results.

But this includes the adoption of an organizational approach that has the various dimensions associated with a functioning intelligence, from processes to architecture/urbanism, from information to knowledge, from IS to ICT. This organizational approach should allow for flexible responses to the complexity of the surrounding environment supported in a real knowledge management practice in the whole of the organization.

Knowledge management means organizing and making adequate policies towards processes and these towards the tools and instruments of management (Costa, 2006), allowing possible to know what is done, why it's done, by whom it's done, when it's done, and what the expected value is. This is the starting point for the development of organizational intelligence. The organizational transformation must be centered in the (re)engineering of the internal processes and organizational structures, as well as in the urban development of the economic networks and value chain (Anunciação & Zorrinho, 2006).

In this sense, intelligence can be seen as a strategic factor for organization competitiveness, making necessary to appeal organizations to redefine and improve their organizational systems, making them «intelligent». This is a current challenge imposed to management professionals that demands creativity and ability. The need of the management to achieve a sketch of the future, with due foreknowledge, constitutes a permanent challenge. Only so, is possible to centralize efforts and focus resources on the priorities and on the identified opportunity targets.

Managers should drive, lead, command or induce, knowing how to have the adequate attitudes, developing the necessary actions, initiatives for market leadership (proactivity), and not only as a response or as a consequence to factors (reaction). In the background, the vision, reorganization and management proactivity skills express anticipation capacity, which estimates the skills of perception, interpretation, understanding of clients' needs and habits, mobilizing resources in the face of market signals and strengthening positions for competitors. Therefore, it is important that perceived perceptions, through the ability of vision, match actual needs and market changes so that resource mobilization and investment can yield positive results and confidence for the organization to make decisions and actions proactive towards the market.

The latter goal of intelligence should be excellence. This last level of organizational performance requires the integration of all previously referenced conditions, such as the availability of quality information (precise, quick, reliable and appropriate concerning facts, events, tendencies and relationships), the transition of information to knowledge, the existence of urban flexibility at level of the organizational systems (Anunciação & Zorrinho, 2006), the exploration of the integrated potential between technologies and economic activities, the interconnection and integration of activities with partners, the commercial responsiveness to the time-to-market.

The relevance of these characteristics emerges when we consider hyper-competition, a synonym of enormous turbulence and inconsistency, making more complex to maintain an appropriate performance achieving unique and inimitable strategies.

PROPOSED MODEL

The model shown in Figure 2, follows four dimensions considered central to the generation of organizational intelligence, namely (Anunciação, 2014):

- **Guidance:** Consists of the identification and characterization of the vision. Vision and strategy express desires and ambitions that must be consensual in acceptance, strong guidance, inspiring motivation, and achievable.
- **Posture:** A set of knowledge, skills, techniques and values that determine the organizational reaction to the various internal and external stimuli.
- **Organization:** Integrates the model, characteristics and requirements of the articulation of functions, resources, activities, processes, technological infrastructures, organizational logics, operating dynamics and relationships with economic partners.
- **Optimization:** Expresses the ability to evaluate and monitor organizational performance, through the institutionalization of systematic processes of introspection, and evaluation of the feasibility of eventual changes.

For a more detailed analysis, these dimensions must be conjugated or integrated with five dimensions, which represent the organizational value cycle for the institution:

- **Identity:** Corresponds to the set of characteristics that allow identifying and characterizing, from an internal or external perspective, a particular institution, differentiating it from others in the same economic sector;
- **Direction:** Orientation that allows to lead and motivate employees, as well as the execution of activities and tasks in order to achieve the defined objectives;
- **Roadmap:** Encompasses the conscious and planned specification of a route and respective steps for changing a given organizational situation based on a process of change;
- **Resources:** Represent the elements and means made available to the organization that are necessary for the performance of its activities and the fulfillment of its mission and strategies;
- **Performance:** Corresponds to the identification of the results achieved or obtained in the economic performance of the organization.

Figure 2. Key dimensions of organizational intelligence model

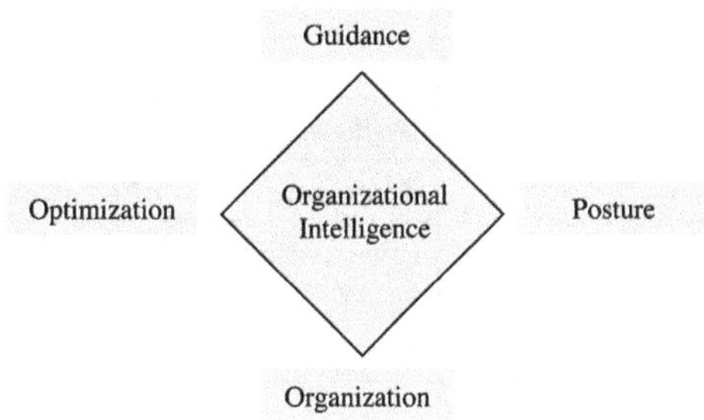

Organizational intelligence must be measured through the conjugation of the axes and cycle previously presented, namely through the measurement of the sub-dimensions resulting from the interception of both. From this conjugation results an array of sub-dimensions, as evidenced in Table 1.

Considering the previous table, we have the following sub-dimensions:

- On the Guidance dimension:
    - **Mission:** Corresponds to the purpose of a company or institution and must express its raison d'être evidencing the essence of its economic activity.
    - **Vision:** Set of aspirations and intentions for the future of the company, and should serve as a reference for the whole company.
    - **Objectives:** Objectives or goals that are intended to be achieved and which should justify the allocation of resources.
    - **Strategies:** Includes defining the desired objectives and identifying and selecting the means and resources to achieve them.
    - **Competitiveness:** The ability of an organization, within its mission, to more successfully meet the needs and expectations of customers.
- On the Posture dimension:
    - **Intellectual Capital:** Corresponds to the organizational value deriving from the competencies and capacities of its human resources, the efficiency and effectiveness of the organizational structure and model and the synergies obtained in the relational domain with the various stakeholders.
    - **Governance:** A set of activities related to political and strategic decisions at the Top Management level, concerning the organization.
    - **Critical Success Factors:** Integrate the variables that provide the most value to customers and that best differentiate the organization from the competition.
    - **Task Force:** Identification of a team with specific competencies, according to the purpose behind its creation and the problems or situations to be solved.
    - **Targeting:** Orientation of the performance of the activities and functions to the goals outlined by focusing on the achievement of the objectives of the area, in order to ensure the achievement of the expected results.
- On the Organization dimension:

Table 1. Sub-dimensions of organizational intelligence model

	Guidance	Posture	Organization	Optimization
Identity	Mission	Intellectual capital	Structure	Information Systems
Direction	Vision	Governance	Culture	Capabilities
Roadmap	Objectives	Critical success factors	Processes	Cycles/Timings
Resources	Strategies	Task Force	Urbanism	Audits
Performance	Competitiveness	Targeting	Contingency Plans	Key performance indicators

  - **Structure:** Represents the ordered set of functions, responsibilities and authority corresponding to groups of activities, which may be associated with nature, product, business, region, etc.
  - **Culture:** Understood as the set of values, behaviors and organizational attitudes that are adopted and felt by the diverse collaborators that integrate the organization and that condition, and distinguish the form of development of the organizational activities.
  - **Processes:** Processes are collections of identifiable and repeatable activities developed directly in the provision of the service.
  - **Urbanism:** Encompasses the set of reflections and measures aimed at the organization and management of the different elements that make up the organization or the information systems.
  - **Contingency Plans:** A document that establishes a set of strategies to deal with a possible threat, or a situation that is not part of the normal functioning of an organization.
- On the Optimization dimension:
  - **Information Systems:** A set of elements (physical, logical, human and procedural / organizational) that, through appropriate relationship rules and previously defined purposes, aim at the production and availability of Information.
  - **Capabilities:** Capabilities are the organizational attributes required for the execution of the processes and for the provision of the service.
  - **Cycles/Timings:** Temporal periods of execution, response or organizational reaction based on the physical, technical, or technological conditions necessary for the adequacy, integration or availability of a given output.
  - **Audits:** Correspond to the systematic analysis or examination of the activities, rules/procedures or systems in a company, in order to ascertain if they are in accordance with the dispositions planned or established previously, and if they have been implemented effectively or if they are adequate.
  - **Key Performance Indicators:** Integrate the set of indicators or metrics that are most relevant to the strategy and the organization, and that allow to determine the existence of success or failure in the options taken or the activities developed.

CONCLUSION AND FUTURE WORK

Intelligence, although a concept closely associated with the human condition, can be, by analogy, adopted for the management and organizations. The main objective is to optimize the functioning and competitiveness of organizations. The concept of intelligence presupposes the capacity of information, selection or choice of the best options and optimization of the performance associated with the resolution of a certain situation or problem

To this end, organizations will have to develop models and tools that, supported by information systems and technologies, allow the transformation of information into knowledge for the development of skills and competences that can optimized, not only the operation, but also the responses to market demands and changes.

The model proposed here seeks to demonstrate a pedagogical approach to the development of organizational intelligence, emphasizing core and transversal axis of organizational functioning, correlating them with a value cycle for the conduct of economic activities. Through the identification and specification of the several dimensions mentioned in the framework, it was sought to cover the set of core aspects of organizational functioning and development. These dimensions constitute a roadmap for the evaluation of the generation or existence of organizational intelligence.

In future works, it is expected to be able to apply this framework to a case study with the aim of being able to validate it and eventual adjustments.

REFERENCES

Alves, R. P., & Falsarella, O. M. (2005). *Conceptual model for Oganizacional Intelligence Applied to Function Maintenance. National conference of Research in Information Science*, Florianópolis.

Anunciação, P. F. (2014). *Ethics, Sustainability and Information and Knowledge Society*. Lisbon: Chiado Editora. (in Portuguese)

Anunciação, P. F. & Esteves, F. M. (2012). Sustainability in the Information Society: A Proposal of Information Systems Requirements in View of the DPOBE Model for Organizational Sustainability. *Procedia Technology*, *5*, 599-606.

Anunciação, P. F., Esteves, F. M., & Santos, J. R. (2013). Some Information Systems Requirements in View of Organizational Sustainability in an Information Society. *Information Resources Management Journal*.

Anunciação, P. F., & Zorrinho, C. D. (2006). *Organizational Urbanism – Who to manage the technological impact in economic organizations*. Sílabo Editions. (in Portuguese)

Bartlett, C. A., & Ghoshal, S. (2000). Going Global: Lessons from late movers. *Harvard Business Review*, *78*(2), 133–142. PMID:11183975

Berry, L. L., Carbone, L. P., & Haeckel, S. H. (2002). Managing the Total Customer Experience. *MIT Sloan Management Review*, *43*(3), 85–89.

Briones-Peñalver, A. J., & Poças-Rascão, J. (2014). *Information tecnhologies (ICT), network organizations, and information systems for business cooperation: a focus on organization and strategic knowledge management. Rethinking the Conceptual Base for New Practical Applications in Information Value and Quality.* IGI Global.

Briones-Peñalver, A. J., Roca-Gonzalez, J. L., & Martinez-Martinez, I. J. (2015). *Innovation management based on knowledge. Analysis of technology-based Defence Companies. Handbook of Research on Effective Project*. IGI Global.

Calof, J. L., & Wright, S. (2008). Competitive intelligence: A practitioner, academic and inter-disciplinary perspective. *European Journal of Marketing*, *42*(7/8), 717–730. doi:10.1108/03090560810877114

Choo, C. W. (1999). *The intelligent organization: The use of information to create awareness and give Meaning Decision Making*. Oxford University Press. (in Spanish)

Costa, L. E. (2006). *Knowledge Management and Competitive Intelligence as instrument of Organizational Sustainability*. Belo Horizonte: Pitágoras Faculty. (in Portuguese)

Dishman, P. L., & Calof, J. L. (2008). Competitive intelligence: A multiphasic precedent to marketing strategy. *European Journal of Marketing*, *42*(7/8), 766–785. doi:10.1108/03090560810877141

Gomes, E., & Braga, F. (2002). *Competitive Intelligence: How to turn information into a lucrative business*. São Paulo: Campus. (in Portuguese)

Guarda, T., Santos, M., Pinto, F., Augusto, M., & Silva, C. (2013). Business Intelligence as a Competitive Advantage for SME's. *Internacional Journal of Trade. Economics and Finance*, *4*(4), 187–190.

Liebowitz, J. (2006). *Strategic Intelligence – Business Intelligence, Competitive Intelligence, and Knowledge Management*. Auerbach Publications. doi:10.1201/9781420013900

López-Cano, V. C., Briones-Peñalver, A. J., & Cegarra-Navarro, J. G. (2015). *Absorptive Capacity and Technology Knowledge: Enhancing Relational Capital. Knowledge and Process Management, 22*(4), 305–317.

McConnell, J., & Ward-Perkins, D. (1998). *The advantage of the Internet for companies: competitive challenges and successful strategies*. Lisboa: Dom Quixote. (in Portuguese)

Murphy, C. (2005). *Competitive Intelligence – Gathering, Analysing, and Putting into Work.* Gower e-Book.

Oliveira, P. H., Gonçalves, C. A., & Paula, E. A. M. (2013). Vision based on Competitive Intelligence Resources. *Management Science Magazine*, *15*(35), 141–151.

Oxman, J. A., & Smith, B. D. (2003). *The Limits of Structural Change. Sloan Management Review, 45*(1), 77–82.

Quinn, J. B. (1992). *Intelligent enterprises*. New York: The Free Press.

Rodrigues, L. C., Rechziegel, W., Esteves, G., & Riscarolli, V. (2012). Competitive Intelligence as Innovation tool to facility Business Processes. [in Portuguese]. *RGO Organizational Management Magazine*, *5*(2), 230–241.

Sapiro, A. (1993). Business intelligence: The Informational Revolution to the Competitive Action. [in Portuguese]. *RAE-Enterprise Administration Magazine*, *33*(3), 106–125.

Senge, P. M. (1992). Mental models. *Planning Review*, *20*(44), 4–10. doi:10.1108/eb054349

Setia, P., & Patel, P. C. (2013). How information systems help create OM capabilities: Consequents and antecedents of operational absorptive capacity. *Journal of Operations Management*, *31*(6), 409–431. doi:10.1016/j.jom.2013.07.013

Silber, K. H., & Kearny, L. (2010). *Organizational Intelligence – A guide to understanding the business of your organization for HR, Training, and Performance Consulting*. Pfiffer.

Tarapanoff, K. (2001). *Organizational and Competitive Intelligence*. Brasília: UNB. (in Portuguese)

Tarapanoff, K.; Júnior, R. H. A.; & Cormier, P. M. J. (2000). Information Society and Intelligence in Information Units. *Ci. Inf. Brasília*, *29*(3), 91-100. (in Portuguese)

Thomas, A., Krishnamoorthy, M., Singh, G., & Venkateswaran, J. (2015). Coordination in a multiple producers–distributor supply chain and the value of information. *International Journal of Production Economics*, *167*, 63–73. doi:10.1016/j.ijpe.2015.05.020

Tsai, J. Y., Raghu, T. S., & Shao, B. B. M. (2013). Information systems and technology sourcing strategies of e-Retailers for value chain enablement. *Journal of Operations Management*, *31*(6), 345–362. doi:10.1016/j.jom.2013.07.009

Zambon, A. C. & Anunciação, P. F. (2014). Competitive Intelligence – Value perceptions in the sector of custom jewelry. *Portuguese and Brazilian management Magazine, 13*(2), 41-60. (in Portuguese)

Zambon, A. C., Sousa, C. S., Andrade, F. J., & Anunciação, P. F. (2015). *Information and value chain.* The Limeira Gross Domestic Product Sector – Brazil, Annual Spain Business Research Conference, Barcelona, Spain.

Chapter 13
The Contribution of Information Science Through Scientific and Technological Knowledge in Intellectual Property

Jorge Lima de Magalhães
Oswaldo Cruz Foundation, Brazil & NOVA University of Lisbon, Portugal

Flavia Maria Lins Mendes
Oswaldo Cruz Foundation, Brazil & Federal University of Rio de Janeiro, Brazil

Adelaide Maria de Souza Antunes
National Institute of Industrial Property, Brazil & Federal University of Rio de Janeiro, Brazil

Zulmira Hartz
NOVA University of Lisbon, Portugal

ABSTRACT

The more than 100 million patents registered in the European Patent Office provide an unprecedented source of scientific and technological information in the history of mankind. The technological management of this information is exploited to develop technological advances in scientific, technological, and educational organizations and companies. New standards of product and process safety and effectiveness have been introduced across the world, and public and private business strategies are under constant review to comply with the prevailing paradigm. The health sector releases more than 1 million papers a year on scientific progress, while technological (patents) advance 10% per year. Therefore, updating the contribution of the information science through scientific and technological knowledge in intellectual property, a case study, will provide a contribution to reflection for the business in research, development, and innovation in health. These facts lead to constant adjustments of business in companies, universities, and government actions. In 2017, three lists of strategic products for the Brazilian Health System were changed. Using new intelligence systems, the government has adopted new strategic partnerships with the private sector, and were conceived in 2017 (others replaced) with budgets of more than US$ 2 billion. This chapter explores the contribution of information science through scientific and technological knowledge in intellectual property.

DOI: 10.4018/978-1-5225-6225-2.ch013

BACKGROUND

The activities of companies, research groups, institutions and national governments are effective when they attribute value and quality to their information. These critical factors are crucial for organizations' success in their domestic and international planning, whatever their long- and short-term strategies (Magalhaes & Quoniam, 2015).

Hence, identify and analyze the amount of scientific information and state of the art with respective correlations have become hard work. The world's technological per-capita capacity to store information has roughly doubled every 40 months, since the 1980s. As of 2012, every day 2.5 quintillion (2.5×10^{18}) bytes of data were created (Lynch, 2008). Therefore, the challenge for science and enterprises is managing Big Data so that this scientific data can be retrieved and processed into strategic information for decision makers.

Intellectual property is an important asset for businesses, and knowledge is becoming increasingly crucial for competitiveness, technology, and economic development. This is particularly true for technology-intensive sectors, where knowledge is regarded as a company's most important asset (Miller, 2000).

Therefore, it is crucially important for organizations to invest in research, development and innovation, if they are to remain active and competitive. Information Science has tools that can help organizations to produce, store and manage data on any activities or processes, resulting in more effective management for innovation. With the increasingly turbulent, complex and competitive conditions in the markets in which companies operate, the use of industrial/intellectual property has become a way of assuring the continuation of their activities in the future by protecting innovations and restricting how their competitors can act. The industrial property information contained in patents identifies the latest science and technology developments, making it a powerful competitive weapon (Pierret, 2006).

The mechanisms for mining information have developed from "manuals" to dedicated portals or websites (from Web 1.0 to Web 2.0), where mass information can be obtained by automated means. This new paradigm allows huge quantities of data to be downloaded in different formats, but it cannot process this data to produce indicators that can actually help decision-makers. Therefore, studies are required using information science, such as technology trends (Quoniam, 2011).

Knowledge Management

Knowledge management requires the involvement and support of all a company's stakeholders to preserve, transmit and develop knowledge. Indeed, it is the individuals who are at the center of the creation of value and hold the keys to the success of such a project. The management of a company's knowledge and know-how is not universal, but depends strongly on the culture of the country in which it is undertaken (Balmisse, 2006).

To motivate individuals to share their knowledge, cultural and human factors must be taken into account, rather than just tools and procedures. Pesqueux et al (2011) define "culture" as the set of values, norms, habits and customs specific to a society that influence the personality of the individual (Pesqueux & Ferrary, 2011).

According Morin (1982), it is necessary to elevate the concept of "system" of this theoretical level to the paradigmatic level. A paradigm, for the author, is the set of fundamental relationships of association and/or opposition between a small number of "master concepts" that command/control all existing knowledge in all the speeches and theories. Thus, to sustain the action cycle (application) of the knowledge to the action progress, it is necessary to try to identify/define what are these notions, featuring the new paradigm and to resume as the (re)definition of the system where the logic of complexity in modeling the actual performance interact with a posture of the researcher and the actors involved in evaluative process (Hartz, 2013; Morin, 1982; Piaget & Garcia, 1983).

In this sense, to identify, classify, measure, disseminate and training competencies are essentials to aid strengthen to the R,D&I in Management of Science, Technology and Innovation in health. According to Hartz (1997), evaluation and health are characterized by phases. The first, is the "systemize" and refers to a "structuralism and cybernetic" practice supported on mathematical theories which stands the analysis of hard-systems (Le Moigne & Jean-Louis, 1994). This approach is still "analytic in nature and positivist in attitude (...). Thus, knowledge needs to be thought as a process that takes place within a complex system of interactions between researchers and knowledge users which may vary in intensity, complexity and level of engagement, depending on the nature of the research and the findings, as well as the needs of the knowledge user – knowledge translation (Brouwers, 2010).

Knowledge translation increasingly assumes critical importance for health research, since it is recognized that the creation of new knowledge, often alone, does not lead to its application or to effective impacts on the health of populations (Canadian Institutes of Health Research, 2016; Who I Special Program on Research and Training in Tropical Diseases, 2006).

Considering the global technological advancement in the knowledge era, the public health arises the need to find better treatments and quality of life to people. Then, to cogitate technology transfer in healthcare by PDP can be understood as a customization for the local reality of a nation. This process can be considered as a translation of knowledge.

Graham et al. (2006) proposed the knowledge-to-action (KTA) process conceptual framework that could be useful for facilitating the use of research knowledge by several stakeholders, such as practitioners, policymakers, patients, and the public. The KTA process has two components: (1) knowledge creation and; (2) action. Each component contains several phases. The authors conceptualized the KTA process to be complex and dynamic, with no definite boundaries between the two components and among their phases. The phases of the action component may occur sequentially or simultaneously, and the knowledge-creation-component phases may also influence the action phases (see Figure 1). In the similar approach, Zicker (2007) and Carvalho et al (2004) highlight some benefits of knowledge translation:

1. Endemic countries involved in setting the research agenda and also on how to respond to their priorities;
2. Strengthen the research capacity of individuals, institutions and networks in the countries whether endemic or low-income;
3. Promote the application of innovations and strategies in political and health practices; and,
4. Serve as a platform to establish consensus on priorities, funding and management processes in research.

According to the World Economic Outlook Database maintained by the International Monetary Fund (IMF), and the official Brazilian statistics agency, Instituto Brasileiro de Geografia e Estatística (IBGE), Brazil is one of the emerging nations set to stand out most on the international stage for its development and economic growth (IBGE, 2011). Brazil's gross domestic product grew by 250 percent, from 2006 to 2012, reaching around R$ 2.5 trillion. Yet despite this promising scenario, one of the many parameters of domestic consumption, or development, in the Brazilian trade balance is a "family" of health-related

Figure 1. Knowledge-to-action (KTA) process conceptual framework
Source: (PIMJAI SUDSAWAD, 2007).

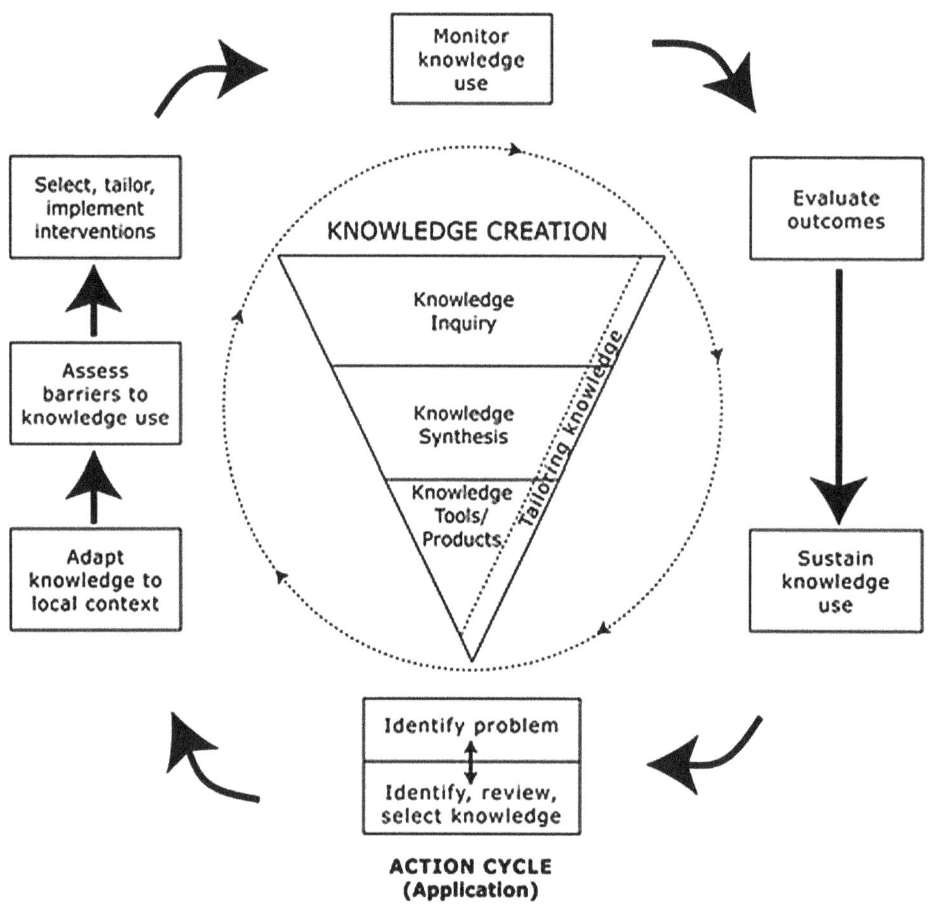

items, which, in 2012, reached a deficit of R$ 12 billion. This group of products includes drugs, medications, equipment and diagnoses for public health (Gadelha, Carvalho, & Pereira, 2012).

Brazil, as well as the global trend, has undergone processes of demographic, epidemiological and nutrition transition with significant reduction of fertility and birth rates besides a steady increase in life expectancy. This process results in changes in the patterns of occurrence of pathologies and increased significantly the prevalence of chronic diseases. This changing in epidemiological profile of the country causes an increased demand for new medical technologies to ensure continuous improvement in the health population (Gadelha, Carvalho, & Pereira, 2012).

In this context, health products are expanding and occupying an important space in the Brazilian economy. It consists of an overload of items with different levels of complexity, ranging from a simple infrared lamp, until the MRI equipment to a culture medium to a reagent kit for detection of HIV and medicines. Therefore, these products are used in conducting medical, dental and physiotherapy procedures, as well as the diagnosis, treatment, rehabilitation and patient monitoring.

It is one of the most vital and dynamic sectors of the economy. According to the Ministry of Health (MoH), in Brazil there was a 72% growth in the number of devices in health services in the National

Health System (SUS - Brazilian term) over the past five years. Sector revenues reached more than 100%, from 2005 to 2013, reaching the order of US$ 5.5bn.

The expansion of access to health care has led to a significant increase in Industrial Economic Health Complex (CEIS) trade deficit in the last decade, from a level of US $ 3 billion in 2003, to US $ 11.5 billion in 2014 (Gadelha & Costa, 2007; Gadhelha, Costa, & Maldonado, 2012).

The Brazilian National Policy guidelines state that all actions should strengthen innovation in the Brazilian health sector, and helps mitigate the R$ 12 billion deficits in drugs and medications in the Brazilian trade balance (Costa, Gadelha, & Maldonado, 2012). According to the Oslo Manual, published by the Organisation for Economic Cooperation and Development (OECD), there are four kinds of innovation: product innovation, process innovation, organizational innovation and marketing innovation, as well as a combination of any of the above. This means that not just knowing about, but monitoring scientific publications, patents and networks of authors and institutions may help a country to develop its science, technology and economic development (Magalhaes, Antunes, & Boechat, 2012). In this Brazilian case study, we will highlight product and process innovation opportunities for the list of medications used by the Unified Health System (Sistema Único de Saúde, SUS), thereby helping to reduce the trade balance, since innovation can impact even established and approved markets.

Concerns about the quality of life of the Brazilian population are on the government's agenda, with public health programs being established to foster healthy aging, linking economic development to improved sanitation conditions and, therefore, yielding more sustainable lifestyles. Consequently, the new applications on how to use the data adding information value to governments, companies and/or institutions, so that can make the right decision are increasingly needed in everyday life.

The Knowledge Management of the Scientific and Technological Information in Intellectual Property: A Case Study of Antiretrovirals

To help reduce the health sector trade deficit and foster technological innovation and development in Brazil, the government has, over the last ten years, spearheaded many actions and investments for different actors from universities, the business sector and government. Thus, the quality of information has been rethought in order to improve the health of the population (Castro, 2002). One of the key actions has been the policy to harness the technological competency and drug and medication production knowhow the country boasted in the mid 1980's (Antunes, & Magalhaes, 2008; Magalhaes, Antunes, & Boechat, 2012). This will encourage the development of the domestic economy, diminish the pharmaceuticals trade balance deficit and promote national science. Directives by the Ministry of Health in Brazil (strategic list for strategic medicines), is part of policy and is based on several key Brazilian issues, including the following provisions:

- Whereas the provisions of article 170 of the Federal Constitution, concerning economic order, founded on the importance of human labor and freedom of enterprise, with the purpose of guaranteeing a dignified existence for all, in accordance with the precepts of social justice;
- Whereas the provisions of article 197 of the Federal Constitution concerning the regulation of health actions and services:
- The fragility and dependency of the Brazilian health industry trade balance, which lacks significant international competitiveness, contributing to the vulnerability of social policies and with a serious impact on sanitation and the SUS budget;

- Observing that any pharmaceutical welfare, immunization programs and healthcare activities, including diagnoses, amongst others, should have their supply guaranteed irrespective of international market oscillations;
- Whereas the health sector plays a key role in orienting policies with an impact on Brazilian social and economic development;
- Whereas the Brazilian health sector has great potential for development, for which reason it is benefitted as a strategic sector in the Brazilian policy for industry, technology and foreign trade; and
- Whereas the cooperation and technical assistance agreement, signed by the Ministry of Health and the Brazilian Development Bank (Banco Nacional de Desenvolvimento Econômico e Social, BNDES) on December 5th, 2007, with the aim of introducing actions, programs and studies designed to develop the Brazilian health sector in Brazilian territory, by which the Ministry of Health is responsible for compiling a list of strategic products to subsidize BNDES in the support of profit-share operations, as set forth in its Health Sector Development Support Program (PROFARMA).

Theses directives establishes a list of strategic products for the SUS, with the purpose of contributing to the development of the health sector. In the same document, it states that this list will be reviewed every two years to keep at the forefront and guide the country towards the most innovative and strategic solutions for the good of the nation. All the actions pertaining to this matter are managed by the Secretariat for Science, Health and Strategic Inputs, which is directly answerable to the Minister of Health.

The aim of the list is not just to divulge the strategic products for the Ministry of Health, but to indicate to the country's main agents which products should be the subject of specific initiatives for boosting local production, for innovation and technology transfer, and mechanisms for their regulation, thereby providing strategic support for the Brazilian health sector. Examples of these agents are public[1] and private producers, regulatory and funding agencies, e.g. the public health regulatory agency (Agência Nacional de Vigilância Sanitária (ANVISA – Brazilian term), BNDES, and the science, technology and innovation funding agency (FINEP – Brazilian term).

In this sense, the antiretroviral (ARV) drugs are listed in the list of essential drugs of the Brazilian Ministry of Health. At each review of the list, new drugs are incorporated as well as some are excluded. The process of updating the strategic medicine list should be annual, in accordance with the recommendations issued by the Executive Group of the Industrial Health Complex (GECIS – Brazilian term), and coordinated by Science, Technology and Strategic Inputs Office - SCTIE with the other offices of the Ministry of Health. The other offices integrate GECIS and Entities of the Competitiveness Council of the sector (Brasil MCTI, 2012). The current list was published in March 2017, where it contemplates 56 strategic medicines (Brasil MCTI, 2017).

Antiretrovirals Case Study

As understanding of the life cycle of HIV has improved, it has led to a greater comprehension of how antiretrovirals can combat the infection. The drugs available on the market are designed to protect the immune system by suppressing HIV viral replication and, transmitting the spread of the virus to new cells. In theory, such treatments should allow the parts of the immune system that are still healthy to expel all the infected cells and wipe out the infection. However, due to certain factors, such as the virus's capacity to acquire resistance, the drugs used are only able to maintain the immune system of patients.

There are 28 drugs used to fight HIV, which are produced by several different companies, and a further 88 that are under development (Wilsdon, Attridge, Sauri, & Kirkpatrick, 2011). When combined correctly, the antiretroviral agents already on the market are capable of suppressing viral replication and maintaining viral levels so low that they are sometimes not detectable using regular blood tests.

With the dynamism of the sector and the resulting emergence of increasingly effective drugs, the treatment strategies available for patients have multiplied, offering increased tolerability, fewer side effects and more simplified dosing regimens. These developments have yielded good results, and it is estimated that the life expectancy of a person with HIV is now virtually the same as that of someone without the virus (Meiners-Chabin, Constance, Protopopescu, Camelia, Chauveau, Julien, & Moatti, 2013).

As new discoveries have been made about the morphology and life cycle of the virus, the drugs used to fight HIV/Aids have also developed, offering a wider range of treatments by new therapeutic classes, such as fusion inhibitors (FIs) and integrase inhibitors (INIs).

Figure 2 shows the emergence of the new classes of antiretrovirals over time, according to their mechanism of action.

From Figure 2, it can be seen that NRTIs were the first antiretrovirals on the market and the only drugs available between 1987 and 1995. Aids treatments started to be more effective since 1995, with the emergence of the first PIs and NNRTIs, which were thenceforth the target of consistent research. The results of this research led to the launch of 11 PIs, and three NNRTIs by 2006. In 2003, the first drugs designed to inhibit the virus's entry into host cells started to be produced. More recently, in 2012-13, a new class of drugs has been produced that are designed to inhibit the integration of the viral DNA with the host DNA.

Antiretroviral therapy is the combination of different kinds of drugs to reduce the likelihood of the virus multiplying in the organism. The combination of three or more drugs is more effective than the use of one (monotherapy). The combination of three or more drugs, also known as the HIV drug cocktail, has become the standard treatment for people who HIV-positive. So far, this is the treatment that offers

Figure 2. Launch of antiretrovirals over time per therapeutic class according to their mechanism of action
Source: Adapted and updated from Wilsdon, Attridge, Sauri, and Kirkpatrick, (2011).
Key: NRTI - nucleoside reverse transcriptase inhibitor; NNRTI - non-nucleoside reverse transcriptase inhibitor; PI - protease inhibitor; INI - integrase inhibitor; FI - fusion inhibitor
**Some authors mention zidovudine as early as 1986.*

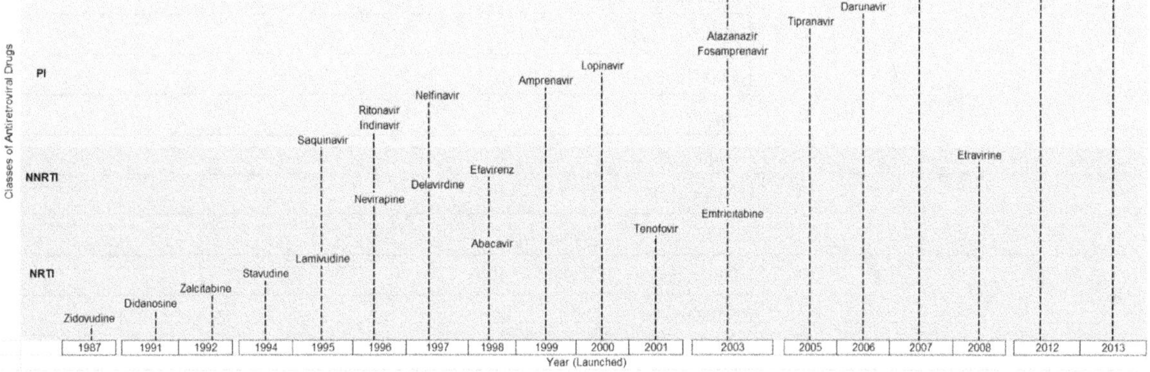

the best changes of preventing the virus from multiplying and keeping the patient's immune system working healthily.

In Brazil, the Department of STDs, AIDS, and Viral Hepatitis, under the aegis of the Disease Vigilance Secretariat, part of the Ministry of Health, is responsible for procuring and distributing drugs free of charge to patients with HIV.

The Brazilian government's policy of providing free access to antiretroviral drugs began in 1991 with the distribution of zidovudine to patients. The policy was boosted in 1996, when a law (#9313) was passed that assured people with HIV/Aids free access to the medicines essential for their treatment through the public health system. Although this sparked much debate and even criticism in the early years, the success of the national HIV/Aids program has gained recognition as an international benchmark in the treatment of the infection/disease. Its benefits include the improved quality of life of patients to greater control of the epidemic in the country (Grangeiro, Teixeira, Bastos, & Teixeira, 2006).

Today, there are 20 antiretrovirals used in Brazil to treat HIV that are listed on the national list of essential medications (Relação Nacional de Medicamentos Essenciais, RENAME), issued by the Ministry of Health. Six are NRTIs, three are NNRTIs, seven are PIs, two are INIs, and there are two FI (see Table 1).

This study is designed to present an overview of world patenting for drugs used in HIV/Aids treatment in Brazil. The temporal evolution of patent applications for different classes of drugs is described,

Table 1. Prescriptions medicines used in anti-HIV therapy in Brazil by class

Class	Medicines
NRTI	Tenofovir
	Abacavir
	Lamivudine
	Stavudine
	Didanosine
	Zidovudine
NNRTI	Etravirine
	Efavirenz
	Nevirapine
PI	Darunavir
	Tipranavir
	Atazanavir
	Fosamprenavir
	Lopinavir
	Ritonavir
	Saquinavir
FI	Enfuvirtide
	Maraviroc
INI	Dolutegravir
	Raltegravir

Source: Own research based on RENAME list.

identifying the main countries of origin of the technologies, the main countries where patent applications are filed, the main technological areas, and the most representative patent holders in numerical terms.

Given that this research is based on information contained in patent documents for drugs and medicines filed around the world, the database chosen to access this data was SciFinder Scholar®, from the Chemical Abstracts Service (CAS, a division of the American Chemical Society), which is accessible in Brazil through the CAPES periodicals portal.[2] This database contains patent documents filed since 1907, with the main patent offices around the world. The main bonus of using SciFinder Scholar® for a study of this kind is that searches can be made based on the molecules, using the CAS RN (CAS Registry Number), or any synonyms of the drugs.

The database was mined in November 2017, using the names of the drugs of each of the five classes. The data retrieved were then exported to VantagePoint® software for treatment and analysis.

The first stage of the data treatment was to remove duplicate documents, using the *remove duplicate records* function. This was done by using the numbers of the priority applications, so that just one patent document remained from each patent family.[3] Next, the number of patent documents belonging exclusively to each class was identified (see Table 2).

The temporal evolution of the patent applications per priority year is shown in Figure 3. There is an upward trend in the number of patents for all classes of antiretrovirals. The highest number of patents is for NRTIs, which were the first class of antiretrovirals developed to treat HIV. It was only much later, in 2000, that the patents for integrase inhibitors and fusion inhibitors increases, which is why the total number of patents for these classes is lower.

The first application for a patent, or priority patent, tends to be filed in the country where the technology is developed, so the country of origin of a technology in a claim can be inferred from the priority country. The analysis revealed that the United States was the country where around 50% of the priority patent applications for antiretrovirals were filed, as shown in Table 3.

The technologies described in the claims from the patents for antiretrovirals were obtained from the first International Patent Classification (IPC) code cited in the documents. The IPC was established in the Strasbourg Agreement (1971) to provide a standard method of organizing the claims in patent documents to facilitate access to the information they contain, and to serve as a tool for searching and retrieving these documents (Wipo, 2017). Every year, the IPC is reviewed, taking into account new information access and retrieval methods, incorporating new technologies and eliminating errors when

Table 2. Total of documents retrieved by antiretroviral class

Class	Active Pharmaceutical Ingredient	Number of Different Patent Documents	Number of Patent Documents for Just One Class of Patents
NRTI	Tenofovir, Abacavir, Lamivudine, Stavudine, Didanosine, and Zidovudine	3,167	2,175
NNRTI	Etravirine, Efavirenz, and Nevirapine	1,097	256
IP	Darunavir, Tipranavir, Atazanavir, Fosamprenavir, Lopinavir, Ritonavir, and Saquinavir	1,655	732
IF	Enfuvirtide and Maraviroc	733	186
INI	Dolutegravir and Raltegravir	309	106
Total of different documents		4,554	3,455

Source: Own research using SciFinder Scholar®.

Figure 3. Temporal evolution of patent applications per class of antiretrovirals
Source: Own research using SciFinder Scholar®; figure produced using VantagePoint®.

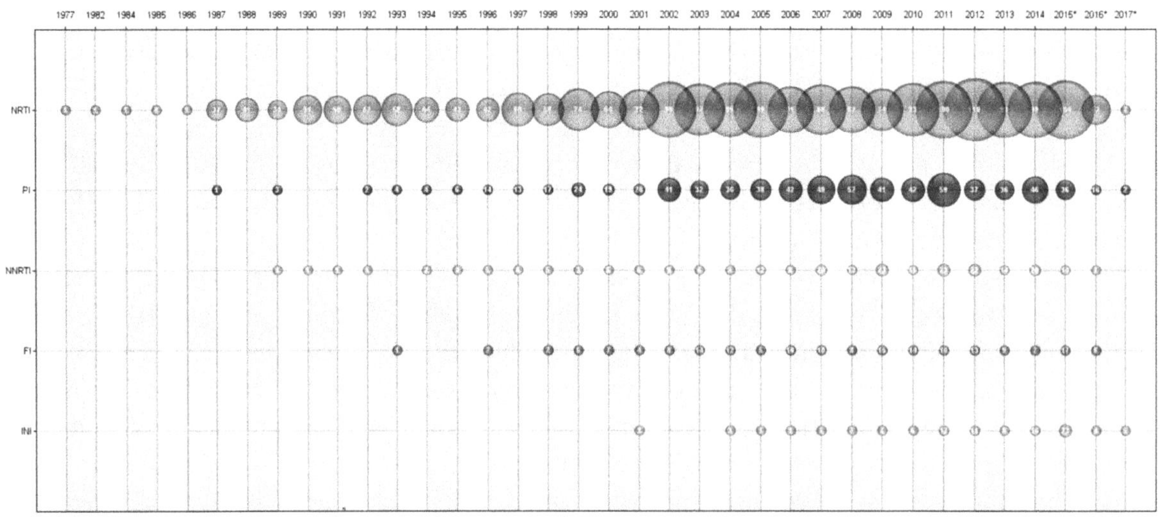

Table 3. Main priority countries per class of antiretroviral

Priority Country	Number of Patent Documents per Class of Antiretroviral				
	NRTI	PI	NNRTI	FI	INI
USA	990	391	107	113	49
China	466	80	56	12	9
India	98	103	42	6	28
Europe (EPO)	84	58	21	17	11
United Kingdom	93	15	5	10	0
Germany	90	21	5	1	1
Japan	52	9	5	0	2
South Korea	50	8	1	17	0
France	53	10	1	0	1
Australia	43	2	2	2	2

Source: Own research based on patent documents retrieved using SciFinder Scholar®.

applicable. It uses a taxonomic structure of sections that are subdivided down into classes, subclasses, groups, and subgroups.

Figure 4 shows the sections under which the patent documents retrieved in this study are classified (A, C, and G), and the main classes and subclasses. As can be seen, the most frequent classification is A61K, which includes preparations for medical purposes. The second most common classification is C07, with 717 patent applications, covering claims for the synthesis of pharmaceutical compounds.

The analysis of the actors involved in the research and development of the technologies cited in patent applications for just one class of antiretroviral revealed the involvement of a wide range of market players. Table 4 shows the patent holders with over 15 patent applications and the classes of antiretrovirals

Figure 4. Technology areas of patent documents citing only one class of antiretroviral[4]
Source: Own research using Vantage Point®.

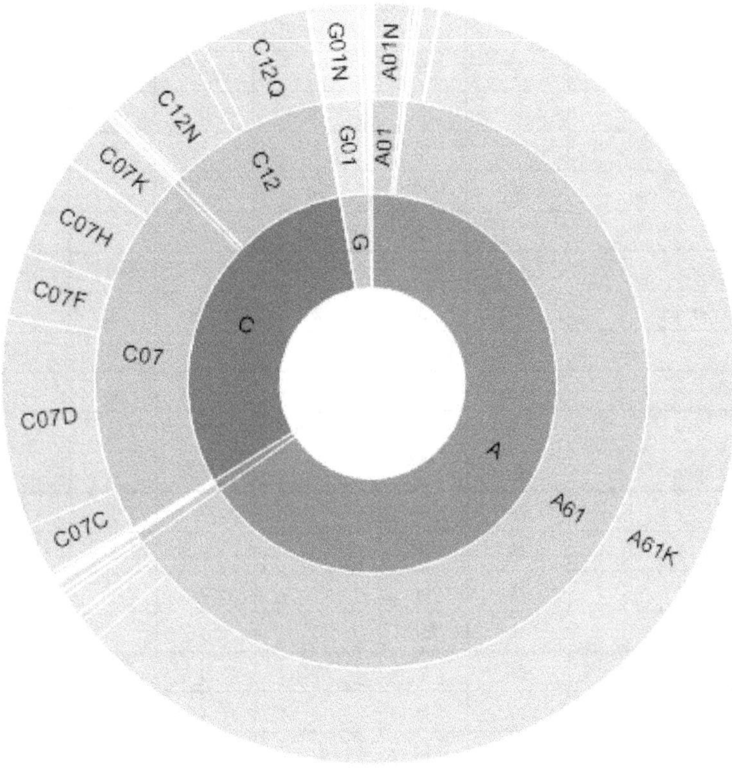

concerned. Two American laboratories, Bristol-Myers and Abbott, are the most active in patenting in this area, with 57 and 56 patent applications, respectively. While Bristol-Myers's claims are mostly for NRTIs, Abbott's are primarily for protease inhibitors. Another curious case is ViiV Healthcare, a company specialized in HIV that has filed 16 patent applications, all for integrase inhibitors.

CONCLUSION

The contribution of information science at the turn of the 21st century has radically changed the way scientific and technological knowledge is addressed, including in intellectual property. There are over 100 million patents filed with the EPO, and it is a great challenge to manage this data and extract useful information from it. Knowledge management, using IT tools, could become a great ally in helping decision-makers in the private and public sectors alike.

In Brazil, public policies could be assisted by an information system that enabled the identification, extraction and management of the technological know-how contained in patents. It would be a useful decision-making tool and help improve healthcare provision for the population.

The intensity with which companies and scientists devote their energy to developing and filing patents to protect their innovations demonstrates how strategically important certain technologies are. In this sense, by observing the progress of certain products over the years, it is possible to deduce the

Table 4. Applicants with over 15 patent applications

Patent Holders	NRTI	PI	NNRTI	FI	INI	No. of Patent Documents
Bristol-Myers Co	35	10	11	0	1	57
Abbott Lab	2	53	1	0	0	56
Glaxo Group Ltd	37	6	1	0	6	50
Gilead Sci Inc	33	11	4	0	0	48
Tibotec Pharm Ltd	9	27	10	1	0	47
Merck & Co Inc	20	7	9	0	9	45
Ranbaxy Lab Ltd	13	21	5	0	0	39
Us Dept Health & Human Services	27	5	2	3	1	38
Janssen Pharma	4	23	10	0	0	37
Hoffmann La Roche Inc	13	14	1	8	0	36
Hetero Res Found	9	12	7	2	4	34
Boehringer Ingelheim Int Gmbh	0	25	8	0	0	33
Cent Nat Rech Sci	29	3	0	0	1	33
Cipla Ltd	18	9	2	0	2	31
Matrix Lab Ltd	11	11	5	2	1	30
Enanta Pharm Inc	0	27	0	0	0	27
Shanghai Desano Chem Pharm Co Ltd	11	7	7	0	0	25
Univ California	19	0	1	4	1	25
Novartis Ag	10	14	0	0	0	24
Wellcome Found Ltd	24	0	0	0	0	24
Mylan Lab Ltd	7	6	3	0	7	23
Teva Pharm Inc	2	4	6	1	10	23
Vertex Pharm Inc	3	19	0	0	0	22
Aurobindo Pharma Ltd	7	8	1	0	4	20
Smithkline Beecham Corp	13	2	0	0	5	20
Inst Pharm & Toxicology Acad Military	15	2	0	2	0	19
Pharmacia & Upjohn Co	5	13	0	0	0	18
Univ Emory	16	2	0	0	0	18
Lupin Ltd	7	8	1	0	1	17
Univ Fudan	13	1	2	1	0	17
Laurus Labs Private Ltd	6	2	4	0	4	16
Viiv Healthcare Co	0	0	0	0	16	16

Source: Own research based on information from patent documents retrieved using SciFinder Scholar®.

technological importance of the products/processes in question, which indicates where most effort is concentrated. However, technologies that are not protected in each country or which expire may then be produced generically and incremental innovations may be made, etc.

In the example given to demonstrate how patent documents can serve as a source of information, it was possible to identify some common indicators in an analysis of patenting per class of antiretrovirals

– the temporal evolution of patenting, the priority countries and the main patent holders. This analysis indicates the emergence of new classes of drugs for HIV/Aids treatment, the new companies, universities and research centers developing these technologies, and the overwhelming concentration of technological developments in the United States and China. As for the content of these documents, inventions related to new formulations and new synthesis pathways are the main claims.

REFERENCES

Antunes & Magalhaes. (Eds.). (2008). Patenteamento e Prospecação Tecnológica no Setor Farmacêutico. Interciencia.

Balmisse, G. (n.d.). *Guide des Outils du Knowledge Management: panorama, choix et mise en oeuvre*. Academic Press.

BRASIL. (2017). *Estratégia Nacional de Ciência, Tecnologia e Inovação 2012-2015*. Ministério da Ciência, Tecnologia e Inovação (MCTI). Disponível em: <http://www.mct.gov.br/upd_blob/0218/218981.pdf>

Brouwers, M. C., Kho, M. E., Browman, G. P., Burgers, J. S., Cluzeau, F., Feder, G., ... Zitzelsberger, L. (2010, December 14). AGREE II: Advancing guideline development, reporting and evaluation in health care. *Canadian Medical Association Journal, 182*(18), E839–E842. doi:10.1503/cmaj.090449 PMID:20603348

Canadian Institutes Of Health Research. (n.d.). *Knowledge Translation and Commercialiation*. Disponível em: http://www.cihr-irsc.gc.ca/e/29418.html#1

Carvalho, A. I., Bodstein, R. C., Hartz, Z., & Matida, Á. H. (2004, September). Concepts and approaches in the evaluation of health promotion. *Ciência & Saúde Coletiva, 9*(3), 521–529. doi:10.1590/S1413-81232004000300002

Castro, A. L. S. (2002). O valor da informação: um desafio permanente. *Revista de Ciência da Informa, 3*(3).

Costa, L. S., Gadelha, C. A. G., & Maldonado, J. (2012). A perspectiva territorial da inovação em saúde: a necessidade de um novo enfoque. Revista de Saúde Pública.

Gadelha, C. A. G., & Costa, L. (2007). Integração de fronteiras: A saúde no contexto de uma política nacional de desenvolvimento. *Cadernos de Saude Publica, 23*(suppl 2), S214–S226. doi:10.1590/S0102-311X2007001400011 PMID:17625648

Gadelha, C. A. G., Costa, L. S., & Maldonado, J. (2012). O Complexo Econômico-Industrial da Saúde e a dimensão social e econômica do desenvolvimento. Revista de Saúde Pública.

Gadelha, P., Carvalho, J. N., & Pereira, T. R. (2012). *A Saúde no Brasil em 2030: diretrizes para a prospecção estratégica do sistema de saúde brasileiro*. Editora FIOCRUZ.

Hartz, Z. (2013, May). From monitoring of performance to the performance of monitoring: New opportunities for evaluation in the management of Health Surveillance. *Ciência & Saúde Coletiva, 18*(5), 1221–1222. doi:10.1590/S1413-81232013000500005 PMID:23670448

IBGE. (2013). *Instituto Brasileiro de Geografia e Estatística*. Brazil: IBGE. Disponível em: http://www.ibge.gov.br/home/estatistica/populacao/tabuadevida/2010/

Le Moigne. (n.d.). *La théorie du Système Géneral: théorie de la modélisation*. Academic Press.

Lynch, C. (2008, September 4). Big data: How do your data grow? *Nature*, *455*(7209), 28–29. doi:10.1038/455028a PMID:18769419

Magalhaes. (2012). *Technological Trends in the Pharmaceutical Industry: the matter of neglected tropical diseases – An overview of the Research, Development & Innovation in Brazil*. Synergia Editora.

Magalhaes. (2015). *Percepção do Valor da Informação por meio da Inteligência Competitiva 2.0 e do Big Data em Saúde in Análise da Informação para tomada de decisão - desafios e soluções* (Vol. 1). Editora Intersaberes.

Magalhães, J. L., Antunes, A., & Boechat, N. (2011). Laboratórios farmacêuticos oficiais e sua relevância para saúde pública do Brasil. *Revista Eletrônica de Comunicação, Informação & Inovação em Saúde*, *5*(1).

Meiners-Chabin. Protopopescu, Chauveau, & Moatti. (2013). Estrutura de mercado e evolução dos preços dos medicamentos antirretrovirais no Brasil. In Propriedade intelectual e políticas públicas para o acesso aos antirretrovirais nos países do Sul. Rio de Janeiro: E-papers.

Miller, J. P. (2000). *Millenium Intelligence: Understanding and Conducting Competitive Intelligence in the Digital Age*. Information Today, Inc.

Morin, E. (1982). Le système: Paradigme ou/et théorie. Science Avec Conscience.

Pesqueux, Y., & Ferrary, M. (2011). Management de la connaissance. *Economica*.

Piaget, J., & Garcia, R. (1983). Article. *Psychogenèse et Histoire des Sciences*, *19*(1), 304.

Pierret, J.-D. (2006). *Methodologie et structuration d'un outil de decouverte de connaissances base sur la litterture biomedicale : une application basee sur le MeSH*. Université du Sud Toulon Var.

Pimjai Sudsawad, S. (n.d.). *Center on Knowledge Translation for Disability and Rehabilitation Research (KTDRR) KT Library - Research Quality*. Disponível em: http://ktdrr.org/ktlibrary/articles_pubs/ktmodels/

Quoniam, L. (2011). *Competitive Intelligence 2.0*. ISTE Ltd and John Wiley & Sons Inc. doi:10.1002/9781118603079

WHO. (2006). *Bridging the "Know-Do" Gap: meeting on knowledge translation in Global Health*. WHO. Disponível em: https://www.measureevaluation.org/resources/training/materials/high-impact-research-training-curricula/bridging-the-know-do-gap.pdf

WHO. (2013). *TDR Performance assessment framework*. Special Programme for Research and Training in Tropical Diseases (TDR). Disponível em: http://www.who.int/tdr/publications/about-tdr/strategy/framework/en/

Wilsdon, T., Attridge, J., Sauri, L., & Kirkpatrick, H. (2011). *Evidence on access to essential medicines for the treatment of HIV/AIDS. Charles River Associates*. CRA.

Wipo. (2017). *Guide to the International Patent Classification.* Retrieved from http://www.wipo.int/export/sites/www/classifications/ipc/en/guide/guide_ipc.pdf

Zicker, F. (2007). *TDR'S Research and Capacity Development Strategy: 30 Years of Research and Capacity Building in Tropical Diseases*. WHO. Disponível em: http://www.edctp.org/forum2007/presentations/Workshop_on_Enhancing_capacity_-_Fabio_Zicker.pdf

ENDNOTES

1 In Brazil, there are national and transnational private and public drugs laboratories. The public ones are called Official Pharmaceuticals Laboratories and are maintained by the federal or state governments. There are 21 such official laboratories in the country (Magalhaes, Antunes, & Boechat, 2011).

2 Coordenação de Aperfeiçoamento de Pessoal de Nível Superior, a Brazilian federal government agency, under the Education Ministry.

3 A patent family is a set of patents filed in different country for the same claim. They are all filed under the same priority number and the priority rights are granted to the original inventor (SOARES & CORREA, 2010).

4 All the descriptions of the classifications can be found at http://web2.wipo.int/classifications/ipc/ipcpub

Chapter 14
Organizational Urbanism: A New Proposal for Information Systems Management

Pedro Fernandes da Anunciação
Instituto Politécnico de Setúbal, Portugal

Francisco Madeira Esteves
Instituto Politécnico de Setúbal, Portugal

Fernando Miguel Gonçalves
Instituto Politécnico de Setúbal, Portugal

ABSTRACT

The challenges of the information society require frequent strategic reorientations that, in the context of the relational economy, emphasize the importance of IS/ICT in adapting to change, competitiveness, and economic sustainability. The easy adoption of the new technologies by the people, the paradigm of the availability of the new distribution channels, the reduction of the time-to-market, among other examples, have demanded to the responsible ones of the organizations and the SI dynamic of management, and adequate framework of the stated challenges. The preparation and management of the associated changes presuppose a coordinated and coherent combination of four vital axes for organizational success: business performance, information management, systems architecture and information technologies, and relational positioning. In this area, the development of an urban view of the value chain and its participants and the adoption of a logistics perspective on the momentum appear to be the primary adjustment in the support of organizational models to business.

INTRODUCTION

The contemporary world is a vast interactive system in which economic organizations seek permanently a balance between the economic, technological and social forces. This balance, which is difficult to achieve, is a permanent challenge for the organization`s management. Technological dimensions, silently and seemingly, harmlessly infiltrate the structures and modus operandi of organizations, generating new

DOI: 10.4018/978-1-5225-6225-2.ch014

management challenges associated with innovation, dynamism, interaction, integration and volatility of activities and agents.

It is a permanent context of turbulence where the continuity and sustainability of organizations are assessed by the weakness of existing certainties regarding the stable maintenance of conditions or market factors. Economic uncertainties put pressure on management to seek responses to change requirements, leading to the search for new models and/or more appropriate instruments or adjustment of existing models. Adjustments are the easiest and immediate solution to combat market volatility, minimizing the costs of change. However, these adjustments often increase the complexity of organizations, especially as their positioning involves greater participation in the economic networks. This participation requires the integration of activities, the sharing of strategies and resources, the adequacy of information systems and technologies and, above all, the enhancement of commitments.

The current economic system requires collaboration and cooperation between various stakeholders and emphasizes the importance of creating common platforms for the integration of different systems and for the joint development of approaches to the market (functional, informational and technological) (Anunciação, 2015; Anunciação & Zorrinho, 2006; Anunciação, 2004). Some of the factors that reflect this new paradigm are evidenced by the current economic drivers, which are associated with different levels of management and organizational difficulties, namely: business globalization, digital services anywhere and anytime, mobile and distributed workforce, new computational tools, outsourcing activities, e-Commerce, automation of operations, strategic use of ICT to obtain competitive advantages, extension of organizations as an organizational and operational model, among others (Benbasat & DeSanctis, 2000).

These key drivers reflect new standards for how organizations operate and, consequently, new levels of efficiency and effectiveness. Organizational adaptation does not seem easy. The opportunities offered by innovation and information systems and technologies have changed the traditional logic of business, allowing organizations to experience a wide range of strategic alternatives and organizational forms (Venkatraman & Henderson, 1998). Organizations, as they integrate economic chains that grow and multiply, constitute links between them, increasing their complexity. In these chains, technological integrations play an important role, insofar as it they enable an economic, commercial and informational approach between the different agents. The participation of companies in collaborative networks influences, conditions and reshapes the respective assets and flows.

The dynamics of the flows, the different structures and forms of organization originate and influence asymmetric competitive behaviors in the networks in which they participate (Gnyawali & Madhavan, 2001). This asymmetry requires organizational and structural flexibility and high standards of competence in the management of the various change projects with a view to facilitate the adaptation to the market, which is required quickly and safely.

The competitiveness of any company is always the result of the ability to build and operate the right mix of resources. A competitive advantage will only be sustainable when, through the continuous development of new capacities, skills and organizational resources, is possible to generate market value by changing the respective conditions (Empson, 2000). In this context, it is necessary that, in the current relational context, organizations seek to adjust management models in order to optimize commercial responses to market needs. Whatever the conceived model, it must anticipate an integration of organizational resources and collective capacities in close articulation with market needs. It is important to keep in mind that these needs are increasingly global, that businesses and consumers communicate and transact products and services in high speed networks, anytime, anywhere. This scenario is a challenge for any manager.

THE POTENTIAL OF THE URBANISTIC APPROACH TO INFORMATION SYSTEMS

Information systems, based on the new information technologies available and adopted by strategic imperatives, constitute a model of representation of organizations and business that, consciously or unconsciously, condition the perception of the respective operations and objectives (Zorrinho, 1991). Their impact goes beyond the internal scope of the organizations and their management, allowing to stimulate the functioning, organization and strategic development of agents and businesses.

This expansion of organizational and economic activities is directly associated with information and its assumption as the main economic asset. Its importance is also reflected at the organizational level, where information also assumes itself as an important organizational asset comparable to any other traditional asset. It should be noted that in the corporate assets universe, information is particularly important insofar as it covers all organizational areas, integrates all information systems necessary to support operations and supports all existing management functions (Anunciação & Esteves, 2014). For its value and relevance, it is important to ensure that the systems that support their treatment and storage are duly architected, managed and safeguarded.

Information systems must support the economic and commercial activities (transactional systems) as well as the quality of decisions (decision support systems), including the strategy, which depends on the organization's own future development (Anunciação & Esteves, 2013). The importance of information systems in the support of organizational activity context stresses the inevitability of management responsibility in understanding the universe or organizational information space inherent in the exercise of their economic activities.

The understanding of the universe of information space depends on the existence of an architectural reference. Architecture is the concept that has best demonstrated the role of information systems in supporting organizational activities and decision-making. Only through architecture, it is possible to understand the level of support of information systems to internal operations, the adequacy of information to the requirements of decisions at different levels of management, the options taken at the technological level, the adequacy of economic performance to the defined objectives, among other important aspects. But it is mainly through architecture that it is possible to assess the adequacy of the requirements of the processing, storage and provision of information to the information space.

However, this concept has evolved from the set of economic and market conditions through, which is evident, either an internal or external context of economic organizations, an either relational or collaborative dimension. In the actual context of society and relational economy, the concept of architecture appears associated with the concept of organizational and information systems Urbanism (Anunciação & Zorrinho, 2006; Zorrinho & Anunciação, 2004; Anunciação, 2004; Anunciação, 1997). This concept of Urbanism of information systems represents a significant analytical and qualitative evolution compared to the concept of architecture. The need of a new relational dimension to the organizational and information systems requires the coherent integration of different systems and, consequently, a new perspective of economic dynamics and dependence about different economic agents.

In an Urbanistic approach, the presence of various players and different systems is evident and the need of a referential activity and anchorage points is critical to support and stabilize economic and organizational complex information systems (Anunciação & Zorrinho, 2006; Zorrinho & Anunciação, 2004; Anunciação, 1997).

It is true that in the economic field information technologies have brought a number of significant benefits to organizations, such as facilitating access to data and information when necessary, whatever the channel, the place or the time, among others aspects. However, this reality becomes a little more complicated when it fits the participation of different economic agents, organizational realities and different systems in a collective perspective of organizational and economic functioning (Domínguez & Garrido, 2010).

The current management is more complex now than in the past. On the one hand, it is necessary a bigger scope and coverage analysis relative to the market, competitors, customers, etc. On the other hand, increases the specificity and depth of analysis, in the identification of market segments, competitive products, the customer preferences, etc. In each of management levels, quality of information is required and its availability at the time of decision making was a critical success factor, according to the specificity of their duties. The development of economic organizations tends to be supported on knowledge operating based logic and the main task of knowledge management is to create an organizational context that encourages the development of new knowledge through exploratory learning (Anunciação, 2015; Gonzalez & Martins, 2014; March, 1991).

Managers focus their analysis only on relevant information and knowledge, adopting technological solutions, such as Business Intelligence, to increase the degree of detail about their goals, objectives and performance. With the Big Data phenomenon, given the exponential growth of data and information, the challenge for organizations corresponds to obtaining qualitative information with economic value to the organization. It is based on information, knowledge and good architectures that more complex organizations seek to adjust the organizational performance to competitive requirements through optimization of the levels of efficiency and effectiveness in the performance of their activities.

THE RELATIONAL DIMENSION AS A SUSTAINABILITY FACTOR IN INFORMATION SOCIETY

Actually, the competition of economic organizations is not in the ability to produce and transform raw materials to market or in a privilege technological position or good financial resources. Competitive advantage is created with the ability to generate new knowledge, as well as in the agility and versatility to information processing and transmission in a relational and collaborative action context. The network concept in the economic sphere, more or less evident to all economic environments and actors, demand that established relationships between organizations and rigor and commitment in the economic relations of all those who are internally set in each of the organizations. In this context, it is necessary to define structures of network organizations, specify the set of relations between the different actors and conceptualize a new collective perspective about the shared activities (Anunciação & Zorrinho, 2006). This conceptualization requires the optimization of resources, the improvement of processes and organizational activities and the perception of the new required levels of efficiency and effectiveness to the participation in the collective activities.

These relationships and connections can't be seen as a set of sporadic or occasional transactions that will settle or consolidate over time by their peers, but, rather, should be framed on organizational principles taking into account technical constraints and management. The solution, which must meet the economic requirements, should allow companies or business groups to develop diversification strategies related and extensive vertical integration. The development of collective activities between various strategic business units, companies or divisions, should build synergies earnings by sharing among themselves, resource, information, systems, taking each of the parties to share in the respective strategic and operational coordination of activities undertaken responsibilities. All organizations that position themselves in organizational networks play both a competitive and cooperative role (Venkatraman & Henderson, 1998).

In this area, Christensen and Overdorf (2000) report that the organizational adjustment should focus, in general, three factors: resources, processes and values. Regarding the first, most managers seek answers from their own resources. However, the possibility of access to other resources through relationships with other economic stakeholders (suppliers, distributors, customers, etc), necessarily leads to organizational adjustments accordingly. With regard to processes, understanding them as the patterns of interaction, coordination, communication and decision used to transform resources into products and services, involves the expression of the joint implementation of capacity on an ongoing basis tasks with full awareness and knowledge of the interdependence logic and responsibility associated with each participation and participant. Values, the third factor, refer to the set of organizational principles to be implemented and promoted by management to guide their organizational and business practices, constituting the foundations of functioning, image and organizational behaviour among all those which interact. Their importance is directly related to the practices and organizational decisions. The bigger, larger and more complex the value chain for the production of products or the provision of the most important services, the bigger the importance of the definition of a consistent set of values in supporting the strategic directions and organizational model.

The network structures are a form of response of business to increase the complexity of their strategic guidelines (Freire, 1997). These structures, that normally develop when there is a wide range of integrated operations under the jurisdiction of an organization that leads the activities and objectives, seek to bring together a number of external companies and internal units that complement each other in achieving goals or common objectives which may correspond to the development, production and marketing of products and/or services. The various stakeholders are no longer seen as a set of units or different elements, hierarchically defined by levels of management or chains of command with their own autonomy and well-defined borders, but as integral components of larger organizational units, an organization (economic) wider and more dynamic, which generates and imposes new performance rates. Their structures and strategies are no longer isolated and are now interrelated between different organizations, sharing and participating in common objectives and giving rise to more significant economic dimension structures. The network value is the number of connections that can be established between members and increases in proportion to the number of people or organizations that can use it (McAfee & Oliveau, 2002).

The impacts of these new organizational forms have not been felt only in large organizations or groups of companies. With economic development and, specially, with the support of information and communication technologies, including through the Internet, there has been a demand or adoption of such structure, also by small organizations. Generally, these small businesses operate in well-defined market niches, subcontracting the most operations to other entities that, through these networks, allow the combination of resources and the focus of competencies in their core business. This enables, through a limited number of functions, the generation of synergies and the achievement of higher levels of productivity and efficiency. To obtain synergies, Karnani (2000) states that it is necessary that different economic sectors or organizations to coordinate and work together. He adds: *Synergy is like Heaven: there are more people who speak it than to actually go there to stop.*

Stacey (1993) mentions that integration versus differentiation is one of three elements that most concern the management of organizations in this context. The concerns of management, regarding the need for integration, understood as the ability to sync and control models, methods and user interfaces, support the full coordination of activities and organizations, resulting from the growing need of the business world interconnection. Increased competition, the emergence of new products, technological differentiation, the crushing of trade margins, among other factors, have led a trend towards outsourcing of activities set that do not belong to the core business and looking for partnerships. This new economic approach resulting therefore needs of interconnection and integration of activities, processes and information systems. The need for activities harmonization in the operations of the various economic agents results from an effort of integration, more and more universal, which gives rise to new requirements in the organizational domain and their support systems, in order to generate response capabilities through the intersection activities, procedures and objectives.

This economic and organizational complexity and the organizational difficulties, materialized in a network share, grow to the same extent of the requirements related to the integration of activities in the various value chains, assuming a share of objectives, strategies, resources, information, systems, technologies and, especially, commitments a critical success factor. In this economic context, with reference to the key axes indicated by Massey, Wheeler and Keen (2000), in particular the ubiquity, mobility, complexity and interactivity, make possible to think the organization far beyond the formal and singular perspective. Necessarily, it is going to fit it in a plural and systemic perspective, emphasizing the logic of order, organization and discipline inserted in a broader context of economic and trade interrelationship. It is also important to know to frame and articulate a coherent and harmonious way the various components of the organizational systems in accordance with existing challenges, not only in functional scope, as well as in information and technology. This requires a shared attitude of cooperation between the various entities, internal and external, in order to create a common platform for the integration of different systems (Krovi, 2001), in order to develop an appropriate management for the articulation of the three systems organizational support (functional, informational and technological).

All these challenges "propose" a careful management and, above all, a capacity for analysis and understanding through learning. The needs of efficiency and effectiveness of "know-how" organization require a constant search for a balance between technological innovation and economic and organizational development. To this end, the systemic requirements of design and harmonious organization of the organizational system in a relational context, through its functional, informational and technological subsystems, require appropriate models to emerging challenges and a deep knowledge of them in order to envisage the development of business activities (s) and organization (s), avoiding a specific logic characteristic treatment of the activities and challenges that arise. Knowing that many of the profound change initiatives or large scale often collapse under the weight of their complexity, it is able to manage and govern the systems so that they are not stuck in the principles of the industrial age, preventing new developments (Manville & Ober, 2003; Hirschhorn, 2002).

The perception of the diversity of factors that more, directly or indirectly, influence this new relational paradigm can be expressed, among others, through the drivers that affecting businesses and require strategic responses (regulatory environment, competitive threats, globalization, customer service, product quality, market share, competitive position, revenues and costs of goods and services, redesign process of business, power of end-users, flattening of the organization and time-to-market, among others); the organizational strategies (one single product, rapid development, minimization of the initial cost and continuous improvement, among others); and the process approach (multi-product distribution, economies of scale, quality and minimization of the costs of adaptation, among others).

All relational changes are complex. New forms of collaboration exert pressures that destabilize the traditionally established routines, their structures and put an extra charge on the development and control mechanisms necessary for adaptation to these situations. Are these new forms of collaboration that have been overcome borders that traditionally delimited functional areas of organizations and even markets, requiring a sharp focus on generating knowledge. The quality of individual performance, traditionally based on planned tasks and accumulated experience, is now focused on the quality of interactions and communications, shared experience and coordination established between the different actors. Organizations collaborate with others towards the joint future construction. This is especially true when we consider the organizational participation in the networked economy (Vandenbosch & Dawar, 2002).

In this context of economic performance, hardly traditional organizational structures will have the capacity to respond to challenges that currently organizations are subject. So, this kind of organizations, will feel the need to gradually create structures that share more democratic values, with greater sharing of responsibilities, autonomy, etc., as opposed to traditional lines more totalitarian. Krog and Cusumano (2001) report the existence of three strategies to manage rapid growth, which increase the organizational complexity and, therefore, the response of difficulty of traditional structures: scaling, duplication and granulation. Many companies have been aware that a periodic overhaul of hierarchical model, which inevitably becomes obsolete in rapidly changing environments, is a losing battle. In management, flexibility must supplant the structure as a principle of governance, just as the organization's design, or its physiology, must supplant the anatomy.

The level of excellence in the management of organizations is the effectiveness and timeliness to market answers. In both capacities, performance and responsiveness are at the core of organizational competencies, which requires timely or "real time" posture. All these new challenges, questioning the way of "knowing" and "doing" things, require the development of new forms of organization, management styles and innovative approaches to change. These, of course, will influence the development of business organizations, whatever the sector of activity in which they are.

The drivers that affect the functioning of the activities supported by new information technologies in relational context are: intensity information, customization, electronic distribution, aggregation effect, operating costs, real-time interface, risk sharing, effects network, standardization and loss of skills (Andal-Ancion, Cartwright & Yip, 2003).

The value created by the intra and inter-organizational collaboration resulted in benefits earned in five main categories (Hansen, Nohria & Tierney, 2004): reduction in costs resulting from the adoption of best practices; improving the decision-making as a result of access to information and knowledge of other entities; increase in income and benefits through sharing of experience and products between the various participants; innovation by combining and pollination of ideas; and increased capacity for collective action resulting from the collaboration of dispersed units.

If the creation of value cannot be done individually, then it is obvious that organizations require collaborative work at some level (Nunamaker, Briggs & Vreede 2000).

A NEW PROPOSAL OF ORGANIZATIONAL URBANISM MODEL

The model presented by Anunciação and Zorrinho (2006) for organizational Urbanism integrates two Submodels. A director model and a detailed model. The director model, also called Metavision, integrates a proposal for organizational management and information systems in complex projects. The detailed model, also known as MRAFITU (Model of Relational Alignment for Functional, Informational and Technological Urbanism), integrates a more specific intervention proposal through a close articulation between the business, information systems and information.

However, from the analysis of the nature and focus of the two models, it is considered that there is a limitation associated with the adequate treatment of the different realities existing or generated within an economic and informational relational context.

The first model explores very well the perspective of the roadmap or the project to be developed when it is intended to initiate an urban analysis of the information systems of the organizations that position themselves in the collaborative or relational domain. And the second provides a very specific analysis of the articulation between activities (business), functionalities (IS) and operations (ICT).

However, it is considered that, between both models, there is a need to define other dimensions of analysis related to information 'logistics'. This new dimension should emphasize a more dynamic aspect associated with the concept of Urbanism, for example, who are the users and the users of information space, who need information for the development of their activities and responsibilities, or who are the information providers that add information to the 'information space'.

Returning to the previous models and as previously mentioned, we have the master model and the detail model. The director model allows ordering and sequencing of steps, activities, critical success factors, risks and responsibility. This director model must support a methodology of intervention to ensure the success of the efforts. This model, designated Metavision, constitutes a management benchmark and intervention plan that incorporates a methodology for managing change from organizational information systems.

The proposed Metavision model, as shown in Figure 1, has the following stages:

1. **Vision:** Predefine the determinants of central guidelines and objectives to be achieved by the organization;
2. **Task Force:** Selection of a "pivot" team to pilot the change process, including all elements of the organizational structure for validation or reformatting the vision and process monitoring;
3. **Critical Success Factors (CSFs):** Identifying the critical success factors in the change process;
4. **Urbanism:** Definition of organizational and informational representations that meet the objectives of change and adhere to the need to respond to the critical success factors identified;
5. **Architecture:** Dynamic comparison between the functional, informational and technological architecture and the urban dimensions identified in the previous step;
6. **"Gap" IS/ICT:** Identification of the differential between the initial availability of IS / ICT and its functional organization and design the new functional model required to support the corresponding Urbanism, defined as the response to FCS identified by the "Task Force" as critical to the realization of the vision;
7. **Migration Plan:** Set of coherent policies among all subsystems or components in order to facilitate the elimination of this gap, looking to provide quality information (relevant, accurate, on time and right place, among other characteristics), according to the diverse needs, as well as ensure proper

monitoring of all activities associated with information management (acquisition, processing, storage, distribution, protection, etc.)

The Figure 1 presents the different phases of Metavision model.

This model, based on the starting formulation of an objective view, allows pilot processes of evolutionary continuous (Auer & Ruohem, 1996; Luftman, 2001), processes of organizational "creative destruction" (Nolan & Croson, 1995), organizational efficiency (Hammer, 2001) and organizational change (Day, 1999), incorporating a component inducing potential changing dynamics.

The change is the normal context of performance of modern organizations and involves an ongoing process of adjustment between organizational models and support models, particularly between the functional, informational and technological models. If designed properly, management change can increase the predictability of the necessary infrastructure while minimizing negative impacts of changes in the objectives and organizational operations (Keyworth & Kirk, 1998). As regards the detail modeled, the alignment model proposed is based on the concepts of organizational and informational Urbanism, and the role of information architecture as the 'Trojan Horse' for the induction and driving organizational change processes (Zorrinho & Anunciação, 2004).

Figure 1. Metavision model
Anunciação & Zorrinho, 2006; Zorrinho, Serrano & Lacerda, 2003.

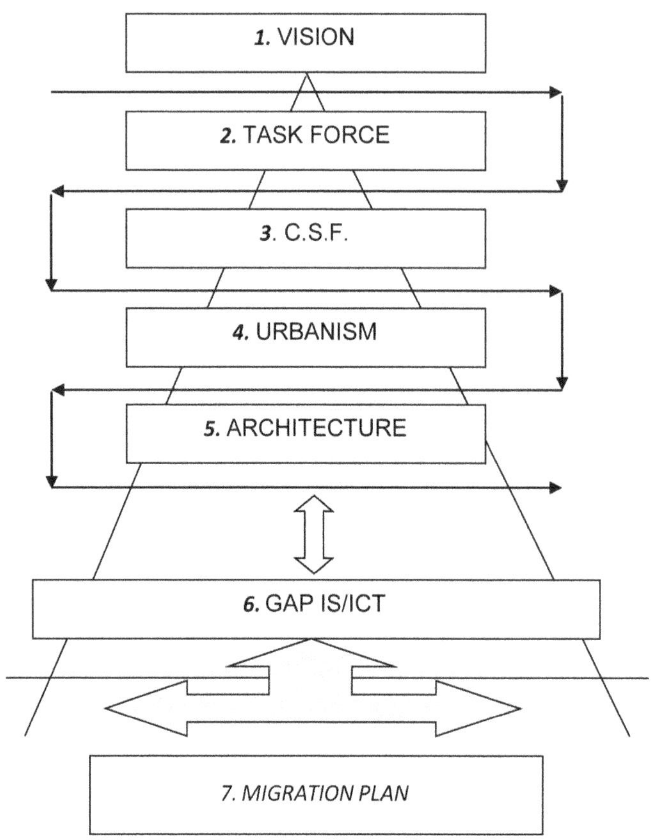

The MRAFITU model, as model of Urbanism, must constitute an intervention framework that incorporates a methodology for managing change, integrated and piloted from the management of the information platform, using the dynamic representation of organizational realities (business/activities, IS and ICT), through a model of urbanization, as shown in Figure 2 (Zorrinho & Anunciação 2004). This model aims at a common logic of value generation based on the three references inherent to organizational functioning: functional, informational and technological.

In this sense, it is intended that the model allows the frame of:

- The direct relationship between the business (s) and the organizational dynamics, with regard to the activities carried out, the responsibilities, the hierarchies, etc., knowing that the organizational management aspect of the business is responsible for obtaining better or worse indices of efficiency and effectiveness in the performance of their respective activities;
- The direct relationship between organizational activities, information systems and information and communication technologies, in the sense that technologies are an integral part of information systems, conditioning the design or organizational redesign, as well as the development of new organizational perspectives, without forgetting that the omission of this relationship compromises any organizational approach made;
- The direct relationship between business and support systems in a relational context, which presupposes functional, informational and technological integration (activities, information and technologies), in an integrated logic of joint creation of value.

THE NEW LOGISTIC PERSPECTIVE TO ORGANIZATIONAL URBANISM

The adoption of a logistics perspective of information systems facilitates the understanding and definition of their architecture. Based on the logistics concept, that including the specification of information, the delivery timings and the role players identification, according to the quality requirements previously

Figure 2. MRAFITU model (Model of Relational Alignment for Functional, Informational and Technological Urbanism)
Anunciação & Zorrinho, 2006.

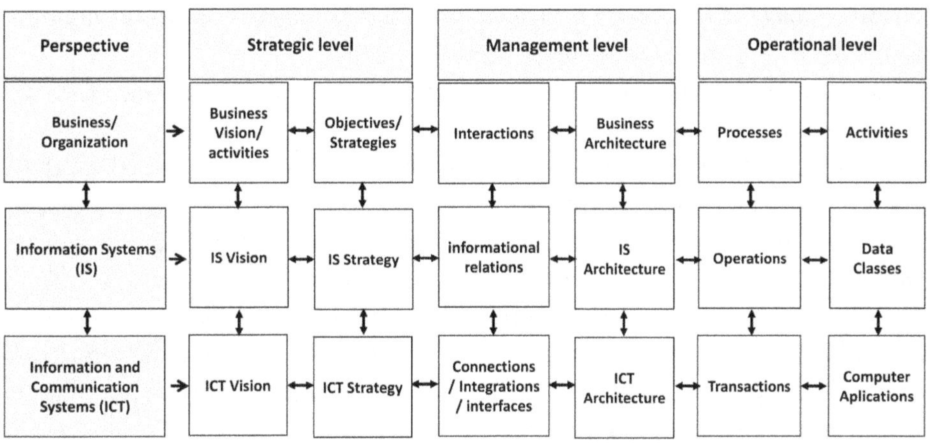

defined, it is possible to obtain a systemic and a global vision of organizational functioning, understand the extended operation with other partners, and define the information system support to all functional processes and requirements.

The logistics concept brings out the dimension of the information space and map it to fit all the information exchange, supporting the organizational dynamic and economic activities of the all agents involved. The dimensions presented in this paper, to frame the information logistics dimension, are based on the dimensions referenced by Linderman et al. (2005). These dimensions look forward to identify the players to interact within the mission of the organization, identify the information space and infrastructure in which their activities are developed, the existing management practices and information services that support the activities of information management. The actors interact through the information space, producing and consuming information. It is therefore important to understand some dimensions that seem essential to the information system management:

- **Referential of Activities:** The adoption of a relational posture or the participation in a network`s structure presupposes the existence of a set of universal activity frameworks capable of framing and justifying the logic of the harmonious functioning of the system. These referential of activities can be set up as a set of activities, products or services, which are essential for the development of activities and cross different areas or organizations.
- **Actors (Players):** Identification of agents that integrate and interact in the informational space associated with the activities developed (clients, suppliers, partners, etc.):
    - **Consumers:** Consumers are clients and users of the information space, who need information for the development of their activities and responsibilities. Using access credentials and the definition of these transactions they order the information they need;
    - **Producers:** Producers are information space providers that add information. This information may be shared temporarily or permanently in the collective information space;
    - **Spaces Federation:** Identification of producers and customers that have specific interests in information spaces, inherent to the nature and specificity of their activities. In this sense, information sharing and consumption in public spaces should provide a coherent integration, according to the scope and value chain is needed through federation services;
    - **Managers:** Managers are responsible for monitoring and controlling information, as a product, and the information space, as a system of processing, storage and provision of information. The management of information services includes a set of processes that ensure the identification of actors, functions (input, processing, storage and availability of information) and performance support systems to these functions.
- **Anchoring Points:** Set of functional, informational and technological characteristics necessary to enable the integration of the respective components or systems.
- **Information Objects:** The identification of information objects processed by the system that which will give meaning and coherence to the information space. The organizational and management performance (efficiency, effectiveness, productivity, etc.) depends on them. These information entities should be formally defined by all actors.
- **Service Information Layers:** The activities of information management are organized into service layers. Each service has a specific set of requirements that enable the information management system to effectively manage and provide value added processing of information within the information space. The service layers include:

  - **Security Layer:** Definition of safety and operation conditions to be applied to the information space;
  - **Workflow Layer:** Identification of enterprise workflows that are patterns of information flows and supporting organizational processes;
  - **Quality of Service Layer:** Specifying the quality of services should seek to ensure a management system space and information that best serves the needs of customers or users and producers;
  - **Transformation Layer:** Has the function to adapt the information, according to the input conditions and output information space either to specific customers or suppliers, or to the management policies defined;
  - **Brokerage Layer:** Intermediation which aims to ensure the link between the available information and existing needs;
  - **Competitive Layer:** Perception of the importance of different characteristics of information that are essential to competitiveness of each actor;
  - **Performance Layer:** Definition and controlling of information systems processing characteristics according to the user necessities and business requirements.
- **The Information Space:** Domain or information space which comprises the set of information that falls within the scope and the universe of the operation of economic activities and actors that comprise it, regardless of where they physically reside. The information space integrates:
  - **Information Catalogue:** Includes the maintenance of the all-meta information related to the needs of the information space and the information contained therein.
  - **Information Repositories:** Includes the identification of repositories needed to receive, maintenance and provision of information.

CONCLUSION

The complexity of information management in the dynamics of relational economic activity originates some difficulties of traditional instruments in organizational, architectural and project management. Organizations and management of IS require rapid, complete and satisfactory answers to the changes imposed by market needs in a collaborative context. The organizational sustainability requires, now and in the future, an understanding of the importance that information plays in the relational context and the capacity to adjust their systems of support.

The need to articulate different structures, functional processes, flows, locations, people, responsibilities, time cycles, organizational dynamics, objectives, etc., justify the introduction of a logistic approach to urbanization model as an important component do better manage the organizational change and the answers to market and different stakeholders. Historically, the concept of Urbanism of IS arises in banking institutions because there are different business and different information systems and technologies. Actually, this is the context of the economy. The success of any organization, in this context, will depend on the way as organizations and their information systems and information and communication technologies will be able to integrate and articulate with other systems of diverse stakeholders, with which they will have to share resources, activities, information, knowledge, etc. This is a proposal that we consider relevant as a critical success factor for the development of economic activities through information systems.

REFERENCES

Andal-Ancion, A., Cartwright, P. A., & Yip, G. S. (2003). The Digital Transformation. *MIT Sloan Management Review*, *44*(4), 34–41.

Anunciação, P. F. (1997). *The Urbanism of Information System*. CETOP Publishing. (in Portuguese)

Anunciação, P. F. (2004). The Urbanism of Banking Information Systems - The Example of Subsystem "Capital Actions". *Management Studies – Portuguese Journal of Management Studies, 9*(1), 37–51.

Anunciação, P. F, (2015). Organizational Change through Information Systems: Metavision-Project Management Model in Internet Banking. In *Handbook of Research on Effective Project Management through the Integration of Knowledge and Innovation*. Academic Press.

Anunciação, P. F., & Esteves, F. M. (2014). Some Information Systems Requirements in View of Organizational Sustainability in an Information Society. *Information Resources Management Journal*, *27*(1), 21–35. doi:10.4018/irmj.2014010102

Anunciação, P. F., & Zorrinho, C. D. (2006). *Organizational Urbanism – Who to manage the technological impact in economic organizations*. Sílabo Publishing. (in Portuguese)

Auer, T., & Ruohem, M. (1996). Understanding The Human Side of Information Systems Management and Use: A Theorethical Framework and Two Evaluation Approaches. In *The Human Side of IT Management*. Idea Group Publishing.

Benbasat, I., & DeSanctis, G. (2000). *Communication Challenges – A Value Network Perspective. In Information Technology and the Future Enterprise* (pp. 144–162). Prentice Hall.

Christensen, C. M., & Overdorf, M. (2000). Meeting the Challenge of Disruptive Change. *Harvard Business Review*, *78*(2), 66–76.

Day, G. S. (1999). Creating a Market-Driven Organization. *Sloan Management Review*, *41*(1), 11–21.

Domínguez, S. J. A., & Garrido, J. A. M. (2010). *Information Systems Management in the company - theory and case studies*. Pirámide Publishing. (in Spanish)

Empson, L. (2000). The Challenge of Knowledge Management, The Strategy Domain. *Economic Daily – Financial Times*, *2*(2294), 32. (in Portuguese)

Freire, A. (1997). *Strategy*. Lisbon: Verbo Publishing. (in Portuguese)

Gnyawali, D. R., & Madhavan, R. (2001, July). Cooperative Networks And Competitive Dynamics: A Structural Embeddedness Perspective. *Academy of Management Review*, *26*(3), 431–445.

Gonzalez, R. V. D., & Martins, M. F. (2014). Mapping the organizational factors that support knowledge management in the Brazilian automotive industry. *false. Journal of Knowledge Management*, *18*(1), 152–176. doi:10.1108/JKM-08-2013-0300

Hammer, M. (2001). The Superefficient Company. *Harvard Business Review*, *79*(8), 82–91. PMID:11550633

Hansen, T. H., Noria, N., & Tierney, T. (2001). What is your strategy for knowledge management? - Organizational Learning. Campus Publishing.

Hirschhorn, L. (2002). Campaigning for Change. *Harvard Business Review*, *80*(7), 98–104.

Karnani, A. G. (2000). Five Strategies for Growth and Value Creation, The Field of Strategy. *Diário Económico – Financial Times*, *5*(2297), 76 – 80.

Keyworth, B. & Kirk, T. (1998). *Changing Perspectives on Change Management.* GartnerGroup, COM-03-3534.

Krogh, G. V., & Cusumano, M. A. (2001). *Three Strategies for Managing Fast Growth. Sloan Management Review, 42*(2), 53–61.

Krovi, R. (2001). Surveying The E-Landscape: New Rules Of Survival. *Information Systems Management, 18*(4), 23–30.

Liderman, M., Haines, S., Siegel, B., Chase, G., Ouellet, D., O'May, J., & Brichacek, J. (2005). *A Reference Model for Information Management to Support Coalition Information Sharing Needs.* Air Force Research Laboratory. Retrieved from www.dtic.mil/cgi-bin/GetTRDoc?AD=ADA464244

Luftman, J. (2001). Assessing Business – IT Alignment Maturity. In *Strategic Information Technology: Opportunities for Competitive Advantage.* Idea Group Publishing. doi:10.4018/978-1-878289-87-2.ch006

Manville, B., & Ober, J. (2003). Beyond Empowerment: Building a Company of Citizens. *Harvard Business Review*, *81*(1), 48–53. PMID:12545922

March, J. G. (1991). Exploration and exploitation in organizational learning. *Organization Science*, *20*(: 1), 71–87. doi:10.1287/orsc.2.1.71

Massey, A. P., Wheeler, B. C., & Keen, P. G. W. (2000). *Technology Matters. In Information Technology and the Future Enterprise.* Prentice Hall.

McAfee, A., & Oliveau, F. X. (2002). Confronting the Limits of Networks. *MIT Sloan Management Review*, *43*(4), 85–87.

Nolan, R., & Croson, D. C. (1995). *Creative Destruction: A Six-Stage Process for Transforming the Organization.* Harvard Business School Press.

Nunamaker, J. F. J., Briggs, R. O., & Vreede, G. (2000). Value Creation Technology – Changing the Focus to the Group. *Information Systems Management*, *18*(4), 102–124.

Stacey, R. (1993). *Learning Organizations and Emerging Strategies, Strategic Thinking and Change Management.* Dom Quixote Publishing. (in Portuguese)

Vandenbosch, M., & Dawar, N. (2002). Beyond Better Products: Capturing Value in Customer Interactions. *MIT Sloan Management Review*, *43*(4), 35–42.

Venkatraman, N., & Henderson, J. C. (1998). Real Strategies For Virtual Organizing. *Sloan Management Review*, *40*(1), 33–47.

Zorrinho, C. (1991). *Information Management*. Lisbon: Editorial Presence. (in Portuguese)

Zorrinho, C., & Anunciação, P. F. (2004, April/June). Metavision: Model of 'urbanization' for Organizational Change [in Portuguese]. *Portuguese and Brazilian Management Journal*, *3*(2), 28–34.

Zorrinho, C., Serrano, A., & Lacerda, P. (2003). *Managing in Complexity - A New Management Paradigm*. Sílabo Publishing. (in Portuguese)

Chapter 15
How Fablabs Manage the Knowledge They Create

Sérgio Maravilhas
Universidade Salvador, Brazil

Joberto Martins
Universidade Salvador, Brazil

ABSTRACT

A collaborative space for stimulating creativity is a place of learning through the exchange and sharing of knowledge and experience among its members. It allows the leveraging of innovation through the use of technological resources available in the space, stimulating the creativity of its participants, enabling the development of products and solutions based on personal projects—do it yourself (DIY)—from ideation, or the construction supported on knowledge developed by other elements together, collaboratively, enhancing the final result—do it with others (DIWO). A research project is being held to create a new lab, or transform and adapt one of the existing lab's, in a Fab Lab or a Maker Space to let students, teachers, and staff give wings to their imagination and develop innovative solutions to solve real problems while they interact and exchange tacit knowledge, making it explicit after concluding their projects when they share their research reports.

INTRODUCTION

Information and knowledge, along with natural and economic resources, proved to be an unprecedented social and strategic expedient (Beuren, 1998; Choo, 1996; 2003; McGee & Prusak, 1995).

The importance of knowledge for organizations is now universally accepted, being, if not the most important, at least one of the resources whose management influences the success of organizations (Davenport, & Prusak, 1998).

Information management relates to the organizational ability to make the right information available for use in decision making (Davenport, 1997), transforming the informational chaos into useful and practical knowledge, leading to benefits for the organizations (Maravilhas, 2014b).

DOI: 10.4018/978-1-5225-6225-2.ch015

Maker Spaces, Hacker Spaces, Tech Shops and Fab Labs are collaborative spaces for stimulating innovation, through the exchange and sharing of information, knowledge, and experience among its members (Blikstein, 2014; Troxler, 2014).

They leverage innovation through the use of technological resources available in the space, stimulating the creativity of its participants and enabling the development of products and solutions based on personal projects from ideation, or the construction supported on knowledge developed by other Makers, collaboratively, enhancing the final result (Gershenfeld, 2005; 2012).

With the motto "Learn, Make, Share" (http://www.forbes.com/2008/08/13/diy-innovation-gershenfeld-tech-egang08-cx_ag_0813gershenfeld.html), these spaces aim to empower its members for the realization of sustainable solutions, local and community-based, using open source tools and equipment's whenever possible (open software, open hardware, open design, open learning), promoting Open Innovation (Chesbrough, 2003) to allow all the possibility of creating low cost products, with the ability to very quickly show the viability of these ideas through the acceptance by the community, leveraging improvements that will make these solutions evolve collaboratively (Anderson, 2010; 2012; Gershenfeld, 2005; 2012).

In these collaborative spaces the participation of all community members is nurtured, promoting equality of race and gender, benefiting from cross-knowledge, shared by every culture and subculture, which will enrich the final result.

Teachers, researchers and students, young and more experienced, men and women of all races and religions, small business owners, inventors and entrepreneurs, members of the local community, all in a horizontal relationship, without titles or awards, just competence and mutual respect, working and learning from each other in a common space.

The purpose is to enhance the entry of women in more technical fields and Engineering, but also to attract students and professionals of Arts and Humanities, Design and Architecture, allowing them to materialize their ideas based on available and affordable technology, supporting creative inventions and aesthetic processes that will enrich the research and development results (R&D) (Blikstein, 2014; Troxler, 2014). Youngsters and adults that abandoned formal education can find here the resources to start their own job and company.

It will be analyzed and described how Fab Labs, which are laboratories of digital fabrication, with broad educational, social and economic advantages, manage their knowledge in a formal and informal way, allowing every member to learn by watching and participate in bigger communal projects. ARHTE project from UNIFACS Laureate University in Salvador, Bahia, Brazil will be also described to show how a future Fab Lab is being constructed and be born in a near future.

The methodology used for the successful development of the project consisted, to begin with, on bibliographic analysis from monographs, journal articles, websites, theses and reports, allowing understanding of the topic, its stakeholders and participating entities. Some conversations with key players in the Fab Lab world have helped to structure the project and the knowledge shared helped avoiding some major obstacles and implementation problems that have been dealt with by other Labs in Brazil, Portugal, Italy, Spain, France, and the United States of America (USA).

The analysis of other existing Fab Labs makes possible to propose the introduction of best practices in the collaborative space, through benchmarking, avoiding mistakes and leaping steps for best performance.

BACKGROUND AND LITERATURE REVIEW

Like the inventors and entrepreneurs of the nineteenth and early twentieth century, who worked in their spare time in a shed, basement or, more recently, garage, creating prototypes of what later would become a successful product in a given market (Rifkin, 2011), Maker Spaces provide access to technologically advanced tools and machines, inserted in a network of participants, called Makers or Tinkerers, that can help answering questions and overcoming obstacles, faster than occurred with the lone inventor (Anderson, 2010; 2012).

A collaborative space for stimulating innovation is a place of learning through the exchange and sharing of knowledge and experience (Hargadon, & Sutton, 2002; Piscione, 2014) among its members, the Makers.

At the same time, allows to leverage innovation through the use of technological resources available in the space, stimulating the creativity of its participants and enabling the development of products and solutions based on personal projects from ideation, or the construction supported on knowledge developed by other users, together, collaboratively, enhancing the final result.

These spaces have several designations and typologies, like Makerspaces, Hackerspaces, Techshops and Fab Labs. In this project we focus in a model, widely tested and in use in several places of the world, the Fab Lab, or fabulous laboratory (Gershenfeld, 2005; 2012), which is a laboratory of digital fabrication, serving as a prototyping platform of physical objects (Eychenne & Neves, 2013), with broad educational (Mandavilli, 2006; Blikstein, 2014) social and economic advantages (Anderson, 2010; 2012; Troxler, 2014).

These Labs are used by everybody to enrich their projects, in a creative manner, inducing the creation of prototypes to leverage innovation, giving wings to their imagination and develop sustainable, social, local, economic innovative solutions to solve real problems.

Created in 2001 in the Massachussets Institute of Technology (MIT) Center for Bits and Atoms (CBA), directed by Neil Gershenfeld, linked to the famous MIT Media Lab, created by Nicholas Negroponte in 1985, the first Fab Lab was funded by the National Science Foundation (NSF) from the USA, and started based on the success of the course taught by Gershenfeld himself titled "How to Make (Almost) Anything" (Gershenfeld, 2012).

Eychenne and Neves mention that these Fab Labs are the *educational component of awareness to digital and personal fabrication, democratizing the conception of techniques and technologies and not just the consumption* (2013, p.10).

To quickly realize the viability of the solution, the machines and tools available in the space will allow developing a prototype that, if it's not feasible, will lead to the search for new solutions (http://www.makerinnovation.cc; http://wefab.cc/), learning with the community how to develop solutions that will be accepted and wanted in the market.

Fail early, fail cheap, fail always, continuing to learn and evolve so that entrepreneurship is encouraged and emulated by others (http://www.instructables.com/id/FabYearBook-2010/).

Students are encouraged to be producers of knowledge and not mere passive recipients (http://studentasproducer.lincoln.ac.uk/).

Fab Labs build bridges between the engineers and fabrication of high-tech products, and other actors usually more averse to technical and manual manufacturing.

The purpose is to enhance the entry of women in more technical fields and Engineering, but also to attract students and professionals of Arts and Humanities, Design and Architecture, allowing them to materialize their ideas based on available and affordable technology, supporting creative inventions and aesthetic processes that will enrich the research and development (R&D) results.

When someone has a difficulty operating the equipment's or machines, they can learn by observing other experienced users working, or they can ask for help, guidance or training with them or a Lab technician, enhancing the capabilities of every member of the space. Tacit knowledge is shared among the members of the Lab, working together, watching and learning, with the ones who own it.

The typology of academic Fab Lab, created in universities or research centers, aims to develop a culture of learning by doing, giving students, teachers, independent inventors and entrepreneurs the opportunity to learn by doing, creating a multidisciplinary space open to the outside to receive different insights and inputs (http://fab.cba.mit.edu/about/faq/).

In such cases, funding depends on the university or research center where they are installed, as well as the purchase of equipment and materials necessary for their operation, having its educational aspect assured by teachers and Postdoctoral Fellows (Eychenne & Neves, 2013) that support the management and maintenance of the space and its dynamics (http://www.fabfoundation.org).

Working in a network, like the Internet supporting them, there are currently 1215 Fab Labs worldwide (https://www.fablabs.io/labs), facilitating the sharing of information and knowledge, connecting people and organizations and, thus, enabling the collaborative innovation (Hatch, 2013; Troxler, 2014).

These spaces aim to develop access to knowledge of science and engineering, democratizing the practice of using the technic on the proposed projects (Blikstein, 2014), providing training courses to the community on the use of the equipment available in the space (http://makercity.wpengine.com/docs/makercity-preview-chapter.pdf), allowing the use of machines to carry out participants own projects or to participate in collaborative projects of the Fab Lab network (Walter-Herrmann & Büching, (Eds.), 2014).

Other users of the space, like students or experts in some technical areas, share their know-how through courses or just by participating in collaborative projects.

CONDITIONS TO START A FAB LAB

For the opening of such a space, the initial investment is around USD$100.000, according to the inventory required and proposed by the CBA, plus the monthly cost of salaries and maintenance of the machinery and location.

Usually, the university or research center provides the space and buys the machines, accessories and consumables, trains the responsible team and pays their salaries, and can provide scholars who collaborate in the organization of the space and scheduling times, while taking their projects and internships.

All Fab Labs must function according to the Fab Charter that sets the rules for the maintenance of the spaces, with its guidelines and policy (http://fab.cba.mit.edu/about/charter/).

On the website of Fab Foundation (http://www.fabfoundation.org/fab-labs/setting-up-a-fab-lab/ideal-lab-layout/) instructions can be found of how a Fab Lab could be implemented in relation to its size, organization and necessary equipment.

Eychenne and Neves (2013), after visiting several spaces in different countries, found some common points, such as: an area between 100 and 250m^2; at least one separated and closed room for the use of the large computer numeric control (CNC) milling machine; a large open central area with side stands containing on one side the less noisy machines and in the other the potentially dangerous machines or the ones who generate dust; computer equipment for work and conference or meeting tables; space for quick meals equipped with coffee machine and refrigerator; exhibition space for the completed projects and a place to store materials and tools.

Regarding the machinery and equipment to start the space, five CNC machines make up its base: a vinyl cutter, a laser cutter, a 3D printer, a precision milling machine, and a large format milling machine.

Being its concept based on the philosophies of open software and open hardware (Troxler, 2014), this reduces its cost, as there are several machines that can be built by themselves, allowing to replicate the equipment at a very low cost (Gershenfeld, 2005; 2012; Troxler, 2014).

The open software and other open source projects like Arduíno – a printed circuit having a microcontroller that allows you to control chips and sensors – also enable the realization of several low-cost projects (Walter-Herrmann & Büching (Eds.), 2014).

Fab Foundation (http://www.fabfoundation.org) also lists the necessary staff for the proper functioning of the space, and the pattern is constituted of: A Director, a Fab Manager, a Guru and three Interns (Eychenne & Neves, 2013).

Economic and Financial Return

The big challenge is to find industrial partners who want to use the space, financially maximizing the continued acquisition of materials and the maintenance of the equipment.

Ford Motors developed a partnership with a Techshop, another type of Maker Space more commercially focused, and the results are astonishing with an increase of about 50% of patent applications from the Ford employees in one year (https://media.ford.com/content/fordmedia/fna/us/en/news/2013/06/01/techshop-and-ford-celebrate-one-year-of-innovation-in-metro-detr.pdf).

This solution allows a very quick and effective market research with the possibility of increasing the ultimate solution based on acceptance and critics received.

Technology transfer between Fab Lab and industry can also be enhanced globally as a product developed in one Lab can solve a problem or need in another place in the world where the network is present.

It's important to remember that in a collaborative space with a conducive environment for innovation, the most important are not the machines and equipment, but people and their ideas (Dodgson & Gann, 2014), that should be encouraged and cherished for the production of new solutions that can be used to solve local community problems and empower its creator and the ones benefiting with the solution. Technology is only an enabler for this purpose.

If successful, the practices introduced can be replicated, creating a new model of innovation support, which will have a significant impact on the social and economic development of countries.

KNOWLEDGE: CREATION AND SHARING

Knowledge derives from data and information. Organized and contextualized data becomes information through contextualization, categorization, calculus, correction and condensation (Davenport, & Prusak, 1998). Information interiorized and applied to a practical task or function, once dominated by the user, becomes knowledge through comparison, consequences, connections and conversation (Davenport, & Prusak, 1998). Knowledge generation occurs when information is compared, combined, analyzed, and rearranged by people.

Knowledge is inside people's head, not in computers and databases. In a book or scientific paper there's the knowledge of the author, what he/she knew and could made explicit to others, which is only a small part of what he/she really knows. For the reader there is only data and information there, until it's possible to act upon that information and integrate it in a process where we can apply what was read and do it ourselves with proficiency (Davenport, 2007; Davenport, & Prusak, 1998). Knowledge is related with action. Someone can read all the books and articles about riding a bicycle. It will be helpful undoubtedly. But, only when someone act with the information received and pedal while maintaining equilibrium, is common sense to accept that the knowledge to ride the bicycle exists, and, curiously, will never be forgotten. That's why learning mathematics implies solving problems and perform hundreds of exercises. No one learns mathematics just by reading books or reports, or watch the teacher solving problems in the blackboard. Although useful, it's not enough. Medical Doctors perform dissections to learn about the human body. They don't just read books. That part is also needed, but we must do it ourselves to learn and internalize the sequences and mechanisms to adopt.

That type of knowledge is called tacit. Socrates, the Greek Philosopher, used to say that a person only knows something when it's possible to teach it to another person, with the same results.

Tacit knowledge can only be learned through practice and experience and is what we call know-how, that knowledge that is acquired through life, but very difficult to explain to others. Constituted by subjective insights, intuitions and hunches of individuals, is not easily communicated or shared. To gain access to such knowledge one may have to be practicing in other related areas of knowledge. What is held in someone's head and includes facts, stories, biases, and insights.

According to Davenport and Prusak we shouldn't learn anything without relating that apprenticeship with practice. They believe that a healthy tension between knowledge and action is the key to the organizational success and, probably, the individual success also (1998).

Trying to make that knowledge available to others, since it is ingrained in people's heads and attitudes, it's imperative to turn it explicit. Explicit knowledge is obtained from facts, from information, almost always through formal education. It is expressed through metaphors, analogies, concepts, hypothesis or models. It's the key for the creation of new knowledge. It is codified, communicable through formal and systematic language. It can be stored in manuals, documentation, patents, blueprints, reports and other accessible sources.

Tacit knowledge is subjective. It is knowledge obtained from experience, with the body, simultaneous, here and now, and analogous, related with practice.

Explicit knowledge is objective. It's a rational knowledge, with the mind, sequential, there and then, digital, related with theories (Nonaka, & Takeushi, 1997).

The process of generating and converting knowledge has four phases. Explicit and tacit knowledge complement each other and, through their interaction, create new knowledge. According to Nonaka and Takeushi (1997), the four phases are:

1. **Socialization:** From tacit to tacit knowledge. Creates shared knowledge. It's a process of sharing experiences, like mental models or technical competencies. This is what happens when someone in an apprenticeship learns with its master. Can be obtained directly from others, without using language, through observation, imitation, and practice.
    a. In Fab Labs, people learn by watching, imitating and practicing with their peers. Different backgrounds and experiences allow internalizing different procedures and activities.

2. **Externalization:** From tacit to explicit knowledge. Creates conceptual knowledge. Tacit knowledge becomes explicit through dialog and collective reflection using metaphors, analogies, concepts, hypothesis and/or models. Books, reports and journal articles are examples of explicit knowledge through externalization. It combines deduction and induction.
    a. In Fab Labs, Makers document their projects, especially on open days when the use of the space is free, and share it so others can learn and improve upon that knowledge, avoiding mistakes and adopting the best practices that conducted to those results so their projects can be successful too.
3. **Combination:** From explicit to explicit knowledge. Creates systemic knowledge. Combining different sets of explicit knowledge allows the creation of some new knowledge. Through education and formal training in schools and universities, different explicit knowledge from the combination of books, texts, activities and the different disciplines combined, generates some new knowledge that can be further developed with time. Computer databases can help categorize some explicit knowledge conducting to new knowledge.
    a. In Fab Labs, the users can search and retrieve documents from different projects from all the Fab Lab world network, in databases involving different disciplines like mechanics, robotics, electrical and computing engineering, design and other aspects that can help solving their own problems and give insights for the construction of new solutions or surpassing technical blocks.
4. **Internalization:** From explicit to tacit knowledge. Creates operational knowledge. The process of incorporating explicit in tacit knowledge, it is related with learn by doing. When internalized in the knowledge base of every individual as mental models or technical know-how, those experiences through socialization, externalization and combination become valuable assets.
    a. In Fab Labs, Makers learn from the analysis of technical documentation and the projects of their peers worldwide, internalizing the new knowledge and consolidating it in its practical use in their projects, learning by doing. That way new knowledge and procedures are integrated to the way of doing things by Makers that can advance to new levels of know-how.

Nevertheless, to make viable the creation of organizational knowledge, the tacit knowledge accumulated needs to be socialized with other members of the organization, starting, that way, a new spiral of knowledge creation that will pass the four phases again, generating new knowledge in an increasing successful non-stopping activity (Nonaka, & Takeushi, 1997).

A new vision must be considered in relation to knowledge and the role it plays in organizations. The efficient exploitation of knowledge as an economic resource and one of the production sectors has become a factor of strategic, economic, social and political importance (Choo, 1996; 2003).

If, on the one hand, an organization cannot function without information and knowledge, on the other it is important to know how to use this resource to improve its use.

Thus, the faster the identification of the relevant knowledge to the organizations and the quicker the access to it, more easily their goals will be achieved (Terra, 2001).

Intellectual Capital can be seen as a competence, skill and/or entrepreneurial intelligence and is recognized as an intangible asset of superior value for organizations (Edvinsson, & Malone, 1998; Stewart, 1998; 2002; Sveiby, 1998).

But, most important is that information and knowledge play a key role in business since they affect the competition at three levels:

1. Modifies the industrial structure and, therefore, changes the rules of the competition (Stewart, 1998; 2002);
2. Creates competitive advantage for organizations offering new ways to overcome their rivals (Sveiby, 1998; Terra, 2001);
3. Create new business opportunities, most often from the organization's internal processes (Almeida, Freitas, & Souza, 2011; Neto, 2013; Rodriguez, 2013).

Fab Labs should follow some recommendations in order to successfully achieve their goals, such as: Value and manage knowledge as a resource so or more important than any other that it needs to function (Maravilhas, 2015b); Pay attention not only to the internal knowledge generated within the Lab, necessary to carry out the organizational tasks it undertakes, but also external knowledge, from various Fab Labs worldwide, in order to maintain their activity cost-effective.

Effective Fab Lab managers should not ask about the cost of obtaining the knowledge needed to increase the successful projects of the Lab. They should ask instead how much will be lost if they don't have it (Maravilhas, 2013a).

Knowledge Management, Creation and Transfer in Fab Labs

In this definition, all the practices that organizations use to represent, create, identify, and distribute knowledge, usually with the support of the computer and Internet, come together. In many organizations a Chief Knowledge Officer (CKO) exists, and mediate all the programs between the workforce and the Board of the organization.

Almost all the for-profit companies depend on technology and software for their knowledge management projects like: knowledge repositories, knowledge bases, expert systems, corporate intranets and extranets, collaborative websites, like wikis, and document/content management software, that allow and promote the knowledge sharing and transfer process between its members.

Knowledge management focuses mainly in the mechanisms used to share and transfer the knowledge assets.

Currently, innovative products are developed based on rapid prototyping in universities R&D departments and research institutes, and in some larger companies.

Only a small group of experts has the possibility of making prototypes in a short period of time and using simple means and resources (Anderson, 2010; 2012). In a Fab Lab this process is democratized and any new technologies are taught so that everyone can enjoy the space and equipment's. Knowledge shared is new knowledge created.

My view is that people are creative animals and will figure out clever new ways to use tools that the inventor never imagined (Steve Jobs in Isaacson, 2011, p. 241).

In relation to its effectiveness, since 2001 at MIT and since 2005, when the first Fab Lab was created outside of MIT, the model has proved to be a facilitator for the creation of regional innovation, building bridges and relationships between experts in technology, design, education, small business owners and entrepreneurs, architects, artists, non-profit organizations, etc.

The idea of a Fab Lab rests on social interaction, in projects involving both academics and craftsmen, the handyman and garage skilled inventors, bringing to the manual and practical learning the ones that in recent years have distanced themselves from the technology and have chosen a more intellectual and less physical training, less hands-on.

The interaction between people with such diverse skills and features, along with the acquired training on the use of the available equipment's, will create a creative and stimulating environment thanks to the power of intellectual and cultural diversity (http://www.thefreelibrary.com/Fabricating+dreams+in+3D%3A+FabLab+Luzern.-a0336489127).

These spaces aim to develop access to knowledge of science and engineering, democratizing the practice of using the technic on the proposed projects (Blikstein, 2014), providing training courses to the community on the use of the equipment's available in the space, allowing the use of machines to carry out participants own projects or to participate in collaborative projects of the Fab Lab network (Walter-Herrmann & Büching, (Eds.), 2014). All the projects are registered and shared so other Makers can replicate them freely, all around the world, making this new explicit knowledge free to the Fab Lab community.

As for the potential of transfer of the generated knowledge, the Fab Lab benefits from an extensive worldwide network which promotes the adoption of knowledge created in several laboratories spread across several continents, allowing to test the acceptance of a huge number of potential users and adapt, improve or complement the initial versions with the return, the feedback, obtained in this way.

There is a special ability in playing with creative ideas developed for a segment or project to explore their innovative application in others. Playing in this environment enables both cross inspiration as the connection, and happy and unexpected combination of unrelated ideas (Dodgson & Gann, 2014).

Regarding the interdisciplinary collaboration, this is enhanced, as mentioned above, by the number of technical, academic and skillful handy men that will cross the space and contribute with tips, advices, warnings and suggestions.

Several areas of knowledge are present in these spaces, such as: electronics, mechanics, computer science, design, chemistry, administration, fine arts, and humanities.

This mix turns the space into a melting pot of cultures and sciences that will enable all to teach and to learn, enriching each of the worldviews involved and profiting all with the multiplicity of the knowledge obtained (http://www.instructables.com/id/FabYearBook-2010/).

Several examples demonstrate the importance of these places for science and technology education, like learning concepts of Engineering and Mathematics (Blikstein, 2014), stimulating creativity and the development of inventions that allow to solve local problems of the communities where these Labs are located, promoting innovation and social economy, empowering people who are part of these networks allowing them an autonomy never imagined before (Mandavilli, 2006; Troxler, 2014).

Because all materials from the projects are made available to the entire network, the potential of dissemination of information allows building on prior knowledge, leveraging innovation and maximizing the previous research (Nonaka, & Takeushi, 1997). That way, the open innovation and the ascent innovation are privileged (Eychenne & Neves, 2013), transforming the Do It Yourself (DIY) model in Do It With Others (DIWO), or Do It Together (DIT) (http://makercity.wpengine.com/docs/makercity-preview-chapter.pd), maximizing the educational and research function, with social and local impacts.

Working and learning collaboratively enhances the final result of the projects, because it's possible to do a sort of market research immediately, getting feedback on how to improve your idea and prototype, failing fast and cheap, allowing to make a better model with the suggestions of the community, using crowdsourcing and the wisdom of the crowds (https://www.diplomacy.edu/resources/books/reviews/wisdom-crowds-why-many-are-smarter-few)

With a markedly educational and research side, interdisciplinary, multidisciplinary and intradisciplinary (Blikstein, 2014; Troxler, 2014), it will allow to develop innovative projects of high scientific quality and high social relevance, following the model "faster, higher, better, more precise", determined by the Fab Lab network. This will be achieved by following the criteria of effectiveness, transfer of knowledge potential, originality and interdisciplinary collaboration (http://fablab-luzern.ch/).

The advantage of being based on an international model that has been tested, offers a place with an innovative atmosphere that will make possible the exchange of knowledge based in fortuitous but fruitful encounters among its members (http://www.thefreelibrary.com/Fabricating+dreams+in+3D%3A+Fab Lab+Luzern.-a0336489127), similar to what happens in the more innovative companies like Google, IDEO, Idealab, Pixar, Apple, among others (Dodgson & Gann, 2014; Isaacson, 2011; 2014; Kahney, 2009; 2013; Majaro, 1990).

Creativity comes from spontaneous meetings, from random discussions. You run into someone, you ask what they're doing, you say 'Wow' and soon you're cooking up all sorts of ideas. So he [Steve Jobs] had the Pixar building designed to promote encounters and unplanned collaborations (Isaacson, 2011).

A Fab Lab attracts more actors from companies than the university itself can do. With its innovative DIY concept, opens innumerous possibilities for universities and ensures a productivity index that will be relevant to the increased volume of innovations, and the consequent creation of wealth resulting from it (http://www.instructables.com/id/FabYearBook-2010/).

The idea of a Fab Lab rests on social interaction, in projects involving both academics and craftsmen, the handyman and garage skilled inventors, bringing to the manual and practical learning the ones that in recent years have distanced themselves from the technology and have chosen a more intellectual and less physical training, less hands-on.

The interaction between people with such diverse skills and features, along with the acquired training on the use of the available equipment's, will allow a creative and stimulating environment thanks to the power of intellectual and cultural diversity (http://www.thefreelibrary.com/Fabricating+dreams+in+3D%3A+FabLab+Luzern.-a0336489127), and the knowledge exchanged this way.

In almost all the Labs, users check other Makers files for inspiration, but usually just in the files from their own Lab. Only in bigger projects, conducted by more experienced users, and involving several Labs, the files from other Labs are retrieved for the purpose of getting insights for the new project at hand or for problem solving. In others, there is no habit of doing so, letting creativity flow only from the participants of the projects, without inspiration from their peers' previous projects.

Usually, users record their projects in their own language for easiness of the task (in English, in Dutch, in Spanish, in French, in Afrikaans, in Tsonga, in Xhosa, in Portuguese, in Italian, in several Chinese dialects, etc.). This makes very difficult the retrieval of information and the appropriation of knowledge, because the linguistic differences will make very hard to analyze and interpret the information gathered by other interested Makers from other countries. Photos, designs, videos, and schemes, are very helpful and should be always included.

In almost all the Labs examples of Makers that had benefited from the information sharing in the Lab exist, but especially in oral form. The written documentation is not as valued as the oral transmission and the observation experience.

The privileged form of information and knowledge exchange goes to videoconference solutions available in all the Labs. Seems to be easier for Makers to watch and debate with their peers, than to read what has been done previously. It's a form of transforming the tacit in explicit knowledge.

All this information should be in English for the use of other Makers in other countries or, at least, have an English Abstract for international user's aid. The Abstract can be done with the Fab Director or Fab Guru help, or using a web translator. A Lab colleague can help performing this task.

This information can be made available in a Cloud Server maintained by the Fab Foundation, allowing the access to everyone involved to improve results and knowledge sharing.

Extensible Markup Language (XML) and databases can be used to assist in the information visualization and finding from all the Labs, along with other technology solutions that allow to search by keyword, subject, title, name or codename of the Makers, subject area, country, city, etc., mixing them all to filter the results. That way, the knowledge sharing will be improved.

Although a lot of computer specialists frequent the Labs, their core business is not managing knowledge. They are more concerned in developing new products or solutions, than in maintaining and sharing the knowledge created in the space.

Examples of Solutions Created in Fab Labs Using Collaborative Knowledge

There are several examples of original products created in Fab Labs that originated innovations shareable amongst all the Labs. *Once prototyped the object and tested the processes, the project can easily be replicated by other Fab Labs in the network* (Eychenne, & Neves, 2013).

Some examples of innovations produced in these spaces include (Gershenfeld, 2012; Mandavilli, 2006):

1. Monitoring sheep herds using Global Positioning System (GPS) and radio frequency developed in Norway, and now used by anyone who needs a similar solution in any location of the world;
2. From the Boston Fab Lab, a project was started to make antennas, radios, and terminals for wireless networks, to provide Internet access. The design was refined at a Fab Lab in Norway, tested at another one in South Africa, deployed from one in Afghanistan, and now is running on a self-sustaining commercial basis in Kenya;
3. Circuit boards and computer chips produced by an eight-year-old girl in Ghana using an MIT design and methodology;
4. Also in Ghana, villagers built large 'collectors' to harness solar energy and built machines to grind seeds into fufu powder for their nourishment;
5. Puzzles 2D, convertible to 3D, were also produced by an eight-year-old girl in the USA (Neil Gerschenfeld's own daughter);
6. In Pabal, India, also based in a MIT model, farmers built sensors to measure the fat content in milk, allowing them to charge for a fair price to industrial buyers;
7. In South Africa, women that never had used a computer, now use it daily to design and send projects to the vinyl cutter, to create decorative products and accessories. Some are girls who left school due to a teenager pregnancy and here can find a profitable and dignifying activity;
8. Also in South Africa other Makers produced a light switch controlled by cellphone, a motion-sensor light and an alarm system, very useful for their unsafe neighborhoods;
9. In the USA, a student produced a sensor that protects women from attackers that can come from behind them, opening a sharp edge spear.

Another way of exchanging knowledge developed from Maker Spaces is the Maker Faire. *The Greatest Show (& Tell) on Earth.* Maker Faire *is part science fair, part county fair, and part something entirely new!* (http://makerfaire.com/). Disseminated all around the world, the Maker Faire allows the exchange of knowledge through the exhibition of bigger projects developed by several Labs in the community.

All the examples show, and several others exist, that the common element in this new creative class is the fact that the creators have been consumers that wanted something previously inexistent. So, instead of being satisfied with the options available, they did something better for themselves (Anderson, 2012).

THE ARHTE PROJECT IN UNIFACS

The Academic Interdisciplinary Program Archimedes, Robert Hooke and Thomas Edison (ARHTE) is a set of educational activities developed within the academic coordination's, involving teachers and students of all courses from the School of Engineering and Information Technology (IT) at UNIFACS – Laureate University, Salvador, Bahia, Brazil, for the development of interdisciplinary activities, involving Computer Science courses, Environmental and Sanitary Engineering, Civil Engineering, Computer Engineering, Production Engineering, Electrical Engineering, Materials and Manufacturing Engineering, Mechanical Engineering, Mechatronics Engineering, Petroleum Engineering, Chemical Engineering, Information Systems, Computer Networks, Internet Systems and Oil and Gas for the development of interdisciplinary activities in an attempt to overcome the fragmentation of knowledge of some specialized areas and the lack of its relation with the problems of modern life (http://www.arhte.unifacs.br/).

The main objective of this program is to create a relationship between technical knowledge and the reality of the student, encouraging the practical application of theoretical concepts covered in class and lectures, making it the most playful and challenging course for the student, as well as prepare much better professionals for the job market (http://www.arhte.unifacs.br/doc/Manual_do_ARHTE_2015.2.pdf).

ARHTE is a mandatory curricular activity to all graduate students enrolled in the school of Engineering courses and IT, from the first to the third semester, and optional for the fourth semester students.

The program started nine years ago and involves about 4.000 students each year. There are 38 teachers involved as Coordinators of the participating teams. The University has about 30.000 students and about 10.000 of them have been enrolled in the program.

The methodology of the ARHTE program operates with the following procedures:

1. In the first semester of the course, all students from Engineering, Information Systems, Computer Science, Networks, and Internet Systems have to participate and be approved in an Entrepreneurship course, which is mandatory to all these students. They also have to attend one of the two days of the project presentation works from other students from the second semester.
2. In the second semester the students will form teams and they need to choose a Coordinator and make their inscription on the program website (http://www.arhte.unifacs.br/index.php). They propose a project and teams must think and have a big idea, plan it accordingly and then present it to an examination board to check if there is a chance of creating a business with their idea (this will be watched by the first semester students to gain knowledge and insights). The goal is to evaluate if there is a market for the idea proposed and start a business to explore it.
3. In the third semester teams must develop the idea proposed previously in the second semester or start a new one if the previous didn't get support to be followed. The goal here is to build a pro-

totype, always with the support of the Coordinator, write a paper and present the results, in slides and video format, to a new examination board.

4. The previous three semesters are mandatory to all the students involved but this fourth semester is optional. The team with four to six students that had good reviews in the previous steps and have a good idea and product to start a business will continue developing their project. The teams will have the support and Coordination of the University Business Incubator team and the support of the business developing team to evolve their entrepreneur vein and learn how to make a business plan, CANVAS plan, and a prototype and present all the results to a specialized business examination board.

The students receive two points in each discipline of their semester. If the teams have four or more students from different courses, they will receive a 10% surplus in the program final grade. The purpose of this advantage is to stimulate the multidisciplinary knowledge exchange that will occur between the team members.

Students have several well-equipped Labs to develop their ideas and prototypes, which can be reserved in the dates they need to fulfill their work and projects.

It is possible to have students from other courses outside of the ones described above in the teams, but they only operate as volunteers and don't have a grade in their courses for that, as happens with these students.

In a near future, the possibility of starting a scientific initiation program that will direct student's projects to start a business or to research and development, with a scientific paper publication in a top journal with peer review, will be implemented.

Becoming a Fab Lab

Although students have Labs at their availability, with all the digital fabrication machines and software that are necessary to fulfill their projects, we cannot call those a Fab Lab yet because they are not open and available to the public, as recommended by the Fab Foundation. Furthermore, the Labs can only be used outside class time and after a request was made and accepted with the reason for using the Lab's and its tools.

Security reasons prevent the access to people not enrolled with the university, because the machines and other equipment's are very expensive and the entrance is not available from the outside. To get to the Labs its necessary to pass through several rooms and it's not secure for the University and their people (teachers, students, and staff) to let outside people walking around everywhere in the building because that can pose several security threats, especially in a city like Salvador where high crime rates occur.

The previous situation prevents students from the contact with external Makers with their knowledge and expertise, that could enrich their own projects. Likewise, prevents the general public in the nearby community from learning with these high skilled students, not stimulating the inventive and innovative vein of the external community.

Steps are being done to analyze the possibility of constructing a new Lab, with the necessary security conditions to operate with everyone interested in participating, or to adapt one of the existing Labs so it can be accessible directly from the outside, avoiding the wrong people can enter the all building.

That's the only measure needed to propose the Lab to the Fab Foundation because all the other elements necessary to become a Fab Lab are already in place.

FUTURE RESEARCH DIRECTIONS

Further research should be performed to analyze other solutions in place in other Fab Labs around the world, to adopt those best practices in all the Labs to improve their management, knowledge sharing, and sustainability.

Also, measures should be done to verify the benefits of that knowledge diffusion for the users. If it is beneficial for the knowledge creation of the Makers and allows them to find solutions to solve problems, and give them insights for new projects, then it should be suggested to all the Labs as a way of reducing the time to market for their products. All the help is welcome and the network can think of it collaboratively to find new ways to improve their results.

The managerial significance and applications of this research, resides on the importance that knowledge creation and diffusion plays on the advancement of the economy, generation of employment, increasing life conditions of the population involved, making all the community beneficiary of these solutions offered by the developments conceived in Fab Labs.

CONCLUSION

After describing the concept of Maker Spaces, the analysis focused in a specific type: The Fab Lab. Several educational advantages promoted by these spaces were described, mainly in the learning of concepts related with Engineering, Calculus, Mathematics, etc., and described the benefits from them to the attraction of women for these specific areas, and in the training of youngsters that didn't pursue their formal education, so they can have useful technical competencies for the labor market.

The different backgrounds of the Makers present in the space will be potentially advantageous for the share and exchange of experiences and knowledge that will enrich all the participants. The Maker movement encourages Master Makers to transfer their knowledge of production techniques to Makers who are less experienced (http://makercity.wpengine.com/docs/makercity-preview-chapter.pdf)

The knowledge created is shared among the network, allowing comments and improvement suggestions that will enrich the result of the projects.

The basic necessary conditions to open a Fab Lab are described, allowing understanding the possibility of creating such a space in a community allowing designing useful solutions for the residents and users of the Lab.

Some examples are presented to demonstrate the viability of these learning spaces and its creative and innovative advantages to empower the people involved, motivating them to create solutions based in rapid prototyping that allow to evaluate the potential of their invention and its acceptance by the community of Makers that can motivate the creation of a new business or improve the competencies at their current jobs.

The Fab Lab aims to develop a culture of learning by doing, giving students, teachers, independent inventors and entrepreneurs the opportunity to learn by doing it themselves (DIY), and learn together (DIWO or DIT) with other Makers from their Lab or another from the network, creating a multidisciplinary space, open to the outside, to receive different insights and inputs (Gershenfeld, 2005; 2012).

Information from projects, collective and individual, and from the courses to learn how to operate all the machines and equipment's in the Lab, together with rules and regulations about the use of the space, can be a knowledge disseminator task.

All this internal and external knowledge needs to be managed, to improve the knowledge creation among the users of the space (Sordi, 2008).

Although, without a general strategy implemented in all Labs, small steps adequate to the Labs publics and technology competence are being taken, being a starting point for a future best practice to be implemented in the entire Fab Lab network.

ARHTE project can be a huge disseminator of new knowledge amongst the students and other publics participating in the activities of the Lab.

REFERENCES

Almeida, M., Freitas, C., & Souza, I. (2011). *Gestão do Conhecimento para a Tomada de Decisão*. Atlas.

Anderson, C. (2010). In the Next Industrial Revolution, Atoms are the New Bits. *Wired Magazine, 1*(25).

Anderson, C. (2012). *Makers: The New Industrial Revolution* (1st ed.). New York: Crown Business.

Beuren, I. (1998). *Gerenciamento da informação: um recurso estratégico no processo de gestão empresarial*. São Paulo: Atlas.

Blikstein, P. (2014). Digital Fabrication and 'Making' in Education: The Democratization of Invention. In J. Walter-Herrmann & C. Büching (Eds.), *FabLabs: Of Machines, Makers and Inventors*. Bielefeld: Transcript-Verlag.

Chesbrough, H. (2003). *Open Innovation: The New Imperative for Creating and Profiting from Technology*. Boston: Harvard Business School Press.

Choo, C. (2003). *Gestão de informação para a organização inteligente: A arte de explorar o meio ambiente*. Lisboa: Editorial Caminho.

Choo, C. W. (1996). The knowing organization: How organizations use information to construct meaning, create knowledge and make decisions. *International Journal of Information Management, 16*(5), 329–340. doi:10.1016/0268-4012(96)00020-5

Davenport, T. (1997). *Information ecology: Mastering the information and knowledge environment*. New York: Oxford University Press.

Davenport, T. (2007). *Profissão: Trabalhador do conhecimento: Como ser mais produtivo e eficaz no desempenho das suas funções*. Amadora: Exame.

Davenport, T., Marchand, D., & Dickson, T. (2004). *Dominando a gestão da informação*. Porto Alegre: Bookmann.

Davenport, T., & Prusak, L. (1998). *Working knowledge: How organizations manage what they know*. Boston: Harvard Business School Press.

Davenport, T., Prusak, L., & Wilson, J. (2003). *What's the big idea: Creating and capitalizing on the best management thinking*. Boston: Harvard Business School.

Dodgson, M., & Gann, D. (2014). *Inovação* (1st ed.). Porto Alegre: L&PM Pocket Encyclopaedia.

Earl, M. (1998). *Information management: The organizational dimension*. New York: Oxford University Press.

Edvinsson, L, & Malone, M. (1998). *Capital Intelectual: Descobrindo o Valor Real de sua Empresa pela Identificação de seus Valores Internos*. S. Paulo: Makron Books.

Eychenne, F., & Neves, H. (2013). *Fab Lab: A Vanguarda da Nova Revolução Industrial*. Associação Fab Lab Brasil.

Gershenfeld, N. (2005). *Fab: The Coming Revolution on Your Desktop: from Personal Computers to Personal Fabrication*. New York: Basic Books.

Gershenfeld, N. (2012). How to Make Almost Anything: The Digital Fabrication Revolution. *Foreign Affairs*, *91*(6).

Hargadon, A., & Sutton, R. (2002). Como Construir uma Fábrica de Inovação. In H. B. Review (Ed.), *Inovação na Prática: On Innovation* (pp. 59–79). Rio de Janeiro: Campus.

Harvard Business Review (Ed.). (2001). *Gestão do Conhecimento: On Knowledge Management*. Rio de Janeiro: Campus.

Harvard Business Review (Ed.). (2002). *Inovação na Prática: On Innovation*. Rio de Janeiro: Campus.

Hatch, M. (2013). *The Maker Movement Manifesto: Rules for Innovation in the New World of Crafters, Hackers, and Tinkerers* (1st ed.). McGraw-Hill Education.

Hinton, M. (2006). *Introducing information management: The business approach* (1st ed.). Burlington: Elsevier Butterworth-Heinemann.

Isaacson, W. (2011). *Steve Jobs* (1st ed.). New York: Simon & Schuster.

Isaacson, W. (2014). Os Inovadores: Uma biografia da revolução digital. S. Paulo: Companhia das Letras.

Jorge, M. (1995). *Biologia, informação e conhecimento*. Lisboa: F. C. Gulbenkian.

Kahney, L. (2009). *A cabeça de Steve Jobs* (2nd ed.). Rio de Janeiro: Agir.

Kahney, L. (2013). *Jony Ive: O gênio por trás dos grandes produtos da Apple*. S. Paulo: Portfolio-Penguin.

Majaro, S. (1990). *Criatividade: Um passo para o sucesso*. Lisboa: Europa-América.

Mandavilli, A. (2006). Make Anything, Anywhere. *Nature*, *442*(8). PMID:16929273

Maravilhas, S. (2013a). A web 2.0 como ferramenta de análise de tendências e monitorização do ambiente externo e sua relação com a cultura de convergência dos media. *Perspectivas em Ciência da Informação*, *18*(1), 126–137. doi:10.1590/S1413-99362013000100009

Maravilhas, S. (2014b). Information Quality and Value. In M. Khosrow-Pour (Ed.), Encyclopedia of Information Science and Technology (3rd ed.; vol. 4). Hershey, PA: IGI Global.

Maravilhas, S. (2015b). Managing an information strategy project: The case of a real estate broker organization. In *Handbook of Research on Effective Project Management through the Integration of Knowledge and Innovation* (pp. 19–43). Hershey, PA: IGI Global. doi:10.4018/978-1-4666-7536-0.ch002

McGee, J., & Prusak, L. (1995). *Gerenciamento estratégico da informação: Aumente a competitividade e a eficiência de sua empresa utilizando a informação como uma ferramenta estratégica*. Rio de Janeiro: Campus.

Neto, R. (2013). Gestão do Conhecimento em Organizações: Proposta de Mapeamento Conceitual Integrativo. S. Paulo: Saraiva.

Nonaka, I., & Takeushi, H. (1997). *Criação de Conhecimento na Empresa: Como as Empresas Japonesas geram a dinâmica da Inovação*. Rio de Janeiro: Campus.

Petit, P. (2001). *Economics and information*. Dordrecht: Kluwer. doi:10.1007/978-1-4757-3367-9

Piscione, D. (2014). *Os Segredos do Vale do Silício: O que Você Pode Aprender com a Capital Mundial da Inovação* (1st ed.). São Paulo: HSM.

Rifkin, J. (2011). *The Third Industrial Revolution: How Lateral Power is Transforming Energy, the Economy, and the World*. New York: Palgrave Macmillan.

Rodriguez, M. (Org.). (2013). Gestão do Conhecimento e Inovação nas Empresas. Rio de Janeiro: Qualitymark.

Sordi, J. (2008). *Administração da Informação: Fundamentos e Práticas para uma Nova Gestão do Conhecimento*. S. Paulo: Saraiva.

Stewart, T. (1998). *Capital Intelectual*. Rio de Janeiro: Campus.

Stewart, T. (2002). *A Riqueza do Conhecimento: O Capital Intelectual e a Organização do Século XXI*. Rio de Janeiro: Campus.

Sveiby, K. (1998). *A Nova Riqueza das Organizações: Gerenciando e Avaliando Patrimónios do Conhecimento*. Rio de Janeiro: Campus.

Terra, J. (2001). *Gestão do Conhecimento: O Grande Desafio Empresarial*. S. Paulo: Negócio Editora.

Troxler, P. (2014). Making the 3rd Industrial Revolution: The Struggle for Polycentric Structures and a New Peer-Production Commons in the Fab Lab Community. In J. Walter-Herrmann & C. Büching (Eds.), *FabLabs: Of Machines, Makers and Inventors*. Bielefeld: Transcript-Verlag.

Walter-Herrmann, J., & Büching, C. (Eds.). (2014). *FabLab: Of Machines, Makers, and Inventors*. Bielefeld: Transcript-Verlag.

Ward, J., & Griffiths, P. (1996). *Strategic planning for information systems* (2nd ed.). Wiley.

Wilson, D. (2002). *Managing information: IT for business processes* (3rd ed.). Woburn: Butterworth-Heinemann.

ADDITIONAL READING

Best, D. (1996a). Business process and information management. In Best (Ed.). The fourth resource: Information and its management. Hampshire: Aslib/Gower.

Best, D. (1996b). *The fourth resource: information and its management*. Hampshire: Aslib/Gower.

Brown, E. (2012). *Working the Crowd: Social Media Marketing for Business* (2nd ed.). British Informatics Society.

Burgelman, R., Christensen, C., & Wheelwright, S. (2009). *Strategic Management of Technology and Innovation* (5th ed.). Singapore: McGraw-Hill.

Christensen, C., Anthony, S., & Roth, E. (2004). *Seeing What's Next: Using the Theories of Innovation to Predict Industry Change*. Harvard Business School Press.

Cleveland, H. (1985). *The Knowledge Executive: Leadership in an Information Society* (1st ed.). New York: Dutton.

Cruz, R. (2011). *O Desafio da Inovação: A Revolução do Conhecimento nas Empresas Brasileiras*. S. Paulo: Senac.

Dantas, J. (2001). *Gestão da Inovação*. Porto: Vida Económica.

Dantas, J., & Carrizo Moreira, A. (2011). O Processo de Inovação: Como Potenciar a Criatividade Organizacional Visando uma Competitividade Sustentável (1ª ed.). Lousã: Lidel.

De Brabandere, L. (2000). *A Gestão das Ideias: da Criatividade à Inovação*. Lisboa: Instituto Piaget.

Eppler, M. (2006). *Managing Information Quality: Increasing the Value of Knowledge-Intensive Products and Processes* (2nd ed.). New York, Heidelberg: Springer.

Freitas Filho, F. (2013). *Gestão da Inovação: Teoria e Prática para Implantação*. S. Paulo: Atlas.

Gleick, J. (2011). The information: A history, a theory, a flood (1st ed.). St. Ives: 4th Estate.

Jamil, G., Maravilhas, S., Malheiro, A., & Ribeiro, F. (Eds.). (2015). *Handbook of Research on Effective Project Management through the Integration of Knowledge and Innovation*. Hershey: IGI Global. doi:10.4018/978-1-4666-7536-0

Kelley, T., & Littman, J. (2001). *The Art of Innovation*. Hammersmith: Harper Collins.

Kelley, T., & Littman, J. (2007). *As Dez Faces da Inovação: o Poder da Criatividade e da Inovação na Empresa* (1st ed.). Lisboa: Presença.

Leonard, D., & Swap, W. (1999). *When sparks fly: igniting creativity in groups*. Boston: Harvard Business School Press.

Leonard-Barton, D. (1995). *Wellsprings of Knowledge: Building and Sustaining the Sources of Innovation*. Harvard Business School Press.

Liautaud, B., & Hammond, M. (2001). e-Business Intelligence: Turning Information into Knowledge into Profit. New York: McGraw-Hill.

Libert, B., & Spector, J. (2009). *Muitas Cabeças Pensam Melhor: Como mobilizar o poder das multidões para o seu negócio* (1st ed.). Porto: Lua de Papel.

Maravilhas, S. (2013b). Social media tools for quality business information. In Talburt, Yeoh & Zhou (Eds.), Information quality and governance for business intelligence. Hershey: IGI Global, 167-193.

Maravilhas, S. (2014a). Competitive Intelligence from Social Media, Web 2.0, and the Internet. In Khosrow-Pour (ed.). Encyclopedia of Information Science and Technology. 3rd Ed. Vol. I, Hershey: IGI Global, 558-566.

Maravilhas, S. (2015a). Social Media Tools for Quality Business Information. In *IRMA - Information Resources Management Association (USA). Social Media and Networking: Concepts, Methodologies, Tools, and Applications* (Vol. II). (Vols. 1–4). pp. 636–662). Hershey: IGI Global.

Maravilhas, S. (2015c). Vantagens Competitivas da Informação de Patentes. In Jamil & Pessoa. Estratégias Defensivas: Assegurando Vantagens Competitivas já Conquistadas. Rio de Janeiro: NovaTerra, 319-347.

Maravilhas, S. (2016). Social Media Intelligence for Business. *International Journal of Organizational and Collective Intelligence*, *6*(4), 100–125. doi:10.4018/IJOCI.2016100102

McGee, J. V., & Prusak, L. (1995). *Gerenciamento estratégico da informação: aumente a competitividade e a eficiência de sua empresa utilizando a informação como uma ferramenta estratégica.* Rio de Janeiro: Campus.

Mohr, J., Sengupta, S., & Slater, S. (2010). *Marketing of High-Technology Products and Innovations* (3rd ed.). New Jersey: Pearson.

Rogers, E. (1995). *Diffusion of Innovations* (4th ed.). Free Press.

Schilling, M. (2013). *Strategic Management of Technological Innovation* (4th ed.). New York: McGraw-Hill.

Silva, R., & Neves, A. (2003). *Gestão de empresas na era do conhecimento.* Lisboa: Sílabo.

Surowiecki, J. (2007). *A Sabedoria das Multidões: Como a inteligência colectiva transforma a economia e a sociedade* (1st ed.). Porto: Lua de Papel.

Tapscott, D., & Williams, A. (2008). Wikinomics: A Nova Economia das Multidões Inteligentes (1ª ed.). Matosinhos: Quidnovi.

Terra, J. (Coord.) (2012). 10 Dimensões da Gestão da Inovação: Uma Abordagem para a Transformação Organizacional. Rio de Janeiro: Campus

Trott, P. (2008). *Innovation Management and New Product Development* (4th ed.). Essex: Prentice Hall.

Ward, J., & Peppard, J. (2002). *Strategic planning for information systems* (3rd ed.). Chichester: Wiley.

KEY TERMS AND DEFINITIONS

Creativity: Reasoning that produces imaginative new ideas and new ways of looking at reality. Creativity is an individual process, arises from the idea that popped into someone's head. Relates facts or ideas without previous relationship and is discontinuous and divergent. No creative process exists if there is no intention or purpose. The essence of the creative process is to seek new combinations.

Explicit Knowledge: Knowledge of the facts obtained from information, almost always through formal education. Expressed through metaphors, analogies, concepts, hypothesis or models. The key for the creation of new knowledge. Codified, communicable through formal and systematic language. Can be stored in manuals, documentation, patents, blueprints, reports and other accessible sources. Explicit and tacit knowledge complement each other.

Fab Lab: A laboratory of digital fabrication, serving as a prototyping platform of physical objects, with broad educational, social and economic advantages. These spaces aim to empower its members for the realization of sustainable solutions, using open source tools and equipment's, to allow all the possibility of creating low cost products which meet the need for one, one hundred, or a thousand people.

Information: A set of data arranged in a certain order and form, useful to people to whom it is addressed. Reduces uncertainty and supports decision-making. Information is considered to support human knowledge and communication in the technical, economic and social domains. Results from the structuring of data in a given context and particular purpose.

Innovation: The application of new knowledge, resulting in new products, processes or services or significant improvements in some of its attributes. A new solution brought to the market to solve a problem in a new or better way than the existent solutions.

Invention: The creation or discovery of a new idea, including the concept, design, model creation or improvement of a piece, product or system. Even though an invention may allow a patent application, in most cases it will not give rise to an innovation.

Knowledge: Is a fluid composed of experiences, values, context information and apprehension about their own field of action that provides a cognitive apparatus for evaluating and incorporating new experiences and information. It originates from data and information and allows acting upon it.

Tacit Knowledge: Knowledge that can only be learned through practice and experience. Know-how. Knowledge that is acquired but difficult to explain to others. Subjective insights, intuitions and hunches of individuals. Not easily communicated or shared. To gain access to such knowledge one may have to be practicing in related areas of knowledge. What is held in someone's head and includes facts, stories, biases, and insights. Tacit and explicit knowledge complement each other.

Chapter 16
Reviewing Information Quality:
The Challenge of the "Analytics" Trend

George Leal Jamil
Informações em Rede Consultoria e Treinamento Ltda., Brazil

ABSTRACT

Revisiting a concept is a mandatory approach when new technologies emerge. This chapter presents a reviewed study, based on a previous publication, about information quality. As it happened in the first attempt to study and develop this concept, it remains challenging, as this context contemplates both tangible and intangible observations for two unlimited parts—information and quality—researched isolated, with its own alternatives, methods and objectives for several years and from different points of view. This chapter's approach is to bring additional work published since the original one and develop conclusions reached at that time, providing a new level of comprehension, which allows the continuity of this debate of information quality, considering the emerging strong trend of analytics processing and informational service offering.

INTRODUCTION

This chapter brings a reflection based on Jamil (2014a), updating the challenges there announced for Information quality and value concepts, specially when addressing the first aspect – Quality – considering new trends of information analysis, management and application, as the "analytics" services and market offers.

The original proposal of this chapter was to discuss information quality and value. These factors must be re-evaluated, as data and information processes face challenging times by the increase in the offer and implementations of new technologies, as those which support big data, internet of things and analytics. Analytics were chosen as the central point for this review of a published chapter, updating its results and appreciation. As the main proposition, Quality and Value regarding information and information management were analyzed, considering those new technologic impacts.

It was previously discussed the fundamentals of quality and value as factors frequently regarded to tangible items, processes or even approached by areas of managerial sciences when evaluating systems

DOI: 10.4018/978-1-5225-6225-2.ch016

performance, integration and external relationships with other systems (Huber, 1990; Davenport & Marchand, 2000; SAA, 2013). Authors persist on debating if quality and value are applicable, relevant when analyzing information, data processing, knowledge management and other related factors. It is my point of view that this discussion produces a context of guidance on observing these factors – quality and value – because both are more relevant now, when thinking information management, than three years ago, when this study was started (Amaral & Souza, 2011; Jamil, 2014b; Jamil, 2015).

This chapter intends to reassess the aspect of information quality, focusing specifically analytics trend, as a focal point chosen to bring light to several themes, proposed mainly for technological emerging offers. This conceptual and practical perspectives were addressed in Jamil (2014a), were we posed an invitation to:

the reader to participate in this research, invoking the real need to update - or rethink - this important conceptual base and the opportunity to do so, by bringing case studies and applications which depicts how information quality and value can be perceived in experiments and real work environments.

This study is a comprehensive attempt to precisely rethink that context, as it was proposed in the original research, both justifying its methodological principles, using that base to promote the discussion update and, finally, not reaching a conclusion, as this is a discussion provoked by undeniable practical arenas, expressively present as a trend in nowadays competitive scenarios.

This chapter reviews the original conceptual discussion about quality and value aspects, inserting analytics and other participative concepts to promote the study update. At the end, some reflections, comparing the original results from the previous publication, aiming to promotion of this permanent discussion – information quality – are produced, offering to the reader a continuation perspective for these analysis, towards considering practical applications.

It was kept the fundamental of information importance, both as an isolated concept and as a practical communication process for organizations, as it continues to be one of the most discussed arguments in the last years (Davenport & Marchand, 2000; Choo, 1996; Choo, 2005; Jamil, Santos, Alves, & e Furbino, 2012; Albesco & Pugna, 2014). Its valuable relationship as a concept, as a "thing" and as "a process" (Allen, 1996), when relating to data, knowledge and intelligence, also produces more impacts as a contributive theoretical part for information systems, information technology management, strategic information, marketing information, big data processing and analytics, among several other fields, (Davenport & Prusak, 2000; Cao; Zhang & Liu, 2006; Laudon & Laudon, 2009; Kimball & Ross, 2010; SAS, 2014; SAS, 2017, Yonce, Taylor, Kelly, & Gnau, 2017).

Information remains a topic that requires attention from researchers, with implication also on the practical, application side, as we need to increase the comprehension about its insertion in organizational systems, both as a concept to be determined, and a practical subject to be precisely assessed, additionally when we assess new waves of technologic events (Buckland, 1991; Castells, 2000; Tuomi, 2001; Barbosa, 2002; Yonce, Taylor, Kelly, & Gnau, 2017).

As it was assessed previously, the conceptual debate about information persists. It is important to perceive that it is a fundamental debate, an original conceptualization that is attempted, with various and significant results. A research basics, exposed to any interested analyst, is: if the basic concept is still so debated, is it possible to precisely understand its composition to other waves, trends and interrelated, systemic integration of information to organizational arrangements? To other systems? To immediate, market-driven implementations? Is it enough to promote a sufficient comprehension? (Jamil, 2001; Jamil, 2013).

With this research context defined, invoking new investigations around information and its remarkable conceptual and practical relationships, a challenging situation to produce this chapter is redefined from the previous study.

Exploring the fundamental proposition of this chapter, quality is a concept, or even considered a complete context, generally related as a level of perception of reliability of a system or a process. It is defined as a composition of factors that composes: a) if the final result of a process fits exactly to what was specified as the initial objective of the complete set of tasks (an interplanetary trip, a surgery, one building or infrastructure work, as a road and its supporting structures – does the final result fit to the original scope?); b) if the result was obtained according to the planned cost; c) if the result was obtained in the precise scheduled time for its conclusion (not less, not more) and; d) if there is any learned lesson and / or estimative which allows to make better previsions or improve our overall performance in the next processes (good practices must be retained, risks must be treated, its factors potentially eliminated)? (Crosby, 1979; ASQ, 2013).

It is a demanded approach, understand quality concept as an indisputable applicative concept, associated with industrial productive systems, aiming to contribute for continuous performance and "zero errors manufacturing" processes and projects. After several years of industrial application, it was then reviewed for service sectors, as it was considered the final user´s perception about the result of a task previously contracted, regarding the same set of basic items previously defined.

Quality reached the situation of being a complete discipline, which is taken as a requirement for new competitive markets and organizations, as those from software development, massive services such as airlines, healthcare systems, entertainment, technology, among many others. As a competitive demand and relevant item, these organizational environments have, in their propositions, Quality as a principle, adopting it as a support for their usual plans, actions and formation (Crosby, 1979; Deming, 1986; Setia & Joglekar, 2013).

Originally planned, the relationship of quality principles to information and its management, constituted a provocation to understand how to promote it, contributing for the evolution to better perspectives of information application in modern organizational arrangements, mainly when perceiving emerging technologies as definition of new business models around information (Krstevski & Manchevski, 2016). Quality represents, this way, a concept which can be not only a complementary issue for information, but also an implicated, interrelated aspect, which can contribute and must be analyzed to together with information.

In the basic work, value was the other conceptual aspect that formed the question to be discussed. In this chapter, as a review, quality is promoted to a first-level of attention, as value was observed as a collateral element, just for methodological reasons, not because it did not deserve the focus anymore. It is a practical posture, to approach quality, but keeping value as a contributive topic for further studies. Along with a potential dilemma regarding the implications and relevance of both concepts itself, it would be advisable for the reader to consider that value is not obscured by quality, as a conceptual base for this study. Otherwise, it should deserve a separate essay, an individual development just as this done to analyze perspectives for quality.

Reflecting on value, it is possible to relate its definition to the strict conception of a property rigorously associated with an amount of a countable metric, a comparative, quantitative aspect. It is regarded as a financial or monetary value, physics metric (temperature, moisture, pressure, liquid flow) measurement, performance of production or operation, among others. This approach, obviously, is essential and cannot be forgotten, as it introduces an indispensable, basic, critical task: to measure evolution or accumulation of something. But it is considered restrictive, limited, as it is the intention of this chapter´s approach to advance in the value concept analysis, as it turns possible to understand its depth and contribution for dynamic organizational processes comprehension (Howard, 1966; Kotler & Keller, 2005).

From Marketing field, it is possible to understand value as a broader and powerful concept. There, value is taken as a fundamental concept, where, in an initial evaluation, it is a perception associated with a relationship of exchange, as it is proposed as the difference between the benefits perceived by a customer when he obtains something, discounting the costs this customer faced to get those benefits (Kotler & Keller, 2005; Laudon & Laudon, 2009). As a decisive concept, it is regarded, eventually to produce Marketing concept itself, as some authors simply add that Marketing is, *a set of disciplines organized to add value to products and services*.

For both quality and value, it was provocative on understand how both concepts were related to information. As unfolds of this observation, understand how information is related to data, knowledge, organization and systems concepts, how it potentially contributes for quality and value for one organization strategic positioning and, on another specific context of observation, how it also presents quality and value, as this fact additionally implicates on how information – and its related concepts – can become more relevant for potential analysis and development in practical applications. Modern organizations, as the proposed startups, some of them aligned to information management perspectives towards knowledge generation (Maital & Seshadri, 2012; Bereznoi, 2014), attempt to encompass these concepts, sometimes intuitively, opening another perspective for this chapter application and future research. It is a motivation to evaluate how these concepts are associated or contribute to each other and to the whole informational context, increasing their comprehension for application in different areas and for detailed studies, applying other methodological definitions and techniques.

Concluding this introduction, it is interesting to analyze how analytics-related process and market offers were proposed as a potential situation to advance the original studies. Analytics can be simply considered as a coordinated way where users´ final data is collected and anonymously treated to produce results that inform about potential interactions dynamics (Chen, Chiang & Storey, 2012; SAS, 2014). As it was discussed by authors, such as Mahrt and Scharkow, (2013) and El-Gayar and Timsina (2014), analytics is becoming both a research field and a profitable commercial sector, where players can offer services for final users. As a comment, these markets are becoming stronger nowadays, as analytics are performing the customized, oriented application of information management for knowledge production and decision-making in organizations. This way, analytics processes were chosen as the background for this study, as it produces interactions with other opportune fields as strategic and marketing planning, information technology management, information systems and several other, also encompassing perspectives to study new technologies insertion in corporative strategies and tactics.

It is recalled that this chapter, adopting its fundamental propositions, methodological support and proposed focus, intends to promote additional discussion, bringing the original approach of value and quality, related to information, originally developed in Jamil (2014a), attempting to productively advance that study to the level of comprehension around the analytics phenomena. Considering this strategy, the chapter evolves with a theoretical background review, updating the original approach and enhancing with opportune developments, then advancing for conceptual integration and, finally, with some reflections on practical cases, therefore supported by the previous conceptual base.

THEORETICAL BACKGROUND

In this section, Quality, Analytics, information and other concepts are approached to leverage our comprehension towards an analytical base. Also, other collateral and contributive concepts are defined, composing this theoretical background.

Quality

A previous definition for quality concept was issued in this chapter introduction, when it was presented as an aspect of a product or service which was thoroughly proposed – not more, not less – in a deliberated plan, after a negotiation with customer or final user, in terms of the exact time and cost for production, adequacy to specifications, finally, allowing the needed managerial knowledge which emerges or is created in each phase, subtask or characteristic, used or implemented to be replicated or improved in the further projects or works. As it can be thought, it is an application-oriented, mainly tangible, industrial concept which presents real challenges and opportunities to be evaluated as to work with an intangible asset, as information.

Walter Shewhart is a researcher who remarkably worked with quality, as it became a relevant concept, specially after the World War II. At this time, industrial systems had an expressive demand for complex evolution and faced the emergence of a worldwide economic stability and expansion, which not only required increased production levels, but its optimized management in various aspects. A classified set of techniques were proposed and developed, as those supported by Statistics oriented to produce an important subset of methods guided to manage production quality through direct statistical control. Shewhart´s work, for example, was associated with a modern attempt to plan, manage and predict production aspects and risks, relating to mathematical models, tools and techniques, reintroducing quantitative managerial principles to adjust productive sites (ASQ, 2013).

It was a remarkable work that defined an interesting principle of quality adjustment, generally identified as measurement. Among quality techniques offered, it is possible to cite: (a) Event mapping and relationships; (b) Strategic and tactical goals proposition and monitoring; (c) Adoption of outsourcing, as a strategic alternative, and; (d) Error sampling, treatment and eliminations and many others. These are fundamentally based on measuring actions, which produced quantitative sampling, as production levels, results, product acceptance and rejection, which could be historically followed, compared and managed, with tactical implications. Broadly speaking, measurement precedes more detailed qualitative analysis, which will complete information and knowledge generation about a project or process, enabling the comprehension on what performance level the current process is being held and how to implement optimization techniques, to evolve it for the next level where any critical variable can be adjusted to develop, correspondently, quality.

Crosby (1979) structured the conceptual base for quality, when he sat a "zero defects" level objective for general processes, which had to produce a result exactly as it was expected. Additionally, a process should operate precise and correctly for the first time (it was created the acronym DIRFT – "do it right for the first time"). This method usually proposed to show a comparison of a measurement of quality deviation based on the principle of nonconformance, taken as the cost of posterior adjustment of a result to what was expected, due to perceived differences between the prediction and the result that was obtained. His works were proposed and diffused during a critical period for the American industrial sector, when it faced tough competition by Japanese corporations which outperformed American in various critical aspects for commercial products supply. Quality, as it was studied and proposed by Crosby, could constitute in an effective competitive characteristic for American enterprises, even reaching competitive and historical conceptions.

William Edward Deming (Deming, 1986; ASQ, 2013) is another researcher and consultant that must be cited when approaching quality concept development. Deming was a remarkable contributor for Quality concept comprehension, study and diffusion, citing, among many of his works the proposition of PDCA – Plan/Do/Check/Act – cycle, a continuum which relates planning, execution, verification and improvement action for a process. Also, his work for Japanese government in the country´s reconstruction after the Second World War is identified along his remarkable work. Deming based his propositions also in Statistics applied to management, announcing the "fourteen principles of quality", considered as a set of principles for quality management in his book, *Out of Crisis*, of 1986.

As Japanese corporations became more and more competitive, especially in a decade from the middle of 1970s to the middle of 1980s, quality concept spread in the world, for scientific studies which captured it from empirical evidences. When these firms started to sell competitively in markets such as industrial machinery, automotive, electronics, computing and many others, in continuity to that cited cycle, scholars and practitioners tried to relate quality principles application to Japanese corporation development (Nonaka, 2008; ASQ, 2013). Nowadays, it is widely perceived in sectors as different as services – medicine, biotechnology, logistics, etc. – and industrial – automotive, electronic, domestic appliances, among others, became no other cases for production or services offering that benefitted from quality studies, but is considerable that quality works as a basic principle that must be implemented for businesses survival and respective development.

It is affirmed, in the end of this section, that Quality can be understood as a basic principle of continuous improvement considering an actual production scenario and the associated capability of obtaining better level of results for a process based on rigorously configured and registered variables and its related measurements, with the application of methods and techniques that will allow to identify the actual level of performance and its related evolutionary aspects.

Analytics

In the last couple of years, one trend affirmed: analytics modelling and processing. As massive amounts of data are being produced and stored every day, as it comes from human basic interactions – such small purchases, services consulting and contract, products acquisitions, etc. – data is created and stored dynamically. First, it can be understood that this data could be monitored only for the control of the specific task itself. But, at the end, as several studies proved, this data and associated information could be processed to promote knowledge about simple or complex phenomena, as, for example, customer behavior, supplier associations, price maneuver in a competitive chain, etc. Advancing this analysis, adding the aspects of unstructured data also processed, as that asset created and processed through social media forums, we reach the point of a more comprehensive analysis of market actions (Chau & Xu, 2012). As these advances continues, we reach the level of Analytics, as approached by SAS (2017), among several sources, where it is possible to understand it as the capability of producing insights that enables decisions based on real facts. According to the same source, analytics is a set of organized resources, as tools, techniques, algorithms and devices which aims to develop these decision-making capabilities. For example, taking one of the existent massive services, such as google, Facebook, Instagram or other social sources platforms, it can be seen that structured and unstructured data can be collected from a focused context, allowing this intended development. It is a way to implement better decision capabilities around user´s objectives and interactions, promoting the application of customer behavior analysis to another higher level of conclusions, as it as previewed in former works about information management (Choo, 2003).

From a commercial point of view, analytics is a term that identifies possibilities and perspectives of integration of software-based tools and methods, analytical processes, techniques and other related resources, which can be applied to collect evidences about one continuous interaction, regarding a phenomenon, then developing structured relations – as percentages of mutual correlation – among variables and conclusions. This way, analytics help a planner to a more useful generated knowledge and information produced by one user, by a group of users and by groups of users of a specific service, becoming a roadmap of user behavior along time and events associated with this specific context. This can contribute for information systems design and other technologic application for final processes in any scale, as it justifies the value associated with this trend – analytics – in actual markets (SAS, 2014; SAS, 2017).

Here, in this chapter, is desired that a comprehension of analytics can implicate in the information quality parameters previously studied. Analytics extend the perspectives of information usage for decisions, expanding also the possibility of requirements for one system design. Additionally, it can result in additional levels of cost and time to manipulate, associated with its complexity. It is a "trade-off" problem, as it also increases the perspectives for knowledge generation to develop comprehension for this specific process. Finally, the good practices, as this new era is still at its beginning, are still being evaluated, with good perspectives ahead. Analytics extend the capabilities of information systems and for the design of informational solutions.

Information Quality

Two broadly accepted concepts – information and quality – are discussed together in this section. The objective is to produce a conceptual leverage that could promote the intended "information quality" aspect, which will guide the future work for this chapter. It is a potential association, which could result in a wide spectrum of research and practical evaluation of information applied in the workplace and markets.

First, it is opportune to discuss one alternative for information concept. As worked by Buckland (1991) and other authors, we could consider:

- "Information-as-a-thing" as one element, related to data and knowledge, as proposed in Davenport and Prusak (2000), Jamil (2012) and Jamil (2014a). In this conceptualization, it is an item, potentially produced from data and used to develop knowledge.
- "Information-as-an-action" or "information-as-a-process" which is usually referred to the communication theory, defining information as an act of inform someone about something (Jamil, 2001; Kearns & Lederer, 2003; Jamil, Santos, Alves, & e Furbino, 2012).

This definition already produces questions to be answered, as, in the first case we face the production of something, in the second, additionally, we have a process, indeed. For example, analyzing the quality fundamentals, of scope adequacy, time and cost management, it is just possible to identify diverse issues to be observed and composed from a potential study on quality (Crosby, 1979; Marchand, 1990; Buckland, 1991; Campos, 2008; Jamil, Santos, Alves, & e Furbino, 2012).

The initial part considers it as an effort to answer those aspects of quality – scope delimitations (user demands), exact cost value, exact time value and perceptions of best practices – when the information is produced and can be delivered to the intended receiver. Information scope can be defined as a definition for its immediate application. For instance, it can be a business concept definition, as "budget limit" or "production level" or a concept, defined in one specific scientific area, as "brand equity" or "logistics transportation level".

Consider the aspect of information quality, it should be analyzed if the resulting content exactly answers to the user needs. This way, it shows a level close to that specified for material items, for produced objects. Some specific points that should be accessed for this relationship are:

- If there is ability to define information scope;
- If there is ability also to recognize its completeness and other essential characteristics;
- If the conditions associated to the scope are – as "application" – are well defined

This first level of this relationship is therefore understood as a perception on how satisfied is a final user with the information produced or recovered from a complete context (Marchand, Kettinger & Rollins, 2001).

The practical, application-oriented previewed by Marchand (1990) and Marchand, Kettinger and Rollins (2001), enlightens new considerations about information quality. Information, as it is generated, stored, shared, processed, gathered, collected and transferred to a user, can be qualified. It is also perceived as more suitable for the user needs, with cost and time management addressed, for instance, when we have application of this considerations of quality in a process that focus information, such as one information system project.

As cited in Jamil (2013), one example can be worked if a user is considered eligible to receive a table with data that consolidates a quantitative parameter, such as consumption of a specific commercial item in a region of a country. Thinking in a regular way to recover this information, using a spreadsheet-like table, one can think a simple structure, showing the sales evolution of this good, by geographical area (or administrative area, such as a state or town), detailed by month. In columns, one dimension, in lines, another dimension, with a cell showing the value and measurements or indicators (sums, averages, etc.) on the last lines and columns of the useful part of the table. A map could also be used to present this information, which, interactively – through a mere "click" in a computerized system – could show the same set of values, spread over the regions considered and associated with small graphs (shown as graphical "balloons" over the region map boundaries) in a disposition that is regularly supported by programming languages available to develop Internet-based systems. It can be affirmed, if analyzing from the user perspective, that the second way to show the same data is more attractive and can even open more possibilities for its application, generating another level of quality, taken as the final user perception of answers (Ravichandran & Rai, 1999; Campos, 2008).

Analyzing implicating factors to evaluate information quality, storage is to be noticed. From information science sponsored or native studies, such as Davenport and Prusak (2000) and Jamil (2005), it is possible to define storage not only as placing information constituents in a physical device, as a magnetic or optical registering, but also the proposition of methods of classification to store it, enabling, through addressing methods, its optimal retrieval. Information technology (IT) and information systems are decisive for these aspects and are to be considered as tools and configurations to promote such quality factor perception for information. It is possible, from computer science works and researches, to identify how storage principles, schemes and organization are strictly related with information usage, reflecting on the user´s appreciation for one automated device or code that deals with information (or, one information system). Additionally, it is interesting to note that these aspects also have potential impacts over time and cost quality factors, as they correspond to impacts on recording, organizing and using information for any organizational purpose (Jamil, 2001; Turban, Mc Lean & Wetherbe, 2002; Lucas Jr., 2005; Turban, Rainer Jr., & Potter, 2007).

Additionally, to this aspect, it must be noticed that analytics processes tend to be, at least at its beginning, a bit chaotic on producing data and information (SAS, 2014; SAS, 2017). Some situations define even the application of automated devices communicating with others (Yonce, Taylor, Kelly, & Gnau, 2017), as Internet of things, became a state-of-the-art solution to be adopted, configured and implemented in organizational situations. This perspective recalls the fundamentals of information technology management and information systems design, as it prescribes situations where information need to be produced coherently to its application, being treated and kept integral for usage (Turban, Jr., & Potter, 2007; Stair, & Reynolds, 2009). Information retrieval, an immense and dynamic research field by itself, is always mentioned because of this fact, and we strongly believe that this situation, without the perspective on managing information from the start, faces a trend of turbulence as those technologies became easier applied.

Cost is another one of the basic aspects for quality that deserves needed study, as information quality is focused as one main theme. Cost can be defined as the sum of efforts and prices to produce and supply information as needed by the final user, calculated from a summary of values that encompasses people costs (wages, for instance), information technology infrastructure, software (for example, to store and retrieve), transmission, codification, classification, recovery systems, among many other information system components.

Some authors, as Marchand, Kettinger and Rollins (2001), Prado (2006), and Campos (2008), comment about information markets or situations where information can be treated, for some purpose or objective, as an item – this way aligned with Buckland (1991) principles – and negotiated for some practical, financial or quantitative value. There is a growing towards a revival of "information broker" services, as startups and traditional firms are dealing, nowadays, with data processing and market intelligence (Jamil, 2014b) to produce informational sets and associated knowledge. As this situation prevails and develops in the market, we are on the verge to define information cost management as one critical topic to promote its application and share throughout marketing value-aggregation chains. These services can be noticed when one can buy – at a predicted cost and time, answering to its scope definition – information in a form of a market research, for instance.

Cost is a factor to be considered as an impacting measurement to calculate the return on investment made by a research company which produced it, or the payment to be made for one specific information supply for a tactical planning decision. Cost is impacted by the adoption of optimal information technology infrastructure, defining the ability on designing IT systems as one related aspect (Leidner & Elam, 1995). This scenario can be found in the modern information services companies which sell prepared, validated and documented information for decision-making for market sectors and strategic planning (Kaplan & Norton, 2007; Porter, 2008). As stated in Jamil (2009), soon it will be possible to define the information market, along with information-as-a-thing and information-as-a-process, as a growing and imposing concept, maybe even pressuring the conceptual redefinition held by authors some years ago (Buckland, 1991; Marchand, Kettinger & Rollins, 2001; O´Brien & Marakas, 2008).

Time, another component of quality, in this case, for information, will be related to a perception that information was delivered to the final user in the expected time. Scope delays and other time-related inadequacies, will impact user´s perception on the quality of the related informational process. It is commonly regarded that an obsolete information starts to lose its related value. In a case of a project, time discrepancies can lead to an additional risk, as it can be applied for decision-making, although referring to a non-existent situation or scenario, implicating in an incoherent decision.

Again, the modern information usage scenario imposes conditions that can bring some implications to its generation and usage. For example, an immediate decision to be taken, in a logistics organization, with an information related to traffic and transportation costs which is generated with a relevant delay, can imply on a manager simply adopting the first set of collected information and implement a precarious transportation arrangement. This incomplete decision, without analyzing other situations, can result in financial and performance losses, as it does not offer alternatives and clear contexts for decisions. Clearly, this is a conflicting aspect of information quality, as, obviously, it cannot be considered the best one, optimal, for any criteria. Attention is called to the precarious situation of the user – the logistics manager, responsible for this decision – to perceive that information delivered is well fitted to organizational needs. So, the estimates must be precise, and the time spent for the information to reach its user is precisely adjusted and reached (Ravichandran & Rai, 1999).

As the last aspect approached in the quality concept, it is approached the potential continuous improvement, best practices learning. Here, one aspect to immediate consider is the role of information in processes, called knowledge management (Jamil, 2005; Jamil, 2014a). A question that arises, when it is possible to think about the organizational life, is what is to be improved in the next context, for information producing and application, in the next execution of a process? So, if information was correctly delivered before, in faster and cheaper way, one can consider that an improvement has been reached for this new cycle of an information supply. As the knowledge management process intends, information becomes an intermediate agent, not a mere object, but an active layer, that helps produce knowledge and all the related process.

It is to consider that cheaper, or excessively simple systems also can produce poor quality contents or even some of difficult access, implicating in corresponding loss of quality perception, as a "confuse" result. So, there is a commitment among these and other factors that must be analyzed to plan the desired level of improvement (Miller, 1996; Kahaner, 1998). This aspect represents a strong difference as quality is studied associated to an intangible asset as information, when we compare it to analysis of tangible production schemes in industries, or even to services-oriented processes.

The intricate and complex relation among cited factors – scope, time, cost and continuous improvement – create several opportunities for studies about information quality: maybe some new factors / aspects should be observed, the user influence – as studied by the valuable "user centered studies" promoted by Information Science – on information systems, the real role of information technology, other factors that are perceived as improvements by the final users along with many others compose a provocative research scenario, similar to those that always were seen along the years when information is studied. A special word must be said relating to analytics, as this trend is irreversible and will result, first, in a wave of information production, then on adoption of new processes definition and tools to implement an optimal information management in organizations.

This section presented information quality as an association of classical approaches, both from technical and theoretical sources from information and quality studies. Together, they produce an opportune scenario for learning, organizational structures, process management, among several other, numberless areas that can contribute from this development. Concluding this section, a discussion about related topics – strategy and information value – will be conducted to produce a base for the final announcements of this chapter.

Contributive Concepts: Information Value and Strategy

Value is a concept regularly defined as a quantifiable measurement of something. But its modern perception, proposed in the Marketing theories, can become significantly provocative, offering interesting ways to study its dynamics relationships with information (Kotler & Keller, 2005; Sheth-Voss & Carreras, 2010). In an objective way, value, as it is proposed by Marketing, can be better understood as a result of a simple equation: the perception of the set of benefits that someone can get from a material item or service minus (discounted) the corresponding cost to obtain such benefits. When this difference results positive, it can be affirmed that the negotiation is valuable or showed a higher aggregated value for the final user. A negative result produces a situation of uninteresting offering relationship. Improving this initial conceptualization, it is opportune to understand both elements of this difference: benefits and costs.

It is important to notice, in this definition, that both evaluations – perception of benefits and its associated costs – are done by the final customer. Value is a concept perceived externally, although it can be promoted by internal, intrinsic characteristics on a product or service, based on the findings of an external user who will appreciate the benefits. These facts, when considered for information relationships, management and application, allowed some discussions over the years, as, for some researchers, it is an unsuccessful attempt on materialize aspects for information, from the consideration of strict measurable components to the complete denial of possibility to announce something as "information value". For others, with whom we align in this chapter and book, this is a well-designed and proved parameter, although no disregard is paid to the eventual opposite views, as they can contribute to review these findings anytime. So far, information value is regarded as a reliable parameter to classify and understand information management with wider comprehension.

It is possible to affirm that information can be considered itself as a benefit – observing those two basic definitions from Buckland (1991) – when we analyze the reduction of uncertainties, precision on decisions, knowledge formation and other organizational and professional benefits (Jamil, Santos, Alves, & e Furbino, 2012; Jamil, 2014a). As a good result from an information management process, regarding all its conceptual work, the results can be applied for decision making, better understanding of that process and its relationship to many others, scenario clarification and knowledge generation from raw and dispersed data in a potential market intelligence arrangement (Jamil, 2013; Yonce, Taylor, Kelly, & Gnau, 2017). Information, as an intangible good, is then related to several organizational results, such as processes definitions and structuration, decision-making and its related coordination and implementation and other unfolding. But its associated benefits perception – easiness of comprehension, learning, application, update, storage/registering and share – compose the notion of benefits associated with information and its management (Jamil, 2005).

Propositions to measure information range from the quantification of bits or electrical signs to money values, reaching this deeper Marketing-oriented conceptualization. It is possible to exert the limited, restrict way, considering information more valuable, in quantitative categorization, comparing it to other, eventually analyzing its perspectives of an index return (cost, price, performance, etc.). There were various attempts to associate the financial analysis of information with other systems and organizations, considering its future sale or buying, as it was in a case of information needed for a critical process. These frequent attempts resulted in model propositions, legal recommendations about ownership, safety and diffusion and costs related also to the infrastructures needed to deal with information (Weiss & Verma, 2002; Kotler & Keller, 2005; Sheth-Voss & Carreras, 2010).

As commented in Jamil (2009), it was there proposed the analysis of a daily, essential managerial parameter: "productivity". Here, we observe this indicator as an information. Productivity can be announced regarded to an industrial or factor line can, as it can be a composition of individual factors, as quantities produced, accepted products, definitively rejected products, cost of performance, industrial performance *per capita* or taken as a ratio from a unity of time (days, weeks, months, etc.), among many others. Exhibiting this information as a "final", concise, isolated number is a routinely task, but some doubts remain: Will this numeric value conduct the informed agent to a more precise decision? Is it correct to think about improving overall industrial performance, based on this index, or is better keep it at this level? It is possible to apply the data and information produced this way, at this level of acceptance and analysis, as it can be regarded to the usage of it to an objective decision, for instance, to adjust or adequate production systems. According to Jamil (2009), *if it is considered too simple and incomplete, demanding new studies for the user, it can be understood as useless*. Otherwise, considered complete, reliable and objective, this final number is applied as soon as it is obtained and is valuable (Ahkbar, 2003).

The cost to get this benefit composition is also complex, as it considers measurement infrastructure, equipment, devices applied, processes definitions (which involve several components *per se,* as skilled people, organizational preparedness, documentation, etc.), site sampling criteria, among many other factors. As presented by Brynjolfsson and Hitt (2000) and Basnuevo (2005), an automated process, such as those associated with analytics, can present a very low cost, implicating in a better perception of informational value, especially for low-margin processes and offers. But in a higher, top-level aggregated value process, with a sophisticated "software layer" information system, for example, it is possible to identify a higher cost, implicating in other calculation or perception for value. This way, information value shows also additional restrictions if it is applied alone, in an isolated managerial decision (Ahkbar, 2003; Kearns & Lederer, 2003).

Jamil (2009) also approached the relationship of information to value aggregation for products and services. Also based on Marketing analysis, value proposition for commercial and technological can be impacted by information availability and its associated quality. As information quality was developed regarding cost, time, scope and learning perspectives, according to the theoretical work previously conducted, this relationship has the potential of better comprehension on how information is associated to quality, reinforcing our findings. It leads to the need of a brief discussion of strategy.

Strategy is defined as a discipline to design a future position for the future of an organization (Hitt, Ireland & Hoskisson, 1999; Mintzberg, Lampel & Ahlstrand, 2009). This short definition composes important issues on if this future scenario and actions can be designed as a documented plan or will it correspond to informal decisions? Is there any strict definition to be adopted as organizational standards for procedures or activities? How long this plan must be kept without changes? These and many other questions concerning to strategy conception are studied in the field or research around this topic and its associated roles to planning tactics and operations. Its repercussions are, organizationally speaking, fundamental.

Porter (1980) studied the case of formal versus informal strategic planning, and this analysis is still actual, demanding more researches and practical observations. As considering information quality and analytics, this is an authentic field where changes are expected to appear, as data and information will be more produced in a potentially disarranged way, posing threats for formal planning structures and processes. For strategic purposes, it can also produce a competitive advantage for one player that can apply analytics dynamically, proposing final scenarios for simulations, studies and implementations, leading to a better capacity for strategic planning. This situation recalls some aspects identified by Mintzberg,

Lampel and Ahlstrand (2009), aligning one organization to a process of composition of deliberated and emerging strategies, in an optimized strategic responsive system.

Approaching an opportune influence of information quality for strategic planning, it is interesting to evaluate some structured methods which have been applied for strategic studies along the last decades as the Product / mission Matrix (Ghemawat, 2006); SWOT model (Andrews, 1980; Barney, 1995); Competitive forces analysis – Porter (1980) and BCG Matrix (Barney, 1995; Writh, Kroll & Parnell, 1998).

Along with endless discussions around its validity, objectives and perspectives, these remarkable propositions have been applied for strategic analysis and studies for many years, generating knowledge, trends and historical evolution, being considered as standards for emerging and established markets (Venkatraman, 1989; Barney, 1995; Mata, Fuerst, & Barney, 1995; Wright, Kroll & Parnell 1999; Hitt, Ireland & Hoskisson, 2001).

Henry Mintzberg and his associated authors presented significant reflections about strategy formulation and planning, and its design for new market challenges, reaching an actual level of discussion (Mintzberg, Lampel, Quinn, & Ghoshal, 2003). In these and other works, they reviewed those planning fundamentals cited earlier (models, techniques, etc.), in an approach that is correct and precise for nowadays scenario, as changes introduced by analytics and associated infrastructure, such as big data and social media platforms can present new demands and realities for strategic formulation and planning.

As other valid strategic formulation principle, it is possible to reach the ideas of resource based view (RBV) theory, which was proposed in a basal work by Edith Penrose, entitled *The theory of the growth of the firm*, published in 1959. There, Penrose assessed the proposition of observing a firm as a collection of resources, which are dimensioned as tangible and intangible components, involving, for example, capabilities, domains, processes, information and knowledge, all under managerial control of the firm (Wernerfelt, 1984; Barney, 1991; Barney, 1995). As intended by this theory, strategic planning is understood as a systematic process on identifying resources and positioning it to obtain potential sustainable competitive advantage. This is a dynamic context where data processing, analytics and other modern trends can result in various unfolding perspectives, as data and information generation are capabilities, or resources, that can be a differential and can produce this advantage through any planning ability, as to foresee strategic and tactical scenarios and the related actions to reach any alternative.

Former resource based view studies approached, initially, information technology resources and its planned application, both to identify and classify critical resources, but also as a valuable item itself, to be considered a potential producer of competitive advantage, to be checked for its possibilities for this offer (Porter, 2001; Kearns & Lederer, 2003; Liao, Kickul & Ma, 2009). Competitive advantage was interpreted as differentials offered by an organization, which are identified by a customer in a competitive industry as a decisive component for choosing one negotiation alternative (Porter, 1980). If these differentials are evident and strong enough for a significant period, all rivals eventually will give up competing for industry leadership, becoming possible followers when we identify this situation as a proposition of sustainable competitive advantage, a desired level of strategic planning (Porter, 1980; Porter, 1996; Barney, 1991; Barney, 1995).

As instructed from these works, all information technology and management can be applied for opportune competitive advantage design, both as a resource or element of a strategic planning base or as a component of a complex element, as an intercommunication that gather an effective working team, who, by its turn, will represent this competitive arrangement which will be perceived as a differential by customers and sector stakeholders. As big data, analytics and associated infrastructure are practically offered, this proposition, of competitive resources being generated and position starts to be more dynamic, demanding domain by strategic planners and implementers, leading to situation where the market must be monitored, interpreted and actions planned in faster and risky ways. Knowledge and information management principles must be reassessed by scholars and practitioners, on information quality under this time, to provide the needed structured capability to correctly approach this new paradigm.

CONCLUSION

When approaching technologic evidences and its immediate offers to the market, several concepts must be reevaluated, with its design and conception revisited to induce an update of its practical application. In this chapter, information quality was assessed under new trends of data and information generation, as produced by analytics tools, methods and processes. As practical findings, nowadays, it is possible to identify, even for small organizations, the functioning of a "big data" context or, generally speaking, *how to identify customer reactions in the corporative asset of social media.* These are undeniable facts that announce how technology-based infrastructure, together with its associated resources, reached the final market, producing new situations for a potential gain in knowledge, which must be reviewed carefully.

As it was shown in this text, as the situation become more massive, in terms of potential knowledge generation, various parameters and aspects of management can be threatened, paradoxically leading the manager to a more confuse situation, exactly when prompt solutions were promised. For example, if we faced the promise of dynamic strategic scenarios proposition, it is possible that we reach a point of diffuse, undetermined scenarios, that will produce a riskier condition for strategic choice and decision-making.

This way, topics, such as information quality, as a specification on how information is generated, stored, shared and developed throughout one organization, and provides optimal conditions to produce knowledge, must be continuously revisited, rethought, to enable data, information and knowledge, being correctly applied, in a managed way.

REFERENCES

Akbar, H. (2003). Knowledge Levels and their Transformation: Towards the Integration of Knowledge Creation and Individual Learning. *Journal of Management Studies*, *40*(8), 1997–2021. doi:10.1046/j.1467-6486.2003.00409.x

Albescu, F., & Pugna, I. B. (2014). *Marketing intelligence: The last frontier of business information technologies. Romanian Journal of Marketing.*

Allen, B. (1996). *Information tasks: Toward a user-centered approach to information systems.* San Diego: Academic Press.

Amaral, S. A., & Souza, A. J. F. P. (2011, January/March). Qualidade da informação e intuição na tomada de decisão organizacional. *Perspectivas em Ciência da Informação*, *16*(1), 133–146. doi:10.1590/S1413-99362011000100008

Andrews, K. (1980). *The concept of corporate strategy.* Homewood, IL: Dow-Jones - Irwin.

ASQ. (2013). *American Society for Quality.* Accessed at www.asq.com

Barbosa, R. R. (2002). Inteligência empresarial: uma avaliação de fontes de informação sobre o ambiente organizacional externo. *Datagrama Zero, 6*(3). Disponível em http://www.dgz.org.br/dez02/Art_03.htm

Barney, J. (1991). Firms Resources and Sustained Competitive Advantage. *Journal of Management, 17*(1), 99–120. doi:10.1177/014920639101700108

Barney, J. (1995). Looking inside for competitive advantage. *The Academy of Management Executive, 9*(4), 49–61.

Basnuevo, C. A. (2005). *Antecedentes e situación atual de los conceptos y métodos para el desarollo de la inteligência organizacional.* Disponível em http://bvs.sld.cu/revistas/aci/vol13_4_05/aci02405.htm

Berenzoi, A. (2014, December). Business Model Innovation in Corporate Competitive Strategy. *Problems of Economic Transition, 57*(8), 14–33.

Brynjolfsson, E., & Hitt, L. (2000). Beyond computation: Information technology, organizational transformation and business performance. *The Journal of Economic Perspectives, 14*(4), 23–48. doi:10.1257/jep.14.4.23

Buckland, M. K. (1991, June). Information as thing. *Journal of the American Society for Information Science, 42*(5), 351–360. doi:10.1002/(SICI)1097-4571(199106)42:5<351::AID-ASI53.0.CO;2-3

Campos, V. F. (2008). *Qualidade Total – INDG.* Belo Horizonte: INDG.

Cao, L., Zhang, C., & Liu, J. (2006). Ontology-based integration of business intelligence. *Web Intelligence and Agent Systems. International Journal (Toronto, Ont.), 4*, 313–325.

Castells, M. (2000). The Rise of the Network Society. In The Information Age: Economy, Society and Culture (Vol. 1). Cambridge, MA: Blackwell.

Chau, M. & Xu, J. (2012). Business intelligence in blogs: understanding consumer interactions and communities. *MIS Quarterly, 36*(4), 1189-1216.

Chen, H., Chiang, R. H. L., & Storey, V. C. (2012). Business intelligence and analytics: from big data to big impact. MIS Quarterly, 36(4), 1165-1188.

Choo, C. W. (1996). The knowing organization: How organizations use information to construct meaning, create knowledge and make decisions. *International Journal of Information Management, 16*(5), 329–340. doi:10.1016/0268-4012(96)00020-5

Choo, C. W. (2005). *The knowing organization: how organizations use information to construct meaning, create knowledge and make decisions.* Oxford, UK: Ed. Oxford University Press. doi:10.1093/acprof:oso/9780195176780.001.0001

Crosby, P. (1979). *Quality is Free.* New York: McGraw-Hill.

Davenport, T. H., & Marchand, D. (2000). *Mastering information management.* Boston: Financial Times Prentice Hall.

Davenport, T. H., & Prusak, L. (2000). *Working knowledge: how organizations manage what they know* (2nd ed.). Harvard Business Press.

Edwards, D. W. (1986). *Out of the Crisis*. Cambridge, MA: Massachusetts Institute of Technology, Center for Advanced Engineering Study.

El-Gayar, O., & Timsina, P. (2014). Opportunities for Business Intelligence and Big Data Analytics In Evidence Based Medicine. In *Annals of 47th Hawaii International Conference on System Science*. IEEE. 10.1109/HICSS.2014.100

Ferrel, O. C., & Hartline, M. (2010). *Marketing Strategy*. South Western College Publications.

Ghemawat, P. (2006). *Strategy and the business landscape* (2nd ed.). Boston: Prentice Hall.

Howard, R.A. (1966). Information value theory. *IEEE Transactions on Systems Science and Cybernetics*, 22-26.

Huber, G. P. (1990). A theory of the effects of advanced information technologies on organizational design, intelligence and decision making. *Academy of Management Review*, *15*(1), 47–71.

Inmon, B., Strauss, D., & Neushloss, G. (2008). *DW 2.0: The Architecture for the Next Generation of Data Warehousing*. Morgan Kaufmann.

Jamil, G. L. (2001). *Repensando a TI na empresa moderna*. Rio de Janeiro: Axcel Books do Brasil.

Jamil, G. L. (2005). *Gestão da Informação e do conhecimento em empresas brasileiras: estudo de múltiplos casos*. Belo Horizonte: Ed. Con / Art.

Jamil, G. L. (2013). Sistemas de informação e a tomada de decisões: estudos de caso sobre a inteligência de mercado e a gestão de marketing. In P. Campos & P. Q. Brito (Eds.), *Novas tendências em Marketing Intelligence*. Ed. Actual.

Jamil, G. L. (2014a). Why quality? Why value? Is information related to these aspects? In G. L. Jamil (Ed.), *Rethinking the conceptual base for new practical applications of information quality and value*. Hershey, PA: IGI Global. doi:10.4018/978-1-4666-4562-2.ch001

Jamil, G. L. (2014b). Inteligência de Mercado como um processo de gestão da informação e do conhecimento: proposta de oficinas de capacitação setoriais. Post-doctoral Project at Faculdade de Letras from Universidade do Porto, Porto University, Portugal.

Jamil, G. L. (2015). Market intelligence. In *Encyclopedia of information Science and Technology, 3rd*. Hershey, PA: IGI Global. doi:10.4018/978-1-4666-5888-2.ch508

Jamil, G. L., Santos, L. H. R., Alves, M. L., & Furbino, L. (2012). *A design framework for a market intelligence system for healthcare sector: a support decision tool in an emergent economy. Chapter approved to appear in Handbook on research of ICTs for Social Services and Healthcare: developments and applications*. Hershey, PA: IGI Publishing.

Jeston, J., & Nellis, J. (2007). *Business Process Management: practical guide to successful implementations* (2nd ed.). Butterworth-Heinemann.

Kahaner, L. (1998). *Competitive Intelligence: How to gather, analyse, and use Information to move your business to the top*. New York: Touchstone Books.

Kaplan, R., & Norton, D. (2007). *The execution premium: linking strategies to operation for competitive advantage*. Harvard Business School Press.

Kearns, G. S., & Lederer, A. L. (2003). A resource based view of IT alignment: How knowledge sharing creates a competitive advantage. *Decision Sciences*, *34*(1), 1–29. doi:10.1111/1540-5915.02289

Kimball, R., & Ross, M. (2010). *Relentlessly Practical Tools for Data Warehousing and Business Intelligence*. John Wiley and sons.

Kotler, P., & Keller, K. (2005). *Marketing Management* (12th ed.). Prentice Hall.

Krstevski, D., & Manchevski, G. (2016). SMEs innovation management: Strategy map for innovation-driven company. *Economic Development.*, *1*(1), 195–210.

Laudon, K., & Laudon, L. (2009). *Management Information Systems* (11th ed.). Prentice Hall.

Leidner, D., & Elam, J. J. (1995). The impact of executive information systems on organizational design, intelligence and decision making. *Organization Science*, *6*(6), 645–664. doi:10.1287/orsc.6.6.645

Liao, J., Kickul, J. R., & Ma, H. (2009). Organizational dynamic capability and innovation: An empirical examination of Internet firms. *Journal of Small Business Management*, *47*(3), 263–286. doi:10.1111/j.1540-627X.2009.00271.x

Lucas, H. C. Jr. (2005). *Information technology: strategic decision making for managers*. Hoboken, NJ: John Wiley and Sons.

Mahrt, M., & Scharkow, M. (2013). The value of big data in Digital Media research. *Journal of Broadcasting & Electronic Media*, *57*(1), 20–33. doi:10.1080/08838151.2012.761700

Maital, S., & Seshadri, D. V. R. (2012). *Innovation management: strategies, concepts and tools for growth and profit*. New Delhi: Sage Publications.

Marchand, D. (1990). Managing Information quality. In Information quality: definitions and dimensions. London: Taylor Graham.

Marchand, D., Kettinger, W., & Rollins, J. (2001). *Making the invisible visible: how companies win the right information, people and IT*. Wyley.

Mata, F. J., Fuerst, W. L. & Barney, J. B. (1995). Information technology and sustained competitive advantage: a resource-based analysis. *Management Information Science Quarterly,* 487-505.

Miller, J. P. (1996). Information science and competitive intelligence: Possible collaborators? *Bulletin of the American Society for Information Science*, *23*(1), 1–12.

Mintzberg, H., Alhstrand, B., & Lampel, J. (2009). *Strategy Safari: the complete guide through the wilds of strategic management* (2nd ed.). Pearson Education Limited.

Mintzberg, H., Lampel, J., Quinn, J. B., & Ghoshal, S. (2003). *The strategy process: concepts, contexts, cases* (4th ed.). Prentice Hall.

Nonaka, I. (2008). *The knowledge creating company*. Harvard Business Review Classics.

O'Brien, J., & Marakas, G. (2008). *Management Information Systems*. Irwin: Mc Graw Hill.

Porter, M. (2001, March). Strategy and the Internet. *Harvard Business Review,* 62-78.

Porter, M. (2008). *On Competition*. Harvard Business School Press.

Prado, D. (2006). *Gerenciamento de programas, portfólios e projetos nas organizações*. Belo Horizonte: INDG.

Ravichandran, T., & Rai, A. (1999). Total Quality Management in Information Systems Development:Key Constructs and Relationships. *Journal of Management Information Systems, 16*(3), 119-155.

SAA – Society of American Archivists. (2013). Accessed at http://www2.archivists.org/glossary/terms/i/informational-value

SAS. (2017). *What is Analytics*? Available at https://www.sas.com/en_us/insights/analytics/what-is-analytics.html

SAS. (n.d.). *SAS Enterprise Miner – SEMMA*. Available at http://www.sas.com/technologies/analytics/datamining/miner/semma.html

Setia, P., Venkatesh, V., & Joglekar, S. (2013). Leveraging digital Technologies: how information quality leads to localized capabilities and customer service performance. MIS Quarterly, 37(2), 565-590.

Sheth-Voss, P., & Carreras, I. (2010). How informative is your segmentation? Marketing Research, 22(4), 9-13.

Stair, R., & Reynolds, G. (2009). *Principles of information systems*. Course Technology.

Tuomi, I. (2000). Data is more than knowledge: Implications of the reversed knowledge hierarchy for knowledge management and organizational memory. *Journal of Management Systems, 16*(3), 103-117.

Turban, E., Mc Lean, E., & Wetherbe, J. (2002). *Information technology for management: transforming business in the digital economy* (3rd ed.). Hoboken, NJ: John Wiley and Sons.

Turban, E., Rainer, R. K. Jr, & Potter, R. E. (2007). *Introduction to information systems*. Hoboken, NJ: John Wiley and Sons.

Venkatraman, N. (1989). Strategic orientation of business enterprises: The construct, dimensionality and measurement. *Management Science, 35*(8), 942–962. doi:10.1287/mnsc.35.8.942

Weiss, S. M., & Verma, N. K. (2002). A System for Real-time Competitive Market Intelligence. *Proceedings of of the eighth ACM SIGKDD international conference on Knowledge discovery and data mining*, 360-365. 10.1145/775047.775100

Wernerfelt. (1984). A resource based view of a firm. *Strategic Management Journal, 5*, 171-180.

Yonce, C., Taylor, J., Kelly, N., & Gnau, S. (2017). BI Experts´ perspective: Are you ready for what´s coming in Analytics. *Business Intelligence Journal, 22*(3), 36-42.

Chapter 17
Optimal Locational Analysis for the Targeting of Investments in Specific Businesses and Territories in Brazil:
An Analysis of the State of Minas Gerais

Brenner Lopes
Nous SenseMaking, Brazil

Luander Falcão
Nous SenseMaking, Brazil

ABSTRACT

With the evolving understanding of the role of the supply chain management and its potential to add value, integrated business planning is a key concept in any modern organization. That brings complexity to supply chain management, requiring companies aiming to operate a world-class process to have a strong coordination between internal functions, which is only possible with a highly efficient information management framework. This chapter discusses how companies can extract competitive advantage from the use of available information on the supply side. For that, as based on the supply-side methodology, a new analytical layer is built and applied based on a multivariate statistical approach that makes possible the creation of groups according to characteristics of the capacity to supply goods and services. As analytical locus, the authors selected the Minas Gerais territory, Brazil.

INTRODUCTION

An efficient management of the supply chain can create a sustainable competitive advantage in nowadays global markets. Being a cross-functional discipline, the supply chain management materializes the strategy. For decades, the supply chain management was a synonym for logistics cost control, not including

DOI: 10.4018/978-1-5225-6225-2.ch017

planning activities, information sharing or any value adding whatsoever. The integrated planning rise in adoption by companies came with the understanding that a superior integrated planning, overseeing all business processes from raw material to customer delivery, can create value to the company, not only reducing logistics and inventory costs, but also assuring revenue by reducing out of stock events and increasing customer satisfaction. To assure a global optimization, each link in the supply chain must also have accountability for the supply and demand management and be *able to manage risks, respond to changes in the economic, technological, and competitive environment and exploit new opportunities more effectively than their competitors* (Glatzel & Röhren, 2014).

The discipline is growing exponentially in complexity with the accelerating spread of the supply chain, geographically, where all information is connected and the outstanding availability of data and the fast pace at which scenarios are changing, driving new trends (Glatzel & Röhren, 2014; Handfield, Straube, Pfohl, & Wieland, 2013). Among those trends, which the Supply Chain Management will have to have absorb and adapt, according to Handfield et al. (2013), are the increasing customer expectations, managing supplier-customer interface, increasing savings pressure, soaring risks and disruptions. In order to operate a world class supply chain management, companies will have to have strong cross functional coordination and assertive decision making, implying a superior capability for identifying and correctly assessing trade-offs to explore opportunities, emerging or not, depending on the competitive position of each company.

For that to happen, a great management is paramount. *If you can't measure it, you can't manage it.* And to measure it, you need an ever evolving information flow. Nowadays, each and every link of the supply chain almost certainly has one or more information systems that use software. Those systems can be assessed and the all the information is connected and analyzed together. But before that, the information input must be accurate and comprehensive. With a good information collection process in place, it is possible to identify and create more assertive action plans to structuring the production network, supplier selection and refine the capacity planning.

Looking exclusively to the supply side by mapping the existing suppliers and ranking them by capacity, the supply chain manager can create a more effective planning to assure the highest quality and lowest price. He can also propose the identification and developing of new suppliers that can assure quality supply and low cost in an increasing volatile environment. The company can also anticipate competitors in finding new suppliers and commercial partners creating a strategic competitive advantage.

The strategic pillar of this chapter is based on the Supply-Side methodology (discussed in the book *Information Systems Management for Effective Logistics and Suplly Chains*, IGI Global, 2016) that uses information as a source of competitive advantage. The methodology classifies suppliers in a given geography according to the measure most relevant to the trade question in question (ie, capacity). Just as in this original methodology, we focus here, exclusively, on the supply side. Applying the Supply-Side methodology, the manager could map which suppliers are able to deliver the products and services in the required quantities and quality required, and in which regions they are.

Starting from the base already established by the broad methodology that involved the construction of the idea discussed in the Supply-Side methodology, in this chapter we will explore other analytical possibilities and possible multiple ways of seeing the same question. In order to fulfill this objective, we first carry out a review of the mentioned methodological approach proposing a new approach, but anchored in the basic precepts of the original methodology.

The new analytical logic will be based on the idea of grouping, where through the clusters analysis technique, within the context of data mining, of the companies with high capacity of supply of goods and services in the State of Minas Gerais, we could understand in segments and localities we have groups of companies with high capacity of supply, stratified in the clusters according to this intensity of capacity of supply.

The result of this new approach would enable decision-makers in organizations to answer questions such as: If I am an investor thinking of investing in a particular territory, having, as one of the main vectors of this decision, the supply capacity of the segments that make up my production chain, where would it be more logical to make such an investment (eg, setting up an industry)?

This chapter is divided in five sections: (1) this introduction; (2) the background, where we discuss the Supply Chain Management, information theories and how they interact, in addition to the cluster analysis technique; (3) application of methodology and analysis of results; (4) future research directions, and; (5) the conclusions of the chapter.

BACKGROUND

Among many definitions of *Supply-Chain*, we will adopt the Council of Supply Chain Management Professionals definition (CSCMP, 2013):

Supply Chain Management encompasses the planning and management of all activities involved in sourcing and procurement, conversion and all logistics management activities. Importantly, it also includes coordination and collaboration with channel partners, which can be suppliers, intermediaries, third-party service providers and customers. In essence, supply chain management integrates supply and demand management within and across companies. Supply Chain Management is an integrating function with primary responsibility for linking major business functions and business processes within and across companies into a cohesive and high-performing business model. It includes all of the logistics management activities noted above, as well as manufacturing operations, and it drives coordination of processes and activities with and across marketing, sales, product design, finance and information technology. (p. 187)

This definition was adopted because we believe it is more comprehensive and similar to what other author such as *Global Supply Chain Forum* (GSCF) and Harrison and Hoek (2003) adopted.

One important concept on any definition of the Supply Chain Management is the goal. For Pires and Vivaldini (2014), the goal is to, *maximize and materialize synergies between the links of the supply chain focusing on a more efficient delivery for the customer.* In order to get to that point, it is indispensable to be able to measure and manage each and every link of the supply chain, leveraging information, as we demonstrate in this chapter.

Supply-Chain and Information Management

The definition presented by Ballou (2004), is a more synthetic one compared to CSCMP (2013). It presents Supply Chain Management as a process, developed to assure that the company meets its targets trough activities. This process demands integration of different areas in the business, what makes the business management complex. According to Barros, Ishikiryama, Peres and Gomes (2015), *To support this process, IT can provide a number of tools to facilitate, streamline and increase the reliability of communications and the exchange of information between organizations.*

Barros et al. (2015) present 3 models of Supply Chain Management, focusing on key business integrations, based on five key category management processes and showing the benefits of IT tools in the Supply Chain Management.

The model shown in Figure 1 explicitly states the value of information in the IT tools for the Supply Chain Management.

Matos, Pires, and Vivaldini (2014) in, *Desenvolvimento de Produtos Sob a Perspectiva da Gestão da Cadeia de Suprimentos*, don't include the information among the theoretical outline used although it appears implicitly in the discussion. Information is considered an endogenous factor with no relevance thus creating a gap in the process about how information will flow between the sub processes, neglecting the necessary interface between business processes to support decision making in every level. According to McGee and Prusak (1994), *information creates more information and knowledge creates more knowledge*, in face of an increasing evolution of information management processes.

When that happen, is because the Supply Chain Management is keeping *goods flowing from source to customer* (Harrison & Hoek, 2003) without risking overstock in any link from supplier to customer. Matos et al. (2014) argued that this is made *through cost and reduction as well as through adding value to the product.*

Applying Information to Supply Chain Management

As seen before, Logistics is not supply chain, but logistics works with the input, stocking and output of products, and is definitely among the activites of the supply chain. *Supply chain incorporate processes and activities related to moving products, information and people* (Martins, Lobo, Alves, & Sprosser,

Figure 1. Diagram for the evaluation of the impact/benefits of IT on SCM
Source: Barros et al. (2015, p. 701).

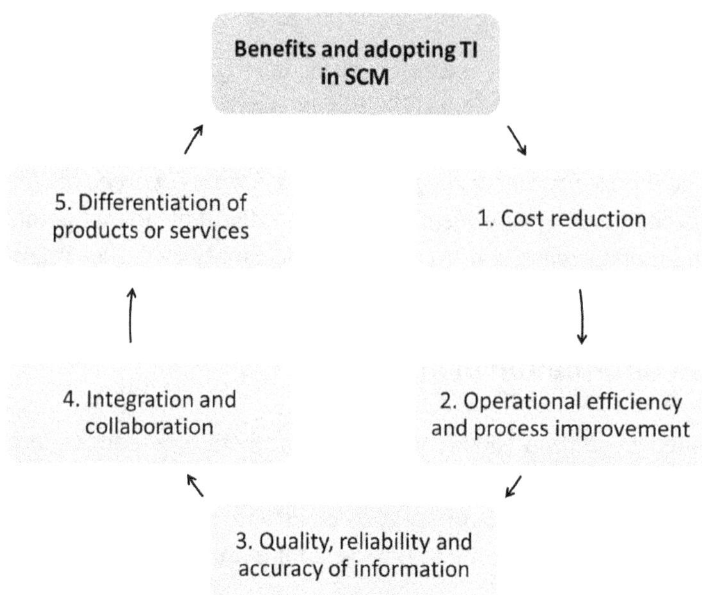

2014). The logistics is also growing exponentially complex, creating trade-offs that demand more analytical and data based decision making.

Amaral and Guerreiro (2014) argued that *the use of information systems and computer technology create the potential to rapidly and precisely identify cost optimization opportunities*. The authors, on the other hand, rely heavily in accounting information to solve trade-offs not considering many other sources of information that can add value to the process. According to Harrison and Hoek (2003), *information is the glue that bond the supply chain processes*.

According to Vaitsman (2001), *information is a known fact, data or event* that supports the decision making process aligned with de goals of the competitive intelligence approach. To Miller (2002), *since intelligence demands information, both internal and external, well managed companies structure their information resources to allow easy access and application*. The debate strength the concept of information relevance on the Supply Chain Management presented by Harrison and Hoek (2003):

In order to make effective supply chain management decision, any organization must put in place an information flow convenient and precise. To achieve that, companies must implement systems that are able to oversee the supply as a whole, from inventory levels in suppliers to demand levels.

Harrison and Hoek (2003) develop the concept of Analytical Intelligence as the *extensive use of data, quantitative and statistical analysis, predictive models and a fact based management* (Davenport & Harris, 2007). In the end, both Analytical Intelligence and Supply Chain Management seek to create a competitive advantage from gains of scale, cost reduction, or any other way to increase profitability.

Strategic Supply Chain Management

According to Ballou (2006), the concept of Supply Chain emerges as an evolution of the logistics with the promise to integrate all surrounding processes. In that context, he considers to "refer to the entire supply channel and suggest that coordination is needed throughout the channel. These are ideas that form the basis for supply chain management as practiced today" (Ballou, 2006). This vision is shown in Figure 2.

Matos et al. (2014) argue that in recent years, the Supply Chain Management has been gaining popularity in the academia and among companies, pursuing greater competitiveness. The authors present one of the key factors for this growing popularity: the emergence of the global supply chain and its impact on customer expectation generating cost pressure, as discussed in Handfield et al. (2013).

In the end, companies expect the Supply Chain Management to overcome those barriers generating positive impacts on revenue and profitability for the company, as pointed out by Christopher (as cited in Harrison & Hoek, 2003):

Strategic management of moving and holding inventory of materials, parts and final products, and the respective information flow through the organization and its channels, assuring current and future profit optimization through cost effective and high quality purchasing. (p. 38)

For Harrison and Hoek (2003, p. 38), the *concept of supply chain suggests a stream of linked processes creating a chain*, and the management of this supply chain *should be aligned to product development decisions and be crated and managed to assure that products are delivered at agreed upon cost, timing and quality* (Matos, Pires, & Vivaldini, 2014, p. 72). Ballou (2006) argues that, *A contemporary view of supply chain management is to think of it as managing a set of processes, where a process is a group of activities relevant to achieve a defined objective, such as filling orders*. Harrison and Hoek (2003) corroborate stating that the processes should be linked to each other disconsidering organizational barriers.

This vision is presented in Figure 3.

Figure 2. Evolution of Supply Chain Management
Source: Ballou (2006, pg. 379).

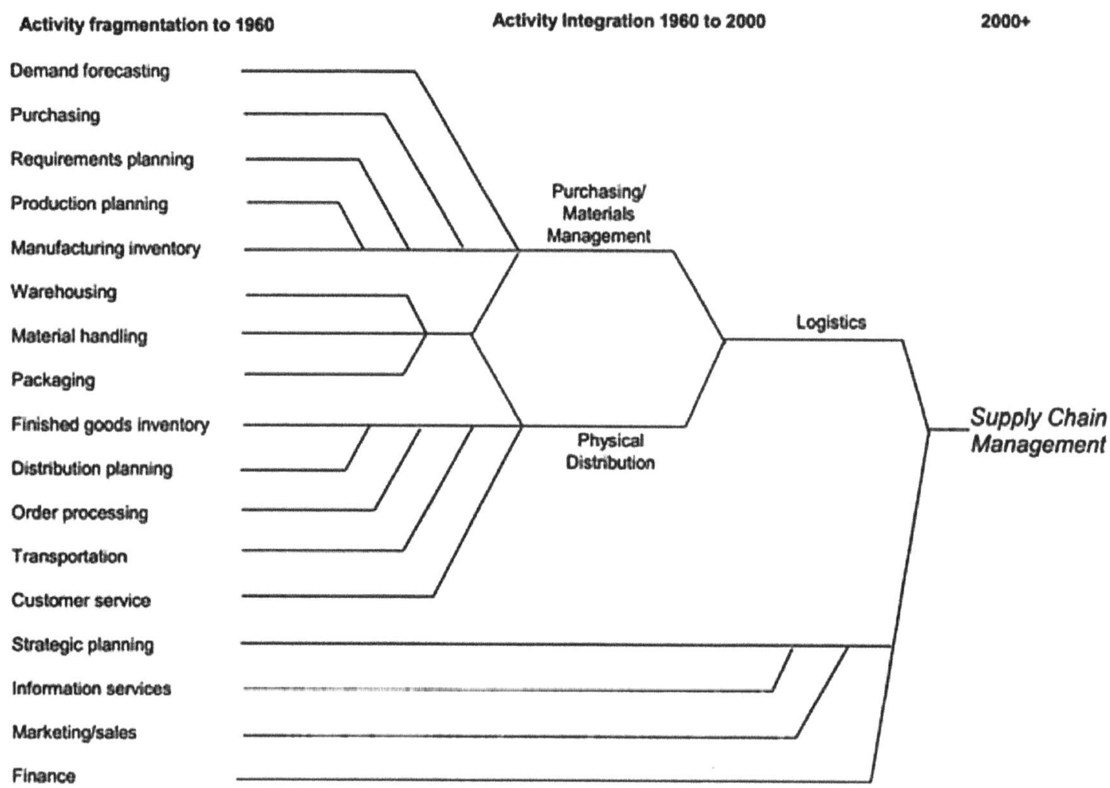

Figure 3. Supply Chain Management overview
Source: Harrison and Hoek (2003).

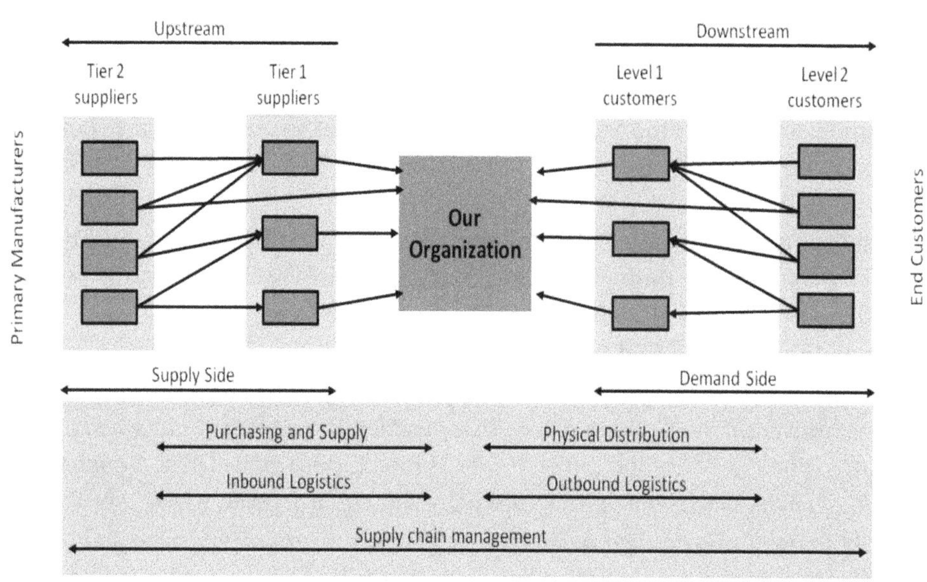

Harrison and Hoek (2003) explain the dynamics as follows:

Material moves downstream. Information, on the other hand, should move upstream. When the supply chain is working effectively, once a customer makes an order, everything else is defined to fulfill that order. The supply chains can be broken down and organized in smaller groups that can be more easily managed. (p. 30)

The integration of processes and information flows makes the supply chain dynamic works effectively. A fundamental part of the integrations depends on the evolution of information systems, more specifically the ERP (Enterprise Resource Planning) systems. According to Harrison and Hoek (2003), *To facilitate information flow, companies are relying on enterprise resource planning systems (ERP).*

IT systems integration has a direct and positive impact on the operation, allowing the information sharing reduce costs. Information, on the order hand, should increase interaction between links on the supply chain. Ballou (2006) identifies the interactions between different areas in the organization, such as marketing, finance, production as Cross-functional Coordination. According to Ballou (2006), Cross-organizational Coordination *has to do with collaborating and coordinating product flows among channel members, i.e., those companies that are not owned or operated by the immediate firm* (Figure 2). Failure in Cross Functional or Organizational Coordination affecting the information flow, can negatively impact the Supply Chain Management goals (Kovacevic, 2013).

According to Matos et al. (2014), *this alignment can promote a better use of the supply chain capacity, increase new product launches effectiveness and improve company performance.* Harrison and Hoek (2003) say that, *the supply chain management is the network management from how the customer coordinates all the links.* For these integration and management to be effective, there must be a strategic decision-making process supported by information and analysis timely available.

Strategic Information Management

For Harrison and Hoek (2003), the *concept of supply chain suggests a stream of linked processes creating a chain.* Each link generates data and information that can be connected providing a comprehensive understanding of the process health. On the other hand, Vaitsman (2001) observes that *not every information available, even if accurate, isn't necessarily valuable and improve the final decision.* Companies aiming at improving the decision-making coordinating different functions should consider the intelligence process to *enable managers to share information* (Miller, 2002). Harrison and Hoek (2003) corroborate that *by sharing demand information with the entire supply chain, we create a chain focused on creating value to the end customer.*

For Mintzberg and Quinn (2001), strategy is *a pattern or plan that integrates goals and the sequence of actions of an organization.* In Ansoff vision, strategy is a matter of *business goals in its essence that is defining the product portfolio.* As pointed out by Mcgee and Prusak (1994), strategy can be divided in three parts: definition, execution and integration *happening in a competitive environment full of information of real and potential value* (Mcgee & Prusak, 1994).

Mcgee and Prusak (1994) identify three generic information strategies: 1) information as competitive advantage; 2) information as production/services, and; 3) information commerce. In the first case, the company builds the capability of internally developing exclusive information on the market and competitors. This capability should be pursued by every company. This conclusion is supported by Harrison and Hoek (2003) when describing that *the goal is to integrate supply and demand data to obtain the most complete and accurate vision of processes, customer and markets.*

Figure 4. Information and competitive strategy
Source: Mcgee and Prusak (1994).

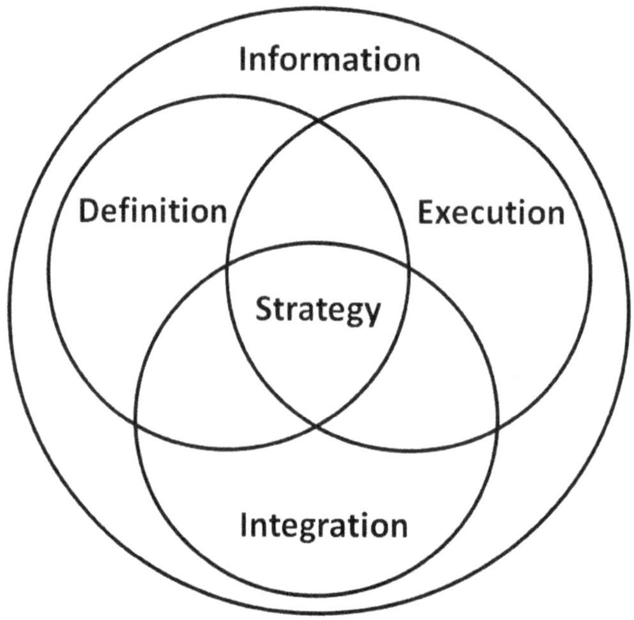

This is possible using IT resources that *can be used to support strategic definitions to variables such ad product differentiation* (Mcgee & Prusak, 1994). For Harrison and Hoek (2003), *technology allows for efficient and fast sharing of information in growing detail and sophistication.*

Finally, *the fundamental reason for the strategic use of information is to obtain competitive advantage* (Mcgee & Prusak, 1994) and to sustain and increase competitive advantage necessary to integrate processes and information flows (Harrison & Hoek, 2003).

Side-Supply Methodology

According to Harrison and Hoek (2003), *the best opportunity to address the market in a timely manner and with the lowest possible inventory is through supply chain integration.* Such integration between suppliers, company and customer processes allows information to flow efficiently along the whole chain, as presented in Figure 5.

In a scenario of rising competitiveness leading to higher price volatility and increasing competition for specialized workers, the understanding of market dynamics is a key strategic asset. The Side-Supply methodology was developed to support and enhance the supply chain strategic planning process, focusing on the supply side and the upstream information flow, impacting the whole organization and downstream stakeholders.

In this study, we will use the Capacity Indicator to rank 629 companies in 27 Brazilian states and map high capacity suppliers. The Capacity Indicator built captures the specificities of different links in the chains and, at the same time, takes into account shared characteristics competitors.

In this study, we must consider three different criteria when building the Supply Indicator. First, the business size represented by the revenue generated. The second criteria is specialization and takes into

Figure 5. Supply Chain information flow
Source: Harrison and Hoek (2003) adapted by the authors.

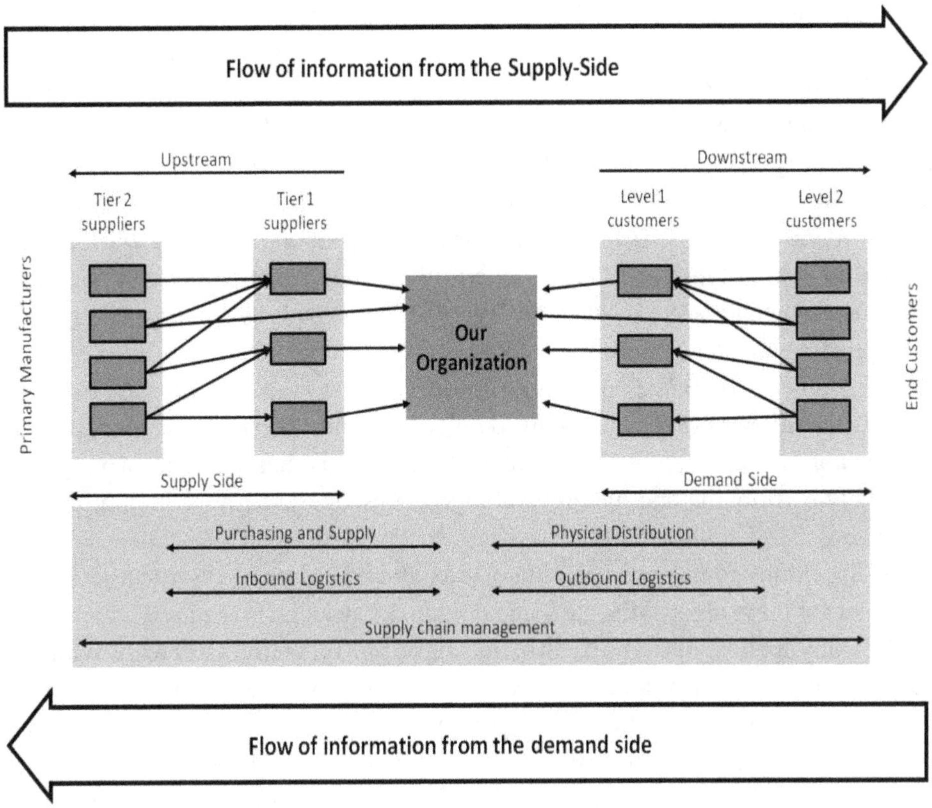

account the activity revenue representativeness in state economy compared to the representativeness of the business in the country. Finally, we considered the growth trend of the business in the 7 years prior to the analysis. All data regarding total wages, number of companies and employees between 2006 and 2014 were extracted from RAIS research from the Ministry of Labor and Employment.

Business Size considers the absolute size of the businesses in the state and country economy. The specialization criteria, also called Location Quotient, aims at identify Local Productive Arrangements in the search for regions that concentrate potential suppliers and deserve a closer look from the company. As the third criteria, was used the Compound Annual Growth Rate (CAGR). Every measure was calculated for the number of companies, employees and total wages for states and country. All data was standardized and weighted equally to create the Capacity Indicator presented.

The methodology and database granularity allow for variable depth of analysis that can be done at the sector, segment or sub-segment levels. It is possible, for example, to analyze the Transformation Industry, zoom in the Food Industry and get further analyzing the Dairy Products market. It can also present analysis broken down by time frame and geography.

The analysis joints out businesses and regions of interest and rise question, such as, *What can explain the growth spike of a certain business in a given region?*. Answering questions like this leads to a better understanding of the underlying conditions for the development of a high capacity supplier, and can help companies gain competitive advantage with a more assertive strategic supply chain management.

Cluster Analysis Technique

For Favero et. al. (2009), cluster analysis *is a technique of clustering, it is an interdependence technique that seeks to group elements according to their natural structure.* Already for Hair et. al. (2009), is a multivariate technique that can receive several names, whose objective is to classify objects into groups according to their relation. Within this perspective, Hair et. al. (2009), cluster analysis as a multivariate analysis, objects (eg cities, people or other entities) should reflect a certain internal homogeneity and external heterogeneity. It works the concept of statistical variable as a central question and *a set of variables that represent the characteristics to compare objects in cluster analysis* (Hair, 2009). It is a technique used in several types of research, due to the strong tradition of grouping individuals and objects, as it can reduce data and / or generate hypotheses. It requires conceptual support, before the technique is applied, to counter criticism as to its use.

According to Maroco (2007), *the analysis of groups or clusters is an exploratory technique of multivariate analysis that allows groups or groups to be grouped into homogeneous groups with respect to one or more common characteristics.* Ester et al. (2016) explains that there are two types of clustering algorithms, namely partitioning and hierarchical algorithms. These authors explain Partitioning algorithms as follows:

Partitioning algorithms construct a partition of a database D of n objects into a set of k clusters, k is an input parameter for these algorithms, i.e some domain knowledge is required which unfortunately is not available for many applications. The partitioning algorithm typically starts with an initial partition of D and then uses an iterative control strategy to optimize an objective function. Each cluster is represented by the gravity center of the cluster (k-means algorithms) or by one of the objects of the cluster located near its center (k-medoid algorithms). (Ester, Kriegel, Sander, & Xu, 1996).

For Maroco (2007), *hierarchical techniques use successive steps of aggregation of the individuals considered individually, each subject being a cluster, and then these are grouped according to their neighborhoods.* According to Ester et al (2016),

the hierarchical decompositionis represented by a dendrograma, tree that iteratively splits D into smaller subsets until each subset consists of only one object. In such a hierarchy, each node of the tree represents a cluster of D.

Application of the Model

As defined by Lopes and Falcão (2016), Supply-Side was built to order the businesses for territories with greater capacity to supply goods and services, allowing managers of Supply Chain to indicate in which regions there is a "pool" of suppliers with delivery capacity, both in volume and quality.

Supply-Side is able to identify states in which businesses have supply capacity, as demonstrated by Lopes and Falcão (2016). However, in doing so, the need arose to know whether in a certain state there would be a group of businesses that had the capacity to provide a range of goods and services. This question is a consequence of the company's need to have multiple suppliers, not just one, even if one of them continues to be the main one. The company being close to a strategic supplier is a factor of competitiveness, but being close to a number of suppliers, with high supply capacity enhances their competitiveness. Thus, with the access to this strategic information, it is possible to optimize the actions between the links of the productive chain, allowing to understand the impact of one action of one supplier on the other and especially on the target company.

To identify which groups these would be, and to understand how they are formed, we opted for the multivariate statistical approach known as cluster analysis, in which business would be grouped. In order to facilitate the understanding and application of the approach, as was done in the Supply-Side methodology, according to Lopes and Falcão (2016), it was decided to select the state of Minas Gerais, and to work with available data for the year 2014. This resulted in 507 selected businesses, distributed across all industries. From these businesses, the General Index median was calculated, with the purpose of selecting only the businesses that had an index value above the median, since these businesses would have high capacity of supply. This procedure sought to select the businesses with the highest indicator value, since this indicates that the higher this result, the greater the capacity to supply goods or services of a given business in a given region. Therefore, it was decided to group only the businesses that had this high capacity of supply.

From this procedure, a total of 127 high-capacity supply businesses were obtained in Minas Gerais. The variables of the sub-indices were selected as the parameters to proceed the grouping. The general index is formed from these three sub-indices, being size, specialization and growth (Lopes & Falcao, 2016).

Data Analysis

To perform the business grouping, the method described in the article Types of Clustering Methods: Overview and Quick Start R Code was used, which defined the procedures as follows:

Partitioning algorithms are clustering techniques that subdivide the data sets into a set of groups k, where k is the number of groups pre-specified by the analyst. There are different types of partitioning cluster methods. The most popular clustering is K-means (MacQueen, 1967), in which each cluster is represented by the center or middle of the data points belonging to the cluster (Source STHDA, 2017).

For the calculations, the statistical package R (2017) was used, and the packages FactoMineR, factoextra and cluster. There were five clusters defined for the calculations.

The Supply-Side methodology was used to identify high-ability supply businesses for companies to improve supplier selection. In making a new selection of this base, we found 127 high-capacity supply businesses in Minas Gerais. These 127 deals were grouped into five clusters, defined a priori by the behavior of the data and be a cluster number that allowed a better division or optimization of the separation of the businesses, according to Figure 6. The businesses were grouped by the K-means method, resulting in five clusters, with 61, 19, 13, 26 and 8 deals, respectively, in clusters 1, 2, 3, 4 and 5.

In this figure, it is possible to observe that the data were divided into two dimensions that support 92% of the variation explained. It is also possible to observe that there are 3 very distinct groups, while two are very close. The clusters were created from the sub-indices, aiming to group the businesses that had the same characteristics of size, growth and specialization, in order to differentiate into groups a set of businesses that have high supply capacity from these sub-indexes. Each cluster was named in order to explain and clarify why they were clustered. This information is given in Table 1.

This table summarizes the main characteristics of each cluster, but does not detail how each cluster is composed and how companies can appropriate this information. That is why each Cluster was detailed in the sequence, with the main characteristics, the businesses that comprise it and what implications for the Supply-Chain sectors.

Figure 6. Supply-Side business cluster
Source: Prepared by the authors, 2017.

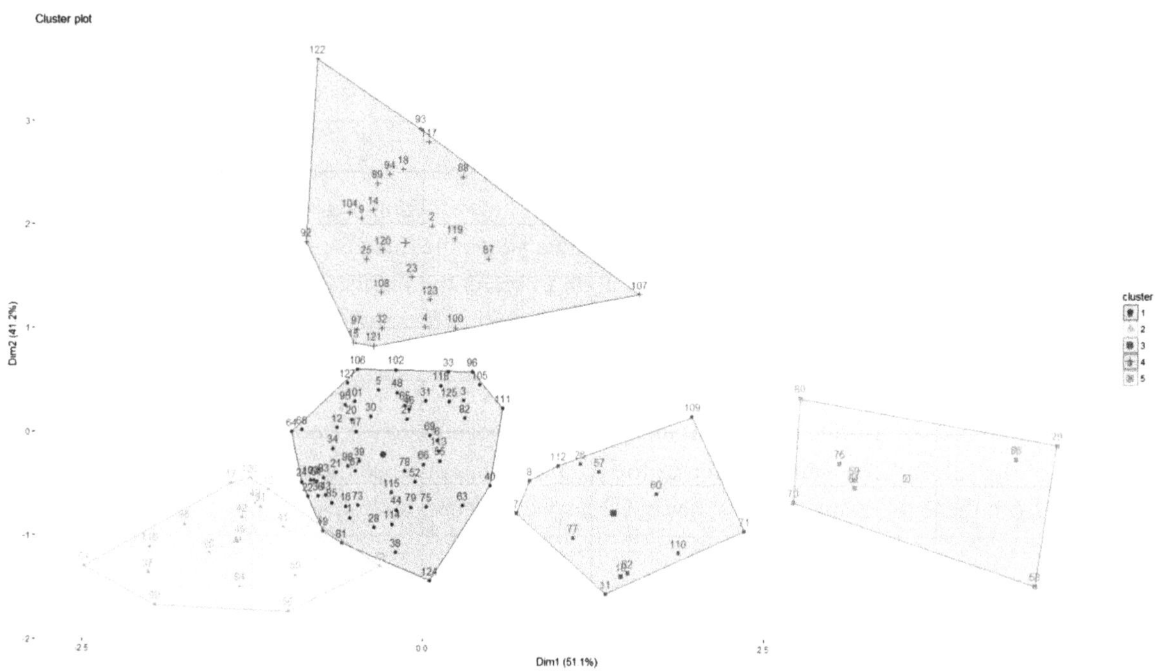

Table 1. Cluster naming

Cluster	Appointment	Description
1	Basics	These are businesses that grow and are specialized, have size and serve most of the business as suppliers - can also be considered as business support
2	Bulky	These are businesses made up of businesses with a large number of companies, employees and salaries, and these last two variables grow over time - they bear some resemblance to cluster 1
3	Specialized	They are businesses that have specialization in the state, such as Extraction of minerals for the manufacture of fertilizers, fertilizers and other chemicals; Drilling and boring; and Manufacture of milk and milk products. The main productive chains of the state of Minas Gerais
4	Service Sector	These are businesses that most belong to the services sector, and those that are not classified as services have service characteristics, such as Other land preparation services, which is classified as Civil Construction (CNAE, 2016) - although the classification has service features
5	APL's of Minas Gerais	The businesses of this cluster are APL's or traditional activities of Minas Gerais

Source: Prepared by the authors, 2017.

Basics

Most of the 61 businesses in this cluster are in the Transformation Industry (36%) and Services (28%), totaling 64% of the cluster's businesses. These deals correspond to 50% of the companies in the cluster, 57% of the number of employees, and 64% of the total wage bill.

The subindex Size among the 3 subscripts is the one that has a greater weight, given the values of each. However, when considering the Specialization and Growth sub-indices, it is noticed that most of the businesses have positive specialization values, which indicates that these businesses have specialization in their business. Half of these businesses have positive growths, and the other half have negative growth. This indicates that not everyone grows. Plotting these data in a dot chart reveals these characteristics.

Of the 61 businesses in this cluster, only one presents characteristics more related to intensive use of knowledge, such as engineering services. The other businesses have a technology already consolidated and of little advance, and many of the techniques that were used to 40 years are still used today, including the businesses related to the transformation industry. An example of this fact are the businesses Manufacture of bakery products, Manufacture of articles of concrete, cement, fiber cement, gypsum and similar materials and Manufacture of prepared foods and plates. These three deals are from Transformation Industry, and are among the 15 largest deals in terms of Indicator. These businesses have low entry barriers, high rivalry, substitute products, and the strength of suppliers over them is high. This justifies the volume of companies, employees and payroll. Because they do not need a specific technology or highly specialized employees, the entry of new entrepreneurs is facilitated. The identification of a differential becomes a factor of competitiveness of the company within these businesses.

But when analyzing the business within this cluster, most have the same characteristics: low entry barriers, high rivalry, high power buyers or suppliers. External environment factors related to economics, legislation, technology and socio-cultural affect directly, but in different intensity the business. Those with high entry barriers, whether due to capital, legislation, or marketing, face high customer pressures, since they have the ability to negotiate.

When analyzing the businesses of this cluster that have specialization and positive growth, it identifies about 30 businesses. These businesses are linked to the growth of the market as a whole: if the market grows, these businesses grow, the reverse being true. Therefore, it can be deduced that these businesses are dependent on the internal consumer market. Any oscillation directly impacts these deals. Here lies a difference of the other cluster: the other clusters are dependent on the external market as well.

These combined factors show that this cluster is the most focused for providing goods and services, to support the operations, and are not vital for the development of the business. Some businesses may have the ability to be strategic providers for some businesses, but most do not. For the Supply Chain area, the importance lies in the fact that in Minas Gerais, in this case, there is a business group with a high supply capacity, when it comes to support business and support activities, which in general are not linked to primary activities, or vital to the operation.

Bulky

Most of the 19 businesses in this cluster are in the Commerce (47%) and Services (26%) sector, totaling 74% of the cluster business. These deals correspond to 78% of the companies in the cluster, 80% of the number of employees, and 81% of the total wage bill.

The fact of concentrating a large number of companies, employees and payroll, justifies its denomination of bulky. But businesses outside these two sectors have the same characteristics. In the case of Transformation, Cement Manufacturing and Manufacture of soft drinks and other non-alcoholic beverages businesses, they have few companies, but generate many jobs and, consequently, generate a significant volume of salary mass. In the case of Agriculture, Cattle Raising, and in Civil Construction, Building Construction and Highway and Railroad Construction, the same characteristic of the cluster follows, with many companies, employees and wage mass, except for railways, which has few companies, when compared to this business with the rest of the cluster.

This cluster has some similarities with cluster 1, especially when evaluating the degree of competition of the business. The difference lies in the details of the business. When observing some business, such as Teaching Activities, it is important to understand how this business is. It is the fifth largest generator of jobs, and the first in generation of payroll, within the cluster. It includes activities of early childhood education and elementary education, secondary education, higher education, vocational education of technical and technological level and activities to support education. Each of these activities has its own dynamics and, in each case, it is still possible to segment by market, as in the case of higher education, which has different entry barriers by market. Another example is the Restaurants business, which initially has low entrance barriers, and several substitute products. But when you look at the market within this business, the barriers tend to dim, depending on the direction and the goal. While in the cluster it was possible to do a competition analysis for the segment as a whole, in this cluster, this possibility is not so viable, given the particularities of each business. They are also very business focused on meeting the domestic demand, and therefore they oscillate according to the consumption power of the economic agents, being the companies, the families and the government.

When plotting the Growth and Specialization sub-indices on a dot chart, the data show that about 10 businesses have positive Specialization but Negative Growth. This indicates that most of the businesses in this cluster are specialized in their activities, even if there are ramifications within each business, but do not grow or have negative growth. This result indicates that these businesses are going through a period of market stalling, limited to market issues, corroborating the dependence of the domestic market.

When considering only the businesses with positive specialization and growth, the following deals are obtained: Construction of buildings, Teaching activities, Restaurants, Freight transport by road and Hospital care activities. These businesses have a characteristic of growth linked to the behavioral change of people, who started to eat more outside the home, to carry more items, whether physical persons or companies, to consume more medical services, of diverse species, as well as to buy more education services.

This cluster also does not have the characteristics of its business being strategic or vital for the supply of goods or services to a company, but they are important when considering the attraction and retention of qualified professionals, interested not only in remuneration or positions, but also in conditions access to entertainment, education and health services. In this case, in Minas Gerais, there is a business group that provides a series of services in quantity and quality. For Supply-Chain units of construction companies, which need to provide some services such as transportation, food and lodging, in Minas Gerais there are companies able to provide.

Specialized

This cluster has basically 13 businesses, distributed in the sectors of Extractive Industry, Construction, Industry of the Transformation and Agropecuária. When plotting these businesses, using only the Growth and Specialization sub-indices, the data show that all of these businesses have positive expertise, and most have negative growth. They serve both the domestic market and the foreign market, and they suffer oscillation of the productive chains which are directly linked.

In the case of the Transformation Industry business, we can mention the businesses Manufacture of milk and derivatives, Production of iron and steel products, Manufacture of footwear in other materials. The state of Minas Gerais is one of the main producers of milk and dairy products in Brazil, a consolidated and traditional production chain in the state. The same happens with the production of iron and steel products, which grew due to the proximity of the iron ore miners, which supply raw material for the production of iron and steel, constituting one of the most traditional businesses in the state. In addition to these there is the Shoe Manufacturing business, very much linked to the fashion productive chain, very present in the state in various businesses.

The drilling and drilling business belongs to the Civil Construction sector, but can be analyzed along with the extractive industry business, which are Extraction of minerals for the manufacture of fertilizers, fertilizers and other chemical products; Activities to support the extraction of minerals, except oil and natural gas; and Extraction of other non-metallic minerals. These businesses are linked to the mining chain in Minas Gerais, which is also traditional and specialized in this state, since in it there is the presence of large deposits of ores, metallic or nonmetallic. In addition to minerals for various purposes. Of the fourth business, only Drilling and drilling shows positive growth, since in the analyzed period the mining sector grew in the state and in the country, as well as the civil construction sector.

The following businesses belong to the Agriculture and Livestock sector: Forest production - planted forests; Support activities for livestock; Breeding pigs; Support activities for forest production; Hunting and related services; and Support activities for agriculture. The state has strong expertise in agribusiness, being one of the main suppliers of inputs to the industry, where several productive plants have been installed in the state, or near the state to take advantage of the proximity with the raw materials as a factor of competitiveness. These businesses are linked to important chains of the state, such as Pulp and Paper and Meat. These chains serve both the domestic and the external markets, not relying on market oscillations, such as the cluster 1 and 2 businesses.

The business of this cluster has low entry barriers, medium rivalry and high customer pressure, as it provides large productive plants that are located at the next link in the chain of production that pressure the margins of the companies in these businesses.

For Supply-Chain areas this cluster becomes more valuable if they are in these businesses or linked to these productive chains, because these businesses are specialized due to the connection with traditional productive chains of the state.

Service Sector

This cluster is basically composed of businesses in the Services sector. Even the businesses that are associated with the Construction, Transformation and Trade sectors correlate with the service sector, such as business Other ground preparation services, Installation of industrial machinery and equipment, and Maintenance and repair of motorcycles, respectively. It is the second largest cluster in term of business number. It has a similar characteristic to that of cluster 3: the businesses of this cluster do not have the subindex Growth with negative value. So, all businesses grow, with some having more expertise than others.

Part of these businesses are geared towards meeting the needs of families, and the other part is aimed at the care of companies. When considering only the businesses focused on the service of the companies, it is perceived that most of them are focused on activities related to civil construction, more specifically infrastructure works and construction of productive plants. In the period analyzed, both Brazil and the state of Minas Gerais show the final stage of a cycle of strong economic growth, still driven by the increase in household consumption power and the growth of the construction sector.

For the Supply-Chain areas, this cluster presents a finding, because in Minas Gerais there is a business group able to provide services in terms of volume and quality, when it comes to construction activities, focused on the Engineering production chain. This chain covers the segments of Builders, Industrial Assembly, Projects and Consultancy and Special Engineering Services. Businesses that demand these segments, or that chain, in Minas Gerais, have a business group available to meet the demands related to this type of activity. Table 2 describes these businesses.

APL's of Minas Gerais

This cluster is the smallest of the five in terms of amount of business. The businesses in this cluster are: Coffee cultivation; Wholesale trade of coffee beans; Mineral extraction; Manufacture of lime and plaster; Manufacture of sneakers of any material; Manufacture of explosives; Manufacture of pharmaceutical preparations; Extraction of gemstones (precious and semi-precious stones).

All businesses, except Manufacture of pharmaceutical preparations, are part of a traditional APL of Minas Gerais, or are sectors in which the state is recognized as a reference.

The state has a strong tradition in the coffee chain, being the main producer and exporter in the country of coffee beans. In the state there are large coffee producing areas, such as in the South, East and the region called Cerrado de Minas. These regions are constantly receiving investment for the improvement of coffee cultivation, aiming at increasing productivity.

In addition to coffee, another sector that supports the state's trade balance is the extraction of ores, more specifically iron ore. In the state there are several deposits, and many municipalities are sustained through these businesses. Due to the abundance of this item, several metallurgical companies settled in the central region of the state, and in the route of iron ore disposal, to take advantage of the proximity to the raw material. As the mining activity is strong in the state, this eventually led to other businesses, such as lime and plaster manufacture.

Table 2. Description businesses of the cluster

Business	Indicator
Other ground preparation servisse	3,113
Incorporation of real estate projects	2,510
Recreational and sporting equipment rental	2,387
Rental of construction machinery and equipment without operator	2,316
Accounting, bookkeeping and auditing activities, accounting and tax	2,233
Installation of industrial machinery and equipment	2,161
Electrical installations	2,140
Activities of monitoring of security systems	2,128
Other specialized construction services	2,123
Telecommunications	2,086
Combined office and administrative support services	2,016
Management and administration of real estate	1,978
Provision and management of human resources for third parties	1,899
Real estate activities on own property	1,863
Assembly of industrial plants and metal structures	1,857

Source: Prepared by the authors, 2017.

In Minas Gerais there is a municipality that concentrates an expressive amount of footwear companies, where one has the configuration of an APL Footwear. This fact justifies the fact that the Tennis Manufacturing business of any material appears in this cluster. Similar to this, in Minas Gerais is another municipality that has a significant amount of explosives manufacturing companies in the local administrative area of Santo Antônio do Monte, which is the only APL this segment in Brazil, the second in the world.

In this cluster, Supply-Chain areas only benefit if they are tied to one of these chains, so this makes this cluster very specific.

FUTURE RESEARCH DIRECTIONS

This chapters goal was to show how informational processes can and should be used to support supply-chain areas offering tools aimed at helping managers in the decision making process. This methodology points out potential suppliers regions and segments allowing managers to plan actions, monitor sectors and segments and deepen the knowledge about the companies in the segment or understand which companies can become new suppliers. Using the methodology presented in this chapter, managers can mitigate impacts and avoid unanticipated issues in material flow both upstream and downstream.

Based on this proposed goal, the authors have chosen a theme that presents a clear gap in the Supply Chain studies, the information management. Supply chain activity and studies generate vast amount of data that is used to support operational and tactical decision, but seldom analyzed to support long term high impact strategic decisions.

The authors suggest that subsequent studies may cover some different aspects of data application for Supply Chain management:

The first and more direct path is to replicate the analysis with increasing geographical granularity, studying high capacity suppliers per city instead of state. Another relevant further development is to study the informations systems that hold information concerning the supply process. Those informations are usually spread among different systems in the company, making it difficult for managers to have a more strategic assessment and have a better understanding of threats and opportunities in the process. Studies aim to create methodologies to unified information systems and the impact of this kind of practice in the results of the companies.

CONCLUSION

According to Harrison and Hoek (2003), *customer satisfaction depends on material and informational flow management along the supply chain*, making the coordination of the supply chain and the understanding of current and potential suppliers strategic to any company.

In this sense, the application of the clusters analysis technique, based on the precepts developed in the supply-side methodology, demonstrates how effective and useful the strategic supply area can be.

From its application and verification of the results, a powerful analytical construct could be structured that would follow the same idea of a filter, from where we could start from a macro analysis, segmenting the similar business clusters according to their potential of supply and understanding of its most relevant characteristics, thus making macro cutbacks for a first level of business prioritization with a high supply potential, regardless of the territorial cut that one wants to analyze. Defining the cluster that deals with the most relevant supply potential, it could be from there already using the suppy-side methodology to apply a new segment-level cut, within a certain business, that we want to analyze; from then on the development of a certain segment and then get to the analysis of a small group of companies, with a very well defined location at the level of cities.

All this high-level information could allow a broad analysis, for example, of the best location for the installation of a new industrial plant in a given country or any other territorial cut, from the point of view of the companies' capacity to supply and places. That is, it could optimization in a rational way and based on high value information, the optimal locality, based on the criteria that characterize the segments and companies that supply this industry. For example, one could choose a location by understanding that proximity to a more strategic type of supply would be important, while also determining a better route criterion for other less priority suppliers according to their importance for the productive process of the industry and the characteristics of your product or services that will be offered.

Finally, based on the characteristics of the analyzed variables, the entire model and its set of information could be plotted in a geoprocessing solution, including for example, the code of the municipalities of a given territory, thus having the visual element as another alternative analytical and decision support.

REFERENCES

ABDI – Agência Brasileira de Desenvolvimento Industrial Engenha¬ria consultiva no Brasil: dcsafios e oportunidades. (2011). *Relatório de Acompanhamento Setorial*. Retrieved December 08, 2015 from http://www.abdi.com.br/Estudo/relatorio_neit_eng-consultiva_final.pdf

Amaral, J. V., & Guerreiro, R. (2014). Conhecimento e Avaliação dos Trade-offs de Custos Logísticos: Um Estudo com Profissionais Brasileiros. *R. Cont. Fin.*, *25*(65), 111–123.

Ansoff, H. I. (1965). *Corporate strategy*. Harmondsworth, UK: Penguin.

Ballou, R. H. (2004). Gerenciamento da cadeia de suprimentos: Logística empresarial. *The Bookman*.

Barros, A. P., Ishikiryama, C. S., Peres, R. C., & Gomes, C. F. S. (2015). Processes and benefits of the application of information technology in supply chain management: An analysis of the literature. *Procedia Computer Science*, *55*, 698–705. doi:10.1016/j.procs.2015.07.077

Christopher, M. (1998). *Logistic and supply chain management: strategies for reduzing cost and improving service* (2nd ed.). Financial Times Pitman.

Classificação Nacional De Atividades Econômicas – CNAE. (n.d.). Retrieved April 18, 2016 from http://cnae.ibge.gov.br/?view=secao&tipo=cnae&versaosubclasse=9&versaoclasse=7&secao=C

CSCMP - Council of Supply Chain Management Professionals. (n.d.). Retrieved December 08, 2015 from http://cscmp.org/resourcesresearch/glossary-terms

Davenport, T. H., & Harris, J. G. (2007). *Competição analítica: vencendo através da Nova Ciência: como decisões baseadas em fatos e business intelligence transformam negócios*. Rio de Janeiro: Elsevier.

Ester, M., Kriegel, H.-P., Sander, J., & Xu, X. (1996). *A Density-Based Algorithm for Discovering Clusters in Large Spatial Databases with Noise.* AAAI Press.

Fávero. (2009). *Luiz Paulo; Belfiore, Patrícia; Silva, Fabiana Lopes da; e Chan, Betty Lilian. Análise de dados: modelagem multivariada para tomada de decisões.* Rio de Janeiro: Editora Elsevier.

Glatzel, C., & Röhren, J. (2014). *Excellence in Supply Chain Management.* Mckinsey & Company. Retrieved December 08, 2015 from http://www.mckinsey.com/~/media/McKinsey/dotcom/customer_service/Operations/PDFS/Supply%20Chain%20Management%20Compendium%202014%20Edition%20-%20Final.ashx

HAIR. (2005). *Joseph F. et al. Análise multivariada de dados. 6ª edição.* Porto Alegre: Bookman.

Handfield, R., Straube, F., Pfohl, H.C. & Wieland, A. (2013). *Trends and Strategies in Logistics and Supply Chain Management - Embracing Global Logistics Complexity to Drive Market Advantage.* Bremen.

Harrison, A., & Hoek, R. (2003). *Estratégia e gerenciamento de logística.* Editora Futura.

IPEA – Instituto de Pesquisa Econômica Aplicada. (2014). *Oportunidades e Desafios para a Engenharia Consultiva no Brasil: Infraestrutura de Transporte.* Retrieved December 08, 2015 from http://www.ipea.gov.br/portal/index.php?option=com_content&view=article&id=21095

Kehrle, L. & Moutinho, L. (2005). Competitividade presente e esperada de arranjos produtivos de calçados na Paraíba. R. *Econ. contemp., 9*(3), 671-697.

Kovacevic, I. (2013). Application of transactional analysis in bullwhip effect analysis. *Anfiteatru Economic*, (33), 210-223.

Lopes, B., Falcão, L., & Canellas, T. (2016). Supply-Side: Mapping High Capacity Suppliers of Goods and Services. In Handbook of Research on Information Management for Effective Logistics and Supply Chains. IGI Global.

MacQueen, J. (1967). Some Methods for Classification and Analysis of Multivariate Observations. In *Proceedings of the Fifth Berkeley Symposium on Mathematical Statistics and Probability* (vol. 1, pp. 281–97). Berkeley, CA: University of California Press. Retrieved from http://projecteuclid.org:443/euclid.bsmsp/1200512992

Maroco, J. (2007). *Análise estatística: com utilização do SPSS. 3.* Lisboa: Edições Sílabo.

Martins, R. S., Lobo, D. S., Alves, A. F., & Sprosser, R. L. (2014). Fatores Relevantes na Contratação de Serviços em Terminais Intermodais para Granéis Agrícolas. [*RESR, Piracicaba-SP, 52*(02), 347–364.

Matos, A. L. T., Pires, S. R. I. & Vivaldini, M. (2014). Desenvolvimento de produtos sob a perspectiva da gestão da cadeia de suprimentos. *Revista de Administração da UEG, Aparecida de Goiânia, 5*(1).

Mcgee, J. V., & Prusak, L. (1994). *Gerenciamento estratégico da informação.* Rio de Janeiro: Campus.

Miller, J. P. (2002). *O milênio da inteligência competitiva.* Porto Alegre: Bookman.

Mintzberg, H. & Quinn, J. B. (2001). *O processo da estratégia* (3rd ed.). Porto Alegre: Bookman.

R Core Team. (2017). *R: A language and environment for statistical computing*. R Foundation for Statistical Computing. Retrieved from https://www.R-project.org/

Silva, C. M., França, M. T. & Oyamada, G. C. (2014). *O setor sucroalcooleiro brasileiro e a competitividade entre os estados de Mato Grosso e Mato Grosso do Sul.* Caderno de Publicações Meio Ambiente, Desenvolvimento Regional e Educação, n. 7.

Vaitsman, H. S. (2001). *Inteligência empresarial: atacando e defendendo*. Rio de Janeiro: Interciência.

Chapter 18
Indian Telecom Industry:
Challenges and Use of Analytics to Manage Customer Churn

Vishal Mahajan
Jaipuria Institute of Management, India

Renuka Mahajan
Jaipuria Institute of Management, India

Richa Misra
Jaipuria Institute of Management, India

ABSTRACT

This chapter begins with an introduction to the telecom sector, communication using mobile phones, and evolution of wireless communication. It presents the current scenario in this sector, the challenges faced by the telecom industry, especially with ever increasing data network traffic. Finally, the possibilities of harnessing the power of big data analytics, new techniques, and technologies that drive innovation in telecom are presented to help service providers make better decisions and react quickly to threats on the competitive horizon.

AN INTRODUCTION TO THE TELECOM SECTOR

In the past couple of years, telecommunication has developed as one of the basic pillars of infrastructure growth. It has become one of the key segments of financial and social growth, which is essential for overall progress of the country. The telecommunications segment, as an industry in India, is one of the most promising in the world and has become the largest telephone network.in the world, second after China (Trai, 2014).

Reforms in the Indian telecom sector, like, carefully, selecting privatization and allowing competition in various sections of the telecom industry, have led to its gradual growth. In India, the competition in the private sector for telecom was introduced through value-added services in the year 1992, and then, slowly, cellular and basic services were opened for private competition (D & B Research, 2009). In 1997, an independent regulator for telecommunication services, the Telecom Regulatory Authority

DOI: 10.4018/978-1-5225-6225-2.ch018

of India (Trai, 2014), was established. Competition in the international long distance (ILD) and in the national long distance (NLD) telephone services were introduced in the beginning of the present decade (Bhattacharya, 2010).

During the early days, only two public sector enterprises, governed by the Government, had dominance in fixed line and land line telecommunication services because of their huge user base. Their dominance can be understood by comparing their figures, as on December 31, 2001. BSNL and MTNL, the two Public Sector Enterprises (PSEs), had 34.73 million Direct Exchange Lines (DELs), while there were only 0.45 million privately owned DELs. These two public sector enterprises were permitted to enter in the cellular sector at the start of the current decade, which is quite late, in comparison to the private players. Hence, their mobile subscriber base is very small as compared to the private operators.

As per TRAI, private operators held 91.45 per cent of the total wireless market share (in terms of the subscriber base), while state-owned operators, such as Bharat Sanchar Nigam Limited (BSNL) and Mahanagar Telephone Nigam Limited (MTNL) held 8.55 per cent market share. However, in the wire-line space, operators like BSNL and MTNL held a share of 72.79 percent (Telecom News India, 2016).

Telecom sector in the country has seen exponential growth due to a series of reforms initiated by the Government, followed by inventions in the area of wireless technology and active participation of the private sector. Government policies and regulatory framework formulated by Telecom Regulatory Authority of India have provided an encouraging atmosphere for the service providers. These initiatives have enhanced the availability of telecom services at a reasonable price to the customers and have made the telecom sector more competitive. In the last two decades, Indian telecommunication sector and mobile telephone segment have got specific attention and focus, which has completely changed the way information and data are shared and communicated. Its stunning growth has helped millions of people stay connected.

This progression, in any case, has and keeps on being at the expense of the atmosphere controlled by an unmanageable and wasteful model of energy generation and utilization. At the same time, this advancement additionally comes with disadvantages, raising crucial questions on the forthcoming operational and financial model of the Telecommunication segment.

EVOLUTION OF TELECOM SECTOR IN INDIA

Telecommunication in India was first introduced in 1851, when the first telephone land lines near Kolkata (which was known as Calcutta then) were laid by the Government. Formal introduction of telephone services in India were made much later, in the year 1881. Considering this, the telecom sector in India is more than 165 years old. Further, Indian telephone services were combined with the postal system in 1883. Post-independence, after 1947, all foreign telecommunication companies were converted to state-owned to create a body, known as the Post, Telephone and Telegraph (PTT), which was administrated by the Ministry of Communication. Until 1984, the Indian telecom sector was completely owned by the Government and the private sector was allowed to manufacture telecommunication equipment only after that. In 1984, the Indian government established Centre for Development of Telematics (C-DOT) - an independent body, to focus all its efforts towards R&D in the telecom segment and to build state-of-the-art telecommunication expertise to meet the rising demands of the telecommunication network in

India. The real growth of the telecom industry began only after 1985 when the government divided the Department of Post and Telegraph to setup two different entities - Department of Telecommunications (DoT) and Department of Posts (D & B Research, 2009).

According to D & B Research (2009), the entire advancement of the telecommunication industry can be categorized into three distinctive periods.

- **Period 1:** Pre-Liberalization Period (1980-89)
- **Period 2:** Post Liberalization Period (1990-99)
- **Period 3:** Post 2000

Due to the Telegraph Act of 1885, until late nineties, the Government of India had complete dominance on all kinds of communication. As stated above, the telecom industry was completely controlled by the government and hence, was categorized as a small sized industry. The New Telecom Policy (NTP-99) framed by the government provided a required push to the progression of the telecom industry and set the much-required drift for liberalization in the industry.

The policies implemented by the government have played an influential and important part to make the telecom industry a big pillar of growth. It allowed the participation by private players virtually in all segments. As a result of this, Indian telecommunication industry is one of the most open market in the world. Today, because of this change in the telecom industry, it has become more competitive. The subscribers are getting benefited due to lower price and availability of advanced technologies. Figure 1 shows the evolution of the Telecom sector in India, spread across last two decades and divided into three phases, summarizing the important events in each phase.

STRUCTURE OF TELECOM INDUSTRY IN INDIA

Telecom industry in India can be largely categorized into basic, cellular and internet services. At present, the private sector players, as well as the public-sector players, are aggressively involved in providing services to the fast emerging telecommunication requirements in India. Involvement of private players is allowed in all segments of the telecom industry, including basic mobile services, DLD (Data Link Discriminator), radio, internet and paging.

The comprehensive organization of the telecommunication industry with respect to the service providers is explained in Figure 2.

PUBLIC SECTOR

Until 2002, VSNL was the only telecommunications infrastructure and service provider in India (16th Annual Report). It was privatized in 2002 and, after that, the two public sector undertakings, namely MTNL (Mahanagar Telephone Nigam Limited) and BSNL (Bharat Sanchar Nigam Limited), were created to provide all the telecom services throughout India. MTNL provides the telecom services to two major metro cities - Delhi and Mumbai, while telecom services to the rest of the country are being provided by BSNL. In the post-liberalization era, these Public Sector Undertakings had gained generous ground and gave intense competition to the telecom players privately operating in the country.

Figure 1. Evolution of telecom sector in India
Source: D&B Research (2009). The Indian Telecom Industry. Retrieved from URL: https://www.dnb.co.in/IndianTelecomIndustry/OverviewTI.asp

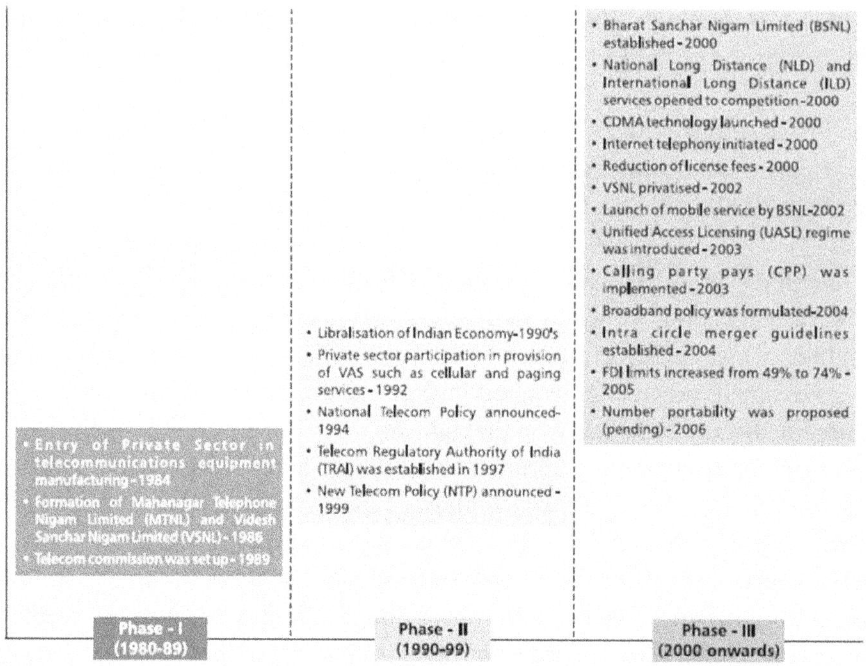

Figure 2. Structure of Indian telecom sector
Source: D&B Research (2009). The Indian Telecom Industry. Retrieved from URL: https://www.dnb.co.in/IndianTelecomIndustry/OverviewTI.asp

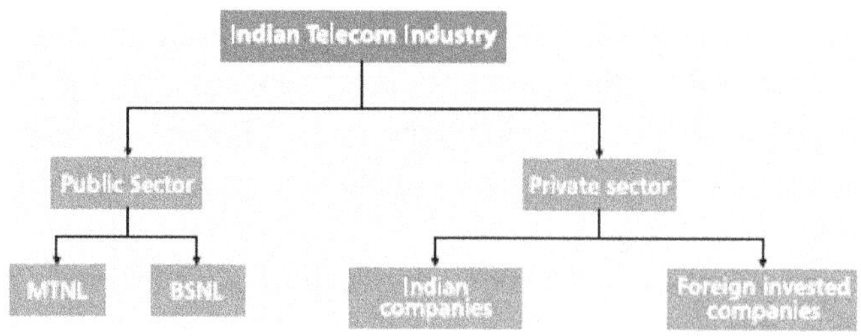

PRIVATE SECTOR

Private operators have played a key role, in the mobile communication segment, in the progress of the telecom industry. The private sector has been growing its base in the telecommunication services segment rapidly after the liberalization of the telecommunication industry. The contribution of the private players towards telecom services segment has been at a very fast pace after the implementation of NTP-99.

Though private organizations are involved in providing both fixed and wireless services. Their major contribution and action has been seen in the wireless segment. The contribution of fixed line subscribers to the private players is only 2%, compared to their overall subscriber base. Some private organizations have their presence across all circles in India, but there are still many circles which are serviced by the regional players.

CONTRIBUTION TO GDP

The open-minded and better policies implemented by the Government of India have been instrumental in the rapid growth of the Indian telecom sector along with the strong consumer demand. The government has enabled easy market access to telecom equipment, and a fair, proactive, regulatory framework that has ensured availability of telecom services to the consumer at affordable prices. The deregulation of the foreign direct investment (FDI) norms has made this sector one of the fastest growing, and among the top five employment opportunity generators in the country (IBEF, 2016).

Figure 3 shows the growing trend of revenue from the telecommunication services.

As per the report by the GSM Association (GSMA) in collaboration with Boston Consulting Group (BSG),

The Indian mobile economy is growing rapidly and will contribute substantially to India's gross domestic product. Driven by the strong adoption of data consumption on handheld devices, the total mobile services market revenue in India is expected to touch US$ 37 billion in 2017, registering a Compound Annual Growth Rate (CAGR) of 5.2 per cent between 2014 and 2017 (YourStory, 2015).

EVOLUTION OF WIRELESS GENERATIONS

The telecommunication services in the world had an incredible jump in the recent couple of years: 6 billion individuals own cellular telephones. It may be interesting to investigate the different generations of cellular systems. Figure 4 shows the timelines and features provided by each generation of technology and examined in the development of mobile communications from first generation to fifth generation.

Figure 3. Telecom sector revenue trend
Source: Researcher's own compilation with data from http://www.dot.gov.in/

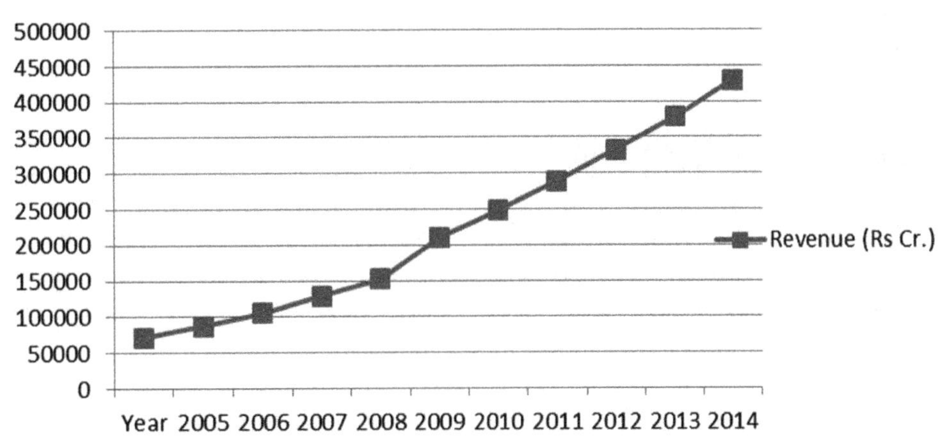

Figure 4. Mobile services provided by each generation
Source: Ahmed, Bilal, and Mustafa, 2015. Comparison between Cellular Generations. International Journal of Engineering, Applied and Management Sciences Paradigms, Vol. 22, Issue 01, 2320-6608

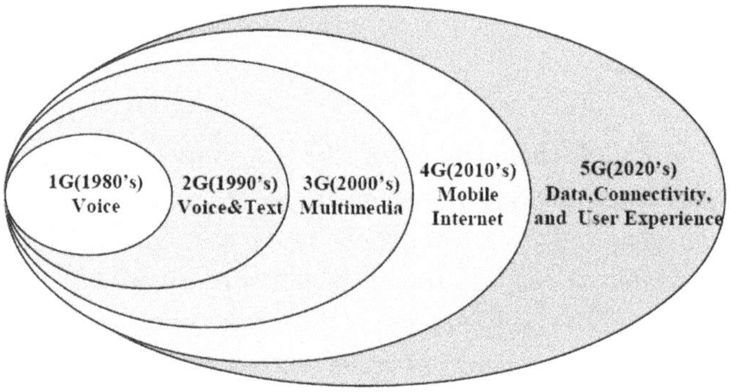

Nearly every telecom service provider, as well as the user, is looking to take the advantage of these third generation and fourth generation services. At present, four generations are prevalent in the mobile industry. These are known as the first generation or 1G, the second generation or 2G, the third generation or 3G, and then the fourth generation or 4G.

FIRST GENERATION (1G) SERVICES OVERVIEW

During early days, when the wireless services were started, analog communication was used that generation is known as First Generation (1G). The first generation of wireless telephone technology, known as 1G mobile telecommunications, was first implemented during 1980s. In 1979, Nippon Telegraph and Telecom (NTT) introduced the first automatic mobile system in Japan, and it was known as first generation. It was capable of speed up to 2.4kbps and permitted the audio calls in one country. Though it was very simple to implement, this technology had many drawbacks like:

- Low Voice Quality
- Battery Life was not adequate
- Phone Size was large
- Minimal Security
- Storage Capacity was limited

SECOND GENERATION (2G) SERVICES OVERVIEW

2G - Second Generation wireless cellular mobile services are based on GSM and is far ahead of First Generation (1G) services. Unlike 1G that had its main emphasis on voiced communication, 2G provides services such as multimedia message (MMS), text messages and picture messages. The bandwidth of 2G is 30-200 KHz.

A typical second generation (2G) uses Global System for Mobile Communications, known as GSM network service. It ranges from 800/900MHz or 1800/1900 spectrum. Second generation was launched in the year 1991, in Finland and used digital signals.

Following Mobile technologies are included into 2G:

- **GPRS:** General Packet Radio Service
- **CDMA:** Code Division Multiple Access
- **GSM:** Global System for Mobile Communication
- **EDGE:** Enhanced Data Rates for GSM Evolution

The greatest advantage of the second generation, as compared to the first generation, is the roaming facility which is available with 2G and the data speed is up to 64kbps. Though 2G service was way ahead of 1G service, it also had drawbacks like:

- Digital signals required for 2G are very strong. If any specific network zone does not have the network coverage, the strength of digital signals would be weak and will interfere with the proper working of mobile phones.
- 2G is not able to support complex data like videos.

2.5G GENERATION SERVICES OVERVIEW

Before the launch of third generation (3G) of mobile technology, 2.5G was introduced between second and third generation telephony. It is occasionally called 2G Cellular Technology. combined with GPRS. Its speed was 64-144 kbps. The main features included in this technology were Audio Calls, Web Browsing, Send/Receive E-mail Messages, and Camera Phones etc. Polyphonic ring tones, color pictures and Java based games all worked on 2.5G.

THIRD GENERATION (3G) SERVICES OVERVIEW

Third generation, which is referred as 3G, was launched in the year 2000. It has an increased data transmission speed from 144Kbps to 2Mbps. The bandwidth of 3G is 15-20MHz and it functions at a range of 2100MHz. The rate at which 3G can transfer data is high and it has an enhanced bandwidth. It can support videos, audio files as well as applications built on the web. 3G has also provided a new opportunity to use service, which was not possible earlier like global roaming. Wide Band Voice Channel was another notable feature provided by third generation, which enabled anyone to send messages, get connected or even call anyone, anywhere in the world and, thus, has shrunk the world into a small place. Another notable feature provided by 3G is the voice clarity: one can now talk without any interference or disruptions. It does not stop here: with 3G phones one can now make video calls, organize video conferences, play 3D games, send or receive Multi Media Messages (MMS).

The features provided by 3G phones are not limited to these, they also enable users to enjoy multiple entertainment options, like 3D gaming, multiple player gaming etc. including communication, Internet,

Mobile T.V, Video Conferencing, Video Calls, Multi Media Messaging Service (MMS) at a faster rate. In spite of all these enhanced features, it has drawbacks like:

- High costs for 3G Licenses Services
- Challenges in building the infrastructure for 3G
- Bandwidth Requirements are high
- High cost of 3G cellular devices
- Size of Cell Phones is large

FOURTH GENERATION (4G) SERVICES OVERVIEW

Successor of 2G and 3G is fourth generation or 4G technologies and it promises downloading at a speed of 100 Mbps and gives ultra-broadband web access. High speed stream is supported on 4G, which promises some additional features, like high definition mobile TV, video conferencing and cloud computing. With 4G, data can be transferred or shared at very fast speed and it is more secure than the previous generations. Following, are the mobile technologies, which are involved in 4G:

- **LTE:** Long Term Evolution Standard (LTE) which is built on the Global System for Mobile communication (GSM)/ Enhanced Data Rates for GSM Evolution (EDGE) and Universal Mobile Telecommunications System (UMTS)/High Speed Packet Access (HSPA),
- **3GPP:** 3rd Generation Partnership Project,
- **MIMO:** Multiple in Multiple Output smart antenna technology,
- **OFDM:** Orthogonal Frequency Digital Multiplexing,
- **802.16e:** Worldwide Interoperability for Microwave Access (WiMAX),
- **802.20:** Mobile Broadband Wireless Access (MBWA)

Though 4G has changed the way mobile technology is being used, it has its own drawbacks like:

- Battery consumption is more
- Implementation is tough due to complexity involved
- Complex hardware requirements
- Cost of instrument required to implement next generation network is very high

FIFTH GENERATION (5G) SERVICES OVERVIEW

Today's wired society is going wireless and if it has a problem, 5G is the answer. 5G technology is going to give tough competition to Computers and Laptops. It will be available in the market by 2020 at an affordable cost with more reliability than previous mobiles (Philip, 2016).

5G Network Model is an All-IP based model for mobile and wireless network interoperability. The All-IP Network (from now on referred as AIPN) has the capacity to satisfy the ever increasing mammoth demands of the burgeoning cellular market (Sapakal & Kadam, 2013).

In addition to this, it also provides a standard framework for all radio access technologies and guidelines. All-IP Network works on packet switching as compared to circuit switching used in its antecedents, and its continuous development offers performance enhancement and cost reduction.

The main objective of 5G networks is to meet the rising demand of telecommunications way beyond 2020, specifically to provide the support for rise in data consumption at an unimaginable growth rate. Users want to connect all their devices to the network and want the data to be available at higher speed. In the coming decade, the data traffic volume is expected to rise multifold, due to the availability of video chats/calls, online streaming and communication among various devices, including cars and homes, as well over the internet. It is expected that fifth generation networks will have to support minimum speed of 100 Megabits per second extendable up to 10 Giga bits per second. Figure 5 shows the services, year they are introduced and key differences between various generations of mobile phone technologies.

COMMUNICATION USING MOBILE PHONES

Mobile phone or cellular telephone is a wireless device which uses radio waves in the band of 800–900 megahertz (MHz) to allow telecommunication inside a definite range of hundreds of square miles. In order to establish a cellular phone system, a geographical zone is divided into smaller zones or areas, typically mapped as uniform hexagrams but, is in fact, overlapping and irregular in shape. Every cell includes a low powered radio transmitter and receiver that allows broadcast of signals between cellular users. The basic function of the mobile phones, when they were invented, was to connect people and enable them to communicate to each other through audio using radio waves within a range called spectrum. Different mobile technologies work on different frequency

Figure 5. Summary of key differences between various mobile technologies
Source: Ahmed, Bilal, and Mustafa, 2015. "Comparison between Cellular Generations". International Journal of Engineering, Applied and Management Sciences Paradigms", Vol. 22, Issue 01, 2320-6608.

Technology	1G	2G	3G	4G	5G
Design Began	1970	1980	1990	2000	
Implementation	1981	1991	2001	2010	2020
Services	Analog voice	Digital voice, short message	Higher capacity, data rates up to 2 Mbps	Higher capacity, completely IP-Oriented, multimedia ,data to hundreds of megabits	Higher capacity, completely IP-Oriented, multimedia, data10 Gb/s
Standards	AMPS, ETACS, NMT etc.	TDMA,CDMA,GSM	WCDMA CDMA-2000	Single standard	
Data Rate	NA	14.4 kbps	2 Mbps	>200 Mbps	10 Gbps
Multiplexing	FDMA	TDMA, CDMA	CDMA	OFDM	
Core Network	PSTN	PSTN	Packet network	Internet	

spectrums. These will be covered in details in the subsequent sections below. The functioning of a mobile phone can be summarized as below:

- The geographical region supported by a cellular network is divided into smaller geographical regions, or cells.
- Within a given region or cell, all communication with a cellular or portable instrument is routed to a base station that supports the cell.
- Since moveable devices are operated using battery, their transmitting power is low. A definite frequency is allotted to a cell for sending and receiving, and can be used by other cells within the broad geographic region. Hence, the spectral efficiency is defined as the usage of a slice of its radio spectrum, enhanced within the area it is serving, by a factor equivalent to the number of times a frequency can be reused.
- As the cellular instrument continues, starting from one cell, then onto the next, over the span of a call, a central controller consequently redirects the call from the old cell to the new cell without any recognizable interference in the reception of signal. This is recognized as "handoff process". The central controller, or "Mobile Telephone Switching Office (MTSO)", in this manner, behaves like a smart central office switch, which follows the movement of a cellular user.

Mobile phones are principally intended for voice communication. But the mobile phones of current generation provide several add-on services and functionalities in addition to the traditional voice communication features like emails, Short Messaging Services (SMS), internet access through packet switching, camera, video player and recording features, Bluetooth functionalities, Multi Media Services (MMS) for sharing and receiving audio, video and image files, Global Position System (GPS), radio and MPEG-2 Audio Layer III (MP3) players.

PRESENT SCENARIO IN THE TELECOM SECTOR

With the passing of time, mobile phones have turned out to be more technology savvy and smart. Earlier, they were manufactured only to offer basic communication services on the move, but with time, they have evolved and are providing capabilities equivalent to small computers including storing and processing data. Besides all this, a smartphone has a cutting edge technology and an ability, that can be used to create applications for collecting, processing and storing data or information for communication, or other purposes, like supporting business and entertainment.

The physical form of a smartphone also varies depending on the type of user. It may come with a modern design, which targets the young or hi-tech users, including touch screen, bigger size of screen, keyboard which is integrated or designed especially for older people with larger visual keypad, SOS button, volume set that can help with the hearing impairment problem and many more (Plaza, Martin, Martin, & Medrano, 2011).

With multiple technologies coming together, like global positioning systems, internet capabilities and the increased capacity of mobile data, smartphones are able to provide innovative, well-designed capabilities to their users, such as gathering information seamlessly, social networking applications implanted into the smartphones including instant messaging, Facebook and Twitter, entertainment applications including multimedia and gaming, and many more. It can also provide value added service applications, such as location based services (LBS) that can provide customized information and application to the user depending on their location and preferences (Alessandro & Trucco, 2011).

User's day to day life, specifically in the developed world, has changed drastically due to the penetration of smartphones. Subscribers are now interacting more with their phone and some spend most of their time managing and using applications and services provided by the smartphones.

During the last decade, the telecom industry has observed incredible growth in user base, known as the subscribers. Increase in network coverage and competitive reduction in prices acted as facilitators for the growth in the subscribers' base. In the next five years, telecom segment is expected to attract investments to a tune of US$ 56.3 billion, and the market capitalization will be beyond the US$ 101 billion benchmark.

Telecom has been the most transformational technology of recent times in India. It needs to be nurtured further for the benefit of the nation. Today, the Indian telecom sector accounts for the 2nd largest private sector investment in infrastructure with INR 1,53,000 crore of investment (during 10th & 11th plan), apart from the two spectrum auctions held in February 2014 and March 2015, causing pay-outs from the industry of INR 61,200 crore and 1,10,000 crores respectively (Coai, 2015).

Telecommunication sector in India has observed incredible growth during the last decade.

Today, the Indian telecom network is the second largest in the world after China. An open-minded policy regime and involvement of the private sector have played a vital role in transforming this sector. The total number of telephones as on 31 January 2014 was 922.04 million, out of which 893.31 million were wireless subscribers (Trai, 2014).

The subscriber base in the telecom industry has grown exponentially over the last decade. This is further fueled by the increase in network coverage and the decline in tariff.

The growth story and the potential have also served to attract newer players in the industry. This resulted in the increase in the intensity of the competition. Internet subscribers in India grew to 243 million as of January 31, 2014 (TRAI, 2014).

According to telecom regulator TRAI, as many as seven out of eight net users in the country are accessing the services through their mobile phones. The total number of mobile internet subscribers stood at 155 million at the end of the last fiscal.

While there was a continuous growth in voice and data related services in the wireless segment, high speed data services were significantly supported by the traditional landline telephony. *The landline telephones are now 27.00 million, and wireless telephones have grown to 944.01 million at the end of December 2014* (TRAI, 2014).

Due to this increase in user base for wireless segment, its share increased to 97.22 percent of the total telecommunication services provided in the country. This growth has been further accelerated due to availability of high broadband landline based connections. The endless demand for wireless services have compelled the telecommunication sector to come up with a solution for generating sustainable resources.to meet requirements of high speed services provided through landline.

In order to attract subscribers and to increase the FDI inflow, FDI limit for the telecommunication sector was increased from 74 per cent to 100 per cent by the Indian government. This has made the sector more attractive and investor friendly. This policy implementation will help the telecommunication service providers consolidate funds and raise debt from foreign as well as the domestic market.

Liberalization and opening of the telecommunication segment has, not only fueled towards swift progress, but also assisted to maximize the consumer benefits in a big way, which is apparent from an enormous drop in charges. Telecommunication sector has observed a continuously increasing trend in subscribers availing telephone services, which resulted in increased teledensity.

Telephone density or teledensity is defined as the number of telephone connections per 100 people living within an area. It varies widely across the countries and also between the urban and rural areas within a country (Gartner, 2012).

Table below shows the trend of teledensity over the years and how it has varied for rural as well as urban subscribers. Table 1 presents the trend of teledensity over the years (Trai, 2016).

Teledensity will be more than 100 when there are more telephone connections as compared to people. Teledensity in third world countries is less than 10. Penetration of telecom services in the country can be observed through the teledensity.

Teledensity has decreased from 76.86% in December 2011 to 74.02% at the end of December, 2013, and again increased to 79.38% as on March, 2015. Teledensity varies across circles and there is a substantial urban-rural divide. While the urban teledensity reached 148.61% at the end of March 2015, the rural teledensity was only 48.37% (Trai, 2016).

The industry will also need to work closely with the Government, international standard bodies and the Civil Society, to promote security and safeguarding the legitimate rights of customers for confidentiality. While the security interests of the Government are supreme, the costs of such programs and initiatives should not only be on the service providers.

Table 1. Teledensity trend: Urban and rural

Year	Rural Teledensity %	Urban Teledensity %	Overall Teledensity %
Dec 2010	147.88	31.18	66.16
Dec 2011	167.85	37.48	76.86
Dec 2012	149.90	39.85	73.34
Dec 2013	144.95	42.67	74.02
Dec 2014	148.06	46.09	77.58
Dec 2015	148.61	48.37	79.38

Source: Trai (2016), The Indian Telecom Services Performance Indicators. Retrieved from URL: http://trai.gov.in/Content/PerformanceIndicatorsReports/1_1_PerformanceIndicatorsReports.aspx

CONVERGENCE OF IT AND TELECOMMUNICATION

Convergence in telecom is the way services are combined into different devices and networks driven by internet. Smartphones is such an example which not only provides voice calling but also provides internet access, access to many applications, videos etc., to name a few. This has helped Internet Protocol define the way information is transmitted to all networks, storage services and cloud computing.

THE RISE OF THE APP ECONOMY

Business Intelligence applications distributed through telecommunication systems, such as video conferencing and integrated communications help employees work together to increase efficiency and optimize all business processes. As per report by Forrester, 226 billion applications were downloaded globally in the year 2015 (Forrester, 2015). They found that, out of 226 billion apps downloaded only 8% were paid applications. They also predicted that, by year 2020, most of the applications will be used for in-app purchasing and advertisements. As per market research perform by App Annie, mobile application economy could touch US101 billion dollars by year 2020 (App Annie, 2016).

CYBER SECURITY IS NO LONGER OPTIONAL

In 2014, 71% of industries, surveyed by security specialists Cyber Edge Group, suffered a cyber-security attack while the average cost of each breach rose 96% and cost $12.7 million per organization (HP, 2015).

In year 2015, the U.S. government has warned that Android devices are a major threat to personal and corporate security. In the current scenario of the uncontrolled expansion of mobile devices, establishments are increasingly implementing mobile applications as a medium to increase efficiency and meet employee demands to work flawlessly anywhere. Serious problem that many organizations and mobile users are overlooking: are mobile applications secure enough and provide protection from any attack?

CURRENT TRENDS

The telecommunications sector continues to be the focus of evolution, revolution and disruption to practically every industry. Mobile and broadband connectivity continue to be integrated into the fabric of today's society and play a key role in driving some key trends, such as streaming video, Internet of Things (IoT) and e–payments using mobile.

Companies operating in the telecom sector will also find new prospects for development in the public sector, as the "smart cities" initiative gets more footing. Digital technologies which are using mobile phones as device to access and pay for common services like transport and parking have increased the consumer demand.

This new found interest and customer needs are helping to take these initiatives forward. Another advantage of these technologies are that the cities can be managed and administered more efficiently. The uses of applications in mobile phones now provide tracking, monitoring, security and many other public facilities like lighting, health and water. Mobile is acting as medium for transformation in organizations and is redefining landscape vendors are operating in. Artificial intelligence and voice based personal assistance are inspiring the conversational applications. The another growing trend, which has been seen due to great focus in Internet of Things (IOT), is the interest in integrated connected home services as per Figure 6.

Data usage has seen drastic upward trend, predominantly due to service like live streaming. It is anticipated to stay on the same track in the years ahead. Wi-Fi usage will continue to be crucial aspect, as service providers are looking to bring more and more mobile traffic onto broadband networks. These networks are using fiber as well as other considerations, like spectrum efficient technologies and potentially unlicensed spectrum solutions (i.e., LTE-U). Voice over LTE (VoLTE) and Voice over Wi-Fi (VoWiFi) services will also be a key focus to help carriers rationalize networks and offer improved and expanded services.

CHALLENGES FOR THE INDUSTRY

Worldwide, there is a growing trend for telecom service provider to follow opportunities outside the traditional telecommunications domain (Dataquest Insight, 2009). Telecom service providers are looking for growth beyond traditional telecom services because the market has matured and has reached the saturation point. With this, the landscape of telecom market is changing at a very fast speed, which is threatening to downgrade service provider's role to data transport/access service providers. Following, are the challenges for the telecom industry:

Revenue Growth

Currently, in the telecommunication sector, there are 15 operators competing for the same revenue share which has seen slow or negligible growth in last couple of years. *The service providers who entered the market during 2009 provided subscriptions at throw away prices loaded with free talk time, which resulted in multiple SIM ownership and reduced realization per minute of use* (Coai, 2015). Due to this

Figure 6. Interest in integrated connected home services
Source: Ericsson ConsumerLab, Connected Homes (2015). Mobile Business trends. Retrieved from URL: http://www.ericsson.com/res/docs/2015/mobility-report/emr-mobile-business-trends-2015.pdf

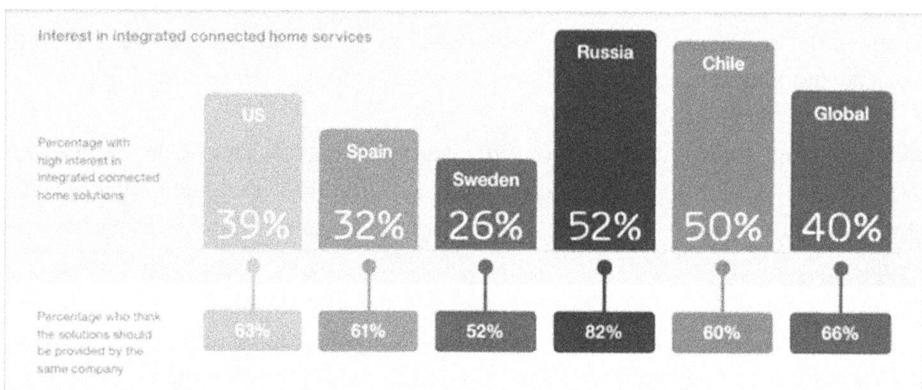

price war to gain the market share, service providers started providing free talk time, which was earlier paid. This resulted in slowdown of revenue growth in telecommunication sector. For many service providers, the revenue from traditional messaging and voice services has reached the peak.

Indeed, instant communication applications, like WhatsApp – which now has 250m users – has reduced providers' revenue share from messaging services faster than everyone's expectations. All this has contributed towards declining average revenue per user (ARPU) over the years (Coai, 2015). Figure 7 shows the trend of average revenue per user.

Lack of Telecom Infrastructure

Infrastructure required to provide telecom services in rural and semi-rural parts of the country is near to nonexistence. It could be one of the key deterrents in taking advantage of the vast rural prospective market. For entering rural areas for providing services, initial fixed cost required is huge and has to be borne by the service providers. Moreover, most of the rural areas are still lacking the elementary infrastructure facilities like electricity and roads, because of this telecom service provides will need more time to provide the telecommunication services and they also have to bear the larger logistic risks as well. As per Cisco (2014) report on trends and analysis of data traffic,

The volume for data traffic on network is expected to reach 132 Exabyte per month by the year 2018. Innovation in smart mobile devices and applications, cloud computing, big data and machine to machine communications (M2M) results in new set of data traffic behavior and features.

All these will force an increased investment in IT and Network infrastructure.

Subscriber Growth

Multi-SIM usage is on increase in the industry. As far as addition of new subscribers is concerned, India is the fastest emerging telecommunication market in the world. Today, subscribers have multiple SIMs, which is driven by availability of devices supporting multi SIM feature. Because of this feature, there

Figure 7. Trend for average revenue per user
Source: Coai (2015). Cellular Operators Association of India's report 2014.Retrieved from URL: http://www.coai.com/statistics/arpu-and-revenue-report

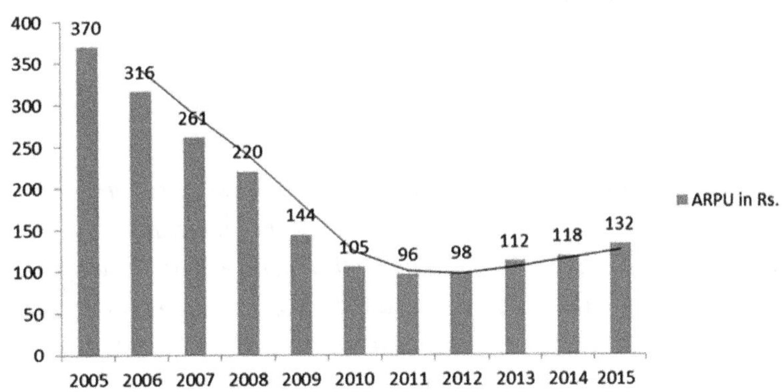

is no meaning of number of subscribers and hence, strength of service providers cannot be measured using subscribers. This impacted the market and forced it to shift their focus towards market share in terms of revenue, instead of market share in terms of customers.

Contribution of dual SIM towards new addition is up to 30%-35% of the newly added subscribers. There is a disparity between revenue market share and customer market share. It has been observed that for some service providers, even if they have higher customer market share, it has not resulted in higher revenue market share for them. The primary reason for this is the low average revenue per user (ARPU) from subscribers. The ARPUs are recently added to the subscriber base due to low tariff, and there is large number of existing subscribers who are inactive. This is forcing service providers to launch new products in the market and come up with new innovative business models to improve their profitability.

4G Roll Out

Service providers have to invest huge amount of money, to pay for spectrum fee and network infrastructure required, for launching fourth generation technologies. Another big challenge faced by service providers for providing Long Term Evolution Standard (LTE) is the use of multiple frequencies by this standard. While all different spectrums or radio frequencies can exist at the same time, it requires service providers to add support for new spectrums, which brings in more complexities and additional cost. Things will be more streamlined and simplified for the service providers and equipment manufactures if the frequencies used by 4G are standardized.

Supporting different frequency ranges and costly equipment required for network are not the only challenges faced by service providers for rolling out the 4G technologies, the cost and availability of 4G enabled mobile phones are also a hindrance. The current cost of 4G enabled handsets is very high. Consumers in India are very sensitive to price and, hence, service providers are coming up with new low-price plans for voice, as well as for data to attract customers. This low-price offer has helped in adoption of 3G in India but for a very brief time. All this is forcing and making service providers to think hard and customize every aspect of 4G roll out to make it suitable for the Indian market.

Reducing Subscriber Churn

In India, the mobile subscriber churn rate is one of the highest in the world. Hence, churn costs a lot to the service provider. With availability of multiple options to choose, subscribers are switching the service providers very easily and this has further increased with number portability. Now operators have to look into retaining these customers, reduce churn, and increase subscriber's lifetime, so that revenue growth can be achieved.

Rising Debt and Consolidation

As per India Ratings and Research firm, *With increase in the cost of spectrum acquisition and heavy cash needed for setting up of infrastructure for mobile internet amid lower profitability, Indian telecom sector is expected to witness consolidation and rising debt.* Apart from spectrum acquisition, the telecom sector is witnessing higher competition in the data space and is forcing infusion of additional capital to maintain the growth. At the same time, cash returns from such investments will be back ended as 4G revenue will gain significance over the next three to five years and, therefore, could be credited negatively in the short term. The stage is set for full-fledged consolidation in the Indian telecom with the recent acquisition of Videocon's 1800 MHz spectrum in two circles of Uttar Pradesh (West) and Gujarat. Reliance Communications in the past acquired Sistema Shyam, which operates under the brand name MTS, in an all-stock merger agreement. Recently,

Reliance Communications and Aircel, which are in talks to merge, have agreed that Reliance Communication will combine its wireless business with the smaller rival owned by Maxis and both parent companies will hold 50% each in what will emerge as India's No. 3 carrier by users (The Economic Times, 2016).

Data Segment

In the past, non-availability of a high-speed broadband network and high cost of computers hindered the reach and availability of internet in India. However, with the rapid spread of high-speed mobile data, that is changing fast. As per Business Standard (2012), *India's internet contribution to gross domestic product could increase from 1.6% in 2012 to between 2.8% and 3.3 per cent by 2015.* As per TRAI (2014) report, there are 155Mn mobile internet users with quarter on quarter growth of 20%, and average data usage per mobile user in India is 61.66 megabyte.

In India, the contribution of voice services is 80% of the total revenue and the rest of the 20% is contributed by data services. Contribution of data services in developed markets, like Japan, is about 50% of the revenue. The primary reason behind this is that internet based cross platform messaging applications had a steady growth. SMS, which was the major contributor to the non-voice revenue segment of the telecom service providers, is gradually losing its importance. Competition is forcing to reduce the rates for voice calling every day. Hence, companies need to focus on data segment to increase the revenues. There were 431 million Internet capable mobile devices in use in India as of December 2012. Mobile data traffic almost doubled between Q2 2012 and Q2 2013, while voice traffic increased by 2% only (Coai, 2013).

According to a report released jointly by Internet and Mobile Association of India and consultancy firm IMRB (2015), mobile internet user base had grown to 306 MN by end of year 2015, growing at a rate of 77% from 2014. They also found that though mobile internet penetration is around 23% in the country, percentage of monthly spent due to data usages has risen from 54% in 2014 to 64% in 2015.

Figure 8 shows the contribution of data towards average revenue per user and shows that it is the lowest and is about 21% of the total revenue in India. Today, entire nation is connected through establishment of high speed Digital Highways and is helping deliver government services to every citizen, which will help in the socioeconomic development of the citizens. But all this is not enough, challenges faced in the development of other areas like health, education, employment generation, and to ensure that economy grows at fast pace requires to strengthen the telecommunication sector further.

Today, world is shrinking because technology has evolved at a very fast pace and its reach is unlimited. This has brought many innovations in the telecom sector as well but all this requires service providers to infuse more capital to be relevant and competitive in the current market scenario.

Figure 8. Data as % of ARPU
Source: Researcher's own compilation.

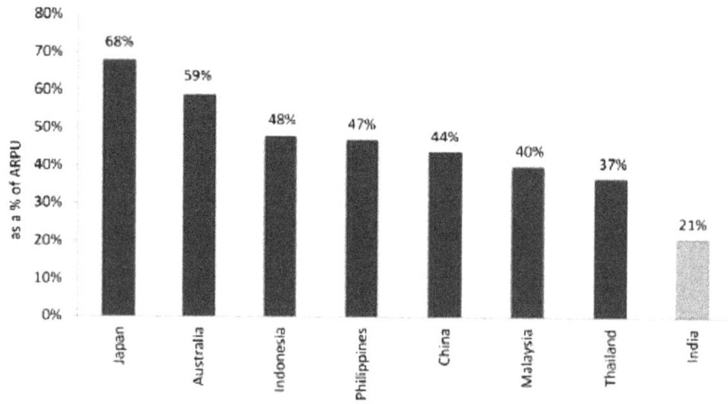

Nationwide Mobile Number Portability

The currently provided mobile number portability functionality is restricted to the same service area. According to the new change proposed by the telecom commission, mobile number portability is implemented across nation by 31st March 2015. This will encourage subscribers to retain their number, even after changing the service provider across country.

Requests for Mobile Number Portability jumped from 109.37 million subscribers at the end of December, 2013 to 111.94 million at the end of January, 2014 (Trai, Press Release No. 13/2014). In the month of Jun-15, a total of 3.68 million subscribers submitted their requests for MNP. With this the collective MNP requests increased from 160.25 million at the end of May 15 to 163.93 million at the end of June 15.

The urban subscribers' share has dropped from 60.03% to 59.86%, whereas in the month of January, 2014 share of rural subscribers has increased from 39.97% to 40.14%. With this, the overall Tele-density in India has seen a growth from 74.02 at the end of December, 2013 to 74.50 at the end of January, 2014. In the month of January, 2014, 2.56 million requests have been made for MNP. This will require certain investments from service providers and will encourage subscribers to change the service provider.

Business environment has become very competitive. There are multiple service providers, which customers can choose from. According to research firm Gartner (2012), churn rate in India is anywhere between 3.5% - 6% per month, one of the highest in the Asia-Pacific region. Also, acquiring a new customer is five to six time costlier than retaining an existing customer. Indian telecom operators are losing more than INR 6000 crore yearly, due to churn and average revenue per user per month (ARPU), which has observed a steep fall and is the lowest in the world (Trai, 2014). To support companies for churn reduction, not only we need to predict which customers are at high risk of churn, but we also need to know how soon these high-risk customers will churn.

Further, with the invention of smartphones, many applications are being developed every day, which have increased the usage of data. Voice usage has decreased as compared to data usage. In less than 7 years, the number of smartphone users has already gone beyond 1 billion, as against 1.5 billion users using Personal Computers. Worldwide, the number of users using internet is more than 2.4 billion, out of which 1.5 billion users are using mobile as a primary or secondary medium to access internet.

More traditional mobile services, such as mobile voice and texting, are being joined by Skype, Facebook, Google+, WhatsApp, ICQ and Twitter. Skype, for example, reported 207 billion minutes of voice and video chats and messages in 2010 between 170 million users (Cisco, 2014).

This has created a big problem for mobile telecom operators as the huge amount of information about new behaviors of the customers is largely opaque to them. With the voice usage, they could track call length, who's calling whom and more, but the data sphere remains an unknown territory. Visibility into customer behavior is declining as the data sphere eclipses the voice sphere (Mishra, Khan & Singh, 2013).

As per Fierce Wireless, (2015),

Data revenue in the wireless services segment will fuel overall telecom growth, driven by the increasing use of smartphones, tablets and other connected devices. Moreover, the growing popularity of mobile video has made larger data plans more alluring. But increasing competition will result in pricing pressure that will prevent substantial growth in wireless margins.

As reported by Nick (2016), revenue from mobile data services of Bharti Airtel, one of the leading service providers in India, grows by 44%. They observed around 76% rise in the data traffic. They also observed that share of revenue from data services has jumped from 12.4% to 23.3% in last one year. They reported that the average data consumption per user has increased from 495 MB to 859 MB in the last 4 quarters.

Another report by Alnoor (2016), presents that every smartphone manufacture in the world is focusing on India currently, because of its growing needs for smartphone and other devices. They have observed an increase of smartphone sales by 23% during first three months of 2016 itself.

Different type of big data generated gathered during telecom customer complete analytic journey suffers from the problems of volume, variety and veracity. Both Structured versus Unstructured Data can be tapped, stored, analyzed for correlations and patterns to arrive at decisions. The structured data set includes customer profile data (including demographic details, number of minutes used, Internet Data Usage etc.) that easily searchable by basic algorithms. Unstructured data is more like human language. It doesn't fit nicely into legacy databases e.g. Call center interaction data, geo spatial data, XDR-extended data records that contains data related to purchase and downloading of apps and the social chatter data which is available on the social media sites.

Further, with nationwide mobile number portability, which was implemented on March 31, 2015, there is a big jump in subscriber churn. This has been further fueled by providing the lifetime free voice services being provided by service providers, such as launch of latest offerings by Reliance Jio. This has forced other service provider to lower their data charges. Hence, it becomes very important to analyze the impact of this change in customer behavior on the "Churn" which will eventually help telecom operators minimize their losses.

RESEARCH OPPORTUNITIES IN THE TELECOM SECTOR

World today has become more competitive with the roll out of technologies such as 5G, M2M and the evolution of Information and Communications Technology (ICT) and digital wallets. Service providers has to invest huge money to provide new technologies to remain relevant in the market.

Mobile data consumption in India surpassed the combined usage of the US and China in 2017. Jio propelled India to the top position in terms of mobile data consumption with 150 crore gigabytes per month, surpassing the combined 130 crore gigabytes mobile data consumption of the US and China.

The current need of the service providers is to keep their customer happy and satisfied by providing excellent services. Mobile companies have the customer care representatives who are trained and educated to interact with customers directly. However, this is not enough because there are many factors that cause customer churns.

Recently telecom providers have consolidated, leaving just three major players in the market by 2017-year end. With Internet of Things, Artificial Intelligence, big data platforms, drones and robots promising to fuel future demand for high-speed data, machine-to-machine communications, instant high definition video transfer as well as its Smart Cities initiative, innovative business models are required to counter the falling revenues in telecom.

THE ROAD AHEAD

Operators have to break their silos, be more collaborative and adopt best practices while simplifying their operations. Analytics and a combination of technologies like predictive analytics, optimization and scoring, recommender system, personalization can help telecom operators capture and analyze data, followed by segmenting their customers to expand their business and gain competitive advantage.

- **Churn Prediction Models:** Predictive analysis techniques are used for developing hypothesis' around churn rates and using data analytics (based on neural networks and decision trees) to assess the causes for churn. The churn prediction model can update risk scores for more accurate and timely retention activities. The input data set includes data usage, call-center history etc. to identify root causes of customer discontent, such as issues at specific call centers as well as dropped calls (Hassouna, 2015).
- **Big Data Management:** Voluminous data, sometimes in petabytes, are gathered and stored throughout the customer lifecycle that ranges from customer acquisition, onboarding to upgrading plans and eventual disconnect (if applicable). It includes customer profiles, offers/promotions, information, usage, data from call centers, social media, web logs, experience and pricing (Senbalc, 2013; Jose, Ramanan, & Kumar, 2017; Kanstren, 2016; Ahmad, 2014). Moreover, API (Application Programming Interface) makes operators' network more programmable.
- **Credit Risk Modeling:** It serves as a scorecard for its customers to classify the good accounts, i.e. the ones with higher likelihood of regular payment, from the bad accounts, i.e. ones with high potential for defaults, for managing credit risk, bad debts and developing a targeted collections approach (Hassouna, 2015).

- **Business Intelligence:** BI tools can help operators find, analyze and share the information with call centers to improve decision-making with query, reporting, analysis, scorecards and dashboards with agile working methods that divide these tasks into short phases with frequent reassessments and adjustments to ensure the best results.
- **Personalization:** The potential of data analytics can be better realized if the organizations can personalize the treatment to the targeted customers having the maximum propensity to churn. A well worked out granular micro-segmentation of customers can be mapped to variety of offers carefully chosen for the customers.
- **Social Media Analytics:** Many social media monitoring and analytic tools can be used to detect sentiment and online reputation from Twitter, feeds, blogs etc. Using keywords, behavioral, demographic, geographic, influencer analysis can go beyond mere social media "listening." Social media offers important insight into consumer opinions and spot buzz related to products and brands.
- **Social Network Analysis:** It helps identify those customers who have a strong sphere of influence, identify groups, group leaders and whether others will churn based on their influence. It consists of two capabilities Group analysis (capability to identify the groups and the leaders of each group from the data) and Diffusion analysis (utilizing the existing churn information of the current churners and to mitigate their influence on others to leave).
- **Recency Frequency Monetary (RFM) Approach:** It helps the organization predict which customers are likely to respond, and which will not respond to their offer, on the basis of how recent customer has made purchase (Recency), how much often they purchase (Frequency) and how much they spend (Monetary) (Li, Lee, & Liu, 2006; Liu & Shih, 2005; Singh & Singh, 2017).

CONCLUSION

Innovation is a continuous process. Hence, service providers must evaluate their existing models and re-invent with strategic use of big data and advanced analytic capabilities, by providing timely and attractive personalized services. Thus, the power of IT and analytics has to be used to deliver the next best offer with analytical decision management to empower the staff, grow revenue and drive profitable customer relationships and profit.

REFERENCES

Ahmad, M., Rashid, T., & Mishra, D. K. (2014). Impact of wireless technology on future of big-data industry. *IT in Business Industry and Government (CSIBIG) Conference on*, 1-4.

Ahmed, S., Bilal, K., & Mustafa, A. (2015). Comparison between Cellular Generations, International Journal of Engineering. *Applied and Management Sciences Paradigms*, *22*(1), 2320–6608.

Alessandro, C., & Trucco, P. C. (2011). Business potential and market opportunities of intelligent LBSs for personal mobility – A European case study. *Procedia Computer Science*, *5*, 906–911. doi:10.1016/j.procs.2011.07.126

Alnoor, P. (2016). India is the next China as appetite for smartphones defy global trend. *Business Standard*. Retrieved from http://www.afaqs.com/news/story/47857_India-is-the-next-China-as-appetite-for-smartphones-defy-global-trend

App Annie. (2016a). *App Forecast*. Retrieved from https://www.appannie.com/insights/market-data/app-annie-releases-inaugural-mobile-app-forecast/

App Annie. (2016b). *App Forecast*. Retrieved from URL: http://go.appannie.com/report-app-annie-market-forecast-2017

Banerjee, A. (2013). *Big Data & Advanced Analytics in Telecom – Huawei, Report by Heavy Reading*. Retrieved from www.heavyreading.com

Bhattacharya, M. (2010). *Telecom Sector in India: Vision 2020*. Department of Telecommunications, Ministry of Communications & IT, and Government of India.

Business Standard. (2012). Retrieved from URL: http://www.business-standard.com/article/companies/internet-technology-to-contribute-up-to-3-3-to-gdp-by-2015-112121900084_1.html

Cisco. (2014). *The Zettabyte Era—Trends and Analysis, Visual Networking Index (VNI)*. Retrieved from URL: http://www.cisco.com/c/en/us/solutions/collateral/service-provider/visual-networking-index-vni/VNI_Hyperconnectivity_WP.html

COAI. (2013). *Cellular Operators Association of India's report 2013*. Retrieved from Cellular Operators association of India: http://www.coai.com/Reports-and-Papers

COAI. (2015). *Cellular Operators Association of India's report 2014*. Retrieved from URL: http://www.coai.com/statistics/arpu-and-revenue-report

D & B Research. (2009). *Indian Telecom Industry*. Retrieved from URL: https://www.dnb.co.in/IndianTelecomIndustry/OverviewTI.asp

Dataquest Insight. (2009). *New Revenue Opportunities for Telecom Carriers, 2013*. Author.

Ericsson Consumer Lab. (2015). *Connected Homes. Mobile Business trends*. Retrieved from URL: http://www.ericsson.com/res/docs/2015/mobility-report/emr-mobile-business-trends-2015.pdf

FierceWireless. (2015). *Wireless industry revenue*. Retrieved from URL: http://www.fiercewireless.com/story/moodys-wireless-industry-revenue-likely-grow-3-4-2016/2015-12-02

Forrester. (2015). *World Mobile Application Spending Forecast, 2015 To 2020 (Global)*. Retrieved from URL: https://www.forrester.com/report/Forrester+Research+World+Mobile+Application+ Spending+Forecast+2015+To+2020+Global/-/E-RES122413

Gartner. (2012). *IT Glossary*. Retrieved from Gartner: http://www.gartner.com/it-glossary/teledensity/

Hassouna, M., Tarhini, A., Elyas, T., & AbouTrab, M. S. (2015). Customer Churn in Mobile Markets: A Comparison of Techniques *International Business Research*, *8*(6). doi:10.5539/ibr.v8n6p224

HP. (2015). *Strategic Focus: Software, Products & Services*. Retrieved from URL: http://www8.hp.com/us/en/hp-news/press-release.html?id=1815969#.VxDdk_l97IV

IBEF. (2016). *Telecom Sector in India*. Retrieved from URL: http://www.ibef.org/industry/telecommunications.aspx

IBM Software Report. (2013). *Minimize customer churn with analytics*. IBM.

Jose, B., Ramanan, T. R., & Kumar, S. D. M. (2017). Big data provenance and analytics in telecom contact centers. *Region 10 Conference TENCON 2017 - 2017 IEEE*, 1573-1578.

Kanstren, T., Liikka, J., Mäkelä, J., Luoto, M., & Prokkola, J. (2016). Preliminary big data in a 5G test network. *Big Data (Big Data) IEEE International Conference on*, 2722-2727.

Li, L., Lee, F., & Liu, W. (2006). Timely Product Recommendation Based on RFM Method. *Proc. The International Conference on Business and Information (BAI 2006).*

Liu, D., & Shih, Y. (2005). Hybrid approaches to product recommendation based on customer lifetime value and purchase preferences. *Journal of Systems and Software*, *77*(2), 181–191. doi:10.1016/j.jss.2004.08.031

Mishra, K. A., Khan, M. I., & Singh, Y. P. (2013). Review: Recent Trends in World and Indian Telecom Market. *International Journal of Management*, *4*(3), 61–74.

Nick, W. (2016, April 28). Bharti Airtel revenue jumps on surging mobile data use. *Total Telecom*. Retrieved from http://www.totaltele.com/view.aspx?ID=493606

PhilipB. (2016). Retrieved from http://www.swinburne.edu.au/news/latest-news/2016/01/the-g-in-5g-how-mobile-generations-have-evolved.php

Plaza, I., Martin, L., Martin, S., & Medrano, C. (2011). Mobile applications in an aging society: Status and trends. *Journal of Systems and Software*, *84*(11), 1977–1988. doi:10.1016/j.jss.2011.05.035

Sapakal, R., & Kadam, S. (2013). 5G Mobile Technology. *International Journal of Advanced Research in Computer Engineering & Technology*, *2*(2), 568–571.

Şenbalc, C., Altuntaş, S., Bozkus, Z., & Arsan, T. (2013). Big data platform development with a domain specific language for telecom industries. In *IEEE High Capacity Optical Networks and Emerging* (pp. 116–120). Enabling Technologies.

Singh, I., & Singh, S. (2017). Framework for Targeting High Value Customers and Potential Churn Customers in Telecom using Big Data Analytics. *I.J. Education and Management Engineering*, *1*, 36–45.

Telecom news India. (2016). *Telecom subscriber base*. Retrieved from URL: http://www.tele.net.in/news/item/18999-telecom-subscriber-base-in-the-country-reaches-1043-billion-in-january-2016-reports-trai&tag=Wireline

The Economic Times. (2016). *Telecom*. Retrieved from URL: http://economictimes.indiatimes.com/industry/telecom/reliance-communications-aircel-rejig-merger-plan-each-will-hold-50-in-new-entity/articleshow/51183146.cms

TRAI. (2014). *Press Release on Telecom Subscription Data as on 31st March, 2014*. Retrieved from Telecom Regulatory Authority of India: http://www.trai.gov.in

TRAI. (2016). *The Indian Telecom Services Performance Indicators*. Retrieved from URL: http://trai.gov.in/Content/PerformanceIndicatorsReports/1_1_PerformanceIndicatorsReports.aspx

VSNL. (2001-2002). *16th Annual Report*. Retrieved from URL: http://www.tatacommunications.com/sites/default/files/AR2002.pdf

YourStory. (2015). *Internet growth, impacts and success: what will India be like with 550 million users in 2018?* Retrieved from URL: http://yourstory.com/2015/02/internet-india-2018/

Chapter 19
U.InovAcelerator: A Model of Innovation Indicators

Fábio Goncalves
University of Porto, Portugal

Maria Pinto
University of Porto, Portugal

Alexandra Xavier
University of Porto, Portugal

ABSTRACT

Following the reflection around the emergence of "research university" in the context of the slow but progressive increase in value of science and technology and research and development in Portugal, a study applied to the knowledge transfer and the process of innovation in the university, in the context of a master dissertation in information science (IS), study area of information management, is presented. The university is one of the most important institutions in the context of the national innovation system (SNI), being part of its mission the creation and transfer of knowledge. At the University of Porto (U.Porto), projects, such as the University of Porto Innovation unit (U.Porto Inovação) and the Science and Technology Park of the University of Porto (UPTEC) seek to support the university's innovation value chain, promoting the reinforcement and solidification of knowledge transfer and of the relations between the university and companies, as well as the promotion and support to the creation of companies with a technological, scientific, and creative base, and the attraction of numerous innovation centers of national and international companies. This chapter points out an informational perspective on I&D+i (research and development and innovation) and entrepreneurship, based on the systemic theory and the quadripolar method, as theoretical and methodological guidance tools, and an information management/knowledge management approach of innovation models for the knowledge economy, the national and international referents, and corresponding set of indicators. An exploratory study, which allowed the identification of internal and external agents, the resources, the relations between actors and institutions, the processes and flows, and the main inputs and outputs, is presented. The most relevant result is embodied in a model of innovation indicators in an academic context and applied to the University of Porto.

DOI: 10.4018/978-1-5225-6225-2.ch019

INTRODUCTION[1]

A National Innovation System (NIS) integrates a group of institutions and organizations from the same country that interact, create, develop, use and share innovation. It includes companies, research and teaching institutes, technology centers, public administration services and the financial system. The main goals of this cooperation are the production of research and development (R&D), the sharing and transfer of knowledge and the development of products, tasks or activities that are identified as essential to innovation.

The University presents itself as one of the most crucial institutions in the context of the NIS, being intrinsic to its mission the production and the transfer of knowledge. At the University of Porto (U.Porto), the focus on R&D and innovation (R&D+i) is reflected in projects such as the creation of University of Porto Innovation (U.Porto Innovation) – a structure aimed at supporting university research, entrepreneurship and the relation between the university and companies – and the University of Porto Science and Technology Park (UPTEC), a structure that supports the transfer of knowledge from the university to the business market, so as to economically and socially enhance the generated knowledge and provide a favorable environment for the incubation of new companies, but also bolster those which were already launched into the market. Their objectives are diversified, with emphasis on supporting the U.Porto's innovation value chain, strengthening and solidifying the transfer of knowledge and the institution's relations with companies, promoting and supporting the creation of technological, scientific and creative companies and the attraction of innovation centers of national and international companies.

It is in this context that the U.InovAcelerator project arises. This project aims to the creation of an Innovation Observatory in the U.Porto ecosystem based upon three key elements:

1. An innovation chain model;
2. A model of innovation indicators and;
3. An informational, technological and information services for innovation model. This observatory will serve as an instrument for the aggregation, collection, systematization and dissemination of information in the context of U.Porto, being supported by an informational accelerator, with the role of monitoring the transfer of knowledge and the innovation value chain.

This study focuses on the development of the innovation indicators model.

The need to develop such model is due to the fact that there is a great dispersion of information and a reductive approach, considering the indicators already identified and obtained. It is, therefore, necessary to implement automated processes that can, from the informational mass that integrates the University's informational system, extract indicators to support the decision making, to reveal the quality of academy action and to guide its strategic definition, making the University an indispensable element in the production of new knowledge and in its transformation into innovation, that is, having an impact on the development of the communities in which it plays an important role.

The study was carried out in the Information Management (IM) area, with a contribution from the Science and Technology Management and Scientific Communication studies, which includes the bibliometric analysis (Silva, 2013), a domain that applies statistical and mathematical methods to analyze and build indicators on the dynamics and evolution of the scientific and technological information of certain disciplines, areas, organizations or countries.

The Systemic Theory and the Quadripolar Method are, respectively, the instruments of theoretical and methodological guidance of the approach of the object of study and work, the infocommunicational flow, under a holistic and dynamic vision that follows the design and execution of a project.

The main results of the study are the:

1. Analysis of the historical evolution of the science, technology and innovation indicators, in an international level and in the context of the U.Porto;
2. The identification, creation and adaptation of R&D+i indicators applied to the U.Porto;
3. The identification and analysis of U.Porto's R&D+i and entrepreneurship areas and processes and;
4. The creation and development of a model of innovation indicators applied to the academic context of U.Porto.

CONCEPTS AND CONTEXT

The combination and evidence of the complementarity between the concepts of "science" and "technology" are at the core of the science and technology activities. These consist of *the set of activities related to the creation, expansion, dissemination and application of scientific and technological knowledge* (Pinto, 2015b).

Innovation is, thus, a key concept in the contemporary world, constituting one of the main advantages for the economic development of a country or region, as well as for the competitiveness of its companies and research institutions. Nevertheless, there's a lot of other concepts that are equally important to understand and concretize innovation, like "invention". Schumpeter (1934) presents one of the oldest, but clearest, definitions of invention – an "idea, outline or model for a new or improved solution" – distinguishing it from innovation, because, although it is indeed the creation of something new, it has no economic relevance. So, it is nothing more than the creation of new knowledge.

Through the process of transfer of knowledge, it is possible to communicate and share research results with the society, since it involves companies, universities and other research institutions. It includes aspects such as legal protection and intellectual property, learning and techniques related to the development of market strategies, marketing and licensing of private companies, or the support for the creation of spin-off companies.

In the Oslo Manual of the Organization for Economic Cooperation and Development (OECD), we find the most consensual definition of innovation:

the implementation of a new or significantly improved product (good or service), process, a new marketing method, a new organizational method in business practices, in the organization of the workplace or in external relations (OCDE, 2005).

Thus, four types of innovation appear: 1) product; 2) marketing; 3) process, and; 4) organizational.

On the other hand, innovation management includes the integration of innovation into the work processes of organizations through new products, services, business models and partnerships. It also supports the implementation of new tools that stimulate innovation, such as value chains, organizational designs, workflows and financing mechanisms.

The relationship between Innovation Management and IM is, nowadays, increasingly close, being seen as a key factor by the various participants of an NIS. Surviving the rapid changes in the market and having a competitive advantage leads to an efficient and effective information management, which directly influences not only the development of business activities and the strategic definition, but also the creation and sharing of knowledge and the consequent innovation.

Pinto approaches IM as the:

integrated management of the entire information lifecycle, including the identification, understanding, logical representation and redesign of organizational processes and physical configurations and/ or technological means that model their production, flow, use, dissemination and preservation, in the context of human and social action (Pinto, 2014; 2017).

She also points out, in a more comprehensive formulation of IM as a transversal and applied study area in Information Science (IS), to the:

study, design, implementation and development of processes and services inherent to the infocommunicational flow, allowing the construction of operational models of maximum efficiency and profitability (Pinto 2015a).

IM, and everything that involves the production of information, its organization and representation, as well as the informational behavior of the various prosumers throughout the infocommunicational flow, play a crucial role in innovation and its management, being the innovation closely linked to information and knowledge (Vick, Nagano & Santos, 2009), from the production of scientific and technical knowledge to organizational knowledge.

In 1985, Lundvall introduced the concept of innovation system, which consisted in a framework projected to understand innovation and encompassing the interaction between the subjects needed to transform an idea into a product, process or service for later launching. Innovation systems are categorized as national, regional, local, sectoral and technological, and can also be classified as micro, meso or macro innovation system. There are several functions, as well as advantages, assigned to innovation systems, namely the support to R&D for the creation of new knowledge, the networking through markets, the financing of innovation processes and other activities to facilitate the commercialization of the knowledge produced (research results), the creation of competencies and the creation and transformation of institutions that influence organizations and innovation processes (Edquist, 2006).

At the national level we have the NIS. It encompasses the flow of technologies and information among individuals, companies and institutions, which presents itself as the key to the creation of innovation processes. It is composed by several actors, including companies, governments, universities and other public and private institutions that play and important role in the development of processes related to the system. Its functions include the creation of human capital (knowledge and experience), the creation and diffusion of technological opportunities, the development and launch of new products, the facilitation of financing, the creation of markets, the expansion of knowledge and the facilitation of regulations for technologies, materials and products which can broaden the market and improve its accesses (Feinson, 2003).

NISs are targeted by numerous evaluations directed to their capacities, to the relationships between their participants, to the diffusion of technology and knowledge, among other factors. For this purpose, there are three levels to evaluate an NIS: micro, meso and macro level (Silva, 2005).

THE UNIVERSITY AND THE INNOVATION SYSTEM

The term University takes us to the community of teachers and students, supported by an administrative structure, which form an institution that exercises teaching, research and interaction with the community, with "academic freedom", presenting itself as an abstract element, as an institution endowed with legal personality that is materially the university community (Pinto, 2015b). It is the responsibility of the University, as an institution, to integrate teaching, research and knowledge transfer, positioning itself both in the Scientific, Educative and Innovative systems, reinforcing its role in the last one, namely in the form of the relationship with the State.

From the second half of the twentieth century, universities play an increasingly prominent and decisive role in economic development – Knowledge Economy – and consequent knowledge capitalization, taking place the so-called second academic revolution (Etzkowitz, 1993). The university stands out in R&D, innovation, professional training, economic development and knowledge creation (Mowery, 2005). The Etzkowitz and Leydesdorff (1998) model of the Triple-Helix reflects this role and the interactions between the main actors: the University, the State and the Companies (Pinto, 2015a).

In Portugal, these changes were felt mainly from the 90's onwards, intensifying the relationship between the university and industry (Gago, 1994). The University is responsible for the production of knowledge, the formation of human capital, the transfer of knowledge for the development of startups, the technological innovation and leadership promotion and for the scientific and technological infrastructures (Guerreiro, 2005). However, this does not prevent the country, when compared with other European countries, from having low impact rates for scientific production (FCT, 2013).

METHODOLOGY

Having as reference the Quadripolar Method, it is here detailed the exploratory study executed. The exploratory research aims to create new ideas, developing a well-founded image about the situation under study and bringing some familiarization about the details related to it, as well as some the determination of its viability in the future.

It was sought to identify studies at a preliminary stage of research that allowed a better understanding of the problem and the identification of references to its approach, namely techniques and tools to be used.

This option is related to the fact that this is a study with few precedents. Similar publications and projects are scarce and either are too recent (to be validated) or too old (outdated). In addition, there are few universities that work with innovation indicators and, even those that do, do not have a very extensive and/or specific set of indicators. Another goal is to develop the model as a multifaceted structure, so it can cover the largest number of U.Porto areas involved with R&D+i.

The approach applied to this study was purely qualitative. It was used for its reasoning many sources, such as monographs, scientific articles and other published documents, as well as a vast amount of field work. So, in the end, there are four unique moments to be identified in this approach:

1. The first one sets itself in a theoretical-pratical context, in order to understand the evolution and the current state of the innovation area, its relation with IM, as well as the University's role in the NSI, through an exploratory documentary research in some specific sources;

2. The second one is characterized by an analysis of the R&D+i and entrepreneurship panorama in the U.Porto context. The exploratory approach was maintained, both through studies carried out in the context of the U.Porto and websites related with it. The direct and participant observation resulted from the fact that the observer participated and had an active role in the environment under observation (U.Porto student), which enabled the contact with numerous specialists, access to important information, as well as facilitated the analysis process. Joint meetings were also held, involving individuals interested in this study and actively participating in the context under analysis;
3. The third one corresponds to the mapping, typification and selection of the indicators, having as sources manuals and guidance documents issued by entities such as COTEC, INE, OECD and EUROSTAT. Indicators in use or mentioned in U.Porto reports, as well as in project data and reference cases, were mapped. Then, came the analysis and classification of the indicators by type (input or output), area (R&D, technology transfer, innovation, etc.) and application (in higher education, companies or government). The next stage looked at the reduction of the indicators set. Indicators repeated and/or similar were eliminated, as well as those that did not apply to the academic context. For a comparative analysis of the remaining indicators, these were regrouped by areas of affectation. Using the information collected in the U.Porto ecosystem analysis, the indicators that adjusted (or could adjust) to its criteria were selected. Finally, adaptations and adjustments were made to determine the final set of indicators;
4. The last moment focused on the development of the model structure and on the distribution of the indicators. A new cycle of meetings was held – with the Pro-Rectory for Innovation and Entrepreneurship (U.Porto) and with Shared Services of U.Porto – whose purpose was to understand what the main stakeholders wanted in terms of the model of innovation indicators and priority processes. These meetings were also important to proceed to the validation of the selected (or created) indicators. Also worth mentioning, is the importance of the analysis of the structure of several reference models, namely the ones from COTEC[2] and the Global Innovation Index (GII), as well as others from proposals such as the ones produced by Heitor (2004) and Manjón (2010).

MAIN FINDINGS

Initially, an analysis was made about the R&D+i and entrepreneurship areas and processes of U.Porto. This analysis was based on the U.Porto innovation value chain (Figure 1). It was concluded that U.Porto has a great development and maturity, being one of the most successful universities in the area. The organization shown through the value chain and the relation obtained between all the entities related to

Figure 1. The innovation value chain
Source: Brito (2013).

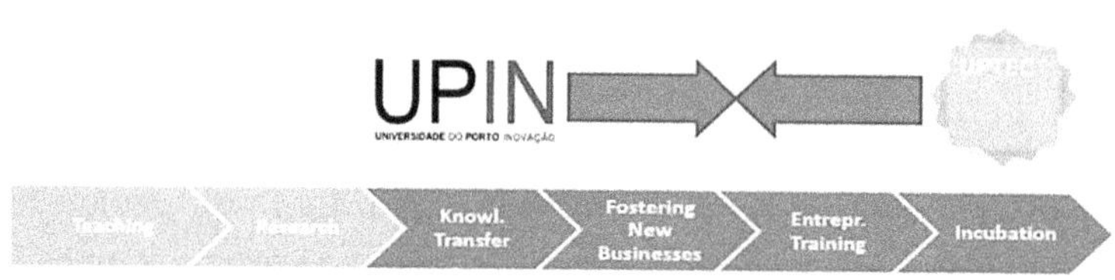

the university, with respect to the execution and continuous improvement of the processes of R&D+I and entrepreneurship, presents itself as a solid base for the future of U.Porto and its entrepreneurs.

In a second phase, the model of innovation indicators was planned and developed.

For the construction of the model, objectives were defined regarding its structure and content, namely:

1. **Conceptual Clarity:** The mapping of the indicators present in the model was based on the structure of the innovation value chain, which is the basis for the organization of the model areas and the information required around each one. The concepts introduced in the new model should not differ from the ones already in use in references such as the value chain;
2. **Indicator Simplicity and Objectivity:** The lack of (or close to) a model, or a set of indicators in use in the U.Porto, as well as concrete examples from other universities at a national level, requires the approach to the R&D+i area to be clear and concrete. The mapping was executed regarding only the information available at U.Porto. The indicators created or adapted were made as simple and objective as possible, since this model is the first approach of its kind to this area;
3. **Proposal of Indicators With Added Value and Possibilities for Benchmarking:** Existing indicators and in use in the U.Porto ecosystem have been identified and selected. These went through a meticulous analysis and development. Some were eliminated, some were adapted. And in some special cases, new indicators were created from scratch. The main goal for this indicators was what they could mean as a description of the R&D+I, but also what kind of information they could retrieve. The indicators should be the main tool to understand where does the R&D+i area of U.Porto stands in the present and what could be improved. Where is it failing? Where is it succeeding? What can we improve? All questions that the analysis of this indicators should be able to answer. And, since this indicators can be a tool for the university to evaluate itself, they should also be suitable for benchmarking with other universities and institutions of its genre;
4. **Broad and Dynamic Model:** The model should provide information that portrays, in some way, the state of all the main R&D+i and entrepreneurship areas and processes of U.Porto, as well as having dynamic indicators, which can be easily adapted or changed in order to bring a bigger efficiency to the model.

Finally, the model was produced (Figure 2), adopting a structure that presents 5 main pillars, subdivided into 13 areas.

The first pillar, Resources, is intended to give an understanding of what resources U.Porto has in its R&D+i and entrepreneurship area, and also how these are allocated. This pillar also has the purpose to provide some context regarding the scope of the model. It is divided into 4 areas:

1. **Human Resources:** Those that integrate the R&D+i and entrepreneurship area;
2. **Infrastructures:** All infrastructures at U.Porto's service;
3. **Investments:** U.Porto investments in R&D+i and entrepreneurship;
4. **Financing:** U.Porto financing in R&D+i and entrepreneurship.

The second pillar is the Transfer of Knowledge pillar, assumed as something indispensable for the U.Porto. It is from this pillar that the model starts its evaluation of the main capabilities of U.Porto's R&D+i and entrepreneurship area. It is divided into 2 areas:

Figure 2. Representation of the model of innovation indicators in an academic context

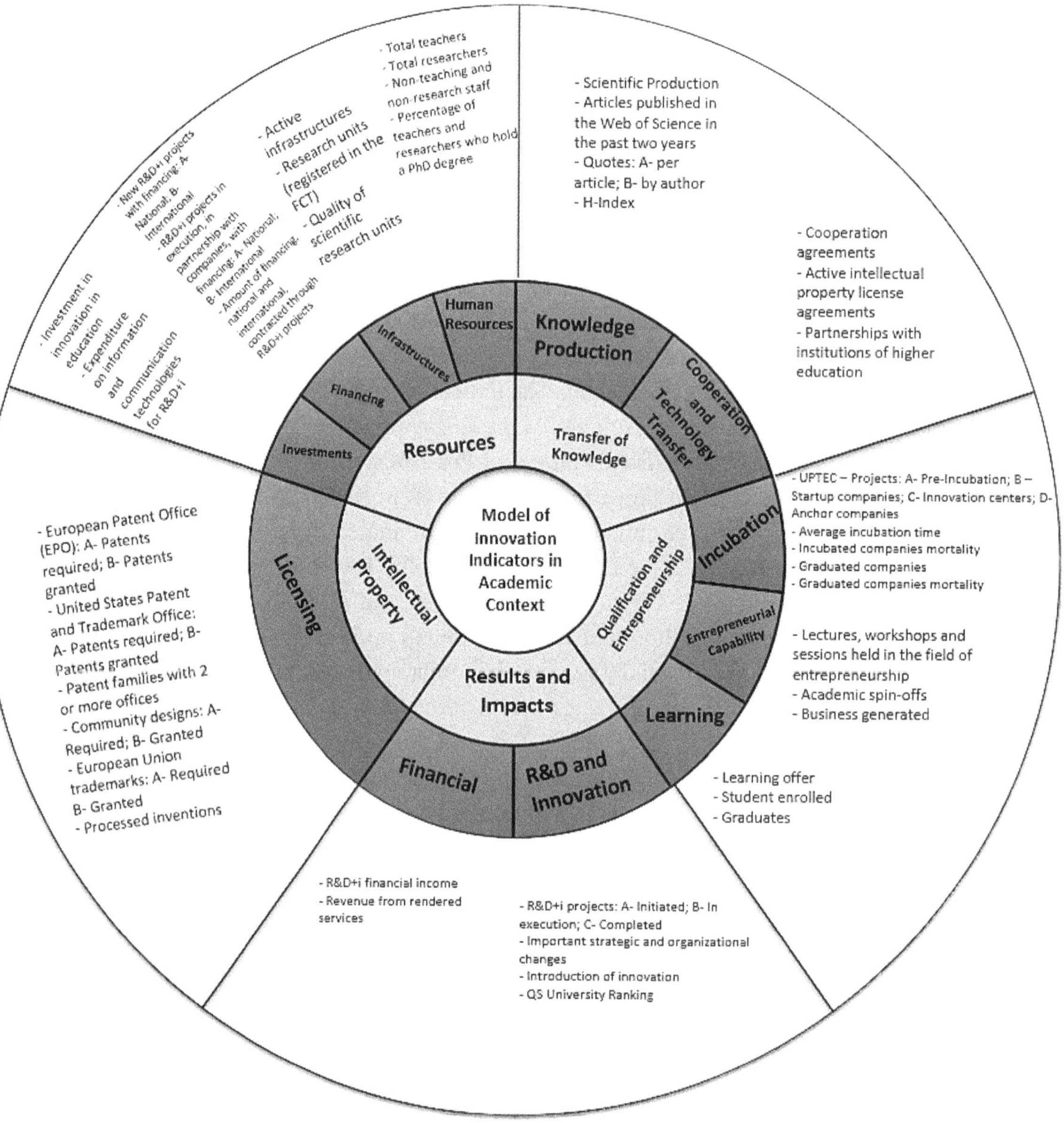

1. **Knowledge Production:** Evaluation of the scientific knowledge produced in the U.Porto;
2. **Cooperation and Technology Transfer:** Coverage of U.Porto's cooperation with other entities and the transfer of technology carried out in this context.

The third pillar is named Intellectual Property and it is aimed at the output of R&D+i in the U.Porto, regarding the licensing of projects and/or technologies. It has only one area, Licensing, that intends to analyze the outputs of intellectual property, namely patents, community designs and European Union trademarks.

The fourth pillar refers to Qualification and Entrepreneurship and intends to understand what U.Porto has to offer in terms of learning and what the current situation of entrepreneurship in the university is. It has 2 areas:

1. **Learning:** Analysis of U.Porto's learning offer, students enrolled and other solutions to promote entrepreneurship in the university;
2. **Entrepreneurial Capability:** Analysis of the capability and expansion of entrepreneurship in U.Porto;
3. **Incubation:** General analysis of the incubation process, with focus on the UPTEC case.

The fifth, and final pillar is Results and Impacts. Here, it is intended to analyze the results and impacts of R&D+i and entrepreneurship in the U.Porto. It is constituted by 3 areas:

1. **Financial:** Financial impact on R&D+i within the U.Porto;
2. **R&D and Innovation:** Impacts on R&D+i in the U.Porto;
3. **Obstacles and Incentives to Innovate (Only Active in a Mapping Developed and Dedicated to UPTEC Companies):** Aims to understand what prevents companies from innovation and what can encourage them to do so.

Each area intends to reflect one or more of the processes previously analyzed in the scope of R&D+i and entrepreneurship in the U.Porto. The areas of Resources and Results and Impacts are considered to affect all areas of the value chain, since the content they present is less specific and broader.

In addition to the main mapping, a set of shorter and business-oriented indicators was also developed. These indicators are applied to the companies incubated in UPTEC (Table 2). Their objective are to analyze innovation and its practices in these companies. Not all the areas of the structure of the main model are used, highlighting a differentiating characteristic: the first 11 indicators apply to a data collection of all companies, with the remaining indicators being more of a survey type indicator (only intended to measure the impacts of innovation in companies in an individual way).

CONCLUSION

This model presents three different types of indicators typology: input, process and output indicators. However, this typology applies much more to the pillars of the model, since it is these that lead to its structure and provide meaning to the model, being properly ordered according to its typology and the analysis performed to the U.Porto (existing only one exception).

So, resources correspond to the inputs, that is, what the university has and what is a reality in its context.

The transfer of knowledge, intellectual property and qualification and entrepreneurship represent the processes. Something that is explicit, since these are the pillars that have the indicators that analyze the information relative to innumerous processes of the U.Porto.

Regarding outputs, these are represented by intellectual property (the exception) and by the results and impacts pillar. These pillars refer to the results produced and conclusions drawn from the processes carried out in terms of R&D+i and entrepreneurship in the U.Porto.

Table 1. Model of innovation indicators in an academic context

	Indicators	Pillar - Area	Source	Frequency
1	Total teachers	Resources – Human Resources		Annual
2	Total researchers	Resources – Human Resources		Annual
3	Non-teaching and non-research staff	Resources – Human Resources	U.Porto (2016)	Annual
4	Percentage of teachers and researchers who hold a PhD degree	Resources – Human Resources	Adapted from U.Porto (2016)	Annual
5	Active infrastructures: A- University Poles; B- Faculties; C- Business Schools; D- Libraries; E- Museums; F- e-learning cafes	Resources - Infrastructures	Adapted from.U.Porto (2016)	Annual
6	Research units (registered in the FCT)	Resources - Infrastructures	U.Porto (2016)	Annual
7	Quality of scientific research units	Resources - Infrastructures	COTEC	Annual
8	Investment in innovation in education	Resources - Investments		Annual
9	Expenditure on information and communication technologies for R&D+i	Resources - Investments		Annual
10	New R&D+i projects with financing: A- National; B- International	Resources - Financing	Adapted from Plano Estratégico U. Porto 2020	Annual
11	R&D+i projects in execution, in partnership with companies, with financing: A- National; B- International	Resources - Financing	Adapted from Relatório de Atividade e Contas U.Porto 2016	Annual
12	Amount of financing, national and international, contracted through R&D+i projects	Resources - Financing	Adapted from Plano Estratégico U. Porto 2020	Annual
13	Scientific production	Transfer of Knowledge – Knowledge Production	Adapted from INE	Annual
14	Articles published in the Web of Science in the past two years	Transfer of Knowledge – Knowledge Production	Adapted from U.Porto (2016)	Bi-annual
15	Quotes: A- per article; B- by author	Transfer of Knowledge – Knowledge Production		Annual
16	H-Index	Transfer of Knowledge – Knowledge Production	Adapted from GII 2016	Annual
17	Cooperation agreements	Transfer of Knowledge – Cooperation and Technology Transfer	U.Porto (2016)	Annual
18	Active intellectual property license agreements	Transfer of Knowledge – Cooperation and Technology Transfer	U.Porto (2016)	Annual
19	Partnerships with institutions of higher education	Transfer of Knowledge – Cooperation and Technology Transfer	U.Porto (2016)	Annual
20	European Patent Office (EPO): A- Patents required B- Patents granted	Intellectual Property - Licensing	Adapted from Heitor et al. (2004)	Annual
21	United States Patent and Trademark *Office*: A- Patents required B- Patents granted	Intellectual Property - Licensing	Adapted from Heitor et al. (2004)	Annual
22	Patent families with 2 or more offices	Intellectual Property - Licensing	Adapted from GII 2016	Annual
23	Community designs: A- Required; B- Granted	Intellectual Property - Licensing	Adapted from Eurostat	Annual
24	European Union trademarks: A- Required B-Granted	Intellectual Property - Licensing	Adapted from Eurostat	Annual
25	Processed inventions	Intellectual Property - Licensing	U.Porto (2016)	Annual
26	Learning offer: A- Degree; B- Integrated Masters; C- Master; D- PhD; E- Continuous Learning	Qualification and Entrepreneurship - Learning	Adapted from U.Porto (2016)	Annual
27	Students enrolled: A- Degree; B-Integrated Masters; C- Master; D- PhD; E- Non-degree courses	Qualification and Entrepreneurship - Learning	Adapted from U.Porto (2016)	Annual

continued on following page

Table 1. Continued

	Indicators	Pillar - Area	Source	Frequency
28	Graduates: A- Degree; B- Integrated Masters; C- Master; D- PhD	Qualification and Entrepreneurship - Learning	Adapted from U.Porto (2016)	Annual
29	Lectures, workshops and sessions held in the field of entrepreneurship	Qualification and Entrepreneurship – Entrepreneurial Capability		Annual
30	Academic spin-offs	Qualification and Entrepreneurship – Entrepreneurial Capability	Lanari (2000)	Annual
31	Business generated	Qualification and Entrepreneurship – Entrepreneurial Capability		Bi-annual
32	UPTEC – Projects: A- Pre-Incubation; B – Startup companies; C- Innovation centers; D- Anchor companies	Qualification and Entrepreneurship - Incubation	Adapted from U.Porto (2016)	Annual
33	Average incubation time	Qualification and Entrepreneurship - Incubation	Lanari (2000)	Annual
34	Incubated companies mortality	Qualification and Entrepreneurship - Incubation	Lanari (2000)	Bi-annual
35	Graduated Companies	Qualification and Entrepreneurship - Incubation	U.Porto (2016)	Annual
36	Graduated Companies Mortality	Qualification and Entrepreneurship - Incubation	Lanari (2000)	Bi-annual
37	R&D+i financial income	Results and Impacts - Financial	Adapted from Relatório Atividade e Contas U.Porto 2016	Annual
38	Revenue from rendered services	Results and Impacts - Financial		Annual
39	R&D+i projects: A- Initiated; B- In execution; C- Completed	Results and Impacts – R&D and Innovation		Annual
40	Important strategic and organizational changes	Results and Impacts – R&D and Innovation	Adapted from Heitor et al. (2004)	Annual
41	Introduction of innovation	Results and Impacts – R&D and Innovation		Annual
42	QS University Ranking	Results and Impacts – R&D and Innovation	Adapted from GII 2016	3 years

The model uses or adapts 18 indicators that were already in use in the U.Porto ecosystem. These proved to be very important for the development of the model, since they gave some much needed context to the research.

The model also has 12 original indicators (between the 2 mappings), which were created with the objective of strengthening weaknesses, but also enriching other strengths.

As for the remaining indicators, most are adapted. It should also be noted that the indicators that meet the designation of "adapted" have undergone considerable changes, in comparison to their original state. This is most often due to the need to adapt the indicators to an academic context and to some specific R&D+i processes.

The model structure evidences conceptual clarity. The inspirations in the U.Porto innovation value chain facilitated the process and allowed the structure to be concise and correct in its ordering and definition.

The simplicity and objectivity of the indicators, coupled with the need for a broad and dynamic model, has been one of the obstacles to its deepening, thus revealing what will be one of its possible weaknesses: its lack of specificity in some content. However, it would not be possible to take another path in covering all areas and processes of R&D+i and entrepreneurship. Specificity in indicators will be something

Table 2. Indicators mapping applied to UPTEC

	Indicators	Pillar - Area	Source	Frequency
1	Expenditure on innovation in the last year: A- Fraction of R&D expenditure; B- Fraction of the acquisition of machinery and equipment expenditure; C- Fraction of the expenditure on the acquisition of other external knowledge, training, marketing, design and other preparations for the production or distribution of innovations	Resources - Investments	Adapted from Heitor et al. (2004)	Annual
2	Level of investment of companies in the training of their employees	Resources - Investments	Adapted from COTEC	Annual
3	Venture capital companies	Resources - Financing	Adapted from Heitor et al. (2004)	Annual
4	Companies with investments made by business angels	Resources - Financing		Annual
5	Companies with investments made by banks or other sources	Resources - Financing		Annual
6	Cooperation in R&D projects: A- Cooperation in R&D projects with other companies; B- Cooperation in R&D projects with institutions of the scientific system	Transfer of Knowledge – Cooperation and Technology Transfer	Adapted from Heitor et al. (2004)	Annual
7	Companies that applied for patents	Intellectual Property - Licensing	Adapted from OCDE	Annual
8	Companies that have registered one or more community designs	Intellectual Property - Licensing	Adapted from OCDE	Annual
9	Companies that have registered one or more European Union trademarks	Intellectual Property - Licensing	Adapted from OCDE	Annual
10	Companies offering formal training	Qualification and Entrepreneurship - Learning	Adapted from GII 2016	Annual
11	Average number of working days expended in training activities	Qualification and Entrepreneurship - Learning	Adapted from Heitor et al. (2004)	Annual
12	Turnover resulting from the introduction of product innovations in the market	Results and Impacts - Financial	Adapted from CIS 2014	Annual
13	High technology products and services: exports	Results and Impacts - Financial	COTEC	Annual
14	Dynamics of business innovation in the past year	Results and Impacts – R&D and Innovation	Adapted from Heitor et al. (2004)	Annual
15	Self-assessment of the company's performance in terms of innovation	Results and Impacts – R&D and Innovation	Adapted from Heitor et al. (2004)	Annual
16	Identification of the type of innovation in which the company's efforts were concentrated	Results and Impacts – R&D and Innovation	Adapted from Heitor et al. (2004)	Annual
17	Identification of the activity in which the management of the company will focus its innovation management activities next year	Results and Impacts – R&D and Innovation	Adapted from Heitor et al. (2004)	Annual
18	Barriers to innovate	Results and Impacts – Obstacles and incentives to innovate	Adapted from Heitor et al. (2004)	Annual
19	Relevant factors to stimulate innovation	Results and Impacts – Obstacles and incentives to innovate	Adapted from Heitor et al. (2004)	Annual
20	Incentives to innovate in the future	Results and Impacts – Obstacles and incentives to innovate	Adapted from Heitor et al. (2004)	Annual

possible, once there is a more standardized situation at the level of indicators and statistics in the area. Although the model encompasses all R&D+i and entrepreneurship areas, there is no guarantee that, within 6 months to 1 year, this aspect will no longer be a reality. The growth momentum of the U.Porto in this area is intense, with more and more infrastructures and resources focused on entrepreneurship and innovation every day.

With respect to the added value of the indicators and the possibility of benchmarking, these have two distinct realities: if, on the one hand, the indicators really guarantee the passage of knowledge about the current state and information available at the R&D+i level in the U.Porto, for another, it is difficult to compare U.Porto with other universities due to the scarcity of similar cases and the difference in the development of indicators and other means of analysis of this area at the international level.

The mapping applied to UPTEC may prove to be a good support for the development of innovation in the companies it hosts. It is hoped that this addition can lead to the development of an Innovation Scoreboard specifically applied to this reality, producing valuable information for the success of companies and the promotion of innovation at various levels.

REFERENCES

Barata, J. M. M. (1992). Inovação e Desenvolvimento Tecnológico: Conceitos, modelos e medidas. Pistas para a investigação aplicada. *Estudos de Economia, XII*, *2*, 147–171.

Barómetro Inovação. (n.d.). Disponível em: http://barometro.cotecportugal.pt/pt/indicadores/modelo-de-indicadores-de-idi/dimensoes-pilares-e-indicadores/dimensao-pilares-e-indicadores.html

Barro Ameneiro, S. (2015). La Transferencia de I+D, la Innovación y el Emprendimiento en las Universidades. Educación Superior en Iberoamérica. Informe 2015. *Centro Interuniversitario de Desarrollo - CINDA Red Emprendia Universia*, 541.

Bartol, K. M., & Srivastava, A. (2002). Encouraging Knowledge Sharing: The Role of Organizational Reward Systems. *Journal of Leadership & Organization Studies*, *9*(1), 64–76. doi:10.1177/107179190200900105

Bossel, H. (1999). *Indicators for Sustainable Development: Theory, Method, Applications*. International Institute for Sustainable Development (Vol. 68). Retrieved from http://www.ulb.ac.be/ceese/STAFF/Tom/bossel.pdf

Breschi, S., & Malerba, F. (1997). Sectoral Innovation Systems: Technological Regimes, Schumpeterian Dynamics, and Spatial Boundaries. In Systems of Innovation: Technologies, Institutions and Organizations (pp. 130–156). Academic Press. doi:10.1016/S0024-6301(98)90244-8

Brito, C. (2013). *The Entrepreneurship Ecosystem of the University of Porto*. UPTEC.

Capuano, E. A. (2015). Informação sobre Conceitos e Indicadores de Inovação. *Revista de Informação, 16*(1).

Carlsson, B. (1995). *Technological systems and economic performance: The case of factory automation*. Economics of science, technology, and innovation.

Condesso, F. (1999). *A autonomia universitária no direito espanhol e português*. Estudo monográfico.

Cooke, P. (1992). Regional Innovation Systems: Competitive Regulation in the New Europe. *Geoforum, 23*(3), 365–382. doi:10.1016/0016-7185(92)90048-9

Cornell University, INSEAD, & WIPO. (2015). *The Global Innovation Index 2016: Winning with Global Innovation. The Global Innovation Index 2016*. Retrieved from http://www.codespring.ro/wp-content/uploads/2012/11/GII-2012_Cover.pdf

da Silva, A. M., Malheiro, A., & Silva, D. (2014). Ciência da Informação e comportamento informacional: Enquadramento epistemológico do estudo das necessidades de busca, seleção e uso. *Prisma*, *21*, 1–61.

Da Silva, C. C. M. (2005). *O Papel do Sector Público na Inovação e na Mudança Tecnológica nas Empresas*. Universidade do Minho.

Deakins, D., & Freel, M. (2003). *Entrepreneurship and Small Firms (3rd ed.)*. McGraw-Hill Education.

DELTCI – Dicionário Electrónico de Terminologia em Ciência da Informação. (n.d.). *Verbete Informação*. Disponível em http://www.ccje.ufes.br/arquivologia/deltci/def.asp?cod=45

Dewar, R. D., & Dutton, J. E. (1986). The Adoption of Radical and Incremental Innovations: An Empirical Analysis. *Management Science*, *32*(11), 1422–1433. doi:10.1287/mnsc.32.11.1422

Direção-Geral de Estatísticas de Educação e Ciência, & Direção de Serviços de Estatística da Ciência e Tecnologia e da Sociedade de Informação. (2016). *Sumários Estatísticos: CIS 2014 - Inquérito Comunitário à Inovação*. Academic Press.

Edquist, C. (1997). Systems of Innovation: Technologies, Institutions and Organizations (Book). *Long Range Planning*. doi:10.1016/S0024-6301(98)90244-8

Edquist, C. (1997). Systems of innovation approaches - Their emergence and characteristics. In *Systems of Innovation* (pp. 1–35). Technologies, Institutions and Organizations. doi:10.1016/S0024-6301(98)90244-8

Edquist, C. (2005). Systems of innovation: Perspectives and challenges. The Oxford Handbook of Innovation, 181–208. doi:10.1093/oxfordhb/9780199286805.003.0007

Erichsen, M., & Borges, N. (1995). A informação como recurso gerencial das organizações na sociedade do conhecimento. *Ciência da Informação*, *24*(2), 1–15. Retrieved from http://revista.ibict.br/index.php/ciinf/article/viewArticle/551

Etzkowitz, H. (1993). Technology transfer: The second academic revolution. *Technology Acess Report*, *6*, 7–9.

Etzkowitz, H., & Leydesdorff, L. (2000). The dynamics of innovation: From National Systems and "' Mode 2 '" to a Triple Helix of university – industry – government relations. *Science And Technology*, *29*(2), 109–123. doi:10.1016/S0048-7333(99)00055-4

Etzkowitz, H., Webster, A., & Healey, P. (1998). Introduction. In *Capitalizing Knowledge*. Albany, NY: State University of New York Press.

European Environment Agency. (2005). Copenhagen: The European Environment - State and Outlook.

Eurostat. (n.d.). Disponível em: http://ec.europa.eu/eurostat/web/science-technology-innovation/data/database

Feinson, S. (2003). National Innovation Systems Overview and Country Cases. *Knowledge Flows and Knowledge Collectives: Understanding The Role of Science and Technology Policies in Development*, 13–38. 10.1787/9789264239012-en

Freeman, C. (1987). *Technology and Economic Performance: Lessons from Japan*. London: Pinter Publishers.

Fundação para a Ciência e a Tecnologia. (2013). *Diagnóstico do Sistema de Investigação e Inovação: desafios, forças e fraquezas rumo e 2020*. Author.

Gabinete de Planeamento Estratégico e Participação Empresariais. (2017). *Universidade do Porto - Relatório de Atividades e Contas - Ano 2016*. Author.

Gago, M. (1994). *Prospectiva do ensino superior em Portugal*. Lisboa: Ministério da Educação.

Gibson, R., & Skarzynski, P. (2008). Innovation to the core - a blueprint for transforming the way your company innovates. Harvard Business Press.

Godinho, M. (2007). Indicadores de C&T, inovação e conhecimento: Onde estamos? Para onde vamos? *Analise Social*, *42*(182), 239–274.

Gonçalves, F. (2017). *U. InovAcelerator: um modelo de indicadores de inovação em contexto académico (Dissertação de Mestrado)*. Universidade do Porto.

Grupp, H., & Mogee, M. E. (2004). Indicators for national science and technology policy: How robust are composite indicators? *Research Policy*, *33*(9), 1373–1384. doi:10.1016/j.respol.2004.09.007

Guerreiro, J. (2005). As funções da universidade no âmbito dos sistemas de inovação. In *Estudos II* (pp. 131–148). Faculdade de Economia da Universidade do Algarve.

Haddad, P. R. (2010). Quatro gerações de indicadores. *Estadão*. Retrieved from http://economia.estadao.com.br/noticias/geral,quatro-geracoes-de-indicadores-imp-,556824

Hagedoorn, J., & Cloodt, M. (2003). Measuring innovative performance: Is there an advantage in using multiple indicators? *Research Policy*, *32*(8), 1365–1379. doi:10.1016/S0048-7333(02)00137-3

Heitor, M., Bóia, M. J., Conceição, P., & Beira, E. (2004). *Mapear Conhecimento e Inovação em Portugal: Uma proposta para um sistema de indicadores e um programa de observação*. Academic Press.

Hwang, V., & Horowitt, G. (2012). *The Rainforest: The Secret to Building the Next Silicon Valley*. Los Altos Hills: Regenwald.

INE. (n.d.). Disponível em: https://www.ine.pt/xportal/xmain?xpid=INE&xpgid=ine_base_dados

INOVA+ - Serviços de Consultadoria em Inovação Tecnológica S. A. (2007). *Conceitos e Modelos de Inovação* (1st ed.). AEP - Associação Empresarial de Portugal.

Janger, J., Schubert, T., Andries, P., Rammer, C., & Hoskens, M. (2017). The EU 2020 innovation indicator: A step forward in measuring innovation outputs and outcomes? *Research Policy*, *46*(1), 30–42. doi:10.1016/j.respol.2016.10.001

Johnson, A., & Jacobsson, S. (2003). The emergence of growth industry: a comparative analysis of the German, Dutch and Swedish wind turbine industries. *Change, Transformation and Development*, (1), 197–227. 10.1007/978-3-7908-2720-0_12

Kayano & Caldas. (2002). *Indicadores para o Diálogo*. GT Indicadores. Plataforma Contrapartes Novib.

Kleinknecht, A., Van Montfort, K., & Brouwer, E. (2002). The Non-Trivial Choice between Innovation Indicators. *Economics of Innovation and New Technology*, *11*(2), 109–121. doi:10.1080/10438590210899

Lanari, L. A. (2000). *Indicadores de desempenho estratégico para uma incubadora de empresas de base tecnológica: uma proposta*. Escola de Biblioteconomia da UFMG.

Laursen, K., & Salter, A. (2004). Searching high and low: What types of firms use universities as a source of innovation? *Research Policy*, *33*(8), 1201–1215. doi:10.1016/j.respol.2004.07.004

Liu, X., & White, S. (2001). Comparing innovation systems: A framework and application to China's transitional context. *Research Policy*, *30*(7), 1091–1114. doi:10.1016/S0048-7333(00)00132-3

Liyanage, C., Elhag, T., Ballal, T., & Li, Q. (2009). Knowledge communication and translation – a knowledge transfer model. *Journal of Knowledge Management*, *13*(3), 118–131. doi:10.1108/13673270910962914

Lundvall, B.-Å. (1992). *National Systems of Innovation: Towards a Theory of Innovation and Interactive Learning. National systems of innovation Towards a theory of innovation and interactive learning*. Retrieved from http://books.google.com/books?id=iDXGwacw-4oC&pgis=1

Lundvall, B.-Å., Johnson, B., Andersen, E. S., & Dalum, B. (2002). National systems of production, innovation and competence building. *Research Policy*, *31*(2), 213–231. doi:10.1016/S0048-7333(01)00137-8

Manjón, J. V. G. (2010). A proposal of indicators and policy framework for innovation benchmark in europe. *Journal of Technology Management & Innovation*, *5*(2).

McLaughlin, P., Bessant, J., & Smart, P. (2005). *Developing an Organizational Culture that facilitates Radical Innovation in a Mature Small to Medium Sized Company: Emergent Findings*. Retrieved from http://hdl.handle.net/1826/858

Metcalfe, J. S. (1995). The Economic Foundations of Technology Policy: Equilibrium and Evolutionary Perspectives. Handbook of the Economics of Innovation and Technological Change, 409–512.

Mowery, D. (2005). *Universities in National Innovation Systems*. Retrieved January 13, 2017, from http://www.aau.org/sites/default/files/urg/docs/Uni _natn_Inovatn_sys.pdf

Mowery, D. C., & Sampat, B. N. (2005). Universities in National Innovation Systems. The Oxford Handbook of Innovation, 209–239. doi:10.1093/oxfordhb/9780199286805.003.0008

Mugnaini, R., Jannuzzi, P. D. M., & Quoniam, L. (2004). Indicadores bibliométricos da produção científica brasileira: Uma análise a partir da base Pascal. *Ciência da Informação*, *33*(2), 123–131. doi:10.1590/S0100-19652004000200013

Nelson, R. R. (1993). National Innovation Systems: A Comparative Analysis. In *National Innovation Systems* (p. 541). Academic Press. http://books.google.co.uk/books?id=YFDGjgxc2CYC

OCDE. (1996). *The Knowledge-Based Economy. OCDE/GD* (Vol. 96). Retrieved from https://www.oecd.org/sti/sci-tech/1913021.pdf

OCDE. (1997). Manual de Oslo: Diretrizes para a Coleta e Interpretação de dados sobre Inovação Tecnológica. *OCDE, Eurostat E Financiadora de Estudos E Projetos*, 184. 10.1787/9789264065659-es

OCDE. (2002). Rumo a um desenvolvimento sustentável: indicadores ambientais. *Cadernos de referência ambiental, 9*.

OCDE. (2016). *Main Science and Technology Indicators 2016-2*. OCDE.

OCDE. (n.d.). Retrieved from https://www.ine.pt/xportal/xmain?xpid=INE&xpgid=ine_base_dados

Okubo, Y. (1997). Bibliometric Indicators and Analysis of Research Systems: Methods and Examples. *OECD Science, Technology and INdustry Working Papers, 1997*(1), 1–70.

Patel, P., & Pavitt, K. (1994). The Nature and Economic Importance of National Innovation Systems. *STI Review, 14*, 9–32.

Pereira, F., Verocai, H., Cordeiro, V., & Gomes, C. (2016). Sistemas de Informação e Inovação: Um estudo bibliométrico. *Revista de Gestão Da Tecnologia E Sistemas de Informação, 13*(1), 81–100.

Pinto, M. M. (2014). *Gestão e Preservação da Informação: o impacto do pensamento sistémico*. Retrieved from http://eiarquivos2014.weebly.com/uploads/2/6/0/6/26068081/p1_4_mp.pdf

Pinto, M. M. (2015a). *A Gestão da Informação nas Universidades Públicas Portuguesas: Reequacionamento e proposta de modelo (Tese de Doutoramento em Informação e Comunicação em Plataformas Digitais)*. Universidade do Porto - Universidade de Aveiro.

Pinto, M. M. (2015b). The Portuguese University: knowledge leverage towards innovation. In Handbook of research on effective project management through the integration of knowledge and innovation (pp. 466–490). IGI Global. doi:10.4018/978-1-4666-7536-0.ch024

Pinto, M. M. (2017). Gestão e Preservação da Informação: o impacto do pensamento sistémico. In N. Vaquinhas, M. Caixas, & H. Vinagre (Eds.), *Da produção à preservação informacional: desafios e oportunidades* (pp. 326–364). Évora: Publicações do Cidehus; doi:10.4000/books.cidehus.2758

Reitoria da Universidade do Porto - U.Porto. (2017). Acedido a 21 de março de 2017. https://sigarra.up.pt/reitoria/pt

Rickne, A. (2000). *New Technology-Based Firms and Industrial Dynamics Evidence from the Technological System of Biomaterials in Sweden*. Chalmers University of Technology.

Schumpeter, J. (1934). The theory of economic development. *Joseph Alois Schumpeter*, 61–116.

Schumpeter, J. A. (1939). *Business cycles*. NBER Books. doi:10.1016/j.socscimed.2006.11.007

Silva, V. (2014). *Gestão da Informação de acidentes de trabalho em profissionais de saúde: proposta de um sistema de gestão da sinistralidade laboral no centrol hospitalar São João*. Faculdade de Engenharia da Universidade do Porto.

Smith, K. (2005). Measuring Innovation. The Oxford Handbook of Innovation, 148–177. doi:10.1093/oxfordhb/9780199286805.003.0006

Tidd, J., Bessant, J., & Pavitt, K. (1997). Managing Innovation - Integratin Technological, Market and Organizational Change. *Technovation, 18*(5), 195–238. doi:10.1016/S0166-4972(98)80033-3

UNESCO. (2009). Definitions of R&D, innovation and S&T activities. In *Training Workshop on Science, Technology and Innovation Indicators*. Cairo, Egypt: UNESCO.

Universidade do Porto. (2016). *U.Porto em Números 2016/17*. Porto: Universidade do Porto. Retrieved from https://sigarra.up.pt/up/pt/web_gessi_docs.download_file?p_name=F1340943912%2FUPorto_em_Numeros_2016-2017.pdf

Universidade do Porto. (2016). *Plano Estratégico U*. Porto: Author.

U.Porto Inovação. (2017). Acedido a 23 de março de 2017. https://upin.up.pt

UPTEC. (2017). Acedido a 23 março de 2017. http://uptec.up.pt/

Vick, T., Nagano, M. S., & Santos, F. C. A. (2009). Aportes da gestão da informação para a criação de conhecimento em equipes de inovação. *Perspectivas em Ciência da Informação, 14*(2), 204–219. doi:10.1590/S1413-99362009000200014

Xavier, A. (2008). *Manual de Identificação e Classificação das Actividades de IDI*. COTEC Portugal. Retrieved from http://www.cotecportugal.pt/imagem/manual_idi_projecto2.pdf

Zaltman, G., Duncan, R., & Holbek, J. (1973). *Innovations and Organizations*. New York: Wiley. Retrieved from http://www.amazon.com/Innovations-Organizations-Gerald-Zaltman/dp/047198129X

ENDNOTES

1 Chapter resulting from the project developed in the scope of the Master in Information Science (Gonçalves, 2017), promoted by the Faculty of Engineering of the University of Porto (FEUP) and the Faculty of Arts and Humanities of the University of Porto (FLUP), under the orientation of Maria Manuela Pinto and co-orientation of Maria Alexandra Xavier. The hosting organization was the Center for Innovation, Technology and Entrepreneurship (CITE) of the Institute for Systems and Computer Engineering, Technology and Science (INESC TEC).

2 See also Xavier (2008).

Compilation of References

Abbott, A. (2004). Pressured staff 'lose faith' in patent quality. *Nature, 429*(6991), 423. doi:10.1038/429493a PMID:15175718

ABDI – Agência Brasileira de Desenvolvimento Industrial Engenha¬ria consultiva no Brasil: desafios e oportunidades. (2011). *Relatório de Acompanhamento Setorial.* Retrieved December 08, 2015 from http://www.abdi.com.br/Estudo/relatorio_neit_eng-consultiva_final.pdf

Ackoff, R. L. (1989). From Data to Wisdom. *Journal of Applied Systems Analysis, 16*(1), 3–9.

Adair, J. (2011). *The art of creative thinking: How to be innovative and develop great ideas.* Norfolk, VA: Kogan Page.

Adeoti-Adekeye, W. (1997). The importance of management information systems. *Library Review, 46*(5), 318–327. doi:10.1108/00242539710178452

AGTMAEL. (2009). *Antoine van. O Século dos Mercados Emergentes.* São Paulo: Cultrix.

Ahmad, M., Rashid, T., & Mishra, D. K. (2014). Impact of wireless technology on future of big-data industry. *IT in Business Industry and Government (CSIBIG) Conference on,* 1-4.

Ahmed, S., Bilal, K., & Mustafa, A. (2015). Comparison between Cellular Generations, International Journal of Engineering. *Applied and Management Sciences Paradigms, 22*(1), 2320–6608.

Akbar, H. (2003). Knowledge Levels and their Transformation: Towards the Integration of Knowledge Creation and Individual Learning. *Journal of Management Studies, 40*(8), 1997–2021. doi:10.1046/j.1467-6486.2003.00409.x

Akbar, H. (2003). Knowledge Levels and their Transformation: Towards the Integration of Knowledge Creation and Individual Learning. *Journal of Management Studies, 8*(40), 1999–2020.

Albescu, F., & Pugna, I. B. (2014, July). *Marketing intelligence: The last frontier of business information technologies. Romanian Journal of Marketing*, 55–68.

Albrecht, M. A., Bosma, R., van Dinter, T., Ernst, J.-L., van Ginkel, K., & Versloot-Spoelstra, F. (2010). Quality assurance in the EPO patente information resource. *World Patent Information, 32*(4), 279–286. doi:10.1016/j.wpi.2009.09.001

Aleixo, J. A., & Duarte, P. (2015). Big data opportunities in healthcare. How can medical affairs contribute? *Revista Portuguesa de Farmacoterapia, 7,* 230–236.

Alessandro, C., & Trucco, P. C. (2011). Business potential and market opportunities of intelligent LBSs for personal mobility – A European case study. *Procedia Computer Science, 5,* 906–911. doi:10.1016/j.procs.2011.07.126

Allen, B. (1996). *Information tasks: Toward a user-centered approach to information systems.* San Diego: Academic Press.

Almeida, M., Freitas, C., & Souza, I. (2011). *Gestão do Conhecimento para a Tomada de Decisão.* Atlas.

Alnoor, P. (2016). India is the next China as appetite for smartphones defy global trend. *Business Standard*. Retrieved from http://www.afaqs.com/news/story/47857_India-is-the-next-China-as-appetite-for-smartphones-defy-global-trend

Alturas, B. (2013). *Introdução aos sistemas de informação organizacionais* (1st ed.). Lisboa: Sílabo.

Alves, R. P., & Falsarella, O. M. (2005). *Conceptual model for Oganizacional Intelligence Applied to Function Maintenance. National conference of Research in Information Science*, Florianópolis.

Amaral, J. V., & Guerreiro, R. (2014). Conhecimento e Avaliação dos Trade-offs de Custos Logísticos: Um Estudo com Profissionais Brasileiros. *R. Cont. Fin.*, *25*(65), 111–123.

Amaral, L., & Varajão, J. (2000). *Planeamento de sistemas de informação* (2nd ed.). Lisboa: FCA.

Amaral, S. A. (1998). *Marketing: abordagem em unidades de informação*. Brasília: Thesaurus.

Amaral, S. A. (2001). *Promoção; o marketing visível da informação*. Brasília: Brasília Jurídica.

Amaral, S. A. (2004). *Marketing da informação na Internet; ações de promoção. Colaboração de S.P. Santos e W.J.Araújo*. Campo Grande: Editora da UNIDERP.

Amaral, S. A. (2008). Marketing da informação: Entre a promoção e a comunicação integrada de marketing. *Informação e Sociedade Estudos, João Pessoa*, *18*(1), 31–44.

Amaral, S. A., & Souza, A. J. F. P. (2011, January/March). Qualidade da informação e intuição na tomada de decisão organizacional. *Perspectivas em Ciência da Informação*, *16*(1), 133–146. doi:10.1590/S1413-99362011000100008

Amaral, S., & Silveira, A. (1993). *Marketing em unidades de informação: estudos brasileiros. Organizado por Amélia Silveira e Sueli Angelica do Amaral*. Brasília: IBICT /SENAI.

American Marketing Association. (2007). *Dictionary of Marketing Terms*. Disponível em: http://www.marketingpower.com/mgdictionary- view1862.php

Andal-Ancion, A., Cartwright, P. A., & Yip, G. S. (2003). The Digital Transformation. *MIT Sloan Management Review*, *44*(4), 34–41.

Anderson, C. (2010). In the Next Industrial Revolution, Atoms are the New Bits. *Wired Magazine*, *1*(25).

Anderson, C. (2012). *Makers: The New Industrial Revolution* (1st ed.). New York: Crown Business.

Andrews, K. (1980). *The concept of corporate strategy*. Homewood, IL: Dow-Jones - Irwin.

Ansoff, H. I. (1965). *Corporate strategy*. Harmondsworth, UK: Penguin.

Ansoff, H. I. (1978). *Strategic Management*. Macmillan.

Antunes & Magalhaes. (Eds.). (2008). Patenteamento e Prospecação Tecnológica no Setor Farmacêutico. Interciencia.

Antunes, A., & Mercado, A. (2000). *A aprendizagem tecnológica no Brasil: a experiência da indústria química e petroquímica*. Editora E-papers.

Anunciação, P. F, (2015). Organizational Change through Information Systems: Metavision-Project Management Model in Internet Banking. In *Handbook of Research on Effective Project Management through the Integration of Knowledge and Innovation*. Academic Press.

Anunciação, P. F. & Esteves, F. M. (2012). Sustainability in the Information Society: A Proposal of Information Systems Requirements in View of the DPOBE Model for Organizational Sustainability. *Procedia Technology*, *5*, 599-606.

Anunciação, P. F. (2004). The Urbanism of Banking Information Systems - The Example of Subsystem "Capital Actions". *Management Studies – Portuguese Journal of Management Studies, 9*(1), 37–51.

Anunciação, P. F. (1997). *The Urbanism of Information System.* CETOP Publishing. (in Portuguese)

Anunciação, P. F. (2014). *Ethics, Sustainability and Information and Knowledge Society.* Lisbon: Chiado Editora. (in Portuguese)

Anunciação, P. F., & Esteves, F. M. (2014). Some Information Systems Requirements in View of Organizational Sustainability in an Information Society. *Information Resources Management Journal, 27*(1), 21–35. doi:10.4018/irmj.2014010102

Anunciação, P. F., Esteves, F. M., & Santos, J. R. (2013). Some Information Systems Requirements in View of Organizational Sustainability in an Information Society. *Information Resources Management Journal.*

Anunciação, P. F., & Zorrinho, C. D. (2006). *Organizational Urbanism – Who to manage the technological impact in economic organizations.* Sílabo Editions. (in Portuguese)

App Annie. (2016a). *App Forecast.* Retrieved from https://www.appannie.com/insights/market-data/app-annie-releases-inaugural-mobile-app-forecast/

App Annie. (2016b). *App Forecast.* Retrieved from URL: http://go.appannie.com/report-app-annie-market-forecast-2017

Arbix, G. (2007). *Inovar ou Inovar. A indústria brasileira entre o passado e o futuro.* São Paulo: Editora Papagaio.

Arnould, E. J., & Wallendorf, M. (1994). Market-oriented ethnography: Interpretation building and marketing strategy formulation. *JMR, Journal of Marketing Research, 31*(4), 484–504. doi:10.2307/3151878

Ashton, W. B., & Klavans, R. A. (1997). *Keeping abreast of science and technology: Technical intelligence for business.* Columbus, OH: Battelle Press.

Asiamah, F. N. (2009). *Exploiting tacit knowledge through knowledge management technologies.* Paper presented at the Learning Forum London Conference (ELFEL). Retrieved: http://bura.brunel.ac.uk/bitstream/2438/3613/4/EXploiting%20Tacit%20Knowledge%20-%20Frank%20Nyame-Asiamah.pdf?origin=publication_detail

ASQ. (2013). *American Society for Quality.* Accessed at www.asq.com

Auer, T., & Ruohem, M. (1996). Understanding The Human Side of Information Systems Management and Use: A Theorethical Framework and Two Evaluation Approaches. In *The Human Side of IT Management.* Idea Group Publishing.

Awasthi, S., Singh, S., Soni, R., & Jaidka, P. (2016). Internet of Things Using Raspberry pi 2. *IJEECS, 5*(5).

Azevedo, A., & Lourenço, A. (2008). *KDD, SEMMA and CRISP-DM: a parallel overview.* Available at http://hdl.handle.net/10400.22/136

Back, W., & Moreau, K. (2001, March). Information management strategies for project management. *Project Management Journal,* 10–20.

Badia, A. (2014). Data, information and knowledge: An information science analysis. *Journal of the Association for Information Science and Technology, 65*(6), 1279–1287. doi:10.1002/asi.23043

Bailey, S. (2002). Do you need a taxonomy strategy? *Inside Knowledge, 5*(5). Accessed at http://www.ikmagazine.com/

Ballou, R. H. (2004). Gerenciamento da cadeia de suprimentos: Logística empresarial. *The Bookman.*

Balmisse, G. (2006). *Guide des Outils du Knowledge Management: panorama, choix et mise en oeuvre.* Recuperado de http://www.leslivresblancs.fr/informatique/applications-pro/knowledge-management/livre-blanc/outils-du-km-panorama-choix-et-mise-en-oeuvre-62.html

Balmisse, G. (n.d.). *Guide des Outils du Knowledge Management: panorama, choix et mise en oeuvre.* Academic Press.

Banerjee, A. (2013). *Big Data & Advanced Analytics in Telecom – Huawei, Report by Heavy Reading.* Retrieved from www.heavyreading.com

Barata, J. M. M. (1992). Inovação e Desenvolvimento Tecnológico: Conceitos, modelos e medidas. Pistas para a investigação aplicada. *Estudos de Economia, XII, 2*, 147–171.

Barbosa, R. R. (2002). Inteligência empresarial: uma avaliação de fontes de informação sobre o ambiente organizacional externo. *Datagrama Zero, 6*(3). Disponível em http://www.dgz.org.br/dez02/Art_03.htm

Barbosa, P., & O'Reilly, A. (2011). *Harvard trends: Tendências de gestão.* Porto, Portugal: Vida Económica.

Barney, J. (1991). Firms Resources and Sustained Competitive Advantage. *Journal of Management, 17*(1), 99–120. doi:10.1177/014920639101700108

Barney, J. (1995). Looking inside for competitive advantage. *The Academy of Management Executive, 9*(4), 49–61.

Barney, J. B. (1986). Strategic Factor Markets: Expectations, Luck and Business Strategy. *Management Science, 32*(10), 1231–1241. doi:10.1287/mnsc.32.10.1231

Barómetro Inovação. (n.d.). Disponível em: http://barometro.cotecportugal.pt/pt/indicadores/modelo-de-indicadores-de-idi/dimensoes-pilares-e-indicadores/dimensao-pilares-e-indicadores.html

Barreto, A. de A. (1994). *A transferência de informação, o desenvolvimento tecnológico e a produção do conhecimento.* Disponível na Internet no endereço: http://www.alternex.com.br/_aldoibct/infor/informa.html

Barreto, A. de A. (1999). Oferta e a demanda da informação: condições técnicas, econômicas e políticas. Ciência da Informação, 28(2), 168-173.

Barreto, A. de A. (2002). Transferência da informação para o conhecimento. In O campo da ciência da informação: gênese, conexões e especificidades. João Pessoa: Editora Universitária, UFPB.

Barreto, A. (2000). O mercado de informação no Brasil. *Informação & Informação, 5*(1), 25–34. doi:10.5433/1981-8920.2000v5n1p25

Barreto, A. de A. (1996). A eficiência técnica e econômica e a viabilidade de produtos e serviços de informação. *Ciência da Informação, 25*(3).

Barro Ameneiro, S. (2015). La Transferencia de I+D, la Innovación y el Emprendimiento en las Universidades. Educación Superior en Iberoamérica. Informe 2015. *Centro Interuniversitario de Desarrollo - CINDA Red Emprendia Universia*, 541.

Barros, A. P., Ishikiryama, C. S., Peres, R. C., & Gomes, C. F. S. (2015). Processes and benefits of the application of information technology in supply chain management: An analysis of the literature. *Procedia Computer Science, 55*, 698–705. doi:10.1016/j.procs.2015.07.077

Bartlett, C. A., & Ghoshal, S. (2000). Going Global: Lessons from late movers. *Harvard Business Review, 78*(2), 133–142. PMID:11183975

Bartol, K. M., & Srivastava, A. (2002). Encouraging Knowledge Sharing: The Role of Organizational Reward Systems. *Journal of Leadership & Organization Studies, 9*(1), 64–76. doi:10.1177/107179190200900105

Basnuevo, C. A. (2005). *Antecedentes e situación atual de los conceptos y métodos para el desarollo de la inteligência organizacional.* Disponível em http://bvs.sld.cu/revistas/aci/vol13_4_05/aci02405.htm

Bateson, G. (1972). *Steps to an Ecology of Mind.* Nova Iorque, Balentine Books.

Baxter, M. (2000). Projeto de produto: Guia prático para o design de novos produtos (2ª ed.). S. Paulo: Edgard Blücher.

Bazeley, P. (2015). Mixed Methods in Management Research: Implications for the Field. *Electronic Journal of Business Research Methods, 13*(1), 27–35. Available online at www.ejbrm.com

Beijerse, R. (1999). Questions in KM: Defining and conceptualizing a phenomenon. *Journal of Knowledge Management, 3*(3), 94–109. doi:10.1108/13673279910275512

Beijerse, R. P. (2000). Knowledge management in small and medium-sized companies: Knowledge management for entrepreneurs. *Journal of Knowledge Management, Bingley, 4*(2), 162–179. doi:10.1108/13673270010372297

Bellini, F., Ascenzo, F., Ulskaia, I., & Savastano, M. (2016). Digital service platform for networked enterprises collaboration: A case study of the NEMESYS project. In IESS 2016, LNBIP 247 (pp. 313–326). Academic Press. doi:10.1007/978-3-319-32689-4_24

Bell, S. (2006). A victim of its own success: Internationalization, neoliberalism, and organizational involution at the business council of Australia. *Politics & Society, 34*(4), 543–570. doi:10.1177/0032329206293643

Benbasat, I., & DeSanctis, G. (2000). *Communication Challenges – A Value Network Perspective. In Information Technology and the Future Enterprise* (pp. 144–162). Prentice Hall.

Bennet, D., & Bennet, A. (2008). Engaging tacit knowledge in support of organizational learning. *The Journal of Information and Knowledge Management Systems, 38*(1), 72-94.

Bennett, R. (1998). Business associations and their potential contribution to the competitiveness of SMEs. *Entrepreneurship and Regional Development, 10*(3), 243–260. doi:10.1080/08985629800000014

Berenzoi, A. (2014, December). Business Model Innovation in Corporate Competitive Strategy. *Problems of Economic Transition, 57*(8), 14–33.

Bernardes, R. (2003). Produção de Estatísticas e Inovação Tecnológica: PAEP 1996 – 2001. *São Paulo em Perspectiva, 17*(3-4), 151–167. doi:10.1590/S0102-88392003000300016

Berry, L. L., Carbone, L. P., & Haeckel, S. H. (2002). Managing the Total Customer Experience. *MIT Sloan Management Review, 43*(3), 85–89.

Berry, M. J., & Linoff, G. (1997). *Data mining techniques: for marketing, sales and customer support.* New York: John Wiley and Sons, Inc.

Best, D. (1996a). Business process and information management. In Best (Ed.), The fourth resource: Information and its management. Hampshire: Aslib/Gower.

Best, D. (1996a). Business process and information management. In D. Best (Ed.), *The Fourth Resource: Information and its Management.* Hampshire, UK: Aslib/Gower.

Best, D. (1996b). *The fourth resource: Information and its management.* Hampshire, UK: Aslib/Gower.

Best, D. (1996b). *The fourth resource: information and its management.* Hampshire: Aslib/Gower.

Beuren, I. (1998). *Gerenciamento da informação: um recurso estratégico no processo de gestão empresarial.* São Paulo: Atlas.

Beuren, I. M. (1998). *Gerenciamento da informação: Um recurso estratégico no processo de gestão empresarial*. São Paulo: Atlas.

Bhattacharya, M. (2010). *Telecom Sector in India: Vision 2020*. Department of Telecommunications, Ministry of Communications & IT, and Government of India.

Birch, P., & Clegg, B. (1999). *Criatividade em negócios: Um guia para gestores*. Viseu: Pergaminho.

Blackstock, M. (2015). Toward a Distributed Data Flow Platform for the Web of Things. *IEEE IoT conference: Developing IoT Applications in the Fog: a Distributed Dataflow Approach. 2015 International Conference on the.*

Blikstein, P. (2014). Digital Fabrication and 'Making' in Education: The Democratization of Invention. In J. Walter-Herrmann & C. Büching (Eds.), *FabLabs: Of Machines, Makers and Inventors*. Bielefeld: Transcript-Verlag.

Boblitz, M. C. (2006). Looking out the window: Market intelligence for a view of the real world. Healthcare financial management, 46-53.

Boisot, M. H. (1998). *Knowledge Assets: Securing Competitive Advantage in the Information Economy*. New York, NY: Oxford University Press.

Boland, W. P., Phillips, P. W. B., Ryan, C. D., & McPhee-Knowles, S. (2012). Collaboration and the Generation of New Knowledge in Networked Innovation Systems: A Bibliometric Analysis. *Procedia: Social and Behavioral Sciences*, *52*, 15–24. doi:10.1016/j.sbspro.2012.09.437

Boléat, M. (2003). *Managing trade associations*. London: Trade Association Forum.

Bollen, J., Mao, H., & Zeng, X. (2011). Twitter mood predicts the stock market. *Journal of Computational Science*, *2*(1), 1–8. doi:10.1016/j.jocs.2010.12.007

Boonstra, A. (2013). How do top managers support strategic information system projects and why do they sometimes withhold this support? *International Journal of Project Management*, *31*(4), 498–512. doi:10.1016/j.ijproman.2012.09.013

Borko, H., (1968), Information science: what is it? *American Documentation*, *19*(1), 3-5.

Bossel, H. (1999). *Indicators for Sustainable Development: Theory, Method, Applications*. International Institute for Sustainable Development (Vol. 68). Retrieved from http://www.ulb.ac.be/ceese/STAFF/Tom/bossel.pdf

Botelho, T. M., Amaral, S. A., & Baptista, S. G. (1994). Information and society: An inteligent society in transformation? *International Journal of Information and Library Research, Londres*, *6*(1/2), 64–76.

Boyd, D., & Crawford, K. (2011). Six provocations for Big Data. *Annals of "A decade in Internet time: Symposium on the dynamics of the internet and society"*.

BPM Institute – Business process management Institute. (n.d.). Access at http://www.bpminstitute.org

Brandenburger, A., & Nalebuff, B. (1996). *Co-opetition: A revolution mindset that combines competition and cooperation: The game theory strategy that's changing the game of business*. New York: Bantam Doubleday.

Branscomb, L. (2004). *Where do tech comercial innovations come from?* Paper presented at the Meredith and Kip Frey Lecture in Intellectual Property at Duke Law School. Retrieved 06-09-2010, from http://www.law.duke.edu/conference/namedlectures.html

BRASIL. (2017). *Estratégia Nacional de Ciência, Tecnologia e Inovação 2012-2015*. Ministério da Ciência, Tecnologia e Inovação (MCTI). Disponível em: <http://www.mct.gov.br/upd_blob/0218/218981.pdf>

BRASIL. Ministério da Saúde. Portaria nº 704, de 08 de março de 2017. Define a lista de produtos estratégicos para o Sistema Único de Saúde (SUS), nos termos dos anexos a esta Portaria.

BRASIL. MS. (2014a). Portaria nº 2.531, de 12 de novembro de 2014. Redefine as diretrizes e os critérios para a definição da lista de produtos estratégicos para o Sistema Único de Saúde (SUS) e o estabelecimento das Parcerias para o Desenvolvimento Produtivo (PDP) e disciplina os respectivos processos de submissão, instrução, decisão, transferência e absorção de tecnologia, aquisição de produtos estratégicos para o SUS no âmbito das PDP e o respectivo monitoramento e avaliação.

BRASIL. MS. (2014b). Portaria nº16, de 15 de maio de 2014. Torna pública a decisão de incorporar a doxicilina injetável e o cloranfenicol suspensão para terapêutica da febre maculosa brasileira e outras riquetsioses no Sistema Único de Saúde - SUS.

BRASIL. MS., D. (2014). *Parceria para o Desenvolvimento Produtivo - PDP*. Accessed may 14, 2014, de http://portalsaude.saude.gov.br/index.php/o-ministerio/principal/leia-mais-o-ministerio/581-sctie-raiz/deciis/l2-deciis/12090-parceria-para-o-desenvolvimento-produtivo-pdp

BRASIL/MS/FIOCRUZ/IPEA. (2012). *A Saúde no Brasil em 2030: diretrizes para prospecção estratégica para o Sistema de Saúde Brasileiro*. Brasil. Recuperado de http://www.isags-unasursalud.org/biblioteca_interna.asp?idArea=33&lang=1&idPai=3024

BRASIL/TCU. (2014). *FISCSAÚDE (No. Relatório Sistêmico de Fiscalização da Saúde - FISCSAÚDE)*. Brasil: TCU - Tribunal de Contas da União. Accessed by http://portal2.tcu.gov.br/portal/page/portal/TCU/imprensa/noticias/noticias_arquivos/032.624-2013-1%20Fisc%20Saude.pdf

Bratianu, C., & Orezea, I. (2013). Emotional Knowledge: The Hidden Part of the Knowledge Iceberg. In *Proceedings Book of the 14th European Conference on Knowledge management* (pp. 82-90). Kaunas, Lithuania: Academic Conferences and Publishing International Limited.

Bregonje, M. (2005). Patents: A unique source for scientific technical information in chemistry related industry? *World Patent Information*, *27*(4), 309–315. doi:10.1016/j.wpi.2005.05.003

Breschi, S., & Malerba, F. (1997). Sectoral Innovation Systems: Technological Regimes, Schumpeterian Dynamics, and Spatial Boundaries. In Systems of Innovation: Technologies, Institutions and Organizations (pp. 130–156). Academic Press. doi:10.1016/S0024-6301(98)90244-8

Briones-Peñalver, A. J., & Poças-Rascão, J. (2014). *Information tecnhologies (ICT), network organizations, and information systems for business cooperation: a focus on organization and strategic knowledge management. Rethinking the Conceptual Base for New Practical Applications in Information Value and Quality*. IGI Global.

Briones-Peñalver, A. J., Roca-Gonzalez, J. L., & Martinez-Martinez, I. J. (2015). *Innovation management based on knowledge. Analysis of technology-based Defence Companies. Handbook of Research on Effective Project*. IGI Global.

Brito, C. (2013). *The Entrepreneurship Ecosystem of the University of Porto*. UPTEC.

Bronowski, J. (1977). O senso comum da ciência. In *Belo Horizonte*. Editora da USP.

Brophy, P., & Coulling, K. (1996). *Quality management for information and library managers*. Guildford: Aslib/Gower.

Brouthers, K., Geisser, K., & Rothlauf, F. (2016). Explaining the internationalization of ibusiness firms. *Journal of International Business Studies*, *47*(5), 513–534. doi:10.1057/jibs.2015.20

Brouwers, M. C., Kho, M. E., Browman, G. P., Burgers, J. S., Cluzeau, F., Feder, G., ... Zitzelsberger, L. (2010). AGREE II: Advancing guideline development, reporting and evaluation in health care. *Canadian Medical Association Journal*, *182*(18), E839–E842. doi:10.1503/cmaj.090449 PMID:20603348

Brown, C. (1994). Unix Distributed Programming. In *Building IoT Applications with Raspberry Pi and Low Power IQRF Communication Modules*. Academic Press.

Brown, J., & Duguid, P. (2000). *The social life of information*. Boston: Harvard Business School Press.

Brum, M. A. C., & Barbosa, R. R. (2009). *Comportamento de busca e uso da informação: um estudo com alunos*. Heidelberg, Germany: Springer.

Brünger-Weilandt, S., Geiß, D., Herlan, G., & Stuike-Prill, R. (2011). Quality: Key factor for high value in professional patent, technical and scientific information. *World Patent Information*, *33*(3), 230–234. doi:10.1016/j.wpi.2011.04.007

Brynjolfsson, E., & Hitt, L. (2000). Beyond computation: Information technology, organizational transformation and business performance. *The Journal of Economic Perspectives*, *14*(4), 23–48. doi:10.1257/jep.14.4.23

Buchanan, B. (2008). Unlocking the value of patent data: Patent informatics services at the UK intellectual property office (UK-IPO). *World Patent Information*, *30*(4), 335–337. doi:10.1016/j.wpi.2008.07.002

Buckland, M. (1995). *Information and Information Systems*. New York: Ed. Praeger.

Buckland, M. K. (1991, June). Information as thing. *Journal of the American Society for Information Science*, *42*(5), 351–360. doi:10.1002/(SICI)1097-4571(199106)42:5<351::AID-ASI53.0.CO;2-3

Budford, S. (2014, October). A grounded theory of the practice of web information architecture in large organizations. *Journal of the Association for Information Science and Technology*, *65*(10), 2017–2034. doi:10.1002/asi.23098

Burgelman, R., Christensen, C., & Wheelwright, S. (2009). *Strategic management of technology and innovation* (5th ed.). Singapore: McGraw-Hill.

Burk, C. Jr, & Horton, F. Jr. (1988). *InfoMap: A complete guide to discovering corporate information resources*. Upper Saddle River, NJ: Prentice Hall.

Business Standard. (2012). Retrieved from URL: http://www.business-standard.com/article/companies/internet-technology-to-contribute-up-to-3-3-to-gdp-by-2015-112121900084_1.html

Butler, J. T. (1995). Patent searching using comercial databases. In Lechter (Ed.), Successful patents and patenting for engineers and scientists. New York: The Institute of Electrical and Electronics Engineers (IEEE) Press.

Byosiere, P., & Luethge, J. D. (2008). Knowledge domains and knowledge conversion: An empirical investigation. *Journal of Knowledge Management*, *12*(2), 67–78. doi:10.1108/13673270810859523

Calof, J. L., & Wright, S. (2008). Competitive intelligence: A practitioner, academic and inter-disciplinary perspective. *European Journal of Marketing*, *42*(7/8), 717–730. doi:10.1108/03090560810877114

Camarinha-Matos, L., & Afsarmanesh, H. (2005). Collaborative networks: A new scientific discipline. *Journal of Intelligent Manufacturing*, *16*(1), 439–452. doi:10.100710845-005-1656-3

Campos, V. F. (2008). *Qualidade Total – INDG*. Belo Horizonte: INDG.

Canadian Institutes of Health Research. (2016). *Knowledge Translation and Commercialiation*. Recuperado 11 de abril de 2017, de http://www.cihr-irsc.gc.ca/e/29418.html#1

Canadian Institutes Of Health Research. (n.d.). *Knowledge Translation and Commercialiation*. Disponível em: http://www.cihr-irsc.gc.ca/e/29418.html#1

Cao, L., Zhang, C., & Liu, J. (2006). Ontology-based integration of business intelligence. *Web Intelligence and Agent Systems. International Journal (Toronto, Ont.)*, *4*, 313–325.

Capuano, E. A. (2015). Informação sobre Conceitos e Indicadores de Inovação. *Revista de Informação, 16*(1).

Caraher, V. (2008). The evolution of the patent information world over the next 10 years: A Thomson scientific perspective. *World Patent Information, 30*(2), 150–152. doi:10.1016/j.wpi.2007.07.008

Carboni, D., Serra, A., Pintus, A., Kro, S., Nati, M., & Gurgen, L. (n.d.). *Specific Targeted Research Projects (STReP) SocIoTal - Creating a socially aware citizen-centric Internet of Things*. Available at: http://cordis.europa.eu/docs/projects/cnect/2/609112/080/deliverables/001-SOCIOTALD41V1.pdf

Cardenas, J. M., & Spinola, M. (2011). The roles of tacit knowledge and knowledge management systems for writing academic papers: a research case. *Proceedings of the 12th European Conference on Knowledge management.*

Carlsson, B. (1995). *Technological systems and economic performance: The case of factory automation.* Economics of science, technology, and innovation.

Carneiro, A. (1995). *Inovação: estratégia e competitividade*. Lisboa: Texto Editora.

Carneiro, L., Soares, A., Patrício, R., Azevedo, A., & Pinho de Sousa, J. (2013). Case studies on collaboration, technology and performance factors in business networks. *International Journal of Computer Integrated Manufacturing*, *26*(1–2), 101–116. doi:10.1080/0951192X.2012.681914

Cartaxo, R. J. (2011). *Metodologia de Priorização para Produção Nacional dos Medicamentos Pertencentes à Lista do Sistema Único de Saúde* (Dissertação de Mestrado). National Institute of Industrial Property (INPI - Instituto Nacional de Propriedade Industrial). Accessed by http://www.inpi.gov.br/portal/

Caru, A., & Cova, B. (2008). Small versus big stories in framing consumption experiences. *Qualitative Market Research, 11*(2), 166–176. doi:10.1108/13522750810864422

Carvalho, A. I., Bodstein, R. C., Hartz, Z., & Matida, Á. H. (2004). Concepts and approaches in the evaluation of health promotion. *Ciência & Saúde Coletiva, 9*(3), 521–529. doi:10.1590/S1413-81232004000300002

Castells, M. (2000). The Rise of the Network Society. In The Information Age: Economy, Society and Culture (Vol. 1). Cambridge, MA: Blackwell.

Castells, M. (1999). *A sociedade em rede*. São Paulo: Paz e Terra.

Castells, M. (2000). *The information age: The rise of the network society* (2nd ed.). London: Blackwell.

Castells, M. (2001). *A sociedade em rede: A era da informação: Economia, sociedade e cultura* (5th ed.). São Paulo: Paz e Terra.

Castells, M. (2004). *A galáxia internet: Reflexões sobre internet, negócios e sociedade*. Lisboa: Fundação Calouste Gulbenkian.

Castillo, J. (2002). A note on the concept of Tacit knowledge. *Journal of Management Inquiry, 11*(1), 46–57. doi:10.1177/1056492602111018

Castro, A. L. S. (2002). O valor da informação: um desafio permanente. *Revista de Ciência da Informa, 3*(3).

Castro, J. (1999). *Invento & inovação tecnológica: Produtos & patentes na construção*. Annablume Editora Comunicacao.

Cattela, R. (1981). Information as a corporate asset. *Information & Management*, *4*(1), 29–37. doi:10.1016/0378-7206(81)90023-9

Cavalcante & De Negri. (2011). *Trajetória recente de indicadores de inovação no Brasil*. IPEA; Texto para discussão, 1659.

Cavalier, T. (2001). Perspectives on machine translation of patent information. *World Patent Information*, *23*(4), 367–371. doi:10.1016/S0172-2190(01)00045-X

Centro de Gestão e Estudos Estratégicos, & Ministério da Ciência, Tecnologia e Inovação. (2013). *Parcerias Estratégicas nº 36* (36º ed, Vol. 18). Brasília: CGEE. Accessed by http://www.cgee.org.br/parcerias/parcerias.php

Certo, S. C., & Peter, J. P. (2010). Administração Estratégica: planejamento e implantação de estratégias (3o ed). Prentice-Hall.

Chakroun, N. (2012). Improving patent information quality: Development and the disclosure requirements. *The Journal of World Intellectual Property*, *15*(3), 199–220. doi:10.1111/j.1747-1796.2012.00438.x

Chan, P. K., Fan, W., Prodromidis, A. L., & Stolfo, S. F. (1999). Distributed data mining in credit card fraud detection. *IEEE Intelligent Systems and their Applications,* 14(6), 67-74.

Chapman, P., Clinton, J., Kerber, R., Khabaza, T., Reinartz, T., Shearer, C., & Wirth, R. (2000). *Step-by-step mining guide*. SPSS.

Chapman, A. S.Tickborne Rickettsial Diseases Working Group. (2006). Diagnosis and Management of Tickborne Rickettsial Diseases: Rocky Mountain Spotted Fever, Ehrlichioses, and Anaplasmosis - United States. A Practical Guide for Physicians and Other Health-Care and Public Health Professionals. Morbidity and Mortality Weekly Report (MMWR). *Recommendations and Reports*, *55*(RR04), 1–27. PMID:16572105

Chau, M. & Xu, J. (2012) Business intelligence in blogs: understanding consumer interactions and communities. *MIS Quarterly, 36*(4), 1189-1216.

Chau, M. & Xu, J. (2012). Business intelligence in blogs: understanding consumer interactions and communities. *MIS Quarterly, 36*(4), 1189-1216.

Chau, M., & Xu, J. (2012). Business intelligence in blogs: understanding consumer interactions and communities. MIS Quarterly, 36(4), 1189-1216.

Chauhan, P., & Kumar, T. (2015). Power Optimization in Wireless Sensor Network: A Perspective. *International Journal of Engineering and Technical Research, 3*(5).

Checkland, P., & Holwell, S. (1998). *Information, systems and information systems: making sense of the field.* John Wiley & Sons.

Chen, H., Chiang, R. H. L., & Storey, V. C. (2012). Business intelligence and analytics: from big data to big impact. MIS Quarterly, 36(4), 1165-1188.

Chesbrough, H. (2003). *Open Innovation: The New Imperative for Creating and Profiting from Technology*. Boston: Harvard Business School Press.

Chesbrough, H. (2010). Business model innovation: Opportunities and barriers. *Long Range Planning*, *43*(2-3), 354–363. doi:10.1016/j.lrp.2009.07.010

Choo, C. W. (2006), Research in information and Knowledge Management. Brasília: PUC/DF, 29 abr. Palestra.

Choo, C. W. (1994). Perception and use of information sources by chief executives in environmental scanning. *Library and Information Science Research. Boston, 16*(1), 23–40. doi:10.1016/0740-8188(94)90040-X

Choo, C. W. (1996). The knowing organization: How organizations use information to construct meaning, create knowledge and make decisions. *International Journal of Information Management, 16*(5), 329–340. doi:10.1016/0268-4012(96)00020-5

Choo, C. W. (1999). *The intelligent organization: The use of information to create awareness and give Meaning Decision Making*. Oxford University Press. (in Spanish)

Choo, C. W. (2003). *Gestão de informação para a organização inteligente: A arte de explorar o meio ambiente*. Lisboa: Editorial Caminho.

Choo, C. W. (2003). *The knowing organization: how organizations use information to construct meaning, create knowledge and make decisions.* Oxford, UK: Ed. Oxford University Press.

Choudary, S. P. (2013). Why business models fail: pipes and platforms. *Wired Magazine*. Accessed at http://www.wired.com/2013/10/why-business-models-fail-pipes-vs-platforms/

Chou, J. S., & Yang, J. G. (2012). Project management knowledge and effects on construction projects outcomes: An empirical study. *Project Management Journal, 43*(5), 47–67. doi:10.1002/pmj.21293

Christensen, C. M., & Overdorf, M. (2000). Meeting the Challenge of Disruptive Change. *Harvard Business Review, 78*(2), 66–76.

Christensen, C., Anthony, S., & Roth, E. (2004). *Seeing what's next: Using the theories of innovation to predict industry change*. Boston: Harvard Business School Press.

Christopher, M. (1998). *Logistic and supply chain management: strategies for reduzing cost and improving service* (2nd ed.). Financial Times Pitman.

Chui, M., Loffler, M., & Roberts, R. (2010). *The Internet of Things*. McKinsey Quaterly.

Churchman, C. W. (1971). *The design of inquiring systems: basic concepts of systems and organizations*. New York: Bencis Books.

Cisco. (2014). *The Zettabyte Era—Trends and Analysis, Visual Networking Index (VNI)*. Retrieved from URL: http://www.cisco.com/c/en/us/solutions/collateral/service-provider/visual-networking-index-vni/VNI_Hyperconnectivity_WP.html

Clark, T. D. Jr, Jones, M. C., & Armstrong, C. P. (2007). The dynamic structure of management support systems: Theory development, research, focus and direction. *Management Information Systems Quarterly, 31*(3), 579–615. doi:10.2307/25148808

Classificação Nacional De Atividades Econômicas – CNAE. (n.d.). Retrieved April 18, 2016 from http://cnae.ibge.gov.br/?view=secao&tipo=cnae&versaosubclasse=9&versaoclasse=7&secao=C

Cleveland, H. (1985). *The knowledge executive: Leadership in an information society*. New York: Dutton.

COAI. (2013). *Cellular Operators Association of India's report 2013*. Retrieved from Cellular Operators association of India: http://www.coai.com/Reports-and-Papers

COAI. (2015). *Cellular Operators Association of India's report 2014*. Retrieved from URL: http://www.coai.com/statistics/arpu-and-revenue-report

Condesso, F. (1999). *A autonomia universitária no direito espanhol e português*. Estudo monográfico.

Cooke, A. (2001). *A guide to finding quality information on the internet: Selection and evaluation strategies* (2nd ed.). Cornwall, UK: Library Association Publishing.

Cooke, P. (1992). Regional Innovation Systems: Competitive Regulation in the New Europe. *Geoforum*, *23*(3), 365–382. doi:10.1016/0016-7185(92)90048-9

Corbin, J., & Strauss, A. (2015). *Basics of qualitative research: Techniques and procedures for developing grounded theory*. Thousand Oaks, CA: Sage Publications.

Cornell University, INSEAD, & WIPO. (2015). *The Global Innovation Index 2016: Winning with Global Innovation. The Global Innovation Index 2016*. Retrieved from http://www.codespring.ro/wp-content/uploads/2012/11/GII-2012_Cover.pdf

Cortada, J. (1996). *TQM for information systems management: Quality practices for continuous improvement*. Singapore: McGraw-Hill.

Costa, E., Soares, A., & Sousa, J. (2015). A new insight in the SMEs internationalization process. In M. L. Camarinha-Matos, F. Bénaben, & W. Picard (Eds.), *Risks and Resilience of Collaborative Networks: 16th IFIP WG 5.5 Working Conference on Virtual Enterprises, PRO-VE 2015, Albi, France, October 5-7, 2015* (pp. 398–410). Cham: Springer International Publishing. 10.1007/978-3-319-24141-8_36

Costa, E., Soares, A., & Sousa, J. (2016b). Situating case studies within the design science research paradigm: An instantiation for collaborative networks. In H. Afsarmanesh, M. L. Camarinha-Matos, & A. Lucas Soares (Eds.), *Collaboration in a Hyperconnected World: 17th IFIP WG 5.5 Working Conference on Virtual Enterprises, PRO-VE 2016, Porto, Portugal, October 3-5, 2016* (pp. 531–544). Cham: Springer International Publishing. 10.1007/978-3-319-45390-3_45

Costa, E., Soares, A., & Sousa, J. (2017a). From data sources to information sharing in SME collaborative networks supporting internationalization: A socio-semantic approach. In L. M. Camarinha-Matos, H. Afsarmanesh, & R. Fornasiero (Eds.), *Collaboration in a Data-Rich World: 18th IFIP WG 5.5 Working Conference on Virtual Enterprises, PRO-VE 2017, Vicenza, Italy, September 18-20, 2017* (Vol. 506, pp. 478–490). Cham: Springer International Publishing. 10.1007/978-3-319-65151-4_43

Costa, L. S., Gadelha, C. A. G., & Maldonado, J. (2012). A perspectiva territorial da inovação em saúde: a necessidade de um novo enfoque. Revista de Saúde Pública.

Costa, L. S., Gadelha, C. A. G., & Maldonado, J. (2012). A perspectiva territorial da inovação em saúde: a necessidade de um novo enfoque. *Revista de Saúde Pública*. doi:10.1590/S0034-89102012005000066

Costa, E., Soares, A., & Sousa, J. (2016a). Information, knowledge and collaboration management in the internationalisation of SMEs: A systematic literature review. *International Journal of Information Management*, *36*(4), 557–569. doi:10.1016/j.ijinfomgt.2016.03.007

Costa, E., Soares, A., & Sousa, J. (2017b). Institutional networks for supporting the internationalisation of SMEs: The case of industrial business associations. *Journal of Business and Industrial Marketing*, *32*(8), 1182–1202. doi:10.1108/JBIM-03-2017-0067

Costa, E., Soares, A., Sousa, J., & Jamil, G. (2017). Information management for network transformation in industrial enterprises associations: The case of the internationalization process. In G. L. Jamil, A. L. Soares, & C. R. M. Pessoa (Eds.), *Handbook of Research on Information Management for Effective Logistics and Supply Chains* (pp. 415–436). Hershey, PA: IGI Global; doi:10.4018/978-1-5225-0973-8.ch022

Costa, L. E. (2006). *Knowledge Management and Competitive Intelligence as instrument of Organizational Sustainability*. Belo Horizonte: Pitágoras Faculty. (in Portuguese)

Coulouris, G., Dollimore, J., Kindberg, T., & Gordon, B. (2013). *Sistemas Distribuídos - Conceitos e Projetos*. Bookman Editora.

Courtney, M. (2013). Puzzling out Big Data. Engineering & Technology, 7(12), 56-60.

Cova, V., & Cova, B. (2001). *Alternatives Marketing: réponses marketing aux évolutions récentes des Consommateurs*. Paris: Dunod.

Covert, A. (2014). *How to make sense for any mess: Information architecture for everybody*. Charleston, SC: CreateSpace Independent Publishing. Retrieved from www.createspace.com

Cox, T. H. (1993). *Cultural Diversity in Organizations*. San Francisco: Berrett Koehler.

Cremona, L., Ravarini, A., & Sutanto, J. (2014). KMS in a cluster of firms: The role of a digital platform. In *Information Systems, Management, Organization and Control. Lecture Notes in Information Systems and Organisation* (Vol. 6, pp. 253–262). Cham: Springer; doi:10.1007/978-3-319-07905-9_18

Cristea, D. S., & Capatina. (2009). *A. Perspectives on knowledge management models*. Analls do "Duranea de Jos". *Universidade de Galati*, *15*(2), 355–366.

Cristea, D. S., & Capatina, A. (2009). Perspectives on knowledge management models. *Annals of "Duranea de Jos". Galati University*, *15*(2), 355–366.

Crosby, P. (1979). *Quality is Free*. New York: McGraw-Hill.

Cruz, E. (1990). *Planeamento estratégico: Um guia para a PME* (3rd ed.). Lisboa: Texto Editora.

CSCMP - Council of Supply Chain Management Professionals. (n.d.). Retrieved December 08, 2015 from http://cscmp.org/resourcesresearch/glossary-terms

Cunha, M. B. (2001). *Para saber mais: fontes de informação em ciência e tecnologia*. Brasília, DF: Briquet de Lemos.

Curbatov, O. (2003). *L'intégration du consommateur par le knowledge Marketing: conception, production et consommation d'un produit personnel* (doctoral dissertation). Université de Nice-Sophia Antipolis en cotutelle avec l'Institut International de Management, Nice, France.

Curbatov, O., Pavlidis, P., & Gay, M. (2006). Le knowledge marketing ou le détournement du marketing (pourquoi le web perfume détourne le marketing vers la "competence clients-entreprises"? (The Knowledge Marketing or Marketing Reloaded: smell web matrix reloaded). *5th International Congress Marketing Trends Venice*.

D & B Research. (2009). *Indian Telecom Industry*. Retrieved from URL: https://www.dnb.co.in/IndianTelecomIndustry/OverviewTI.asp

da Silva, A. M., Malheiro, A., & Silva, D. (2014). Ciência da Informação e comportamento informacional: Enquadramento epistemológico do estudo das necessidades de busca, seleção e uso. *Prisma*, *21*, 1–61.

Da Silva, C. C. M. (2005). *O Papel do Sector Público na Inovação e na Mudança Tecnológica nas Empresas*. Universidade do Minho.

Dahlman. (1998, September). UMTS/IMT-2000 Based on Wideband CDMA. *IEEE Communications Magazine*, 70–80.

Dantas, J. (2001). *Gestão da inovação*. Porto: Vida Económica.

Dataquest Insight. (2009). *New Revenue Opportunities for Telecom Carriers, 2013*. Author.

Davenport, T. (1998). Managing customer knowledge. *CIO Magazine*. Retrieved from http://www.cio.com/archive/060198/think.html,10.5.2004

Davenport, T. H., & Marchand, D. (2000). *Mastering information management*. Boston: Financial Times Prentice Hall.

Davenport, T. (1997). *Information ecology: Mastering the information and knowledge environment*. New York: Oxford University Press.

Davenport, T. (2007). *Profissão: Trabalhador do conhecimento: Como ser mais produtivo e eficaz no desempenho das suas funções*. Amadora: Exame.

Davenport, T. H., & Harris, J. G. (2007). *Competição analítica: vencendo através da Nova Ciência: como decisões baseadas em fatos e business intelligence transformam negócios*. Rio de Janeiro: Elsevier.

Davenport, T. H., & Prusak, L. (1998). *Working Knowledge: How Organizations Manage What They Know*. Boston: Harvard Business School Press.

Davenport, T. H., & Prusak, L. (2000). *Working knowledge: how organizations manage what they know* (2nd ed.). Harvard Business Press.

Davenport, T., Marchand, D., & Dickson, T. (2004). *Dominando a gestão da informação*. Porto Alegre: Bookmann.

Davenport, T., & Prusak, L. (1998). *Working Knowledge*. Cambridge, MA: Harvard University Press.

Davenport, T., & Prusak, L. (1998). *Working knowledge: How organizations manage what they know*. Boston: Harvard Business School Press.

Davenport, T., Prusak, L., & Wilson, J. (2003). *What's the big idea: Creating and capitalizing on the best management thinking*. Boston: Harvard Business School.

David, E., & Thawornwong, S. (2005). The use of data mining and neural networks for forecasting stock market returns. *Expert Systems with Applications*, *29*(4), 927–940. doi:10.1016/j.eswa.2005.06.024

Davila & Epstein. (2007). *As regras da inovação*. Porto Alegre : *The Bookman*.

Davis, K. (1967). Human relations at work: The dynamics of organizational behavior (3rd ed.). Nova Iorque: International Student Edition.

Day, G. S. (1999). Creating a Market-Driven Organization. *Sloan Management Review*, *41*(1), 11–21.

de Carvalho, A. L. B., Souza, M. de F., Shimizu, H. E., Senra, I. M. V. B., & de Oliveira, K. C. (2012). SUS management and monitoring and evaluation practices: Possibilities and challenges for building a strategic agenda. *Ciência & Saúde Coletiva*, *17*(4), 901–911. doi:10.1590/S1413-81232012000400012

De Man, D. (2012). A platform for market intelligence. *International Trade Forum,* (4), 21-21.

De Negri, J. A., & Kubota, L. C. (2008). *Políticas de Incentivo à Inovação Tecnológica*. Brasília: Ipea - Instituto de Pesquisa Econômica e Aplicada. Accessed by http://www.ipea.gov.br/portal/index.php?option=com_content&view=article&id=5569

de Reuver, M., Sørensen, C., & Basole, R. (2017). The digital platform: A research agenda. *Journal of Information Technology*, 1–12. doi:10.105741265-016-0033-3

De Roche. (2016). Five critical aspects to applying data analytics. *Teammate solutions*. Available at http://www.teammatesolutions.com/five-critical-steps-to-applying-data-analytics.aspx

Deakins, D., & Freel, M. (2003). *Entrepreneurship and Small Firms (3rd ed.)*. McGraw-Hill Education.

Dean, J., & Ghemawat, S. (2008). MapReduce: Simplified data processing on large clusters. *Communications of the ACM, 51*(1), 107–113. doi:10.1145/1327452.1327492

Dearlove, D. (1998). *Key management decisions: Tools and techniques of the executive decision-maker* (1st ed.). Wiltshire: Financial Times/Pitman.

Dearlove, D. (1998). *Key management decisions: Tools and techniques of the executive decisionmaker*. Wiltshire, UK: Financial Times/Pitman.

Debelak, D. (2006). *Business models made easy*. Entrepreneur Press.

Deloitte. (2017). *Five questions about applying analytics to risk management*. Available at https://www2.deloitte.com/content/dam/Deloitte/au/Documents/risk/deloitte-au-risk-risk-angles-applying-analytics-risk-management-250215.pdf

DELTCI – Dicionário Electrónico de Terminologia em Ciência da Informação. (n.d.). *Verbete Informação*. Disponível em http://www.ccje.ufes.br/arquivologia/deltci/def.asp?cod=45

Denning, S. (2000). *The Springboard: How Storytelling Ignites Action in Knowledge-era Organizations*. KMCI Press.

Devlin, K. (1999). *Info Sense: turning information into knowledge*. W. H. Freeman and Company.

Dewar, R. D., & Dutton, J. E. (1986). The Adoption of Radical and Incremental Innovations: An Empirical Analysis. *Management Science, 32*(11), 1422–1433. doi:10.1287/mnsc.32.11.1422

Diakantonis, D. (2016, February). Data Matters in Healthcare. *Mergers and Acquisitions*, 27-28.

Dillon, A. (2002). Information architecture in JASIST: Just where did we come from? *Journal of the American Society for Information Science and Technology, 53*(10), 821–823. doi:10.1002/asi.10090

Dimitrios, N. K., Sakas, D. P., & Vlachos, D. S. (2013). The role of information systems in creating strategic leadership model. *Procedia: Social and Behavioral Sciences, 73*, 285–293. doi:10.1016/j.sbspro.2013.02.054

Dinur, A. (2011). *Tacit Knowledge Taxonomy and Transfer: Case Based Research.* Working paper. Institute of behavioral and Applied Management.

Direção-Geral de Estatísticas de Educação e Ciência, & Direção de Serviços de Estatística da Ciência e Tecnologia e da Sociedade de Informação. (2016). *Sumários Estatísticos: CIS 2014 - Inquérito Comunitário à Inovação*. Academic Press.

Dishman, P. L., & Calof, J. L. (2008). Competitive intelligence: A multiphasic precedent to marketing strategy. *European Journal of Marketing, 42*(7/8), 766–785. doi:10.1108/03090560810877141

DoD 5000. (1995). *Modeling and Simulation Master Plan*. Author.

Dodge, J. (2010). Mining for Gold. *Information Management, 10*(5), 38-40.

Dodgson, M., & Gann, D. (2014). *Inovação* (1st ed.). Porto Alegre: L&PM Pocket Encyclopaedia.

Domínguez, S. J. A., & Garrido, J. A. M. (2010). *Information Systems Management in the company - theory and case studies*. Pirámide Publishing. (in Spanish)

Donnelly, J. H., Gibson, J. L., & Ivancevich, J. M. (2000). *Administração: Princípios de gestão empresarial* (10th ed.). Lisboa: McGraw-Hill.

Dosi, G. (1996). The Contribution of Economic Theory in the Understanding of a Knowledge-Based Economy. In Foray & Lundvall (Eds.), Employment and Grow in the Knowledge-Based Economy. OCDE.

Dotzel, T., Shankar, V., & Berry, L. (2013, April). Service Innovativeness and Firm Value. *JMR, Journal of Marketing Research*, *50*(2), 259–276. doi:10.1509/jmr.10.0426

Dou, H. (2004). Benchmarking R&D and companies through patent analysis using free databases and special software: A tool to improve innovative thinking. *World Patent Information*, *26*(4), 297–309. doi:10.1016/j.wpi.2004.03.001

Dou, H., Leveillé, V., Manullang, S., & Dou, J. Jr. (2005). Patent analysis for competitive technical intelligence and innovative thinking. *Data Science Journal*, *4*(31), 209–236. doi:10.2481/dsj.4.209

Drell, L., & Davis, J. (2014, Fall). Getting started with predictive analytics. Marketing Healthcare Services, 23-27.

Dremel, C., Herterich, M. M., Wulf, J., Waizmann, J., & Brenner, W. (2017, June). Digital transformation often requires big data capabilities. *MIS Quarterly Executive*, *16*(2), 81–100.

Dreshler, R., & Smith, K. (2011). Making business model innovation stick. People and Strategy, 34(4), 19-23.

Dresner, S. (2008). *The principles of sustainability* (2nd ed.). Chippenham: Earthscan.

Drucker, P. (2000). *Desafios da gestão para o século 21: Civilização*. Academic Press.

Drucker, P. (1987). *Inovação e gestão* (2nd ed.). Lisboa: Presença.

Drucker, P. F. (2002). *O melhor de Peter Drucker: obra completa. Tradução de Maria L. Leite Rosa, Arlete Simille Marques e Edite Sciulli*. São Paulo: Nobel.

Dubet, F. (1994). *La sociologie de l'expérience*. Paris: Seuil.

Dulken, S. (2000). The future of patent information. *World Patent Information*, *22*(4), 273–275. doi:10.1016/S0172-2190(00)00082-X

Durugbo, C. (2015). Modelling information for collaborative networks. *Production Planning and Control*, *26*(1), 34–52. doi:10.1080/09537287.2013.847217

Dutot, V., Bergeron, F., & Raymond, L. (2014). Information management for the internationalization of SMEs: An exploratory study based on a strategic alignment perspective. *International Journal of Information Management*, *34*(5), 672–681. doi:10.1016/j.ijinfomgt.2014.06.006

Earl, M. (1998). *Information management: The organizational dimension*. New York: Oxford University Press.

Edmunds, A., & Morris, A. (2000). The problem of information overload in business organizations: A review of the literature. *International Journal of Information Management*, *20*(1), 17–28. doi:10.1016/S0268-4012(99)00051-1

Edquist, C. (2005). Systems of innovation: Perspectives and challenges. The Oxford Handbook of Innovation, 181–208. doi:10.1093/oxfordhb/9780199286805.003.0007

Edvinsson, L, & Malone, M. (1998). *Capital Intelectual: Descobrindo o Valor Real de sua Empresa pela Identificação de seus Valores Internos*. S. Paulo: Makron Books.

Edwards, D. W. (1986). *Out of the Crisis*. Cambridge, MA: Massachusetts Institute of Technology, Center for Advanced Engineering Study.

Eisenhardt, K. (1989). Building theories from case study research. *Academy of Management Review*, *14*(4), 532–550. doi:10.5465/AMR.1989.4308385

Eisenhardt, K. M., & Santos, F. M. (2002). Knowledge-Based View: A New Theory of Strategy? In A. Pettigrew, H. Thomas, & R. Whittington (Eds.), *Handbook of Strategy and Management*. Sage Publications.

Eisenhardt, K., & Graebner, M. (2007). Theory building from cases: Opportunities and challenges. *Academy of Management Journal*, *50*(1), 25–32. doi:10.5465/AMJ.2007.24160888

El-Bashir, M. Z., Collier, P., & Sutton, S. G. (2011). The Role of Organizational Absorptive Capacity in Strategic Use of Business Intelligence to Support Integrated Management Control Systems. *The Accounting Review*, *86*(1), 155–184. doi:10.2308/accr.00000010

El-Gayar, O., & Timsina, P. (2014). Opportunities for Business Intelligence and Big Data Analytics In Evidence Based Medicine. *Annals of 47th Hawaii International Conference on System Science*. 10.1109/HICSS.2014.100

Empson, L. (2000). The Challenge of Knowledge Management, The Strategy Domain. *Economic Daily – Financial Times*, *2*(2294), 32. (in Portuguese)

Ênfase em Eficiência Energética. (n.d.). Disponível em: http://www.inf.ufpr.br/aldri/disc/artigos/Relatorio_Tecnico_Mauricio_Barros.pdf/

Eppler, M. J. (2006). Managing information quality: increasing the value of information in knowledge-intensive products and processes (2nd ed.). Heidelberg, Germany: Springer.

Eppler, M. (2006). *Managing information quality: Increasing the value of knowledge-intensive products and processes* (2nd ed.). New York: Springer.

Erichsen, M., & Borges, N. (1995). A informação como recurso gerencial das organizações na sociedade do conhecimento. *Ciência da Informação*, *24*(2), 1–15. Retrieved from http://revista.ibict.br/index.php/ciinf/article/viewArticle/551

Ericsson Consumer Lab. (2015). *Connected Homes. Mobile Business trends*. Retrieved from URL: http://www.ericsson.com/res/docs/2015/mobility-report/emr-mobile-business-trends-2015.pdf

Ernst, H. (2003). Patent information for strategic technology management. *World Patent Information*, *25*(3), 233–242. doi:10.1016/S0172-2190(03)00077-2

Ester, M., Kriegel, H.-P., Sander, J., & Xu, X. (1996). *A Density-Based Algorithm for Discovering Clusters in Large Spatial Databases with Noise*. AAAI Press.

Esteves. (2015). *Tese Mestrado A Internet das Coisas: Avaliação do grau de aceitação da tecnologia RFID pelo cidadão comum UCP*. Available at http://repositorio.ucp.pt/bitstream/10400.14/19429/1/Andr%C3%A9%20Gil%20Capela%20Esteves_355413001_Tese%20Mestrado%20Gest%C3%A3o_A%20Internet%20das%20Coisas-avalia%C3%A7%C3%A3o%20do%20Grau%20de%20Aceita%C3%A7%C3%A3o%20da%20tecnologi~1.pdf

Esty, D., & Winston, A. (2008). *Do verde ao ouro: Como empresas inteligentes usam a estratégia ambiental para inovar, criar valor e construir uma vantagem competitiva*. Cruz Quebrada: Casa das Letras.

Etzkowitz, H. (1993). Technology transfer: The second academic revolution. *Technology Acess Report*, *6*, 7–9.

Etzkowitz, H., & Leydesdorff, L. (2000). The dynamics of innovation: From National Systems and "' Mode 2 '" to a Triple Helix of university – industry – government relations. *Science And Technology*, *29*(2), 109–123. doi:10.1016/S0048-7333(99)00055-4

Etzkowitz, H., Webster, A., & Healey, P. (1998). Introduction. In *Capitalizing Knowledge*. Albany, NY: State University of New York Press.

European Commission. (2014). *Supporting the Internationalisation of SMEs*. Retrieved from http://ec.europa.eu/enterprise/policies/sme/files/support_measures/internationalisation/report_internat_en.pdf

European Environment Agency. (2005). Copenhagen: The European Environment - State and Outlook.

Eurostat. (n.d.). Disponível em: http://ec.europa.eu/eurostat/web/science-technology-innovation/data/database

Evaluation of Restful Web Services in Raspberry Pi. (n.d.). Available at https://www.researchgate.net/publication/283865130_Performance_and_energy_evaluation_of_RESTful_web_services_in_Raspberry_Pi

Eychenne, F., & Neves, H. (2013). *Fab Lab: A Vanguarda da Nova Revolução Industrial*. Associação Fab Lab Brasil.

Farias, Andrade, Freitas, & Junior. (2013). *Metodologias para mensurar o impacto da inovação nas atividades empresariais*. Anprotec, XXIII Seminário Nacional de Parques Tecnológicos e Incubadora de Empresas. Recife, Pernambuco.

Faulkner, D., & Bowman, C. (1995). *The essence of competitive strategy* (1st ed.). Exeter: Prentice Hall.

Fávero. (2009). *Luiz Paulo; Belfiore, Patrícia; Silva, Fabiana Lopes da; e Chan, Betty Lilian. Análise de dados: modelagem multivariada para tomada de decisões*. Rio de Janeiro: Editora Elsevier.

Fayyad, U., Piatetsky-Shapiro, G., & Smyth, P. (1996). The KDD process for extracting useful knowledge from volumes of data. *Communications of the ACM*, *39*(11), 27–34. doi:10.1145/240455.240464

Fedorowicz, J., & Sawyer, S. (2012). *Designing collaborative networks: Lessons learned from public safety. IBM Center for The Business of Government: Collaborating Across Boundaries Series*. Washington, DC: IBM Center for The Business of Government.

Feinson, S. (2003). National Innovation Systems Overview and Country Cases. *Knowledge Flows and Knowledge Collectives: Understanding The Role of Science and Technology Policies in Development*, 13–38. 10.1787/9789264239012-en

Ferreira, A. B. H. (2002). *Novo Aurélio século XXI*. São Paulo: Nova Fronteira.

Ferrel, O. C., & Hartline, M. (2010). *Marketing Strategy*. South Western College Publications.

Fielding. (2000). *Architectural Styles and the Design of Network-based Software Architectures* (Tese de Doutorado). University of California, Irvine, CA. Available at: http://www.ics.uci.edu/~fielding/pubs/dissertation/rest_arch_style.htm

FierceWireless. (2015). *Wireless industry revenue*. Retrieved from URL: http://www.fiercewireless.com/story/moodys-wireless-industry-revenue-likely-grow-3-4-2016/2015-12-02

Figueiredo, T. A., & Pepe, V. L. E. (2016). *A seleção de medicamentos estratégicos para a produção nacional e a saúde coletiva. Saúde Amanhã. Textos para Discussão; n. 21*. Fundação Oswaldo Cruz.

Folkens, J. (n.d.). *Building a gateway to the Internet of Things*. Available at: http://www.ti.com/lit/wp/spmy013/spmy013.pdf

Fonseca, J. J. S. (2002). *Metodologia da pesquisa científica*. Fortaleza: UEC.

Forrester. (2015). *World Mobile Application Spending Forecast, 2015 To 2020 (Global)*. Retrieved from URL: https://www.forrester.com/report/Forrester+Research+World+Mobile+Application+ Spending+Forecast+2015+To+2020+Global/-/E-RES122413

Francke, H. (2009). Towards an architectural document analysis. *Journal of Information Architecture*, *1*(1). Accessed at http://journalofia.org/volume1/issue1/03-francke/jofia-0101-03-francke.pdf

Freeman, C. (1987). *Technology and Economic Performance: Lessons from Japan*. London: Pinter Publishers.

Freire, A. (1997). *Strategy*. Lisbon: Verbo Publishing. (in Portuguese)

Fries, L. (2008). Boundarians that shape market actors: A review of the literature on business associations. In *24th IMP-conference* (pp. 1–17). Academic Press.

Fuld, L. M. (1994). *The new competitor intelligence: the complete resource for finding, analyzing and using information about your competitors*. John Wiley & Sons.

Fundação para a Ciência e a Tecnologia. (2013). *Diagnóstico do Sistema de Investigação e Inovação: desafios, forças e fraquezas rumo e 2020*. Author.

Gabinete de Planeamento Estratégico e Participação Empresariais. (2017). *Universidade do Porto - Relatório de Atividades e Contas - Ano 2016*. Author.

Gadelha, C. A. G., & Costa, L. S. (2012a). Saúde e desenvolvimento no Brasil: avanços e desafios. *Revista de Saúde Pública*. doi:10.1590/S0034-89102012005000062

Gadelha, C. A. G., Costa, L. S., & Maldonado, J. (2012). O Complexo Econômico-Industrial da Saúde e a dimensão social e econômica do desenvolvimento. Revista de Saúde Pública.

Gadelha, C. A. G., Costa, L. S., & Maldonado, J. (2012b). O Complexo Econômico-Industrial da Saúde e a dimensão social e econômica do desenvolvimento. *Revista de Saúde Pública*. doi:10.1590/S0034-89102012005000065

Gadelha, P., Carvalho, J. N., & Pereira, T. R. (2012c). *A Saúde no Brasil em 2030: diretrizes para a prospecção estratégica do sistema de saúde brasileiro*. Editora FIOCRUZ. Accessed by https://portal.fiocruz.br/sites/portal.fiocruz.br/files/documentos/capa_saude_brasil_2030.pdf

Gadelha, C. A. G (2015). Reflexões sobre a saúde na agenda contemporânea do desenvolvimento: Saúde, desenvolvimento e inovação. Rio de Janeiro. *Cepesc*, *1*, 43–69.

Gadelha, C. A. G., & Costa, L. (2007). Integração de fronteiras: A saúde no contexto de uma política nacional de desenvolvimento. *Cadernos de Saude Publica*, *23*(suppl 2), S214–S226. doi:10.1590/S0102-311X2007001400011 PMID:17625648

Gadelha, C. A. G., Quental, C., & Fialho, B. de C. (2003). Saúde e inovação: Uma abordagem sistêmica das indústrias da saúde. *Cadernos de Saude Publica*, *19*(1), 47–59. doi:10.1590/S0102-311X2003000100006 PMID:12700783

Gadelha, P., Carvalho, J. N., & Pereira, T. R. (2012). *A Saúde no Brasil em 2030: diretrizes para a prospecção estratégica do sistema de saúde brasileiro*. Editora FIOCRUZ.

Gago, M. (1994). *Prospectiva do ensino superior em Portugal*. Lisboa: Ministério da Educação.

Gal, Y. (2004). The reward effect: A case study of failing to manage knowledge. *Journal of Knowledge Management*, *8*(2), 73–83. doi:10.1108/13673270410529127

Gammu. (n.d.). *SMSD*. Available at: https://pt-br.wammu.eu/smsd/

Gartner. (2012). *IT Glossary*. Retrieved from Gartner: http://www.gartner.com/it-glossary/teledensity/

Garvin, D. (1988). *Managing quality*. New York: Free Press.

Gashi, P., Hashi, I., & Pugh, G. (2013). Export behaviour of SMEs in transition countries. *Small Business Economics*, *42*(1), 407–435. doi:10.100711187-013-9487-7

Gasik, S. (2011). A model of project knowledge management. *Project Management Journal*, *42*(April), 23–44. doi:10.1002/pmj.20239

Gatlin, L. (1972). *Information theory and the living system*. Columbia University Press.

Gavirneni, S., Kapuscinski, R., & Tayur, S. (1999). Value of Information in Capacitated Supply Chains. *Management Science*, *45*(1), 16–24. doi:10.1287/mnsc.45.1.16

Gawer, A. (2009). *Platforms, markets and innovation*. Edward Elgar Publishing. doi:10.4337/9781849803311

Gershenfeld, N. (2005). *Fab: The Coming Revolution on Your Desktop: from Personal Computers to Personal Fabrication*. New York: Basic Books.

Gershenfeld, N. (2012). How to Make Almost Anything: The Digital Fabrication Revolution. *Foreign Affairs*, *91*(6).

Ghazawneh, A., & Henfridsson, O. (2015). A paradigmatic analysis of digital application marketplaces. *Journal of Information Technology*, *30*(3), 198–208. doi:10.1057/jit.2015.16

Ghemawat, P. (2006). *Strategy and the business landscape* (2nd ed.). Boston: Prentice Hall.

Gibson, R., & Skarzynski, P. (2008). Innovation to the core - a blueprint for transforming the way your company innovates. Harvard Business Press.

Gillespie, G. (2014). How reporting and analytics and metrics affect outcomes and save lives. *Information & Management*, A2–A5.

Ginman, M. (1990). Quality information, and information for quality. In Wormell (Ed.), Information quality: Definitions and dimensions. London: Taylor Graham.

Giudici, A., & Blackburn, R. (2013). *Digitally-enabled network intermediation in the internationalization of SMEs: A dynamic capabilities perspective*. Nemode.Ac.Uk.

Glatzel, C., & Röhren, J. (2014). *Excellence in Supply Chain Management*. Mckinsey & Company. Retrieved December 08, 2015 from http://www.mckinsey.com/~/media/McKinsey/dotcom/customer_service/Operations/PDFS/Supply%20Chain%20Management%20Compendium%202014%20Edition%20-%20Final.ashx

Gleick, J. (2011). The information: A history, a theory, a flood (1st ed.). St. Ives: 4th Estate.

Gluhak, A., Krco, S., Nati, M., & Pfsterer, D. (2011). A Survey on Facilities for Experimental Internet of Things Research. *IEEE Communications Magazine*, *49*(11), 58–67.

Gnyawali, D. R., & Madhavan, R. (2001, July). Cooperative Networks And Competitive Dynamics: A Structural Embeddedness Perspective. *Academy of Management Review*, *26*(3), 431–445.

Godeluck, S. (2000). *A explosão da economia na internet*. Lisboa: Livros do Brasil.

Godinho, M. (2003). *Estudo sobre a utilização da propriedade industrial em Portugal* (Vol. 1). Lisboa: Instituto Nacional da Propriedade Industrial.

Godinho, M. (2007). Indicadores de C&T, inovação e conhecimento: Onde estamos? Para onde vamos? *Analise Social*, *42*(182), 239–274.

Goffman, W. (1970). Information Science: Discipline or disappearance. *Aslib Proceedings*, *22*(12), 589–596. doi:10.1108/eb050268

Golden-Biddle, K., & Locke, K. (2007). *Composing qualitative research* (Vol. 2). Thousand Oaks, CA: Sage Publications. doi:10.4135/9781412983709

Gomes, E., & Braga, F. (2001). *Inteligência competitiva: Como transformar informação em um negócio lucrativo*. Rio de Janeiro: Campus.

Gomes, E., & Braga, F. (2002). *Competitive Intelligence: How to turn information into a lucrative business*. São Paulo: Campus. (in Portuguese)

Gonçalves, A. (2006). Problema de Saúde Pública: Caracterizando e avaliando aplicações. *Revista Brasileira de Epidemiologia*, *9*(2), 253–255. doi:10.1590/S1415-790X2006000200013

Gonçalves, F. (2017). *U. InovAcelerator: um modelo de indicadores de inovação em contexto académico (Dissertação de Mestrado)*. Universidade do Porto.

Gonzalez, R. V. D., & Martins, M. F. (2014). Mapping the organizational factors that support knowledge management in the Brazilian automotive industry. *false*. *Journal of Knowledge Management*, *18*(1), 152–176. doi:10.1108/JKM-08-2013-0300

Google. (2017). *Google Analytics*. Available at https://www.google.com/analytics/

Grant. (1991). The Resource-Based Theory of Competitive Advantage: Implications for Strategy Formulation. *Californian Management Review*, *33*(3), 114-125.

Grant, R. (1991). The Resource-based Theory of Competitive Advantage: Implications for Strategy Formulation. *California Management Review*, *33*(3), 114–135. doi:10.2307/41166664

Grant, R. (2002). *Contemporary strategy analysis: concepts, techniques, applications*. Wiley-Blackwell.

Grant, R. M. (2002). *Contemporary strategy analysis: Concepts, techniques, applications*. Oxford, UK: Wiley-Blackwell.

Greif, S. (1987). Patents and economic growth. *International Review of Industrial Property and Copyright Law*, *18*(2), 191–213.

Grupp, H., & Mogee, M. E. (2004). Indicators for national science and technology policy: How robust are composite indicators? *Research Policy*, *33*(9), 1373–1384. doi:10.1016/j.respol.2004.09.007

Guarda, T., Santos, M., Pinto, F., Augusto, M., & Silva, C. (2013). Business Intelligence as a Competitive Advantage for SME's. *Internacional Journal of Trade. Economics and Finance*, *4*(4), 187–190.

Gubbi, J., Buyya, R., & Marusic, S. (2013). Internet of Things (IoT): A vision, architectural elements, and future. *Future Generation Computer Systems*, *29*(7). Available at: http://ac.els-cdn.com/S0167739X13000241/1-s2.0-S0167739X13000241-main.pdf?_tid=68615f94-c2f6-11e6-856400000aacb361&acdnat=1481827605_39fd450812ee030ddd19162300d8d3f

Gubbi, J., Buyya, R., Marusic, S., & Palaniswami, M. (2013). Internet of Things (IoT): A vision, architectural elements, and future directions. *Future Generation Computer Systems*, *29*(7), 1645–1660. doi:10.1016/j.future.2013.01.010

Guerreiro, J. (2005). As funções da universidade no âmbito dos sistemas de inovação. In *Estudos II* (pp. 131–148). Faculdade de Economia da Universidade do Algarve.

Guinchat, C., & Menou, M. (1994). Introdução geral às ciências e técnicas da informação e documentação (2nd ed.). Tradução de Miriam Vieira da Cunha. Brasília: IBICT.

Gupta, P. (2008). *Inovação empresarial no século 21*. Porto: Vida Económica.

Haberman, M. (2001). The role of intelectual property and patent information in successful innovation, production and marketing: Case study 1: The non-spill drinking vessel. *World Patent Information*, *23*(1), 71–73. doi:10.1016/S0172-2190(00)00105-8

Haddad, P. R. (2010). Quatro gerações de indicadores. *Estadão*. Retrieved from http://economia.estadao.com.br/noticias/geral,quatro-geracoes-de-indicadores-imp-,556824

Hagedoorn, J., & Cloodt, M. (2003). Measuring innovative performance: Is there an advantage in using multiple indicators? *Research Policy*, *32*(8), 1365–1379. doi:10.1016/S0048-7333(02)00137-3

HAIR. (2005). *Joseph F. et al. Análise multivariada de dados. 6ª edição*. Porto Alegre: Bookman.

Hammer, M. (2001). The Superefficient Company. *Harvard Business Review*, *79*(8), 82–91. PMID:11550633

Hampton, J. (2011). *SEMMA and CRISP-DM: Data mining methodologies*. Available at http://jesshampton.com/2011/02/16/semma-and-crisp-dm-data-mining-methodologies/

Handfield, R., Straube, F., Pfohl, H.C. & Wieland, A. (2013). *Trends and Strategies in Logistics and Supply Chain Management - Embracing Global Logistics Complexity to Drive Market Advantage*. Bremen.

Hansen, T. H., Noria, N., & Tierney, T. (2001). What is your strategy for knowledge management? - Organizational Learning. Campus Publishing.

Hargadon, A., & Sutton, R. (2002). Como Construir uma Fábrica de Inovação. In H. B. Review (Ed.), *Inovação na Prática: On Innovation* (pp. 59–79). Rio de Janeiro: Campus.

Harrison, A., & Hoek, R. (2003). *Estratégia e gerenciamento de logística*. Editora Futura.

Hartz, Z. M. de A. (1997). *Avaliação em saúde: dos modelos conceituais à prática na análise da implantação de programas*. Publisher FIOCRUZ. Accessed by de http://books.scielo.org/id/3zcft

Hartz, Z. M. de A. (2012). *Avaliação em saúde: dos modelos conceituais à prática na análise da implantação de programas*. Editora FIOCRUZ. Recuperado de http://books.scielo.org/id/3zcft

Hartz, Z. (2013, May). From monitoring of performance to the performance of monitoring: New opportunities for evaluation in the management of Health Surveillance. *Ciência & Saúde Coletiva*, *18*(5), 1221–1222. doi:10.1590/S1413-81232013000500005 PMID:23670448

Harvard Business Review (Ed.). (2001). *Gestão do Conhecimento: On Knowledge Management*. Rio de Janeiro: Campus.

Harvard Business Review (Ed.). (2002). *Inovação na Prática: On Innovation*. Rio de Janeiro: Campus.

Haslinda, A. & Sarinah, A. (2009). A review of knowledge management models. *Journal of International Social Research*, *2*(9), 187-198.

Haslinda, A., & Sarinah. (2009). A review of knowledge management models. *The Journal of International Social Research*, *2*(9), 187-198.

Hassouna, M., Tarhini, A., Elyas, T., & AbouTrab, M. S. (2015). Customer Churn in Mobile Markets: A Comparison of Techniques *International Business Research*, *8*(6). doi:10.5539/ibr.v8n6p224

Hatch, M. (2013). *The Maker Movement Manifesto: Rules for Innovation in the New World of Crafters, Hackers, and Tinkerers* (1st ed.). McGraw-Hill Education.

Hausenblas, M. (n.d.). *Key Requirements for an IoT Data Platform*. Available at: https://www.mapr.com/blog/key-requirements-iot-data-platform

Hax & Wilde. (2001). *The Delta Project. Discovering New Sources of Profitability in a Network Economy*. Palgrave.

Heavin, C., & Adam, F. (2011). Optimizing customers as knowledge resources and recipients: Cases in Small to Medium Sized Software Enterprises. In M. M. Al Shammari (Ed.), *Customer-centric Knowledge management concepts and applications*. Information Science Reference.

Heitor, M., Bóia, M. J., Conceição, P., & Beira, E. (2004). *Mapear Conhecimento e Inovação em Portugal: Uma proposta para um sistema de indicadores e um programa de observação*. Academic Press.

Henderson, B., & Venkatraman, N. (1999). Strategic alignment: Leveraging information technology for transforming organizations. *IBM Systems Journal*, *32*(1), 472–484. doi:10.1147j.382.0472

Hessels, J., & Terjesen, S. (2010). Resource dependency and institutional theory perspectives on direct and indirect export choices. *Small Business Economics*, *34*(2), 203–220. doi:10.100711187-008-9156-4

Hess, P., & Siciliano, J. (1996). *Management: Responsibility for performance*. New York: McGraw-Hill.

Hildebrandt, D. (1997). *Internet tools of the profession: a guide for information professionals*. Chicago: Special Libraries Association.

Hill, J. (2003). *System Architecture for Wireless Sensor Networks. Spring 2003* (Vol. 1). Berkeley, CA: University of California.

Hinton, M. (2006). *Introducing information management: The business approach*. Burlington, MA: Elsevier Butterworth-Heinemann.

Hirschhorn, L. (2002). Campaigning for Change. *Harvard Business Review*, *80*(7), 98–104.

Hitt, M., Ireland, R. D., & Hoskisson, R. E. (2012). *Strategic management: concepts, competitiveness and globalization*. Boston: Cencage Learning.

Hoffman, L. (2013). Looking back at the big data. *Communications of the ACM*, *56*(4), 21–23. doi:10.1145/2436256.2436263

Homem, M. L. (2000). *Necessidade e desejo*. Disponível na Internet no endereço: www.psico.org.br

Howard, R.A. (1966). Information value theory. *IEEE Transactions on Systems Science and Cybernetics*, 22-26.

HP. (2015). *Strategic Focus: Software, Products & Services*. Retrieved from URL: http://www8.hp.com/us/en/hp-news/press-release.html?id=1815969#.VxDdk_l97IV

Hsu, W., Chen, H., & Cheng, C. (2013). Internationalization and firm performance of SMEs: The moderating effects of CEO attributes. *Journal of World Business*, *48*(1), 1–12. doi:10.1016/j.jwb.2012.06.001

Hu, M., & Liu, B. (2004). Mining and summarizing customer reviews. In *Proceedings of the tenth ACM SIGKDD international conference on Knowledge discovery and data mining (KDD '04)*. ACM. Retrieved from http://doi.acm.org/10.1145/1014052.1014073

Huang, K., Lee, Y., & Wang, R. (1998). *Quality information and knowledge*. Upper Saddle River, NJ: Prentice Hall.

Huber, G. P. (1990). A theory of the effects of advanced information technologies on organizational design, intelligence and decision making. *Academy of Management Review*, *15*(1), 47–71.

Hunsicker, J. (1989, Summer). Strategies for European survival. *The McKinsey Quarterly*, 37–47.

Hwang, V., & Horowitt, G. (2012). *The Rainforest: The Secret to Building the Next Silicon Valley*. Los Altos Hills: Regenwald.

IA Institute – Information Architecture Institute. (2014). *Recommended reading*. Accessed at http://iainstitute.org/en/learn/education/recommended_reading.php

IAB – Instituto de Arquitetos do Brasil. (2014). Accessed at http://www.iab.org

IBEF. (2016). *Telecom Sector in India*. Retrieved from URL: http://www.ibef.org/industry/telecommunications.aspx

IBGE. (2013). *Instituto Brasileiro de Geografia e Estatística*. Brazil: IBGE. Disponível em: http://www.ibge.gov.br/home/estatistica/populacao/tabuadevida/2010/

IBM Software Report. (2013). *Minimize customer churn with analytics*. IBM.

IBM. (n.d.). *Node-RED*. Available at: https://developer.ibm.com/open/openprojects/node-red/

Idris, K. (2003). *Intellectual property: A power tool for economic growth*. Geneva: World Intellectual Property Organization.

IMS Institute for Healthcare Informatics. (2012). *The Global Use of Medicines: outlook through 2016*. Recuperado de www.imshealth.com

INE. (n.d.). Disponível em: https://www.ine.pt/xportal/xmain?xpid=INE&xpgid=ine_base_dados

Inmon, B., Strauss, D., & Neushloss, G. (2008). *DW 2.0: The Architecture for the Next Generation of Data Warehousing*. Morgan Kaufmann.

INOVA+ - Serviços de Consultadoria em Inovação Tecnológica S. A. (2007). *Conceitos e Modelos de Inovação* (1st ed.). AEP - Associação Empresarial de Portugal.

International Telecommunication Union. (2005). *ITU internet reports 2005: The internet of things*. Available at: https://www.itu.int/osg/spu/publications/internetofthings/

IPEA – Instituto de Pesquisa Econômica Aplicada. (2014). *Oportunidades e Desafios para a Engenharia Consultiva no Brasil: Infraestrutura de Transporte*. Retrieved December 08, 2015 from http://www.ipea.gov.br/portal/index.php?option=com_content&view=article&id=21095

Isaacson, W. (2014). Os Inovadores: Uma biografia da revolução digital. S. Paulo: Companhia das Letras.

Isaacson, W. (2011). *Steve Jobs* (1st ed.). New York: Simon & Schuster.

Itami, H. (1987). *Mobilizing Invisible Assets. Massachusetts*. Cambridge, MA: Harvard University Press.

Jacobson, A., & Prusak, L. (2006). The cost of knowledge. *Harvard Business Review*. Accessed at https://hbr.org/2006/11/the-cost-of-knowledge

Jain, C. L. (2016). How to Use Big Data and Predictive Analytics to Improve the Success of New Products. *Review of Business*, *37*(1), 48-55.

Jain, C. L. (2016). How to use big data and predictive analytics to improve success of new products. *Review of Business*, *37*(1), 48–55.

Jamil, G. L. (2013). Approaching Market Intelligence concept through a case analysis: Continuous knowledge for marketing strategic management and its complementarity to competitive intelligence. *Annals of Centeris 2013 – Conference on Technology and Information Systems*.

Jamil, G. L. (2014). Inteligência de Mercado como um processo de gestão da informação e do conhecimento: proposta de oficinas de capacitação setoriais. Post-doctoral Project at Faculdade de Letras from Universidade do Porto, Porto University, Portugal.

Jamil, G. L. (2014b). Inteligência de Mercado como um processo de gestão da informação e do conhecimento: proposta de oficinas de capacitação setoriais. Post-doctoral Project at Faculdade de Letras from Universidade do Porto, Porto University, Portugal.

Jamil, G. L. (2001). *Repensando a Tecnologia da informação na empresa moderna*. Rio de Janeiro: Axcel Books do Brasil.

Jamil, G. L. (2001). *Repensando a TI na empresa moderna*. Rio de Janeiro: Axcel Books do Brasil.

Jamil, G. L. (2005). *Gestão da Informação e do conhecimento em empresas brasileiras: estudo de múltiplos casos*. Belo Horizonte: Ed. Con / Art.

Jamil, G. L. (2013). Approaching Market Intelligence concept through a case analysis: Continuous knowledge for marketing strategic management and its complementarity to competitive intelligence. *Annals of Centeris 2013 – Conference on Technology and Information Systems*.

Jamil, G. L. (2013). Sistemas de informação e a tomada de decisões: estudos de caso sobre a inteligência de mercado e a gestão de marketing. In P. Campos & P. Q. Brito (Eds.), *Novas tendências em Marketing Intelligence*. Ed. Actual.

Jamil, G. L. (2014a). Why quality? Why value? Is information related to these aspects? In G. L. Jamil (Ed.), *Rethinking the conceptual base for new practical applications of information quality and value*. Hershey, PA: IGI Global. doi:10.4018/978-1-4666-4562-2.ch001

Jamil, G. L. (2015). Market intelligence. In *Encyclopedia of information Science and Technology*. Hershey, PA: IGI Global. doi:10.4018/978-1-4666-5888-2.ch508

Jamil, G. L., Jamil, L. C., Vieira, A. A. P., & Xavier, A. J. D. (2015). Challenges in modelling Healthcare services: a study case of information architecture perspectives. In G. L. Jamil (Ed.), *Handbook of research on information architecture and management in modern organizations*. Hershey, PA: IGI-Global.

Jamil, G. L., & Magalhães, L. F. C. (2015). Perspectives for big data analysis for knowledge generation in project management contexts. In G. L. Jamil (Ed.), *Handbook of research on effective project management research through the integration of knowledge and innovation*. Hershey, PA: IGI Global. doi:10.4018/978-1-4666-7536-0.ch001

Jamil, G. L., Santos, L. H. R., Alves, M. L., & Furbino, L. (2012). *A design framework for a market intelligence system for healthcare sector: a support decision tool in an emergent economy. Chapter approved to appear in Handbook on research of ICTs for Social Services and Healthcare: developments and applications*. Hershey, PA: IGI Publishing.

Jamil, G. L., Santos, L. H. R., Alves, M. L., & Furbino, L. (2012). *A design framework for a market intelligence system for healthcare sector: a support decision tool in an emergent economy. In Handbook on research of ICTs for Social Services and Healthcare: developments and applications*. Hershey, PA: IGI Publishing.

Janger, J., Schubert, T., Andries, P., Rammer, C., & Hoskens, M. (2017). The EU 2020 innovation indicator: A step forward in measuring innovation outputs and outcomes? *Research Policy*, *46*(1), 30–42. doi:10.1016/j.respol.2016.10.001

Jaziri-Bouagina, D. (2017). The Tacit Knowledge Through the Customer Experience: Conceptualization, Externalization Methods, and Use Application to Tunisian Thalassotherapy Centers. In Handbook of Research on Tacit Knowledge Management for Organizational Success. Academic Press.

Jaziri-Bouagina, D., & Triki, A. (2013). The customer experiential knowledge management (CEKM) - concept proposition and research framework development. *Proceedings of the 14th European Conference on Knowledge Management ECKM 2013, 1*, 307-314.

Jaziri-Bouagina, D., & Jamil, G. L. (2017). *Handbook of Research on Tacit Knowledge Management for Organizational Success*. Hershey, PA: IGI Global. doi:10.4018/978-1-5225-2394-9

Jeston, J., & Nellis, J. (2007). *Business Process Management: practical guide to successful implementations* (2nd ed.). Butterworth-Heinemann.

Johnson, A., & Jacobsson, S. (2003). The emergence of growth industry: a comparative analysis of the German, Dutch and Swedish wind turbine industries. *Change, Transformation and Development*, (1), 197–227. 10.1007/978-3-7908-2720-0_12

Johnson, J. E. (2012, July). Big data + Big Analytics + Big opportunity. *Financial & Executive*, 51-53.

Jolly, A. (2003). *Innovation: Harnessing creativity for business growth*. New York: Kogan Page.

Jorge, M. (1995). *Biologia, informação e conhecimento*. Lisboa: F. C. Gulbenkian.

Jose, B., Ramanan, T. R., & Kumar, S. D. M. (2017). Big data provenance and analytics in telecom contact centers. *Region 10 Conference TENCON 2017 - 2017 IEEE*, 1573-1578.

Junior. (n.d.). *Instalando o Raspbian no Raspberry Pi*. Available at: http://www.jairjunior.eng.br/artigos/instalando-o-raspbian-no-raspberry-pi/

Kahaner, L. (1997). *Competitive intelligence: How to gather, analyze, and use information to move your business to the top*. New York: Simon & Schuster.

Kahaner, L. (1998). *Competitive Intelligence: How to gather, analyse, and use Information to move your business to the top*. New York: Touchstone Books.

Kahney, L. (2009). *A cabeça de Steve Jobs* (2nd ed.). Rio de Janeiro: Agir.

Kahney, L. (2013). *Jony Ive: O gênio por trás dos grandes produtos da Apple*. S. Paulo: Portfolio-Penguin.

Kalseth, K. (1991). Business information strategy: The strategic use of information and knowledge. *Information Services & Use*, *11*(3), 155–164. doi:10.3233/ISU-1991-11307

Kane, H., Ragsdell, G., & Oppenheim, C. (2006). Knowledge Management Methodologies. *Electronic Journal of Knowledge Management*, *4*(2), 141–152.

Kanstren, T., Liikka, J., Mäkelä, J., Luoto, M., & Prokkola, J. (2016). Preliminary big data in a 5G test network. *Big Data (Big Data) IEEE International Conference on*, 2722-2727.

Kaplan, R., & Norton, D. (2007). *The execution premium: linking strategies to operation for competitive advantage*. Harvard Business School Press.

Karnani, A. G. (2000). Five Strategies for Growth and Value Creation, The Field of Strategy. *Diário Económico – Financial Times*, *5*(2297), 76 – 80.

Kayano & Caldas. (2002). *Indicadores para o Diálogo*. GT Indicadores. Plataforma Contrapartes Novib.

Kearns, G. S., & Lederer, A. L. (2003). A resource based view of IT alignment: How knowledge sharing creates a competitive advantage. *Decision Sciences*, *34*(1), 1–29. doi:10.1111/1540-5915.02289

Kehrle, L. & Moutinho, L. (2005). Competitividade presente e esperada de arranjos produtivos de calçados na Paraíba. R. *Econ. contemp.*, *9*(3), 671-697.

Keller, K. L. (2006). *Administração de marketing*. *12* (p. 750). São Paulo: Pearson Prentice Hall.

Kettinger, W., & Marchand, D. (2005). Leveraging information locally and globally: the right mix for flexibility and standardization. *IMD Bulletin*. Available at https://www.imd.org/research/publications/upload/Marchand_Kettinger_WP_2005_02_Level_1.pdf

Kettinger, W., Zhang, C., & Marchand, D. (2011). CIO and Business executive leadership approaching to establishing company-wide information orientation. *MIS Quarterly Executive*, *10*(4), 157–174.

Keyworth, B. & Kirk, T. (1998). *Changing Perspectives on Change Management*. GartnerGroup, COM-03-3534.

Khanagha, S., Volberda, H., & Oshri, I. (2014) Business model renewal and ambidexterity: structural alteration and strategy formation process during transition to a Cloud business model. Research and Development Management, 44(3), 322-240. doi:10.1111/radm.12070

Kimball, R., & Ross, M. (2010). *Relentlessly Practical Tools for Data Warehousing and Business Intelligence*. John Wiley and sons.

Kim, W. C., & Mauborgne, R. (2015). *Blue Ocean Strategy: Expanded Edition*. Harvard Business Review Press.

King, W., Grover, V., & Hufnagel, E. (1989). Using information and information technology for sustainable competitive advantage: Some empirical evidence. *Information & Management*, *17*(2), 87–93. doi:10.1016/0378-7206(89)90010-4

Kira, T. A. R. A. P. A. N. O. F. F. (org.). (2015). Analise da Informacao para Tomada de Decisao Desafios e Solucoes (Vol. 1). Brasil: Editora Intersaberes. Recuperado de http://www.estantevirtual.com.br/b/kira-tarapanoff/analise-da-informacao-para-tomada-de-decisao-desafios-e-solucoes/158550962

Kleinfeld, R., Doukas, C., & Radziwonowicz, L. (2014). *Glue.things – a Mashup Platform for wiring the Internet of Things with the Internet of Services*. Research Gate.

Kleinknecht, A., Van Montfort, K., & Brouwer, E. (2002). The Non-Trivial Choice between Innovation Indicators. *Economics of Innovation and New Technology*, *11*(2), 109–121. doi:10.1080/10438590210899

Knight, G., & Liesch, P. (2016). Internationalization: From incremental to born global. *Journal of World Business*, *51*(1), 93–102. doi:10.1016/j.jwb.2015.08.011

Koch, A. (1991). Patent information to stimulate innovation in small and medium sized companies. *World Patent Information*, *13*(4), 201–205. doi:10.1016/0172-2190(91)90193-9

Koelen, M. A. (2001). Health promotion research: Dilemmas and challenges. *Journal of Epidemiology and Community Health*, *55*(4), 257–262. doi:10.1136/jech.55.4.257 PMID:11238581

Kostopoulos, C. K., Spanos, Y. E., & Prastacos, G. P. (2002). The Resource – Based View of the Firm and Innovation: Identification of Critical Linkages. *The 2nd European Academy of Management Conference*. Retrieved from http://ecsocman.hse.ru/data/165/663/1219/rb_view.pdf

Kotler, P. (1978). *Marketing para organizações que não visam o lucro*. São Paulo: Atlas.

Kotler, P. (1998). *Princípios de marketing. Tradução Vera Whately* (p. 527). Rio de Janeiro: Prentice Hall do Brasil.

Kotler, P. (2000). *Administração de marketing* (p. 764). São Paulo: Prentice Hall.

Kotler, P. (2003). *Marketing de A a Z: 80 conceitos que todo profissional precisa saber* (p. 95). Rio de Janeiro: Campus.

Kotler, P., Armstrong, G., Saunders, J., & Wong, V. (1999). *Principles of marketing* (2nd ed.). Cambridge, UK: Prentice Hall.

Kotler, P., & Keller, K. (2005). *Marketing Management* (12th ed.). Prentice Hall.

Kovacevic, I. (2013). Application of transactional analysis in bullwhip effect analysis. *Anfiteatru Economic*, (33), 210-223.

Kovac, W. (2010). Documentation and quality: Quo vadis patent system? A patent attorney's view. *World Patent Information, 32*(3), 229–232. doi:10.1016/j.wpi.2009.07.008

Kozinets, R. V. (2002). The field behind the screen: Using netnography for marketing research in online communities. *JMR, Journal of Marketing Research, 39*(1), 61–72. doi:10.1509/jmkr.39.1.61.18935

Kremer, M., Mantin, B., & Ovchinnikov, A. (2013). *Strategic consumers, Myopic retailers.* Darden School of Business at University of Virginia. Available at http://www.darden.virginia.edu/web/uploadedFiles/Darden/Faculty_Research/Research_Publications/Ovchinnikov_StrategicConsumer_MyopicRetailers.pdf

Krieg-Sigman, K. (1995). Kissing in the dark: Promoting and communicating in a public library setting. *Library Trends, 43*(3).

Krogh, G. V., & Cusumano, M. A. (2001). *Three Strategies for Managing Fast Growth. Sloan Management Review, 42*(2), 53–61.

Krols, B. (2010). *Invenções acidentais que mudaram as nossas vidas.* Booksmile.

Krovi, R. (2001). Surveying The E-Landscape: New Rules Of Survival. *Information Systems Management, 18*(4), 23–30.

Krstevski, D., & Manchevski, G. (2016). SMEs innovation management: Strategy map for innovation-driven company. *Economic Development., 1*(1), 195–210.

Krupp, F., & Horn, M. (2009). *Reinventar a energia: Estratégias para o futuro energético do planeta.* Alfragide: Estrela Polar.

Kumar, N. (2004). *Marketing as strategy: Understanding the CEO's agenda for driving growth and innovation.* Boston: Harvard Business School Press.

Küpers, W. (2005). Phenomenology of embodied implicit and narrative knowing. *Journal of Knowledge Management, 9*(6), 114–133. doi:10.1108/13673270510630006

Ladkin, D. (2012). *Perception, Reversibility, "Flesh": Merleau-Ponty's Phenomenology and Leadership as Embodied Practice.* Retrieved from http://integralleadershipreview.com/6280-perception-reversibility-flesh-merleau-pontys-phenomenology-and-leadership-as-embodied-practice

Lanari, L. A. (2000). *Indicadores de desempenho estratégico para uma incubadora de empresas de base tecnológica: uma proposta.* Escola de Biblioteconomia da UFMG.

Landaetta, R. R. (2008). Evaluating benefits and challenges of knowledge transfer across projects. *Engineering Management Journal, 20*(1), 29–38. doi:10.1080/10429247.2008.11431753

Lang, J. C. (2001). Managerial concerns in knowledge management. *Journal of Knowledge Management, 5*(1), 43–59. doi:10.1108/13673270110384392

Laranja, M., Simões, V., & Fontes, M. (1997). *Inovação tecnológica: Experiências das empresas portuguesas.* Lisboa: Texto Editora.

Las Casas, A. L. (2006). *Administração de marketing: conceitos, planejamento e aplicações à realidade brasileira.* São Paulo: Atlas.

Las Casas, A. L. (2010). *Marketing de varejo. 4.* São Paulo: Atlas.

Lattès, R. (1992). *O Risco e a fortuna: A grande aventura da Inovação.* Lisboa: Difusão Cultural.

Laudon, K., & Laudon, J. P. (2013). *Management information systems: Managing the digital firm* (13th ed.). New York: Prentice Hall.

Laudon, K., & Laudon, L. (2009). *Management Information Systems* (11th ed.). Prentice Hall.

Laursen, K., & Salter, A. (2004). Searching high and low: What types of firms use universities as a source of innovation? *Research Policy*, *33*(8), 1201–1215. doi:10.1016/j.respol.2004.07.004

Lautenborn, R. (1990). New marketing litany: 4P's passé; C-words take over. *Advertising Age*, 26.

Lawrence, S., & Giles, C. L. (2000). Accessibility of information on the Web. *Intelligence*, *11*(1), 32–39. doi:10.1145/333175.333181

Le Coadic, Y. (1996). *A ciência da informação.* Brasília: Briquet de Lemos.

Le Moigne, J. L. (1979). Informer la décision ou décider de l'information. *E.S./ Sciences de Gestion,* (1), 863-888.

Le Moigne. (n.d.). *La théorie du Système Géneral: théorie de la modélisation.* Academic Press.

Le Moigne, J. L. (1994). *La théorie du Système Géneral: théorie de la modélisation.* France.

Leidner, D., & Elam, J. J. (1995). The impact of executive information systems on organizational design, intelligence and decision making. *Organization Science*, *6*(6), 645–664. doi:10.1287/orsc.6.6.645

Leonidou, L., & Theodosiou, M. (2004). The export marketing information system: An integration of the extant knowledge. *Journal of World Business*, *39*(1), 12–36. doi:10.1016/j.jwb.2003.08.005

LeRouge, C., Mantzana, V., & Wilson, E. V. (2007). Healthcare information systems research: Revelations and visions. *European Journal of Information Systems*, *16*(1), 669–671. doi:10.1057/palgrave.ejis.3000712

Liao, J., Kickul, J. R., & Ma, H. (2009). Organizational dynamic capability and innovation: An empirical examination of Internet firms. *Journal of Small Business Management*, *47*(3), 263–286. doi:10.1111/j.1540-627X.2009.00271.x

Liautaud, B., & Hammond, M. (2001). e-Business intelligence: Turning information into knowledge into profit. New York: McGraw-Hill.

Liderman, M., Haines, S., Siegel, B., Chase, G., Ouellet, D., O'May, J., & Brichacek, J. (2005). *A Reference Model for Information Management to Support Coalition Information Sharing Needs.* Air Force Research Laboratory. Retrieved from www.dtic.mil/cgi-bin/GetTRDoc?AD=ADA464244

Liebowitz, J. (2006). *Strategic Intelligence – Business Intelligence, Competitive Intelligence, and Knowledge Management.* Auerbach Publications. doi:10.1201/9781420013900

Li, L., Lee, F., & Liu, W. (2006). Timely Product Recommendation Based on RFM Method. *Proc. The International Conference on Business and Information (BAI 2006).*

Linke, R. (2017). Applying analytics – Ideas and examples from leading companies and researches. *MIT Management Newsroom.* Available at http://mitsloan.mit.edu/newsroom/articles/reading-list-applying-analytics/

Liu, D., & Shih, Y. (2005). Hybrid approaches to product recommendation based on customer lifetime value and purchase preferences. *Journal of Systems and Software*, *77*(2), 181–191. doi:10.1016/j.jss.2004.08.031

Liu, X., & White, S. (2001). Comparing innovation systems: A framework and application to China's transitional context. *Research Policy*, *30*(7), 1091–1114. doi:10.1016/S0048-7333(00)00132-3

Liyanage, C., Elhag, T., Ballal, T., & Li, Q. (2009). Knowledge communication and translation – a knowledge transfer model. *Journal of Knowledge Management, 13*(3), 118–131. doi:10.1108/13673270910962914

Lopes, B., Falcão, L., & Canellas, T. (2016). Supply-Side: Mapping High Capacity Suppliers of Goods and Services. In Handbook of Research on Information Management for Effective Logistics and Supply Chains. IGI Global.

López-Cano, V. C., Briones-Peñalver, A. J., & Cegarra-Navarro, J. G. (2015). *Absorptive Capacity and Technology Knowledge: Enhancing Relational Capital. Knowledge and Process Management, 22*(4), 305–317.

Lucas, H. C. Jr. (2005). *Information technology: strategic decision making for managers*. Hoboken, NJ: John Wiley and Sons.

Luftman, J. (2001). Assessing Business – IT Alignment Maturity. In *Strategic Information Technology: Opportunities for Competitive Advantage*. Idea Group Publishing. doi:10.4018/978-1-878289-87-2.ch006

Lundvall, B.-Å. (1992). *National Systems of Innovation: Towards a Theory of Innovation and Interactive Learning. National systems of innovation Towards a theory of innovation and interactive learning*. Retrieved from http://books.google.com/books?id=iDXGwacw-4oC&pgis=1

Lundvall, B.-Å., Johnson, B., Andersen, E. S., & Dalum, B. (2002). National systems of production, innovation and competence building. *Research Policy, 31*(2), 213–231. doi:10.1016/S0048-7333(01)00137-8

Lynch, C. (2008). Big data: How do your data grow? *Nature, 455*(7209), 28–29. doi:10.1038/455028a PMID:18769419

Macedo, B. B. (2014b, outubro). *Parcerias para o Desenvolvimento Produtivo (PDPs) como Estratégia de Acesso aos Medicamentos*. FIOCRUZ/Farmanguinhos, Rio de Janeiro. Accessed by www.far.fiocruz.br

Macedo, B. B. (2014a). *Parcerias para o Desenvolvimento Produtivo (PDPs) como Estratégia de Acesso aos Medicamentos*. FIOCRUZ/Farmanguinhos, Especialização de Gestão da Inovação em Medicamentos Fitoterápicos.

Macedo, M., & Barbosa, A. (2000). *Patentes, pesquisa & desenvolvimento: Um manual de propriedade intelectual*. Rio de Janeiro: Fiocruz. doi:10.7476/9788575412725

MacInnis, D. (2011). A framework for conceptual contributions in Marketing. *Journal of Marketing, 75*(4).

MacQueen, J. (1967). Some Methods for Classification and Analysis of Multivariate Observations. In *Proceedings of the Fifth Berkeley Symposium on Mathematical Statistics and Probability* (vol. 1, pp. 281–97). Berkeley, CA: University of California Press. Retrieved from http://projecteuclid.org:443/euclid.bsmsp/1200512992

Magalhães, J. L., & Quoniam, L. (2013). Perception of the Information Value for Public Health: A Case Study for Neglected Diseases. In *Rethinking the Conceptual Base for New Practical Applications in Information Value and Quality* (p. 345). IGI Global. Recuperado de http://www.igi-global.com/chapter/perception-information-value-public-health/84218

Magalhaes, J. L., Antunes, A. M. S., & Boechat, N. (2012a). *Technological Trends in the Pharmaceutical Industry: the matter of neglected tropical diseases – An overview of the Research, Development & Innovation in Brazil*. Synergia Editora. Accessed by http://www.livrariasynergia.com.br/livros/M39700/9788561325732/tendencias-tecnologicas-no-setor-farmaceutico-a-questao-das-doencas-tropicais-negligenciadas-edicao-bilingue.html

Magalhaes, J. L., Antunes, A. M. S., & Boechat, N. (2012b). *Tendências Tecnológicas no Setor Farmacêutico: a questão das doenças tropicais negligenciadas - uma perspectiva da P,D&I no Brasil* (Vol. 1). Synergia Editora. Accessed by http://www.livrariasynergia.com.br/livros/M39700/9788561325732/tendencias-tecnologicas-no-setor-farmaceutico-a-questao-das-doencas-tropicais-negligenciadas.html

Magalhães, J. L., Antunes, A., & Boechat, N. (2011). Laboratórios farmacêuticos oficiais e sua relevância para saúde pública do Brasil. *Revista Eletrônica de Comunicação, Informação & Inovação em Saúde, 5*(1).

Magalhães, J. L., Hartz, Z., & Antunes, A. M. (2016). An Evaluation of Partnership for Productive Development in Brazil's Healthcare: Measuring the Knowledge Translation from Implementation to the Impact. In Handbook of Research on Information Architecture and Management in Modern Organizations (pp. 525-544). IGI Global.

Magalhaes. (2012). *Technological Trends in the Pharmaceutical Industry: the matter of neglected tropical diseases – An overview of the Research, Development & Innovation in Brazil.* Synergia Editora.

Magalhaes. (2015). *Percepção do Valor da Informação por meio da Inteligência Competitiva 2.0 e do Big Data em Saúde in Análise da Informação para tomada de decisão - desafios e soluções* (Vol. 1). Editora Intersaberes.

Magalhães, J. L., Quoniam, L., & Boechat, N. (2013). Web 2.0 Tools for Network Management and Patent Analysis for Health Public. *Revista de Gestão Em Sistemas de Saúde, 2*(1), 26–41. doi:10.5585/rgss.v2i1.48

Mahoney, J., & Pandian, J. R. (1992). The Resource-Based View Within The Conservation of Strategic Management. *Strategic Management Journal, 13*(5), 363–380. doi:10.1002mj.4250130505

Mahrt, M., & Scharkow, M. (2013). The value of big data in Digital Media research. *Journal of Broadcasting & Electronic Media, 57*(1), 20–33. doi:10.1080/08838151.2012.761700

Maia, J. M. (1996). Propriedade industrial: Comunicações e artigos do presidente do INPI. Lisboa: Instituto Nacional da Propriedade Industrial (INPI).

Maital, S., & Seshadri, D. V. R. (2012). *Innovation management: strategies, concepts and tools for growth and profit.* New Delhi: Sage Publications.

Majaro, S. (1990). *Criatividade: Um passo para o sucesso.* Lisboa: Europa-América.

Makadok, R., & Barney, J. B. (2001, December). Strategic Factor Marketing Intelligence: An application of information economics to strategy formulation and competitor intelligence. *Management Science, 47*(12), 1621–1638. doi:10.1287/mnsc.47.12.1621.10245

Malone, T. W., Weill, P., Kai, R. K., Urso, V., Herman, G., Apel, T. G., & Woerner, S. (2006). *Do some business models perform better than others?* MIT Sloan Research Paper no. 4615-06. Accessed at http://papers.ssrn.com/sol3/papers.cfm?abstract_id=920667##

Mandavilli, A. (2006). Make Anything, Anywhere. *Nature, 442*(8). PMID:16929273

Manjón, J. V. G. (2010). A proposal of indicators and policy framework for innovation benchmark in europe. *Journal of Technology Management & Innovation, 5*(2).

Mansour, O. (2009). Group intelligence: A distributed cognition perspective. In *INCOS'09 - International Conference on Intelligent Networking and Collaborative Systems* (pp. 247–250). IEEE. 10.1109/INCOS.2009.59

Mantovani. (2012). *Estudo sobre computação ubíqua.* Available at: http://www.unifil.br/portal/arquivos/publicacoes/paginas/2012/11/516_867_publipg.pdf

Manual de Frascati, OCDE. (2013). *Metodologia proposta para definição da pesquisa e desenvolvimento experimental.* F. Iniciativas, Assessoria em P&D.

Manual de Oslo. (1997). *Diretrizes para coleta e interpretação de dados sobre inovação. Terceira Edição. OCDE.* FINEP.

Manville, B., & Ober, J. (2003). Beyond Empowerment: Building a Company of Citizens. *Harvard Business Review, 81*(1), 48–53. PMID:12545922

Maravilhas, S. (2009). A informação de patentes: Vantagens da sua utilização como estímulo à criatividade, I&D, inovação e competitividade das empresas portuguesas. In IAPMEI (Ed.), Parcerias Científicas para a Inovação (pp. 91-110). Lisboa: IAPMEI.

Maravilhas, S. (2012). *A informação de patentes como fonte de ideias para produtos e negócios sustentáveis.* Paper presented at the Vth GECAMB, Conferência CSEAR Portuguesa de Gestão e Contabilidade Ambiental, Porto, Portugal.

Maravilhas, S. (2013b). Social media tools for quality business information. In Talburt, Yeoh & Zhou (Eds.), Information quality and governance for business intelligence. Hershey, PA: IGI Global.

Maravilhas, S. (2014b). Information Quality and Value. In M. Khosrow-Pour (Ed.), Encyclopedia of Information Science and Technology (3rd ed.; vol. 4). Hershey, PA: IGI Global.

Maravilhas, S., & Borges, M. (2009). O impacto das bibliotecas digitais de patentes no processo de inovação em Portugal. In Borges & Sanz-Casado (Eds.), A ciência da informação criadora de conhecimento (Vol. 2, pp. 47-63). Coimbra: Actas do IV Encontro Ibérico EDIBCIC 2009.

Maravilhas, S. (2013a). A web 2.0 como ferramenta de análise de tendências e monitorização do ambiente externo e sua relação com a cultura de convergência dos media. *Perspectivas em Ciência da Informação, 18*(1), 126–137. doi:10.1590/S1413-99362013000100009

Maravilhas, S. (2015b). Managing an information strategy project: The case of a real estate broker organization. In *Handbook of Research on Effective Project Management through the Integration of Knowledge and Innovation* (pp. 19–43). Hershey, PA: IGI Global. doi:10.4018/978-1-4666-7536-0.ch002

Maravilhas, S., & Borges, M. (2011a). Os recursos de informação usados na I&D em Portugal: Caracterização dos centros de investigação do ensino superior público das áreas de ciência, tecnologia e medicina. In P. Guerrero & V. Moreno (Eds.), *Límites, fronteras y espacios comunes: Encuentros y desencuentros en las Ciencias de la Informacíon* (pp. 321–333). Badajoz: Actas do V Encontro Ibérico EDICIC.

Maravilhas, S., & Borges, M. (2011b). A utilização da informação de patentes pelos centros de investigação do ensino superior público: o seu impacto no processo de inovação em Portugal. In P. Guerrero & V. Moreno (Eds.), *Límites, fronteras y espacios comunes: Encuentros y desencuentros en las ciencias de la informacíon* (pp. 364–376). Badajoz: Actas do V Encontro Ibérico EDICIC.

Marchand, D. (1990). Managing Information quality. In Information quality: definitions and dimensions. London: Taylor Graham.

Marchand, D. (1990). Managing information quality. In Wormell (Ed.), Information quality: Definitions and dimensions. London: Taylor Graham.

Marchand, D. (n.d.). Value Add through effective use of information. *IMD Bulletin.* Available at http://www.imd.org/research/challenges/tc00110.cfm

Marchand, D., & Davenport, T. (Eds.). (2000). *Mastering Information Management.* New York: Financial Times Prentice Hall.

Marchand, D. (1990). Infotrends: A 1990s outlook on strategic information management. *Industrial Management Review, 5*(4), 23–32.

Marchand, D., & Horton, F. Jr. (1986). *Infotrends: Profiting from your information resources.* New York: Wiley.

Marchand, D., Kettinger, W., & Rollins, J. (2001). *Making the invisible visible: how companies win the right information, people and IT*. Wiley.

March, J. G. (1991). Exploration and exploitation in organizational learning. *Organization Science, 20*(: 1), 71–87. doi:10.1287/orsc.2.1.71

Marcovitch, J. (1983). *Administração em ciência e tecnologia*. São Paulo: Edgard Blücher.

Marcus, D. (1995). Benefits of using patent databases as a source of information. In Lechter (Ed.), Successful Patents and Patenting for Engineers and Scientists. New York: The Institute of Electrical and Electronics Engineers (IEEE) Press.

Marini. (n.d.). *Comparativo da comunicação de dados em dispositivos móveis: WEB Servicese e Sockets*. Available at: http://painel.passofundo.ifsul.edu.br/uploads/arq/20150522102340173874l499.pdf

Marion. (2002). *Monografia para cursos de administração, contabilidade e economia*. São Paulo: Atlas.

Markides, C. (2000). *All the right moves: A guide to craft a breakthrough strategy*. Boston: Harvard Business School Press.

Markides, C. (2008). *Game-changing strategies: How to create new market space in established industries by breaking the rules* (1st ed.). San Francisco: Wiley.

Markovitch, D. G., Steckel, J. H., & Yeung, B. (2005). Using Capital Markets as Market Intelligence: Evidence from the Pharmaceutical Industry. *Management Science, 51*(10), 1467–1480. doi:10.1287/mnsc.1050.0401

Maroco, J. (2007). *Análise estatística: com utilização do SPSS. 3*. Lisboa: Edições Sílabo.

Maroco, J. (2007). *Análise Estatística: com utilização do SPSS*. Lisboa: Edições Sílabo.

Martins, R. S., Lobo, D. S., Alves, A. F., & Sprosser, R. L. (2014). Fatores Relevantes na Contratação de Serviços em Terminais Intermodais para Granéis Agrícolas. [*RESR, Piracicaba-SP, 52*(02), 347–364.

Mascitelli, R. (2000). From Experience: Harnessing Tacit Knowledge to Achieve Breakthrough Innovation. *Journal of Product Innovation Management, 17*(3), 179–193. doi:10.1016/S0737-6782(00)00038-2

Massey, A. P., Wheeler, B. C., & Keen, P. G. W. (2000). *Technology Matters. In Information Technology and the Future Enterprise*. Prentice Hall.

Mata, F. J., Fuerst, W. L. & Barney, J. B. (1995). Information technology and sustained competitive advantage: a resource-based analysis. *Management Information Science Quarterly*, 487-505.

Matos, A. L. T., Pires, S. R. I. & Vivaldini, M. (2014). Desenvolvimento de produtos sob a perspectiva da gestão da cadeia de suprimentos. *Revista de Administração da UEG, Aparecida de Goiânia, 5*(1).

Mattar, F. N. (1996). *Pesquisa de marketing: metodologia, planejamento*. São Paulo: Atlas.

Matthews, C., & Sternberg, R. (2009). Developing experience-based (tacit) knowledge through reflection. *Learning and Individual Differences, 19*(4), 530–540. doi:10.1016/j.lindif.2009.07.001

Maznevski, M. L. (1994). Understanding our diferences: Performance in decision-making groups with diverse members. *Human Relations, 47*(5), 531–552. doi:10.1177/001872679404700504

Mc Gee, J. V., & Prusak, L. (1994). *Gestão estratégica da informação*. Rio de Janeiro: Campus.

Mc Kinsey. (2011). *Big Data: the next frontier for innovation, competition and productivity*. Accessed at http://www.mckinsey.com/insights/business_technology/big_data_the_next_frontier_for_innovation

McAfee, A., & Brynjolfsson, E. (2012). Big data: the management revolution. Harvard Business Review, 90(10), 60-68.

McAfee, A., & Brynjolfsson, E. (2012). Big data: The management revolution. *Harvard Business Review*, *90*(10), 60–68. PMID:23074865

McAfee, A., & Oliveau, F. X. (2002). Confronting the Limits of Networks. *MIT Sloan Management Review*, *43*(4), 85–87.

Mccarthy, E. J. (1978). Marketing básico: uma visão gerencial. Rio de Janeiro: Zahar, 2v.

McConnell, J., & Ward-Perkins, D. (1998). *The advantage of the Internet for companies: competitive challenges and successful strategies*. Lisboa: Dom Quixote. (in Portuguese)

Mcgee, J. V., & Prusak, L. (1994). *Gerenciamento estratégico da informação*. Rio de Janeiro: Campus.

McGee, J. V., & Prusak, L. (1995). *Gerenciamento estratégico da informação: Aumente a competitividade e a eficiência de sua empresa utilizando a informação como uma ferramenta estratégica*. Rio de Janeiro: Campus.

McKinsey Global Institute. (2011). *Big Data: The Management Revolution - Harvard Business Review*. Recuperado de http://hbr.org/2012/10/big-data-the-management-revolution/ar/1

McLaughlin, P., Bessant, J., & Smart, P. (2005). *Developing an Organizational Culture that facilitates Radical Innovation in a Mature Small to Medium Sized Company: Emergent Findings*. Retrieved from http://hdl.handle.net/1826/858

McPherson, P. (1994). Accounting for the value of information. *Aslib Proceedings*, *46*(9), 203–215. doi:10.1108/eb051366

McQueen, R. J. (1998). Four views of knowledge and knowledge management. *Proceedings of Fourth Americas Conference on Information Systems*, 609-611. Retrieved from http://www.mngt.waikato.ac.nz/departments/staff/bob/WEBDOCS/pubs/aiskm98.pdf

Mehmod, H. (2017). *A big list of what Google Analytics can and can not do*. Available at http://marketlytics.com/blog/list-of-things-google-analytics-can-and-cannot-do

Meiners-Chabin. Protopopescu, Chauveau, & Moatti. (2013). Estrutura de mercado e evolução dos preços dos medicamentos antirretrovirais no Brasil. In Propriedade intelectual e políticas públicas para o acesso aos antirretrovirais nos países do Sul. Rio de Janeiro: E-papers.

Melén, S. (2009). *New insights on the internationalisation process of SMEs: A study of foreign market knowledge development. Economic Research Institute*. Stockholm School of Economics.

Melese, J. (1990). *Approches Systémiques des Organisations: Vers l'Entreprise à Complexité Humaine*. Paris, France: Les Éditions d'Organisation.

Mergel, I. (2017). *Digital service teams: Challenges and recommendations for government. IBM Center for The Business of Government: Using Techbology Series*. Washington, DC: BM Center for The Business of Government.

Merriam, S. (1998). *Qualitative research and case study applications in education*. San Francisco: Jossey-Bass Publishers.

Metcalfe, J. S. (1995). The Economic Foundations of Technology Policy: Equilibrium and Evolutionary Perspectives. Handbook of the Economics of Innovation and Technological Change, 409–512.

Michael, A., & Salter, B. (2006). *Mobile marketing: Achieving competitive advantage through wireless technology* (1st ed.). Oxford: Butterworth-Heinemann.

Michalko, M. (1991). *Thinkertoys: A handbook of business creativity for the 90's*. Berkeley, CA: Ten Speed Press.

Mikamo, S. (2013). Business associations and politics in the post-EDSA Philippines: Neither oligarchy nor civil society. *Philippine Political Science Journal*, *34*(1), 6–26. doi:10.1080/01154451.2013.789165

Miller, S. (2002). Competitive Intelligence - an overview *Competitive Intelligence Magazine*, *14*(3), 43-55. Retrieved from http://www.sci.org/library/overview.pdf

Miller, J. P. (1996). Information science and competitive intelligence: Possible collaborators? *Bulletin of the American Society for Information Science*, *23*(1), 1–12.

Miller, J. P. (2000). *Millenium Intelligence: Understanding and Conducting Competitive Intelligence in the Digital Age*. Information Today, Inc.

Miller, J. P. (2002). *O milênio da inteligência competitiva*. Porto Alegre: Bookman.

Minelli, M., Chambers, M., & Dhiraj, A. (2013). *Big Data, Big Analytics*. John Wiley & Sons, Inc. Recuperado de https://books.google.com.br/books/about/Big_Data_Big_Analytics.html?hl=pt-BR&id=Mg3WvT8uHV4C

Ministério da Saúde. M. (2014a). *Parceria para o Desenvolvimento Produtivo - PDP*. Accessed in November 2, 2014. http://portalsaude.saude.gov.br/index.php/o-ministerio/principal/leia-mais-o-ministerio/581-sctie-raiz/deciis/l2-deciis/12090-parceria-para-o-desenvolvimento-produtivo-pdp

Ministério da Saúde. M. (2014b). *SCTIE*. Accessed in November 2, 2014 by http://portalsaude.saude.gov.br/index.php/o-ministerio/principal/secretarias/sctie

Mintzberg, H. & Quinn, J. B. (2001). *O processo da estratégia* (3rd ed.). Porto Alegre: Bookman.

Mintzberg, H. (1987, July). Crafting strategy. *Harvard Business Review*, 66–75.

Mintzberg, H., Alhstrand, B., & Lampel, J. (2009). *Strategy Safari: the complete guide through the wilds of strategic management* (2nd ed.). Pearson Education Limited.

Mintzberg, H., Lampel, J., Quinn, J. B., & Ghoshal, S. (2003). *The strategy process: concepts, contexts, cases* (4th ed.). Prentice Hall.

Mishra, K. A., Khan, M. I., & Singh, Y. P. (2013). Review: Recent Trends in World and Indian Telecom Market. *International Journal of Management*, *4*(3), 61–74.

Mobberley, C. (n.d.). *What is Node-Red?* Available at: https://learn.adafruit.com/raspberry-pi-hosting-node-red/what-is-node-red

Mohr, J., Sengupta, S., & Slater, S. (2010). *Marketing of high-technology products and innovations* (3rd ed.). Upper Saddle River, NJ: Pearson.

Moody, D., & Walsh, P. (1999). *Measuring the value of information: an asset evaluation approach*. European Conference on Information Systems. Disponível: http://wwwinfo.deis.unical.it/~zumpano/2004-2005/PSI/lezione2/ValueOfInformation.pdf

Moody, D., & Walsh, P. (2009). *Measuring the value of information: an asset evaluation approach*. European Conference on Information Systems, 1999. Disponível: http://wwwinfo.deis.unical. it/~zumpano/2004-2005/PSI/lezione2/ValueOfInformation.pdf

Mooradian, N. (2005). Tacit knowledge: Philosophic roots and role in KM. *Journal of Knowledge Management*, *9*(6), 104–113. doi:10.1108/13673270510629990

Morin, E. (1982). Le système: Paradigme ou/et théorie. Science Avec Conscience.

Morin, E. (1982b). Le système: Paradigme ou/et théorie. Science avec Conscience.

Morin, E. (1982a). Le Système: Paradigme ou/et Théorie. In *Sicence avec Conscience*. Paris, France: Libraire Arthème Fayard.

Morin, E. (1991). Da necessidade de um pensamento complexo. In *Para navegar no Século XXI*. Rio de Janeiro: Vozes.

Morin, E. (1996). *O Problema epistemológico da complexidade*. Publicações Europa- América.

Morville, P., & Rosenfeld, L. (2009). *Information architecture for the world-wide-web: Designing large scale web-sites* (3rd ed.). Sebastopol, CA: O'Reilly Media.

Mowery, D. (2005). *Universities in National Innovation Systems*. Retrieved January 13, 2017, from http://www.aau.org/sites/default/files/urg/docs/Uni _natn_Inovatn_sys.pdf

Mowery, D. C., & Sampat, B. N. (2005). Universities in National Innovation Systems. The Oxford Handbook of Innovation, 209–239. doi:10.1093/oxfordhb/9780199286805.003.0008

Mueller, H., & Nyfeler, T. (2011). Quality in patente information retrieval: Communication as the key factor. *World Patent Information*, *33*(4), 383–388. doi:10.1016/j.wpi.2011.06.012

Mugnaini, R., Jannuzzi, P. D. M., & Quoniam, L. (2004). Indicadores bibliométricos da produção científica brasileira: Uma análise a partir da base Pascal. *Ciência da Informação*, *33*(2), 123–131. doi:10.1590/S0100-19652004000200013

Müller, R., Glücker, J., Aubry, M., & Shaun, J. (2013). Project management knowledge flows in network of project managers and project management offices: A case study in the pharmaceutical Industry. *Project Management Journal*, *44*(2), 4–19. doi:10.1002/pmj.21326

Mullins, J. (2014). *The customer funded business*. Hoboken, NJ: Wiley.

Murphy, C. (2005). *Competitive Intelligence – Gathering, Analysing, and Putting into Work*. Gower e-Book.

Naetebusch, R., Schoeppel, H. R., & Fichtner, H. (1994). Patent information in a large electrical company, exemplified by the situation at Siemens. *World Patent Information*, *16*(4), 198–206. doi:10.1016/0172-2190(94)90003-5

Neeley, T. (2015). Equipes globais que funcionam: Uma estrutura para diminuir a distância social. *Harvard Business Review*.

Nelson, R. R. (1993). National Innovation Systems: A Comparative Analysis. In *National Innovation Systems* (p. 541). Academic Press. http://books.google.co.uk/books?id=YFDGjgxc2CYC

Neto, R. (2013). Gestão do Conhecimento em Organizações: Proposta de Mapeamento Conceitual Integrativo. S. Paulo: Saraiva.

Nick, W. (2016, April 28). Bharti Airtel revenue jumps on surging mobile data use. *Total Telecom*. Retrieved from http://www.totaltele.com/view.aspx?ID=493606

Nolan, R., & Croson, D. C. (1995). *Creative Destruction: A Six-Stage Process for Transforming the Organization*. Harvard Business School Press.

Nonaka, I. (1997). *Organizational Knowledge Creation*. Presentation summary at the Knowledge Advantage Conference.

Nonaka, I. (1991). The knowledge creating company. *Harvard Business Review*, *69*, 96–104.

Nonaka, I. (1994). A Dynamic Theory of Organizational Knowledge Creation. *Organization Science*, *5*(1), 14–37. doi:10.1287/orsc.5.1.14

Nonaka, I. (2008). *The knowledge creating company*. Harvard Business Review Classics.

Nonaka, I., & Konno, N. (1998). The concept of 'Ba': Building a foundation for knowledge creation. *California Management Review*, *40*(3), 40–54. doi:10.2307/41165942

Nonaka, I., & Takeuchi, H. (1995). *The Knowledge Creating Company*. New York, NY: Oxford University Press.

Nonaka, I., & Takeushi, H. (1997). *Criação de Conhecimento na Empresa: Como as Empresas Japonesas geram a dinâmica da Inovação*. Rio de Janeiro: Campus.

Nunamaker, J. F. J., Briggs, R. O., & Vreede, G. (2000). Value Creation Technology – Changing the Focus to the Group. *Information Systems Management*, *18*(4), 102–124.

O'Connor, A. (1993). Successful strategic information systems planning. *Journal of Information Systems*, *3*(2), 71–83. doi:10.1111/j.1365-2575.1993.tb00116.x

Obenshaim, M. K. (2004). Application of data mining techniques to healthcare data. Statistics for Hospital Epidemiology, 25(8), 690 – 695.

OCDE. (1996). *The Knowledge-Based Economy. OCDE/GD* (Vol. 96). Retrieved from https://www.oecd.org/sti/sci-tech/1913021.pdf

OCDE. (1997). Manual de Oslo: Diretrizes para a Coleta e Interpretação de dados sobre Inovação Tecnológica. *OCDE, Eurostat E Financiadora de Estudos E Projetos*, 184. 10.1787/9789264065659-es

OCDE. (2002). Rumo a um desenvolvimento sustentável: indicadores ambientais. *Cadernos de referência ambiental, 9*.

OCDE. (2016). *Main Science and Technology Indicators 2016-2*. OCDE.

OCDE. (n.d.). Retrieved from https://www.ine.pt/xportal/xmain?xpid=INE&xpgid=ine_base_dados

Oda, S. (2009). JP-NETe - A English-language search tool for Japanese unexamined patents. *World Patent Information*, *31*(2), 131–134. doi:10.1016/j.wpi.2008.09.002

Ohata, M. & Kumar, A. (2012, September). Big Data: A Boom for Business Intelligence. *Financial Executive*.

Ojala, M. (1989). A patently obvious source for competitor intelligence: The patent literature. *Database*, *12*(4), 43–49.

Okubo, Y. (1997). Bibliometric Indicators and Analysis of Research Systems: Methods and Examples. *OECD Science, Technology and INdustry Working Papers, 1997*(1), 1–70.

Oliveira, J. (1999). *Gestão da inovação*. S. João do Estoril: Sociedade Portuguesa de Inovação |Principia.

Oliveira, J. (2017). Data "government" models for Healthcare. Business Intelligence Journal, 17(4), 34-39.

Oliveira, A. (2004). *Análise do investimento em sistemas e tecnologias da informação e da comunicação*. Lisboa: Sílabo.

Oliveira, P. H., Gonçalves, C. A., & Paula, E. A. M. (2013). Vision based on Competitive Intelligence Resources. *Management Science Magazine*, *15*(35), 141–151.

Oracle. (n.d.). *The Java EE 6 Tutorial*. Available at: http://docs.oracle.com/javaee/6/tutorial/doc/gijqy.html

Orna, E. (1996). Valuing information: Problems and opportunities. In D. Best (Ed.), *The Fourth Resource: Information and its Management*. Hampshire, UK: Aslib/Gower.

Orna, E. (1999). *Practical information policies* (2nd ed.). Cambridge, UK: Gower.

Osterwalder, A., & Pigneur, Y. (2010). *Business model generation: A handbook for visionaries, game changers, and challengers*. Wiley.

Oxman, J. A., & Smith, B. D. (2003). *The Limits of Structural Change. Sloan Management Review, 45*(1), 77–82.

Pacheco, C. G., & Valentim, M. L. P. (2010). Informação e conhecimento como alicerces para a gestão estratégica empresarial: um enfoque nos fluxos e fontes de informação. In M. Valentim (Ed.), *Gestão, mediação e uso da informação* (pp. 319–341). São Paulo: Cultura Academica.

Pandikumar, S., & Vetrive, R. S. (n.d.). *Internet of Things Based Architecture of Web and Smart Home Interface Using GSM*. Available at: https://ijirset.com/upload/2014/iciet/it/8_611.pdf

Paramkusham, R. B., & Gordon, J. (2013, Fall). Inhibiting factors for knowledge transfer in information technology projects. *Journal of Global Business and Technology*, *9*, 2.

Park, S. H., Huh, S. Y., Oh, W., & Han, S. P. (2012). A social network-based inference model for validating customer profile data. MIS Quarterly, 36(4), 1217-1237.

Parker, G., Alstyne, M., & Choudary, S. (2016). *Platform Revolution*. New York: W. W. Norton & Company, Inc.

Parung, J., & Bititci, U. (2008). A metric for collaborative networks. *Business Process Management Journal*, *14*(5), 654–674. doi:10.1108/14637150810903048

Pask, G. (1969). *Learning strategies and teaching strategies*. Surrey: System Research Ltd.

Patel-Campillo, A., & DeLessio-Parson, A. (2016). Why types of operations, trade associations, and production trends matter in the geographic branding of an emerging industry. *Journal of Wine Research*, *27*(3), 242–256. doi:10.1080/09571264.2016.1202218

Patel, P., & Pavitt, K. (1994). The Nature and Economic Importance of National Innovation Systems. *STI Review*, *14*, 9–32.

Pearson, A., Jordan, Z., & Munn, Z. (2012). Translational Science and Evidence-Based Healthcare: A Clarification and Reconceptualization of How Knowledge Is Generated and Used in Healthcare. *Nursing Research and Practice*, *2012*, 1–6. doi:10.1155/2012/792519 PMID:22474583

Penrose, E. (1959). *The Theory of the Growth of the Firm*. Londres: Basil Blakwell.

Penrose, E. (1959). *The theory of the growth of the firm*. New York: John Wiley and Sons.

Penzias, A. (1989). *Ideas and information: Managing in a high-tech world*. New York: W. W. Norton.

Penzias, A. (1995). *Harmony: Business, technology & life after paperwork*. New York: Harper Collins.

Pereira, F., Verocai, H., Cordeiro, V., & Gomes, C. (2016). Sistemas de Informação e Inovação: Um estudo bibliométrico. *Revista de Gestão Da Tecnologia E Sistemas de Informação*, *13*(1), 81–100.

Perera, C., Zaslavsky, A., Christen, P., & Georgakopoulos, D. (2014). *Context Aware Computing for The Internet of Things: A Survey. IEEE Communications Surveys & Tutorials, 16(1).*

Pesqueux, Y. (2005). *Management de la connaissance: un modèle organisationnel?* Management De La Connaissance: Un Modele Organisationnel? Recuperado de https://halshs.archives-ouvertes.fr/halshs-00004005/document

Pesqueux, Y., & Ferrary, M. (2011). Management de la connaissance (2o ed). Economica. Recuperado de /Entreprise/Livre/management-de-la-connaissance.

Pesqueux, Y., & Ferrary, M. (2011). Management de la connaissance. *Economica.*

Pessoa, C., Silva, T., Rosa, M., & Augusto, T. (2016). Internet das Coisas: Estudo de mercado para aplicação em Pet Shops. *XXIII SIMPEP Proceedings.*

Pessoa, C., Silva, T., Rosa, M., & Jamil, G. (2016). A Internet Das Coisas: Conceitos aplicações, desafios e tendências. *XIII CONTECSI Proceedings.*

Petit, P. (2001). *Economics and information.* Dordrecht: Kluwer. doi:10.1007/978-1-4757-3367-9

Petró, B. (2008). Análise do fluxo informacional dos gestores turísticos da unidade de conservação Parque Natural Municipal Mata Atlantica de Atalanta – SC, p. 184. Dissertação (Mestrado em Ciência da Informação). Universidade Federal de Santa Catarina, Florianópolis.

Petroski, H. (2008). *Inovação: Da idéia ao produto.* São Paulo: Edgard Blücher.

PhilipB. (2016). Retrieved from http://www.swinburne.edu.au/news/latest-news/2016/01/the-g-in-5g-how-mobile-generations-have-evolved.php

Philipp, M. (2006). Patent filing and searching: Is deflation in quality the inevitable consequence of hyperinflation in quantity? *World Patent Information, 28*(2), 117–121. doi:.wpi.2006.01.00610.1016/j

Philipp, M. (2005). Why pay for value-added information? *World Patent Information, 27*(1), 7–11. doi:10.1016/j.wpi.2004.06.001

Phua, C., Lee, V., Smith, K., & Gayler, R. (2010). *A comprehensive survey of data mining-based fraud detection research.* Available at http://arxiv.org/ftp/arxiv/papers/1009/1009.6119.pdf

Piaget, J., & Garcia, R. (1983).. . *Psychogenèse et Histoire des Sciences, 19*(1), 304.

Picot, A. (1989). Information management: The science of solving problems. *International Journal of Information Management, 9*(4), 237–243. doi:10.1016/0268-4012(89)90047-9

Pierret, J.-D. (2006). *Methodologie et structuration d'un outil de decouverte de connaissances base sur la litterture biomedicale : une application basee sur le MeSH.* Université du Sud Toulon Var.

Pieth, M. (2012). Collective action and corruption. In *Collective Action – innovative strategies to prevent corruption.* DIKE AG; doi:10.1007/978-3-319-04480-4_6

Pilon. (2009). *Estudo Para Aplicação De Redes Sem Fio No Ambiente Industrial.* Universidade Tecnológica Federal Do Paraná.

Pimjai Sudsawad, S. (n.d.). *Center on Knowledge Translation for Disability and Rehabilitation Research (KTDRR) KT Library - Research Quality.* Disponível em: http://ktdrr.org/ktlibrary/articles_pubs/ktmodels/

Pinto, M. M. (2014). *Gestão e Preservação da Informação: o impacto do pensamento sistémico.* Retrieved from http://eiarquivos2014.weebly.com/uploads/2/6/0/6/26068081/p1_4_mp.pdf

Pinto, M. M. (2015b). The Portuguese University: knowledge leverage towards innovation. In Handbook of research on effective project management through the integration of knowledge and innovation (pp. 466–490). IGI Global. doi:10.4018/978-1-4666-7536-0.ch024

Pinto, M. M. (2015a). *A Gestão da Informação nas Universidades Públicas Portuguesas: Reequacionamento e proposta de modelo (Tese de Doutoramento em Informação e Comunicação em Plataformas Digitais).* Universidade do Porto - Universidade de Aveiro.

Pinto, M. M. (2017). Gestão e Preservação da Informação: o impacto do pensamento sistémico. In N. Vaquinhas, M. Caixas, & H. Vinagre (Eds.), *Da produção à preservação informacional: desafios e oportunidades* (pp. 326–364). Évora: Publicações do Cidehus; doi:10.4000/books.cidehus.2758

Piscione, D. (2014). *Os Segredos do Vale do Silício: O que Você Pode Aprender com a Capital Mundial da Inovação* (1st ed.). São Paulo: HSM.

Plaza, I., Martin, L., Martin, S., & Medrano, C. (2011). Mobile applications in an aging society: Status and trends. *Journal of Systems and Software*, *84*(11), 1977–1988. doi:10.1016/j.jss.2011.05.035

PMBoK. (2013). *Project Management Body of Knowledge* (5th ed.). Project Management Institute.

Polanyi, M. (1966). *The Tacit Dimension*. London: Routledge & Kegan Paul.

Popp, J., MacKean, G., Casebeer, A., Milward, H., & Lindstrom, R. (2014). Inter-organizational networks: A critical review of the literature to inform practice. *IBM Center for The Business of Government*, 123.

Porter & Millar. (1985, July). How Information Gives You Competitive Advantage. *Harvard Business Review*.

Porter, M. (1980). *Competitive Strategy: Techniques for Analysing Industries and Competitors*. Free Press.

Porter, M. (2001, March). Strategy and the Internet. *Harvard Business Review*, 62-78.

Porter, M. (1980). *Competitive strategy: Techniques for analyzing industries and competitors*. New York: Free Press.

Porter, M. (1985). *Competitive advantage: creating and sustaining superior performance*. New York: Free Press.

Porter, M. (2008). *On Competition*. Harvard Business School Press.

Porter, M. E. (1985). *Competitive advantage: Creating and sustaining superior performance*. New York: Free Press.

Porter, M. E. (2008). *Competitive Strategy: Techniques for Analyzing Industries and Competitors*. Simon and Schuster.

Porter, M. E., & Millar, V. E. (1985). How information gives you competitive advantage. *Harvard Business Review*, *63*(4), 149–160.

Postscapes. (n.d.). *IoT Gateways*. Available at: http://www.postscapes.com/iot-gateways/

Prado, D. (2006). *Gerenciamento de programas, portfólios e projetos nas organizações*. Belo Horizonte: INDG.

Prehofer, C. (n.d.). *From IoT Mashups to Model-based IoT*. Available at: https://www.w3.org/2014/02/wot/papers/prehofer.pdf

Premkumar, G., & King, W. (1994). The evaluation of strategic information systems planning. *Information & Management*, *26*(6), 327–340. doi:10.1016/0378-7206(94)90030-2

Pressman, R. (2006). *Software Engineering* (6th ed.). Boston: Makron Books.

Provan, K., & Kenis, P. (2008). Modes of network governance: Structure, management, and effectiveness. *Journal of Public Administration: Research and Theory*, *18*(2), 229–252. doi:10.1093/jopart/mum015

Quinn, J. B. (1992). *Intelligent enterprises*. New York: The Free Press.

Quoniam, L. (2011). *Competitive Intelligence 2.0*. ISTE Ltd and John Wiley & Sons Inc. Accessed by http://www.iste.co.uk/index.php?f=a&ACTION=View&id=409

Quoniam, L. (2011). *Competitive Intelligence 2.0*. ISTE Ltd and John Wiley & Sons Inc. doi:10.1002/9781118603079

R Core Team. (2017). *R: A language and environment for statistical computing*. R Foundation for Statistical Computing. Retrieved from https://www.R-project.org/

Rabelo, R., & Gusmeroli, S. (2008). The ecolead collaborative business infrastructure for networked organizations. In Pervasive Collaborative Networks (pp. 451–462). Academic Press. doi:10.1007/978-0-387-84837-2_47

Raghupathi, W., & Raghupathi, V. (2014). Big data analytics in healthcare: promise and potential. Health Information Science and Systems, 2(3), 1-10. doi:10.1186/2047-2501-2-3

RAIC – Raic Canada. (2014). *Raic / Irac Architecture Canada*. Accessed at https://www.raic.org/

Rajamaran, A. & Ullman, J. D. (2011). *Mining of massive datasets*. Cambridge Editors.

Rappa, M. (2010). Business models on the web. *Managing the Digital Enterprise*. Retrieved 25-05, 2012, from http://digitalenterprise.org/models/models.html

Rascão, J. (2001). *A análise estratégica e o sistema de informação para a tomada de decisão estratégica* (2nd ed.). Lisboa: Sílabo.

Rascão, J. (2004). *Sistemas de informação para as organizações: A informação chave para a tomada de decisão* (2nd ed.). Lisboa: Sílabo.

Rascão, J. (2008). *Novos desafios da gestão da informação* (1st ed.). Lisboa: Sílabo.

Raspberry Pi Foundation. (n.d.). *Raspberry Pi*. Available at: https://www.raspberrypi.org/help/faqs/#introWhatIs

Ravichandran, T., & Rai, A. (1999). Total Quality Management in Information Systems Development:Key Constructs and Relationships. *Journal of Management Information Systems, 16*(3), 119-155.

Ray, T., & Clegg, S. (2007). Can We Make Sense of Knowledge Management's Tangible Rainbow? A Radical Constructivist Alternative. *Prometheus, 25*(2), 161-185. Retrieved from http://opus.kobv.de/tuberlin/volltexte/2010/2789/pdf/CKM_embedded_Innovaiton_Marketing.pdf

Rayome, A. D. (2016). *Seven ways to build trust and analytics at your company*. Tech Republic Blog. Available at https://www.techrepublic.com/article/infographic-7-ways-to-build-trust-in-data-and-analytics-at-your-company/

Redman, T. (1996). *Data quality for the information age*. Norwood: Artech House.

Reich, B., Gemino, A., & Sauer, C. (2008). A temporary model of information technology project performance. *Journal of Management Information Systems, 24*(3), 9–44.

Reich, B., Gemino, A., & Sauer, C. (2012). Knowledge management and Project-based knowledge in its projects: A model and preliminary results. *International Journal of Project Management, 30*(6), 663–674. doi:10.1016/j.ijproman.2011.12.003

Reich, B., & Wee, S. W. (2006). Searching for knowledge in the PMBoK Guide. *Project Management Journal, 37*(2), 11–27.

Reitoria da Universidade do Porto - U.Porto. (2017). Acedido a 21 de março de 2017. https://sigarra.up.pt/reitoria/pt

RIBA – Royal Institute of British Architects. (2014). Accessed at http://www.architecture.com/Explore/Home.aspx

Ribeiro, D. (2007). Propriedade intelectual: Mais de 30% da investigação em Portugal é redundante. *Jornal de Negócios, Quinta-feira, 34*.

Richards, G. (1996). Production and Consumption of European Cultural Tourism. *Annals of Tourism Research, 23*(2), 261–283. doi:10.1016/0160-7383(95)00063-1

Richardson, R. J. (1999). *Pesquisa social: métodos e técnicas*. São Paulo: Atlas.

Rickne, A. (2000). *New Technology-Based Firms and Industrial Dynamics Evidence from the Technological System of Biomaterials in Sweden*. Chalmers University of Technology.

Rifkin, J. (2011). *The Third Industrial Revolution: How Lateral Power is Transforming Energy, the Economy, and the World*. New York: Palgrave Macmillan.

Robredo, J. (2003), Da Ciência da Informação revisitada aos sistemas humanos de informação. Brasília: Thesaurus; SSRR Informações.

Rodrigues, J. (1999). *A conspiração solar do padre Himalaya*. Porto: Árvore - Cooperativa de Actividades Artísticas.

Rodrigues, L. C., Rechziegel, W., Esteves, G., & Riscarolli, V. (2012). Competitive Intelligence as Innovation tool to facility Business Processes. [in Portuguese]. *RGO Organizational Management Magazine*, *5*(2), 230–241.

Rodriguez, M. (Org.). (2013). Gestão do Conhecimento e Inovação nas Empresas. Rio de Janeiro: Qualitymark.

Rodriguez, V., Barcos, L., & Álvarez, J. (2010). Managing risk and knowledge in the internationalisation process. *Intangible Capital*, *6*(2), 202–235. doi:10.3926/ic.2010.v6n2.p202-235

Roederer, C. (2008). *L'expérience de Consommation: Exploration Conceptuelle, Méthodologique et Stratégique* (doctoral dissertation). Centre de recherche en marketing de Bourgogne IAE, Dijon, France.

Rogers, E. (1995). *Diffusion of innovations* (4th ed.). New York: Free Press.

Rollins, M., & Halinen, A. (2005). Customer Knowledge Management Competence: Towards a theoretical Framework. *Proceedings of the 38th Hawaii International Conference on System Sciences*. 10.1109/HICSS.2005.180

Romagni, P., & .... (1999). *10 instrumentos-chave da gestão* (1st ed.). Lisboa: D. Quixote.

Roman, K., Maas, J., & Nisenholtz, M. (2009). *How to advertise: What works, what doesn't – and why* (3rd ed.). London: Kogan Page.

Romero, D., Galeano, N., & Molina, A. (2009). Mechanisms for assessing and enhancing organisations' readiness for collaboration in collaborative networks. *International Journal of Production Research*, *47*(17), 4691–4710. doi:10.1080/00207540902847280

Rosenfeld & Morville. (2002). *Information Architecture for the World Wide Web*. O'Reilly.

Ross, J. W., Weill, P., & Robertson, D. (2006). *Enterprise Architecture as Strategy: Creating a Foundation for Business Execution*. Boston: Harvard Business Review Press.

Rouse. (n.d.a). *Gateway*. Available at http://internetofthingsagenda.techtarget.com/definition/gateway

Rouse. (n.d.b). *Ethernet*. Available at http://searchnetworking.techtarget.com/definition/Ethernet

Rowley, J. (2006). Information marketing (2nd ed.). Ashgate Publishing Limited.

Rowley, J. (2002). Information marketing in a digital word. *Library Hi Tech*, *20*(3), 352–358. doi:10.1108/07378830210444540

Rozas, N. (2004, November). O que é telemetria? *Revista Gás Brasil, São Paulo*, *1*(15), 13–15.

Russel, S., & Norvig, P. (2009). *Artificial intelligence: A modern approach* (3rd ed.). Prentice Hall.

Ryan, D., & Jones, C. (2012). *The best digital marketing campaigns in the world: Mastering the art of customer engagement* (1st ed.). Croydon: Kogan Page.

Ryan, D., & Jones, C. (2013). *Understanding digital marketing: Marketing strategies for engaging the digital generation* (2nd ed.). Croydon: Kogan Page.

Rygielski, C.; Wang, J. C. & Yen, D. C. (2002). Data mining techniques for customer relationship management. *Technology in Society, 24*(4), 483-502. 10.1016/S0160-791X(02)00038-6

SAA – Society of American Archivists. (2013). Accessed at http://www2.archivists.org/glossary/terms/i/informational-value

Sabaté, F. T. (1989). *Fundamentos de Economia de la Empresa (2nd ed.).* Barcelona: Librería Hispano Americana.

Sampaio, L. A. C. (2009). *Mapeamento das competências gerenciais necessárias aos gerentes das unidades de informação dos Tribunais Superiores do Poder Judiciário Brasileiro.* Brasília: Universidade de Brasília. Faculdade de Ciência da Informação. Dissertação de Mestrado.

Santos, M. E. R., Toledo, P. T. M., & Lotufo, R. A. (2009). *Transferência de Tecnologia: estratégias para a estruturação e getão de Núcleos de Inovação Tecnológica.* Komedi. Accessed by http://www.pucrs.br/ett/Livro%20Transferencia%20de%20Tecnologia.pdf

Santos, R. A. S. (n.d.). *Domótica via dispositivos móveis.* Universidade Federal De Ouro Preto.

Sapakal, R., & Kadam, S. (2013). 5G Mobile Technology. *International Journal of Advanced Research in Computer Engineering & Technology, 2*(2), 568–571.

Sapiro, A. (1993). Business intelligence: The Informational Revolution to the Competitive Action. [in Portuguese]. *RAE-Enterprise Administration Magazine, 33*(3), 106–125.

Saracevic, T. (1991). Information Science: origin, evolution and relations. In *Conceptions of library and information science. Proceedings of the international conference for the celebration of 20th anniversary of the Departament os Information Studies.* University of Tampere.

Sara, V. (2008). The Interplay of Reificative and Participative Processes of Customer Knowledge Creation: An exploratory study of commercial lending. *Journal of Financial Services Marketing, 12*(4), 287–298. doi:10.1057/palgrave.fsm.4760082

Sarkar, S. (2009). Empreendedorismo e inovação (2ª ed.). Lousã: Escolar Editora.

SAS. (2017) *What is Analytics*? Available at https://www.sas.com/en_us/insights/analytics/what-is-analytics.html

SAS. (2017). *What is Analytics?* Available at https://www.sas.com/en_us/insights/analytics/what-is-analytics.html

SAS. (n.d.). *SAS Enterprise Miner – SEMMA.* Available at http://www.sas.com/technologies/analytics/datamining/miner/semma.html

Schiffman, L., & Kanuk, L. (2010). *Consumer behavior.* Prentice Hall.

Schoeppel, H. R., & Naetebusch, R. (1995). Patent searching in a large electrical company, as exemplified by the situation at Siemens. *World Patent Information, 17*(3), 165–172. doi:10.1016/0172-2190(95)00016-S

Schumpeter, J. (1934). The theory of economic development. *Joseph Alois Schumpeter,* 61–116.

Schumpeter, J. A. (1939). *Business cycles.* NBER Books. doi:10.1016/j.socscimed.2006.11.007

Schumpeter, J. A. (1982). *Teoria do desenvolvimento econômico.* São Paulo: Editora Abril Cultural.

Schutzius, M. (2016). *Best practices for industry associations.* Retrieved from http://go.euromonitor.com/white-paper-associations-2016-Associations-Research-Strategy-Development-Member-Engagement.html

SCIP. (2012). Strategic and Competitive Intelligence Professionals. *What is competitive intelligence.* Retrieved from http://www.scip.org/content.cfm

Scott, J. (2010). Does the EPO really care about the quality of patent information? *World Patent Information, 32*(3), 233–236. doi:10.1016/j.wpi.2009.08.004

Sedoglavich, V. (2012). Technological imperatives in the internationalization process: Results from a qualitative investigation of high-tech SMEs. *Management Research Review, 35*(5), 441–459. doi:10.1108/01409171211222386

Seidler, R., Hartmann, E., & Gemünden, H. G. (2004). *The role of tacit knowledge in innovation management.* Competitive Paper submitted to the 20th Annual IMP Conference in Copenhagen.

Şenbalc, C., Altuntaş, S., Bozkus, Z., & Arsan, T. (2013). Big data platform development with a domain specific language for telecom industries. In *IEEE High Capacity Optical Networks and Emerging* (pp. 116–120). Enabling Technologies.

Senge, P. M. (1992). Mental models. *Planning Review, 20*(44), 4–10. doi:10.1108/eb054349

Setia, P., Venkatesh, V., & Joglekar, S. (2013). Leveraging digital Technologies: how information quality leads to localized capabilities and customer service performance. MIS Quarterly, 37(2), 565-590.

Setia, P., & Patel, P. C. (2013). How information systems help create OM capabilities: Consequents and antecedents of operational absorptive capacity. *Journal of Operations Management, 31*(6), 409–431. doi:10.1016/j.jom.2013.07.013

Seymore, S. B. (2009). Serendipity. *North Carolina Law Review, 88*, 185.

Shapiro, C., & Varian, H. (1999). *Information rules: A strategic guide to the network economy.* Boston: Harvard Business School Press.

Sheth-Voss, P., & Carreras, I. (2010). How informative is your segmentation? Marketing Research, 22(4), 9-13.

Shin, M., Holden, T., & Schmidt, R. A. (2001). From knowledge theory to management practice: Towards an integrated approach. *Information Processing & Management, 37*(2), 335–355. doi:10.1016/S0306-4573(00)00031-5

Siegel, S. (2016). Applying analytics to clinical data. *Tech Crunch.* Available at https://techcrunch.com/2016/08/31/applying-analytics-to-clinical-data/

Sigala, M., & Chalkiti, K. (2007). Improving performance through tacit knowledge externalization and utilization: Preliminary findings from Greek hotels. *International Journal of Productivity and Performance Management, 56*(5), 456–483. doi:10.1108/17410400710757141

Silber, K. H., & Kearny, L. (2010). *Organizational Intelligence – A guide to understanding the business of your organization for HR, Training, and Performance Consulting.* Pfiffer.

Silva, A. M. (2008). Modelos e modelizações em Ciência da Informação: o modelo e-lit e a investigação em literacia informacional. *Prisma.Com, 13*(1). Accessed at http://revistas.ua.pt/index.php/prismacom/article/viewFile/785/710

Silva, B. L. R. (2012). *Sistema De Controle Do Trio Automotivo Por Meio De SMS.* Centro Universitário De Brasília –UniCEUB.

Silva, C. M., França, M. T. & Oyamada, G. C. (2014). *O setor sucroalcooleiro brasileiro e a competitividade entre os estados de Mato Grosso e Mato Grosso do Sul.* Caderno de Publicações Meio Ambiente, Desenvolvimento Regional e Educação, n. 7.

Silva, S. L. (2002). Informação e competitividade: A contextualização da gestão do conhecimento em processos organizacionais. *Revista Ciência da Informação., 31*(2), 142–151.

Silva, V. (2014). *Gestão da Informação de acidentes de trabalho em profissionais de saúde: proposta de um sistema de gestão da sinistralidade laboral no centrol hospitalar São João*. Faculdade de Engenharia da Universidade do Porto.

Singh, I., & Singh, S. (2017). Framework for Targeting High Value Customers and Potential Churn Customers in Telecom using Big Data Analytics. *I.J. Education and Management Engineering*, *1*, 36–45.

Slater, P., Twyman, P., & Blackman, M. (2000). The smart way for patent information to help small firms. *World Patent Information*, *22*(4), 337–341. doi:10.1016/S0172-2190(00)00071-5

Smith, K. (2005). Measuring Innovation. The Oxford Handbook of Innovation, 148–177. doi:10.1093/oxfordhb/9780199286805.003.0006

Soares, C., & Carvalho, A. C. P. L. F. (2013). *Novas tendências em Marketing Intelligence* (P. Campos & P. Q. Brito, Eds.). Ed. Actual.

Sommer, L., & Haug, M. (2011). Intention as a cognitive antecedent to international entrepreneurship-understanding the moderating roles of knowledge and experience. *The International Entrepreneurship and Management Journal*, *7*(1), 111–142. doi:10.100711365-010-0162-z

Sordi, J. (2008). *Administração da Informação: Fundamentos e Práticas para uma Nova Gestão do Conhecimento*. S. Paulo: Saraiva.

Spagnoletti, P., Resca, A., & Lee, G. (2015). A design theory for digital platforms supporting online communities: A multiple case study. *Journal of Information Technology*, *30*(4), 364–380. doi:10.1057/jit.2014.37

Speigler, I. (2003). Technology and Knowledge: Bridging and generating gap. *Information & Management*, *40*(6), 533–539. doi:10.1016/S0378-7206(02)00069-1

Spender, J. C. (2003). Exploring uncertainty and emotion in the knowledge-based theory of the firm. *Information Technology & People*, *16*(3), 266–288.

Srinivasan, U., & Arunasalam, B. (2013). Leveraging Big data analytics to reduce healthcare costs. *IEEE Computer. org*. doi:10.1109/MITP.2013.55

Stacey, R. (1993). *Learning Organizations and Emerging Strategies, Strategic Thinking and Change Management*. Dom Quixote Publishing. (in Portuguese)

Stair, R., & Reynolds, G. (2009). *Principles of information systems*. Course Technology.

Stewart, T. (1998). *Capital Intelectual*. Rio de Janeiro: Campus.

Stewart, T. (2002). *A Riqueza do Conhecimento: O Capital Intelectual e a Organização do Século XXI*. Rio de Janeiro: Campus.

Stillwell, W. D. (2003). *Tacit Knowledge and the Work of Ikujiro Nonaka: Adaptations of Polanyi in a business Context*. Retrieved: www.missouriwestern.edu/orgs/polanyi/tad%20web%20archive/tad30-1/tad30-1-pg19-23-pdf.pdf

Streeck, W., Grote, J., Schneider, V., & Visser, J. (2006). *Governing interests: Business associations facing internationalism*. London: Routledge.

Sugahara, C. R., & Januzzi, P. M. (2005). Estudo do uso de fontes de informação para inovação tecnológica na industria brasileira. *Ciência da Informação, Brasília*, *34*(1), 45–56.

Suhr, C. (2004). A change of paradigms: Looking back to the pioneer years of patent information management (1960-1990). *World Patent Information*, *26*(1), 41–43. doi:10.1016/j.wpi.2003.10.017

Sun, H. (2010). *CKM-embedded Innovation Marketing as Success Driver for Product Innovation, Theoretical Framework and Empirical Research* (doctoral dissertation). Von der Fakultät Wirtschaft & Management der Technichen Universität Berlinzur Erlangung des akademischen Grades, Berlin, Germany.

Sun, R., Gregor, S., & Keating, B. (2015). Information technology platforms: Conceptualisation and a review of rmerging research in IS research. In *Australasian Conference on Information Systems* (pp. 1–17). Academic Press.

Sveiby, K. (1998). *A Nova Riqueza das Organizações: Gerenciando e Avaliando Patrimónios do Conhecimento*. Rio de Janeiro: Campus.

Tanenbaum & Van Steen. (n.d.). *Sistemas Distribuídos: Princípios e Paradigmas* (2nd ed.). Pearson.

Tapscott, D., & Williams, A. (2008). *Wikinomics: A nova economia das multidões inteligentes*. Matosinhos: Quidnovi.

Tapscott, D. (1995). *The digital economy: Promise and peril in the age of networked intelligence*. New York: McGraw-Hill.

Tapscott, D. (1999). *Creating value in the network economy*. Boston: Harvard Business School Press.

Tarapanoff, K. (2001). *Organizational and Competitive Intelligence*. Brasília: UNB. (in Portuguese)

Tarapanoff, K. (2006). Informação, conhecimento e inteligência em corporações: relações e complementariedade. In Inteligência, informação e conhecimento. Brasília: IBICT: UNESCO.

Tarapanoff, K.; Júnior, R. H. A.; & Cormier, P. M. J. (2000). Information Society and Intelligence in Information Units. *Ci. Inf. Brasília*, *29*(3), 91-100. (in Portuguese)

Telecom news India. (2016). *Telecom subscriber base*. Retrieved from URL: http://www.tele.net.in/news/item/18999-telecom-subscriber-base-in-the-country-reaches-1043-billion-in-january-2016-reports-trai&tag=Wireline

Terra, J. (2001). *Gestão do Conhecimento: O Grande Desafio Empresarial*. S. Paulo: Negócio Editora.

The Economic Times. (2016). *Telecom*. Retrieved from URL: http://economictimes.indiatimes.com/industry/telecom/reliance-communications-aircel-rejig-merger-plan-each-will-hold-50-in-new-entity/articleshow/51183146.cms

Thomas, A., Krishnamoorthy, M., Singh, G., & Venkateswaran, J. (2015). Coordination in a multiple producers–distributor supply chain and the value of information. *International Journal of Production Economics*, *167*, 63–73. doi:10.1016/j.ijpe.2015.05.020

Thusy, A., & Morris, L. (2004, January). From CRM to Customer Experience: a new Realm for Innovation. Business Digest.

Tidd, J., Bessant, J., & Pavitt, K. (1997). Managing Innovation - Integratin Technological, Market and Organizational Change. *Technovation*, *18*(5), 195–238. doi:10.1016/S0166-4972(98)80033-3

Tidd, J., Bessant, J., & Pavitt, K. (2003). *Gestão da inovação: Integração das mudanças tecnológicas, de mercado e organizacionais*. Lisboa: Monitor.

Tidd, J., Bessant, J., & Pavitt, K. (2008). *Gestão da Inovação*. Porto Alegre: Bookman.

Tigre, P. B. (2006). *Gestão da inovação: a economia da tecnologia no Brasil*. Rio de Janeiro: Elsevier.

Tiwana, A., Konsynski, B., & Bush, A. (2010). Platform evolution: Coevolution of platform architecture, governance, and environmental dynamics. *Information Systems Research*, *21*(4), 675–687. doi:10.1287/isre.1100.0323

Toma, C., & Popa, M. (2014). IoT – Internet of Things Architecture for Context Aware Sensors Data Processing in Waste Management Solution. *Journal of Mobile, Embedded and Distributed Systems, 6*(4).

Tomaél, M. I. (2004). Critérios de qualidade para avaliar fontes de informação na Internet. In M. I. Tomaél & M. L. P. Valentim (Eds.), *Avaliação de fontes de informação na Internet* (pp. 19–40). Londrina: Eduel.

TRAI. (2014). *Press Release on Telecom Subscription Data as on 31st March, 2014.* Retrieved from Telecom Regulatory Authority of India: http://www.trai.gov.in

TRAI. (2016). *The Indian Telecom Services Performance Indicators.* Retrieved from URL: http://trai.gov.in/Content/PerformanceIndicatorsReports/1_1_PerformanceIndicatorsReports.aspx

Traxler, F., & Huemer, G. (2007). *Handbook of business interest associations, firm size and governance: A comparative analytical approach.* London: Routledge.

Trigo, M. R., Gouveia, L. B., Quoniam, L., & Riccio, E. L. (2007). *Using competitive intelligence as a strategic tool in a higher education context* (D. Remenyi, Org.). Nr Reading: Academic Conferences Ltd.

Trott, P. (2008). *Innovation management and new product development* (4ª ed.). Essex, UK: Prentice Hall | Financial Times.

Troxler, P. (2014). Making the 3rd Industrial Revolution: The Struggle for Polycentric Structures and a New Peer-Production Commons in the Fab Lab Community. In J. Walter-Herrmann & C. Büching (Eds.), *FabLabs: Of Machines, Makers and Inventors.* Bielefeld: Transcript-Verlag.

Tsai, J. Y., Raghu, T. S., & Shao, B. B. M. (2013). Information systems and technology sourcing strategies of e-Retailers for value chain enablement. *Journal of Operations Management, 31*(6), 345–362. doi:10.1016/j.jom.2013.07.009

Tseng, K., & Johnsen, R. (2011). Internationalisation and the internet in UK manufacturing SMEs. *Journal of Small Business and Enterprise Development, 18*(3), 571–593. doi:10.1108/14626001111155718

Tuomi, I. (2000). Data is more than knowledge: Implications of the reversed knowledge hierarchy for knowledge management and organizational memory. *Journal of Management Systems, 16*(3), 103-117.

Turban, E., Mc Lean, E., & Wetherbee, J. (2002). *Information technology for management: transforming business in the digital economy* (3rd ed.). Hoboken, NJ: John Wiley and Sons.

Turban, E., Rainer, R. K. Jr, & Potter, R. E. (2007). *Introduction to information systems.* Hoboken, NJ: John Wiley and Sons.

U.Porto Inovação. (2017). Acedido a 23 de março de 2017. https://upin.up.pt

Ulrich, K., & Eppinger, S. (2012). *Product design and development* (5th ed.). Singapore: McGraw-Hill.

UNESCO. (2009). Definitions of R&D, innovation and S&T activities. In *Training Workshop on Science, Technology and Innovation Indicators.* Cairo, Egypt: UNESCO.

Universidade do Porto. (2016). *Plano Estratégico U.* Porto: Author.

Universidade do Porto. (2016). *U.Porto em Números 2016/17.* Porto: Universidade do Porto. Retrieved from https://sigarra.up.pt/up/pt/web_gessi_docs.download_file?p_name=F1340943912%2FUPorto_em_Numeros_2016-2017.pdf

UPTEC. (2017). Acedido a 23 março de 2017. http://uptec.up.pt/

US Government. (2007). *FEA – Consolidate reference model document* (Version 2.3). Accessed at http://www.whitehouse.gov/sites/default/files/omb/assets/fea_docs/FEA_CRM_v23_Final_Oct_2007_Revised.pdf

Utterback, J. (1994). *Mastering the dynamics of innovation: How companies can seize opportunities in the face of technological change.* Boston: Harvard Business School Press.

Utterback, J. M. (1996). *Dominando a dinâmica da inovação*. Rio de Janeiro: Editora Qualitymark.

Vaitsman, H. S. (2001). *Inteligência empresarial: atacando e defendendo*. Rio de Janeiro: Interciência.

Valente. (n.d.). *Um middleware para Internet Das Coisas*. Available at: http://repositorio.ul.pt/bitstream/10451/9211/1/ulfc104490_tm_Bruno_Valente.pdf

Valentim, M. P. (2000). O mercado de informação brasileiro. *Informação & Informação*, *5*(1), 35–42. doi:10.5433/1981-8920.2000v5n1p35

Van Kesteren, J. (2012). How a market intelligence system can add value to the decision making process of NDP and sales: a design science approach. Master Theses of School of Industrial Engineering, Eidhoven University, Netherlands.

Vance, D. (1997). Information knowledge and wisdom: the epistemic hierarchy and computer-based information systems. *Proceedings of the Third America's Conference on Information Systems*, 15-17. Retrieved from http://hsb.baylor.edu/ramsower/ais.ac.97/papers/vance2.htm

Vandenbosch, M., & Dawar, N. (2002). Beyond Better Products: Capturing Value in Customer Interactions. *MIT Sloan Management Review*, *43*(4), 35–42.

Varajão, J. (1998). *A arquitectura da gestão de sistemas de informação* (2nd ed.). Lisboa: FCA.

Vargo, S. L., & Lusch, R. F. (2008). Service-dominant logic: Continuing the evolution. *Journal of the Academy of Marketing Science*, *36*(1), 1–10. doi:10.100711747-007-0069-6

Varun, C., Banerjee, A., & Kumar, V. (2009). Anomaly detection: A survey. *ACM Computing Surveys*, *41*(3), 2–15.

Vavra, T. G. (2013). The advantage of aftermarketing. *American Salesman, 58*(5), 13-18.

Venkatraman, N. (1989). Strategic orientation of business enterprises: The construct, dimensionality and measurement. *Management Science*, *35*(8), 942–962. doi:10.1287/mnsc.35.8.942

Venkatraman, N., & Henderson, J. C. (1998). Real Strategies For Virtual Organizing. *Sloan Management Review*, *40*(1), 33–47.

Vick, T., Nagano, M. S., & Santos, F. C. A. (2009). Aportes da gestão da informação para a criação de conhecimento em equipes de inovação. *Perspectivas em Ciência da Informação*, *14*(2), 204–219. doi:10.1590/S1413-99362009000200014

von Hippel, E. (2005). Democratizing innovation: The evolving phenomenon of user innovation. *Journal Für Betriebswirtschaft*, *55*(1), 63–78. doi:10.100711301-004-0002-8

VSNL. (2001-2002). *16th Annual Report*. Retrieved from URL: http://www.tatacommunications.com/sites/default/files/AR2002.pdf

Vujovic, V. (2016). *Raspberry Pi as a Wireless Sensor Node: Performances and Constraints*. ResearchGate.

Wagner, G. (1990). The value and the quality of information: the need for a theoretical synthesis. In Wormell (Ed.), Information Quality: Definitions and Dimensions. London: Taylor Graham.

Walter-Herrmann, J., & Büching, C. (Eds.). (2014). *FabLab: Of Machines, Makers, and Inventors*. Bielefeld: Transcript-Verlag.

Wang, D. (2009). Chinese to English automatic patent machine translation at SIPO. *World Patent Information*, *31*(2), 137–139. doi:10.1016/j.wpi.2008.10.003

Wang, J., & Gooderham, P. (2014). Institutional change and regional development in China: The case of commodity trading markets. *Environment and Planning. C, Government & Policy*, *32*(3), 471–490. doi:10.1068/c11254b

Ward, J., & Griffiths, P. (1996). *Strategic planning for information systems* (2nd ed.). Wiley.

Watson, D. (2005). *Business models: Investing in companies and sectors with strong competitive advantage*. Harriman House Publishing.

Webster, F. (2000). *Theories of the information society*. Cornwall, UK: Routledge.

Weingand, D. E. (1987). *Marketing/planning library and information services*. Littleton, CO: Libraries Unlimited.

Weiss, S. M., & Verma, N. K. (2002). A System for Real-time Competitive Market Intelligence. *Proceedings of of the eighth ACM SIGKDD international conference on Knowledge discovery and data mining*, 360-365. 10.1145/775047.775100

Wernerfelt. (1984). A resource based view of a firm. *Strategic Management Journal, 5*, 171-180.

WHO. (2006). *Bridging the "Know-Do" Gap: meeting on knowledge translation in Global Health* (No. WHO/EIP/KMS/2006.2). WHO. Recuperado de https://www.measureevaluation.org/resources/training/materials/high-impact-research-training-curricula/bridging-the-know-do-gap.pdf

WHO. (2006). *Bridging the "Know-Do" Gap: meeting on knowledge translation in Global Health*. WHO. Disponível em: https://www.measureevaluation.org/resources/training/materials/high-impact-research-training-curricula/bridging-the-know-do-gap.pdf

WHO. (2013). *TDR Performance assessment framework*. Special Programme for Research and Training in Tropical Diseases (TDR). Disponível em: http://www.who.int/tdr/publications/about-tdr/strategy/framework/en/

WHO. (2013a). *TDR | TDR Performance assessment framework*. Special Programme for Research and Training in Tropical Diseases (TDR). Recuperado de http://www.who.int/tdr/publications/about-tdr/strategy/framework/en/

WHO. (2013b). *Priority Medicines for Europe and the World Update Report*. WHO.

Wilsdon, T., Attridge, J., Sauri, L., & Kirkpatrick, H. (2011). *Evidence on access to essential medicines for the treatment of HIV/AIDS. Charles River Associates*. CRA.

Wilson, T. (1994b). The nature of strategic information and its implications for information management. New Worlds in Information and Documentation, 265-269.

Wilson, T. (2001). Information overload: Myth, reality and implications for health care. *ISHIMR - International Symposium on Health Information Management Research*.

Wilson, D. (2002). *Managing information: IT for business processes* (3rd ed.). Woburn: Butterworth-Heinemann.

Wilson, R. (1987a). Patent analysis using online databases—I: Technological trend analysis. *World Patent Information*, *9*(1), 18–26. doi:10.1016/0172-2190(87)90189-X

Wilson, R. (1987b). Patent analysis using online databases—II: Competitor activity monitoring. *World Patent Information*, *9*(2), 73–78. doi:10.1016/0172-2190(87)90131-1

Wilson, T. (1985). Information management. *The Electronic Library*, *3*(1), 62–66. doi:10.1108/eb044644 PMID:2498741

Wilson, T. (1987). Information for business: The business of information. *Aslib Proceedings*, *39*(10), 275–279. doi:10.1108/eb051066

Wilson, T. (1989a). Towards an information management curriculum. *Journal of Information Science*, *15*(4/5), 203–209. doi:10.1177/016555158901500403

Wilson, T. (1989b). The implementation of information system strategies in UK companies: Aims and barriers to success. *International Journal of Information Management*, *9*(4), 245–258. doi:10.1016/0268-4012(89)90048-0

Wilson, T. (1994a). Tools for the analysis of business information needs. *Aslib Proceedings*, *46*(1), 19–23. doi:10.1108/eb051339

Wipo. (2017). *Guide to the International Patent Classification.* Retrieved from http://www.wipo.int/export/sites/www/classifications/ipc/en/guide/guide_ipc.pdf

Wormell, I. (1990). *Information quality: Definitions and dimensions.* London: Taylor Graham.

Wurman, R. S. (2014). *Richard Saul Wurman's website.* Accessed at http://www.wurman.com/

Wurman, R. S. (1995). *Ansiedade de informação: como transformar informação em compreensão.* São Paulo: Cultura.

Xavier, A. (2008). *Manual de Identificação e Classificação das Actividades de IDI.* COTEC Portugal. Retrieved from http://www.cotecportugal.pt/imagem/manual_idi_projecto2.pdf

Xie, Y., & Amine, L. (2009). Social networks and the internationalization of Chinese entrepreneurs. *Global Business and Organizational Excellence*, *29*(1), 61–78. doi:10.1002/joe.20299

Xu, Z., Pu, F., & Fang, X. (2016). Raspberry Pi Based Intelligent Wireless Sensor Node for Localized Torrential Rain Monitoring. Journal of Sensors.

Yakhlef, A. (2005). Immobility of tacit knowledge and the displacement of the locus of innovation. *European Journal of Innovation Management*, *8*(2), 227–239. doi:10.1108/14601060510594684

Yeomans, M. (2006). *Oil - Petróleo: Guia conciso para o produto mais importante do mundo.* Lisboa: D. Quixote.

Yin, R. (2009). Case study research: Design and methods (4th ed.). Thousand Oaks, CA: Sage Publications.

Yonce, C., Taylor, J., Kelly, N., & Gnau, S. (2017). BI Experts´ perspective: Are you ready for what´s coming in Analytics. *Business Intelligence Journal, 22*(3), 36-42.

YourStory. (2015). *Internet growth, impacts and success: what will India be like with 550 million users in 2018?* Retrieved from URL: http://yourstory.com/2015/02/internet-india-2018/

Yunbiao, S. (n.d.). *Internet of Things: Wireless Sensor Networks.* Available at: http://www.iec.ch/whitepaper/pdf/iecWP-internetofthings-LR-en.pdf

Zaltman, G., Duncan, R., & Holbek, J. (1973). *Innovations and Organizations.* New York: Wiley. Retrieved from http://www.amazon.com/Innovations-Organizations-Gerald-Zaltman/dp/047198129X

Zambon, A. C. & Anunciação, P. F. (2014). Competitive Intelligence – Value perceptions in the sector of custom jewelry. *Portuguese and Brazilian management Magazine, 13*(2), 41-60. (in Portuguese)

Zambon, A. C., Sousa, C. S., Andrade, F. J., & Anunciação, P. F. (2015). *Information and value chain.* The Limeira Gross Domestic Product Sector – Brazil, Annual Spain Business Research Conference, Barcelona, Spain.

Zicker, F. (2007). *TDR'S Research and Capacity Development Strategy.* WHO | TDR. Accessed by http://www.edctp.org/forum2007/presentations/Workshop_on_Enhancing_capacity_-_Fabio_Zicker.pdf

Zicker, F. (2007). *TDR'S Research and Capacity Development Strategy: 30 Years of Research and Capacity Building in Tropical Diseases*. WHO. Disponível em: http://www.edctp.org/forum2007/presentations/Workshop_on_Enhancing_capacity_-_Fabio_Zicker.pdf

Zikmund, W. G. (2000). Business Research Methods (6th ed.). Dryden Press.

Zorrinho, C., Serrano, A., & Lacerda, P. (2003). *Managing in Complexity - A New Management Paradigm*. Sílabo Publishing. (in Portuguese)

Zorrinho, C. (1991). *Gestão da informação*. Lisboa: Editorial Presença.

Zorrinho, C. (1991). *Information Management*. Lisbon: Editorial Presence. (in Portuguese)

Zorrinho, C., & Anunciação, P. F. (2004, April/June). Metavision: Model of 'urbanization' for Organizational Change [in Portuguese]. *Portuguese and Brazilian Management Journal*, *3*(2), 28–34.

Zorrinho, C., Serrano, A., & Lacerda, P. (2003). *Gerir em Complexidade*. Edições Sílabo.

About the Contributors

George Leal Jamil is a Post doctoral student at Universidad Politecnica de Cartagena (Entrepreneurship, Business Models), post doctoral title by Universidade do Porto, Portugal. Doctorate in Information Science, Masters in Computing Science, BsC in Electrical Engineering. Consultor, Teacher for Management topics – Strategy, Marketing, Information and Knowledge management, Project and process management. Regular author and editor of books, papers and chapters for IGI Global, member of a board who analyzes proposed works and new titles. George Jamil is a frequent worker for startups accelerators, incubators and funding / investing programs in various countries and sectors, approaching topics of management and applied scientific methods for business success.

* * *

Adelaide Antunes is a Senior Specialist at INPI (National Institute for Industrial Property). Involved in the Master's and Doctorate course in IP and Innovation at INPI, professor emeritus at the Federal University of Rio de Janeiro (UFRJ), and tenured professor for the Graduate Program in Chemical and Biochemical Process Technologies at UFRJ. Holds a Bachelor's degree in Chemical Engineering from UFRJ (1976), Master's (1979) and Doctoral (1987) degrees in Chemical Engineering from PEQ-COPPE/UFRJ, a Post-Doctorate from the French Institute of Petroleum (IFP), France (1988), and an Executive MBA from COPPEAD/UFRJ (1991). Member The Technology Committee of the Brazilian Chemical Industry Association (ABIQUIM), and of the National Institute of Science and Technology for Drugs and Medications (INCT–INOFAR). Areas of interest: chemicals industry, petrochemicals, alternative energy sources, patents, technology foresight. Supervises masters and doctoral theses in management and technological innovation (total of over 105 dissertations and theses). Author of several books.

Pedro Fernandes Anunciação, PhD in Management at the University of Évora – Portugal. Coordinating Professor at the Information Systems Department of Business Science School Information of the Polytechnic Institute of Setubal – Portugal. President of Scientific-Technical Council of School of Business Science School Information of the Polytechnic Institute of Setubal – Portugal. Member of Advisory Council of Business Science School Information of the Polytechnic Institute of Setubal. Member of the Academic Council and the General Council of the Polytechnic Institute of Setúbal – Portugal. General-Secretary of Information Systems Governance European Club – Portugal. Author/co-author of several books (Urbanism of Information Systems; Organizational Urbanism; Ethics, Sustainability and Information and Knowledge Society).

Cássio Batista is a professor and researcher of FUMEC University (Fundacao Mineira de Educacao e Cultura) in Belo Horizonte, Brazil, and consultant at Vivo/Telefonica Brasil, which is responsible for Evolved Core Network planning. Currently in the Doctoral Program at FUMEC University in Information tecnology and Comunication. Master degree on Telecommunications at UNICAMP (Universidade Estadual de Campinas). Telecommunications Engineer graduated on National Institute of Telecommunications-INATEL.

Antonio-Juan Briones-Peñalver is an Assistant Professor of Management at Technical University of Cartagena (Spain) and researcher for the Portugal International Centre (CIEO), mention of European Doctorate (Doctorate with European accreditation). His research interest is focused on the study of entrepreneurs, start-up business projects, firms´ cooperation & strategic alliances, and, in general, business collaboration relationships along the European Union. In the last years he has been leading several Projects of International Cooperation for Development issues between the European Union and Central America, sponsored by the Spanish Ministry of Foreign Affairs. He is currently Postgraduate Coordinator and Permanent Training and Collaborator at the "Office of Creation of Technology-Based Companies, Chair of Entrepreneurs " at Technical University of Cartagena. Before joining the University, he occupied positions as Economist and Executive Director in several Agribusiness companies.

Luiz Fernando Magalhães Carvalho is a Computer Scientist versed in Data Science. He holds a B.Sc and a M.Sc degree, both from Federal University of Minas Gerais – UFMG, where he studied and developed research around data processing, mining and other tasks related to modern data sciences techniques. He has worked in several international and governmental projects, for example, with data management for national information system of healthcare and currently works at one of the top-level competitors at the Financial sector, Itaú Bank, Brazil, also as a data scientist.

Eric Costa is currently a researcher at the Institute for Systems and Computer Engineering, Technology and Science (INESC TEC) in Portugal. He is also a Ph.D. Student in Industrial Engineering and Management at the Faculty of Engineering, University of Porto (FEUP) in Portugal. His doctoral research aims to study and design information management and collaborative network solutions to improve the support provided by industrial business associations to the internationalisation of SMEs, following a Design Science Research approach. He accomplished his M.Sc. degree in 2012, in the field of Industrial Engineering and Management, at the University of Minho in Portugal. His master thesis was performed under the topic of Lean Management, with the aim of analysing and reducing setup times of metal-mechanic machines of an industrial company. His current research interests include design science research, digital platforms, collaborative networks, information management, industrial business associations, and SME internationalisation.

Francisco Esteves is 57 years old, born in Ponte de Sor, Portugal, currently Delivery Manager at Winning Scientific Management, has done his professional career since 1978 in Banking industry, global Consulting companies, and IT multinational companies. With bachelor degrees in Philosophy and Electronics (Universidade Clássica de Lisboa) and Post Graduate in Organizational Systems (ISEG–Universidade Técnica de Lisboa) develops simultaneously Teaching activity at Escola Superior de Ciências

Empresariais (IPS–Setubal) covering knowledge subjects as IT Governance, Information Management, Business Intelligence and Information Systems Architecture. Currently is a Ph.D. Candidate in IMS Management School (Universidade Nova de Lisboa).

Luander Falcão has developed professional and academic activities, aimed at understanding phenomena and understanding what correlations exist between events. He has a Master's degree in Business Administration, with emphasis on Competitive Intelligence, MBA in Finance and Controllership and a bachelor's degree in Economic Sciences from PUC Minas. It also holds a CIP, Competitive Intelligence Professional, level I and II, awarded by the Fuld-Gilad-Herring Academy of Competitive Intelligence (ACI). He is currently developing analytics projects, elaborating sectoral and regional studies, such as Competitive Intelligence Specialist, through geoprocessing tools, building prospective scenarios, intelligence projects, small data and quantitative modeling. He also develops activities as a lecturer in intelligence events, a postgraduate teacher, and has chapters of books already published.

Fábio Gonçalves holds a Master Degree in Information Science from the Faculties of Engineering and Letters of the University of Porto. This degree was conferred following the Degree in Information Science, granted by the same institutions. He is currently a Key Account and Business Development Manager at IPBRICK S.A., where he works directly with document management software and manages and cooperates with several entities related to numerous areas. To this date he has participated in events such as the IV Post-Graduation Workshop on Information Science and the VIII Iberian Encounter EDICIC. His master's dissertation had as its theme the development of a model of innovation indicators applied to an academic context, a topic on which he develops research and study dynamics and which he seeks to deepen.

Fernando Gonçalves has a Master's Degree in Organization and Information Systems and Graduated in Management. Global IT Manager at Grupo Nabeiro - Delta Cafés, the biggest coffee roaster in Iberia. Previous management experiences both at private and public sector. Board of Directors Secretary of GUSP - Portuguese SAP Users Group Association, and previously President and Vice-President of the Board. Lecturer of Information Systems Governance at Undergraduate and Master's Degrees at the School of Business and Administration of Polytechnic Institute of Setubal. Member of ISACA, European Club for Information Systems Governance, Portuguese Economists Association, Portuguese Information Society Promotion and Development Association, and Accountants Portuguese Chamber. Author of papers, articles, and reports on management and information systems.

Zulmira Hartz holds a licence in education from the Federal University of Rio Grande do Norte, a degree in Medicine from FESO-Rio de Janeiro, and an MA and PhD in Community Health from the Université de Montreal. She also completed a post-doc at the École Nationale de Santé Publique-France. She is a full researcher in Epidemiology at ENSP/Fiocruz (retired), where she coordinated the MA and PhD programs in Public Health, having also coordinated the postgraduate program of the FIOCRUZ. She was also an invited professor at the Department of Social Medicine of the University of Montréal, where she is still an associate researcher of the Groupe de Recherche Interdisciplinaire en Santé (GRIS). She was a deputy director of the Escola Nacional de Saúde Pública (Fiocruz) and deputy director at

Instituto de Higiene e Medicina Tropical/University NOVA of Lisbon. She is currently a full professor, and the convenor of the module on Health Evaluation of the PhD Program in International Health at IHMT/NOVA.

Cecilia Carvalho Jamil is a Brazilian fashion researcher and designer with a Master's diploma from Stockholm University, where she had the opportunity to develop an academic research related to the creative process of fashion. She has a background in Fashion Design and has expanded her studies towards the management of international luxury brands at the Institut Français de la Mode in Paris. She had chapters and reviews in former books at IGI. Her past experiences in product development and marketing in the fashion industry have given her the basis for further explorations regarding the structures, systems and processes employed in this sector.

Liliane Carvalho Jamil is a Medical Doctor, undergraduate by Federal University of Minas Gerais. She had training in hospitals and research centers in Brazil, United States and England, in areas such as emergency room services and plastic surgery. Nowadays Liliane works as a professional intern of Plastic Surgery Clinic of Felicio Rocho Hospital in Belo Horizonte, Minas Gerais and is coursing the Masters in Medicine area at Unifesp, São Paulo, Brazil.

Dhouha Jaziri is a Lecturer at FSEGSO, University of Sousse. She obtained her PhD degree in marketing from the high institute of management of Tunis (ISG, Tunis). Her research concerns the experiential marketing, the knowledge management, the customer knowledge management, innovation management, anti-consumption behavior, digital field. She has a set of peer-reviewed publications and communications such as Tunisia (2012, 2013), Algeria (2008), Cardiff, Great Britain (2013), GECSO (2013,2016) Nancy/Paris- France, Lithuania (2013), Spain (2015). She was a member of program committee, in the workshop of BIO- WorldCIST'15 –(Portugal, 2015-2017). Likewise, she is a member of editorial Advisory Board (IGI global) and she is reviewer of books' chapters and articles in high ranked journals in marketing and management field. Moreover, she published her first book entitled: "The Handbook of Research on Tacit Knowledge Management for organizational success", which was released in April, 2017 by IGI Global USA.

Brenner Lopes is a partner in the consulting firm Nous SenseMaking, specializing in Intelligence, Strategy and Advanced Data Analytics, where he serves as director of consulting and new business. Previously in the Brazilian Service of Support to Micro and Small Companies, he oversaw the management of the Business Intelligence Unit; having also worked in several other units such as territorial development, public policies, strategies and guidelines and project office. He was a consultant to the João Pinheiro Foundation in the Statistics and Information Center and a partner in a Market Research Institute. He is a researcher and post-graduate professor at several universities. Economist, with master's degree in administration, also has several specializations, such as: High Management Program in a Globalized Environment (ESADE-ITAM); High Political and Strategy Studies (War Superior School); Strategic Information Management (UFMG); Superior Studies of Strategic Planning (Public Security Secretariat of the State of Minas Gerais).

Sérgio Maravilhas Lopes has two Post-Docs (UFBA in Patents and innovation + UNIFACS in FAB LABS, both with Grants). He has a PhD in Information and Communication in Digital Platforms, a Master in Information Management (FEUP + Sheffield Univ. UK), a Postgraduate Course in ICT, a Specialization in Innovation and Technological Entrepreneurship (FEUP+NCSU, USA) and a Degree in Philosophy. Sérgio is a frequent author, editor and reviewer for IGI Global journals, books and periodicals.

Jorge Lima de Magalhães is a FUll Professor and Researcher in Public Health at FIOCRUZ – Oswaldo Cruz Foundation in Brazil. Coordinator in the Masters' Program in Management, Research and Development in the Pharmaceutical Industry at Fiocruz. Postgraduate by Aix Marseille Université, France in Competitive Intelligence for Public Health. He holds the Doctor and Master's titles in Sciences by Chemical School at Federal University of Rio de Janeiro. Jorge has over 20 years of experience in strategic management in the pharmaceutical Industrial Operations. Last 16 years he works in R, D & Innovation area of the FIOCRUZ which has three published books, 13 book chapters and several articles in journals indexed in the national and international levels. He is leader of the CNPq Research Group Knowledge Management and Prospecting Health. Jorge's research areas are Management and Technology Innovation, Big data in Health, Competitive Intelligence and Knowledge Management, Web 2.0, Technological Trends, Patents and Knowledge Translation.

Renuka Mahajan, PhD, is an academician and researcher with more than 19 years of total experience in Information Technology, including 17 years in academics and 2 ½ years in corporate field. She has published research papers in reputed scopus journals of springer EAIT, Inderscience IJIL, Inderscience IJDATS, Inderscience IJLEG, JIOS and few other conference proceedings. Her area of interests is educational data mining, data analytics and e-CRM.

Vishal Mahajan is a Certified project management professional and seasoned IT professional with 21+ years of experience in IT delivery & Operations management, Program/Project Management, CoE management and Client relationship management. He is Experience in managing distributed global IT accounts in functional and matrix management structure. He is Skilled at establishing operational excellence within culturally diverse environments with cross functional & cross cultural experience of working at various countries across the globe such as US, UK, and Asia Pacific within Publishing, Manufacturing, Retail and Travel domains. Author is a Bachelor of Engineering (BE) in Electronics Engineering, Post Graduate Diploma in Information Technology (PGDIT) and Doctorate in management.

Marco Elísio Marques graduated in Electrical Engineering from the Pontifical Catholic University of Minas Gerais (Brazil) and has a MsC and PhD in Materials Science and Engineering by the Federal University of Minas Gerais. He has twenty-six years of professional experience acquired in leading technology companies: Nokia Institute of Technology, Lucent Technologies and SID Microelectronics, mainly in the areas of quality, research and development and project management. Actually, he is the Electrical Engineering course coordinator at FUMEC University and a researcher in the areas of Information Technology and Communication.

Joberto B. Martins has PosDoc at ICSI/Berkeley, USA (1995) and Paris Saclay, France (2016), PhD in Computer Science at Université Pierre et Marie Curie (UPMC) (1986). Professor at Salvador University (UNIFACS) (since 2000). Invited Professor at Hochschule für Techknik und Wirtschaft des Saarlandes (HTW), Germany (since 2004) and Université d'Evry, France, teaching PhD courses in Computer Science. Head of NUPERC and IPQoS research groups with research interest on Resource Allocation, SDN/OpenFlow, Cognitive Management, Artificial Intelligence, IoT, Smart City and Smart Grid. More than 140 research contributions in journals, conferences and book chapters. Worked as Professor at UPMC and INT in France and as key speaker for international congresses and companies in Brazil, Europe and US. Active presence at R&D international community with IFIP 3.2, IEEE COMSOC Awards Committee, IEEE Smart City and Smart Grid Research Committees, IEEE CGAA, Ad-hoc consultant for INEP, CAPES, CNPq, FINEP, FAPESB and other research funding institutions.

Flávia Maria Lins Mendes holds a Bachelor's degree in Chemical Engineering from Federal University of Rio de Janeiro (UFRJ) (2009) and Doctoral (2014) degrees in Chemical Engineering from Graduate Program in Chemical and Biochemical Process Technologies at UFRJ. In 2016 concluded the MBA in Project Management at IBMEC. Since 2009 works as a researcher in SIQUIM (Chemicals Industry Information System) developing foresight studies in many areas of chemical industry. Specialist in search and analysis of patent documents. In 2016 was invited to integrate a group of foresight study in Strategic Management Center at FIOCRUZ.

Marlede Menezes has a Master's Degree in Management, Research and Development in Pharmaceutical Industry at FIOCRUZ / Farmanguinhos. Graduation in Pharmacy from the Federal University of Rio de Janeiro (UFRJ) (2013). Post-graduation at the Federal University of Rio de Janeiro (UFRJ) in Industrial Pharmaceutical Technology (Conclusion: 2016). She has scientific papers published in congresses and indexed journals from her master's dissertation.

Richa Misra, PhD, is working as an Assistant Professor (General Management) in Jaipuria Institute of Management, Noida. She has done her Doctorate in Management from Amity University. She has more than 15 years of teaching experience in the field of IT and Decision Science. Her research papers and cases has been published in reputed refereed journals like On the horizon Emerald, International Journal of Logistics and Economics (Inderscience), Journal of Information and Organizational Sciences, ECCH, Abhigyaan(FORE), BHU (IT), IBS Business Review, Smart Journal of Management, Fortune Business Review and International Islamic University Malaysia. She has also presented her Cases and Research in various international conferences and forums.

Sérgio Oliveira holds a degree in Business Administration from the School of Business Administration of Fundação Getúlio Vargas at São Paulo (FGV), a Master's degree in Business Administration from the Federal University of Bahia (2000) and a PhD in Business Administration from School of Business Administration of Fundação Getúlio Vargas at São Paulo (FGV). He is currently a full professor of the master's program at Salvador University - Unifacs. He coordinated postgraduate courses and undergraduate courses in business administration. He worked as an executive in the auditing, consulting, construction and education sectors. He is the author of several articles in scientific journals in the fields of marketing knowledge, innovation and business education.

Maria Manuela Pinto is an assistant Professor and member of CIC.DIGITAL (Centre for Research in Communication, Information and Digital Culture), at the University of Porto - Faculty of Arts and Humanities, also collaborating with CITE (Centre of Innovation, Technology and Entrepreneurship), at the associated laboratory INESC TEC (Institute for Systems and Computer Engineering, Technology and Science). She holds a PhD in Information and Communication in Digital Platforms (University of Porto and University of Aveiro), with a thesis entitled "Information Management in Public Portuguese Universities: Rethinking and proposal of a model", a specialization in Documental Sciences (University of Coimbra) and a graduation in History (University of Porto). Previously she was ICT business manager in private companies (1985-1996), information manager in a municipality (1998-2004) and lecturer in Documental Sciences (FLUC, 2000-2005). She is still involved in entrepreneurship extracurricular activities and coordination member of V.IVE! (Living the Innovation and Entrepreneurship at U.Porto).

Jose Poças Rascão (PhD, ISCTE/ULI) is a Professor of Management Information Systems at the Polytechnic Institute of Setúbal, Portugal. From 2010 to 2015 was Director of course management of distribution and logistics. It is research in the Research Unit in the development of enterprise (UNIDE) at the Lisbon University Institute (ISCTE/ULI). Since 2000 has published 8 books and 8 chapters of books on strategy management and management of information and more than 36 publications on strategy and information management. Under his supervision, 17 Master These have been successfully concluded, from 2007 to 2016, with three more coming up for defense in the next two years. He also works the private management consultant for large and medium-sized enterprises. He worked the management consult at Price Waterhouse and Coopers & Lybrand between 1989 and 1995.

Leandro Henrique Rocha dos Santos is an entrepreneur of market intelligence and market research, in topics related to strategic management information and knowledge management (at IN3 Informação, Inovação and Inteligência Company). He had his BsC of Communication Sciences from Pontificia Universidade Catolica and Masters in Management, studying Marketing Information Systems applied for healthcare consultancy at Fumec University, in Belo Horizonte, Brazil.

António Lucas Soares is a researcher at INESC TEC and associate professor at the Department of Informatics Engineering at the Faculty of Engineering - University of Porto. His area of expertise is Information Systems specialised in applications to Collaborative Networks and Information and Knowledge Management particularly in industrial organisations. His research interests include socio-technical design, knowledge representation, digital platforms for collaboration, design science research. He is currently coordinating the Centre for Enterprise Systems Engineering and the Cluster Industry & Innovation of INESC TEC. At University of Porto, he is the director of the master programme in Information Science (FEUP/FLUP). He is member of the executive committee of the European chapter of the iSchools organisation and a founding member of the Portuguese Chapter of the Association for Information Systems. He publishes regularly in scientific journals in the areas of information and knowledge management.

Jorge Pinho de Sousa is full professor in the Faculty of Engineering of the University of Porto (FEUP), at the Department of Industrial Engineering and Management. He is a research coordinator at INESC TEC where he has been coordinator of the Centre for Enterprise Systems Engineering for around 15 years, with a regular participation in European R&D projects. His main research areas are Decision Support Systems, Operations Management, Transportation Systems, Logistics, and Collaborative Networks. He is the director of the Doctoral Program in Transportation Systems at FEUP (in the framework of the MIT Portugal Program) and coordinator of a post-graduation in Operations Management at Porto Business School. He regularly published in journals with high international impact, and has a frequent and active participation in conferences and other scientific meetings. From 2005 to 2009, has been President of the Portuguese Operations Research Society (APDIO).

Augusto Alves Pinho Vieira is a Medical Doctor by Federal University of Juiz de Fora 2006-2012 Juiz de Fora, Brazil. Board Certificated Ophtamologist, independent researcher. Worked at Hilton Rocha Foundation 2013-2015 Belo Horizonte, Brazil in Ophthalmology Services, Medical Science University 2016 Belo Horizonte, Brazil. Glaucoma Fellow for Federal University of São Paulo, where he also is a Phd Student. Worked in researches such as GSK expression in Schizophrenics 2008-2009, Prevalence of lower urinary tract symptoms in individuals with Down Syndrome 2010-2012 and Patient's perception on glaucoma and different types of treatment (Medical versus surgical treatment) 2014-2015. Received the following Awards: Young Scientist Award, World Federation of Societies of Biological, Psychiatry – WFSBP, Paris, France 2009 and Finalist Idea to Product Competition Latin America (I2P) University of Texas 2009.

Maria Alexandra Lobo Xavier studied Chemical Engineering from Faculty of Science & Technology of Coimbra University. MBA from the Porto Business School. Alexandra Xavier joined INESC Porto in January 2004 to found the Canter for Innovation, Technology and Entrepreneurship and Let-In, the Laboratory for Technology Enterprises. Her activity is focus on Innovation Management and Technology Entrepreneurship. Collaborates with the Masters in Innovation and Technology Entrepreneurship from the Faculty of Engineering of the University of Porto/Portugal, where she teaches several lectures. She also teaches for 5 years in a private University of Marketing IPAM. Between 2000 and 2003, she works in Sonae.com, where she was responsible for the implementation of new media projects. She also have an extensive experience as a senior consulting in Accenture, when she was responsible for several projects in the areas of Information Systems, Strategic Management and Organisation with a special focus in Port Wine, Manufacturing, Utilities and Finance sectors.

Index

K

M

O

P

R

S

T

U

Premier Reference Source
Emerging GIS Applications for Emergency and Disaster Management

Premier Reference Source
Managerial Strategies and Green Solutions for Project Sustainability

Premier Reference Source
Comparative Approaches to Using R and Python for Statistical Data Analysis

Premier Reference Source
Solutions for High-Touch Communications in a High-Tech World

www.ingramcontent.com/pod-product-compliance
Lightning Source LLC
LaVergne TN
LVHW060354300926
849162LV00006B/1266